OECD PROCEEDINGS

Boosting Innovation: The Cluster Approach

PUBLISHER'S NOTE
The following texts are published in their original form to permit faster distribution at a lower cost.

ORGANISATION FOR ECONOMIC CO-OPERATION AND DEVELOPMENT

ORGANISATION FOR ECONOMIC CO-OPERATION AND DEVELOPMENT

Pursuant to Article 1 of the Convention signed in Paris on 14th December 1960, and which came into force on 30th September 1961, the Organisation for Economic Co-operation and Development (OECD) shall promote policies designed:

- to achieve the highest sustainable economic growth and employment and a rising standard of living in Member countries, while maintaining financial stability, and thus to contribute to the development of the world economy;
- to contribute to sound economic expansion in Member as well as non-member countries in the process of economic development; and
- to contribute to the expansion of world trade on a multilateral, non-discriminatory basis in accordance with international obligations.

The original Member countries of the OECD are Austria, Belgium, Canada, Denmark, France, Germany, Greece, Iceland, Ireland, Italy, Luxembourg, the Netherlands, Norway, Portugal, Spain, Sweden, Switzerland, Turkey, the United Kingdom and the United States. The following countries became Members subsequently through accession at the dates indicated hereafter: Japan (28th April 1964), Finland (28th January 1969), Australia (7th June 1971), New Zealand (29th May 1973), Mexico (18th May 1994), the Czech Republic (21st December 1995), Hungary (7th May 1996), Poland (22nd November 1996) and Korea (12th December 1996). The Commission of the European Communities takes part in the work of the OECD (Article 13 of the OECD Convention).

FOREWORD

This book is the product of an extensive research effort, undertaken within the framework of the OECD project on National Innovation Systems, to clarify the concept of clusters of innovative firms and to demonstrate its usefulness for policy making. This co-operative work, led by the Netherlands, involved the organisation of two workshops on "Cluster Analysis and Cluster-based Policy: New Perspectives and Rationale in Innovation Policy Making", the first in Amsterdam (October 1997) and the second in Vienna (May 1998), bringing together researchers and policy makers from a number of OECD countries. It contributed to the overall report on *Managing National Innovation Systems* prepared by the OECD Working Group on Technology and Innovation Policy (TIP), under the aegis of the Committee for Scientific and Technological Policy.

The book is published on the responsibility of the Secretary-General of the OECD.

TABLE OF CONTENTS

PART I: THE METHODOLOGICAL DIMENSION

PART II: THE EMPIRICAL DIMENSION

INTRODUCTION

Innovation is not an activity of a single firm; it increasingly requires an active search involving several firms to tap new sources of knowledge and technology and apply these in products and production processes. The systems of innovation approach demonstrates that the competitiveness of companies is becoming more dependent on complementary knowledge acquired from other firms and institutions. Increasing complexity, costs and risks in innovation are enhancing the value of inter-firm networking and collaboration in order to reduce moral hazard and transaction costs, spurring a multitude of partnerships between firms with complementary assets, in addition to traditional market-mediated relations (*e.g.* purchasing of equipment, licensing of technology). As customers, suppliers and subcontractors, firms exchange information and engage in mutual learning. Interactions are also intensifying between firms and a number of other institutions involved in the innovation process, such as universities and other institutions of higher education, private and public research labs, consultancy and technical service providers and regulatory bodies.

In many countries, clusters of innovative firms are driving growth and employment. Innovative clusters of economic activity are becoming magnets for new technology, skilled personnel and research investment. These groups of enterprises tend to be well established and stable, innovating through strong backward and forward linkages with suppliers and customers. Co-operation in clusters has increasingly become a requirement for success. Moreover, co-operation offers a direct way to improve economic performance and reduce costs. Costs can be reduced if new knowledge and technology can be acquired more cheaply outside the firm than if it were to be produced in house. Co-operation also creates greater opportunities for learning (an essential requirement for productivity improvements), enables risks and R&D costs to be shared, and facilitates flexibility. It may also help reduce time-to-market for new products and processes.

Economic clusters emerge most often where there is a critical mass of firms allowing economies of scale and scope, a strong science and technology base, and a culture conducive to innovation and entrepreneurship. Clusters may also be based on factors such as natural resources or geographical advantages. Many successful clusters have long historical roots, and the emergence of new clusters takes time.

This publication contributes to the second phase of the *OECD National Innovation Systems (NIS) Project*. The first phase of the NIS project aimed at measuring and assessing the "knowledge distribution power" of systems of innovation at the national level. A number of country case studies were published comparing countries' NIS, using common indicators on knowledge flows and interactions between the actors in the NIS (OECD, 1999, *Managing National Innovation Systems*). In the second phase of the project several focus groups were formed to conduct in-depth studies of particular aspects of the innovation system, *i.e. i)* institutional linkages; *ii)* human resource flows; *iii)* innovative firm capacities and behaviour; *iv)* the innovation systems of developing and "catch-up" countries; and *v)* industrial clusters. This book summarises the results of the work of the OECD Focus Group on Clusters, based on two conferences, organised in Amsterdam (10-11 October 1997) and in Vienna (4-5 May 1998), where researchers and policy makers from a number of OECD countries presented and discussed the state of the art in cluster analysis and cluster policy making.

The contributions included in this book provide useful lessons for policy analysts and policy makers, some of which are highlighted below:

- Clusters can be interpreted as reduced-scale national innovation systems. The dynamics, system characteristics and interdependencies of individual clusters are similar to those of national innovation systems. With its focus on knowledge linkages and interdependencies between actors in networks of production, the cluster approach offers a useful alternative to the traditional sectoral approach.

- Clusters can be identified at various levels of analysis. Micro-level analysis focuses on inter-firm linkages, industry- (meso-)level analysis on inter- and intra-industry linkages in the production chain, while macro-level analysis examines how industry groups constitute the broader economic structure. Cluster analysis can also be applied at the regional level.

- The analysis of clusters indicates great diversity in innovation paths depending on the knowledge base of the clusters concerned. This calls for differentiation in policy analysis and policy making.

- Addressing systemic imperfections is increasingly seen as the key rationale for innovation and industrial policies, including cluster-based policies. For government policy, cluster analysis can provide insights into identifying economic strengths and weaknesses, gaps in innovation networks, development opportunities for regions, infrastructure needs and targets for enhanced investment in science and knowledge.

- Cluster initiatives originate in a trend towards new forms of governance and incentive structures based on networks and partnerships. A main task for policy makers is to facilitate the networking process and to create an institutional setting that favours market-induced cluster formation.

- Governments can nurture the development of innovative clusters, primarily through the provision of appropriate policy frameworks in areas such as education, finance, competition and regulation. Also valuable are schemes to stimulate knowledge exchange, reduce information failures and strengthen co-operation among firms. Focused R&D schemes, innovative public procurement, investment incentives and the creation of centres of excellence are more direct policy tools. Regional and local policies and development programmes can also play a role in encouraging cluster formation.

In an increasing number of OECD Member and non-member countries, cluster analysis is seen as an important analytical tool that can underpin industrial and technology policy. In this respect, cluster analysis is a core element of the ongoing work on national innovation systems.

The OECD would like to thank all those who contributed to the success of this work. "Practice what you preach" seems to have been the leading principle of Theo Roelandt and Pim den Hertog, the co-ordinators of the OECD Focus Group work and the editors of this book. I would like to thank them for having efficiently managed this "cluster" of experts and policy makers and fulfilled the ambitious goals of the Focus Group. It is hoped that cluster analysis will not only contribute to a better understanding of innovation, but that it will also provoke some innovation in innovation policy making itself.

Jean Guinet
OECD Secretariat

Chapter 1

CLUSTER ANALYSIS AND CLUSTER-BASED POLICY MAKING IN OECD COUNTRIES: AN INTRODUCTION TO THE THEME

by

Theo J.A. Roelandt
Dutch Ministry of Economic Affairs

Pim den Hertog
Dialogic, Utrecht

1. Introduction

Understanding technical change and innovation is crucial for understanding the dynamics of "knowledge-based economies" and "learning economies". Differences in innovation performance and the related institutional setting particular to a country, partly explain variations in economic performance. In modern innovation theory, strategic behaviour and alliances of firms, as well as interaction and knowledge exchange among firms, research institutes, universities and other institutions, are at the heart of the innovation process. Innovation and the upgrading of productive capacity is a dynamic social process that evolves most successfully in a network in which intensive interaction takes place between those "producing" and those "purchasing and using" knowledge. As a result, innovation researchers and innovation policy makers increasingly focus on the efficiency and efficacy with which knowledge is generated, diffused and used, and on the dynamics of the related networks of production and innovation. Increasingly, the notion of National Innovation Systems (NIS) is used as a conceptual framework for discussing these types of linkages and interactions among the numerous actors involved in the innovation process.

This book presents a perspective on the innovation systems approach, an approach that focuses on networks of production and value chains rather than on nation states. It reviews the cluster perspective on innovation as developed by the OECD Focus Group on Cluster Analysis and Cluster-based Policy.[1] Clusters can be characterised as networks of production of strongly interdependent firms (including specialised suppliers) linked to each other in a value-adding production chain. In some cases, clusters also encompass strategic alliances with universities, research institutes, knowledge-intensive business services, bridging institutions (brokers, consultants) and customers. The cluster perspective provides a number of advantages over the traditional sectoral approach in analysing innovation and innovation networks. These advantages are not limited to the analysis of innovation processes, but extend to the realm of innovation policy making as well. Cluster-based policy aims at removing imperfections of innovation systems (systemic imperfections) by facilitating the efficient functioning of these systems. The main aim of this study is to review cluster methodologies and cluster analyses as well as cluster-based policy initiatives in the OECD countries.

This introductory chapter sets out the major elements of the cluster approach, starting with a brief discussion of the notion of national innovation systems (Section 2). Subsequently, the concept of economic clusters as a reduced scale model of the innovation system approach is introduced. The advantages of the cluster approach compared to the traditional sectoral perspective are illustrated and the main characteristics of cluster analysis reviewed (Section 3). The policy implications are briefly introduced as the cluster perspective redefines the role of the government as well as the character of government interference (Section 4). Finally, the main questions that will be addressed in the various contributions included in this book are presented (Section 5).

2. Systems of innovation approaches

Since its launching in the second half of the 1980s, NIS has developed into a widely used theoretical framework, analytical instrument and – increasingly – gained popularity as a framework for innovation policy making. In the literature, however, the notion of NIS is defined differently.[2] Freeman (1987), for example, originally defined it as the "network of institutions in the public and private sectors whose activities and interactions initiate, import, modify and diffuse new technologies". Lundvall (1992), in his major contribution defined NIS as "the elements and relationships which interact in the production, diffusion and use of new, and economically useful, knowledge (…) and are either located within or rooted inside the borders of a nation state". Metcalfe (1995) describes NIS as "that set of distinct institutions which jointly and individually contribute to the development and diffusion of new technologies and which provides the framework within which governments form and implement policies to influence the innovation process. As such it is a system of interconnected institutions to create, store and transfer the knowledge, skills and artefacts which define new technologies". Despite the conceptual variations, the literature on innovation systems underpins two essential dimensions of innovation:

- The interaction between different actors in the innovation process, particularly between users and producers of intermediate goods and between business and the wider research community, is crucial to successful innovation (interdependency).

- Institutions matter, because innovation processes are institutionally embedded in the setting of systems of production (systemic character).

Networking and interdependency

Theory and practice have revealed that the interaction between the different agents involved in the innovation process is important for successful innovation (Morgan, 1997; Lagendijk and Charles, this volume). Firms almost never innovate in isolation (DeBresson, 1996). Networks of innovation are the rule rather than the exception, and most innovative activity involves multiple actors (OECD, 1999). To successfully innovate, companies are becoming more dependent on complementary knowledge and know-how in companies and institutions other than their own (Figure 1). Innovation is not the activity of a single company (like the "heroic Schumpeterian entrepreneur"), but rather it requires an active search process to tap new sources of knowledge and technology and apply them to products and production processes. A firm's competitiveness is becoming more dependent upon its ability to apply new knowledge and technology in products and production processes. At the same time, the rate of specialisation is rising. Companies are developing strategies to cope with their increasing dependency on their environment such as more flexible organisation structures and the integration of various links in the production chain through strategic alliances, joint ventures and consortia. The division of labour between dissimilar and complementary firms is based on the strategic choice that firms have to make between internalising knowledge or sharing information with external actors. The main goal of most strategic alliances has been to gain access to new and complementary knowledge and to speed up the learning process. There has been a shift by firms towards dis-internalising activities along and between value chains and towards specialisation in those activities that require resources and capabilities in which firms already have, or can easily acquire, a competitive advantage. In the literature, the concept of "alliance capitalism" (Dunning, 1997) is used to indicate this new stage in the development of modern economic systems: the co-existence of competition, sharpened by globalisation and liberalisation, with an increasing number of network relations and strategic alliances.

Figure 1. Strategic alliances, 1990-95

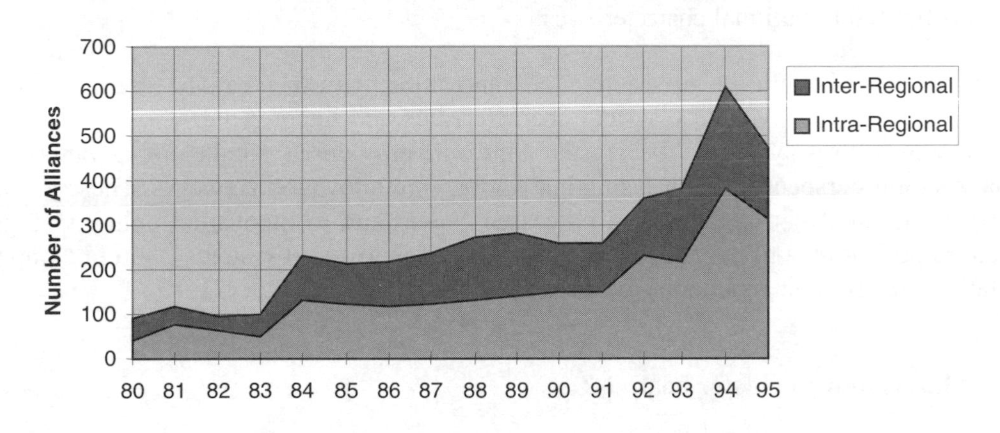

Source: European Commission, *Second European Report on S&T Indicators*, 1997.

Systemic character

A second major factor of the NIS approach is its systemic character. Innovation is no longer perceived as a linear process, but as the result of a complex interaction between various actors and institutions. These actors and institutions and their interconnections constitute a system of strongly interdependent agents. This implies that not only actors, but also institutions, play a major role in innovation. Institutions can be interpreted in a narrow sense, *e.g.* organisations such as universities, research organisations, financial institutions or all kinds of brokerage organisations that are in one way or the other involved in innovation processes. They can also be interpreted more broadly to include "behaviour" as reflected in, for example, routines, norms, rules, laws or, more generally, "the way things are usually done".[3]

Policy addressing systemic imperfections

The systemic character of the NIS concept makes it an appealing and useful tool for understanding the dynamics of innovation, and the NIS approach is increasingly used by policy makers as a framework for industrial and/or innovation policy making. However, firms, organisations and institutions, as well as their interactions, differ substantially across countries. This implies that policy responses to systemic imperfections will be country-specific, as will be shown in the contributions to this book.

Edquist pointed out that the NIS concept is, in fact, a component of a larger family of "systems of innovation" approaches, which all have systems analysis as their common starting point but which differ in the objective and level of analysis (supranational, regional, sectoral or technological systems of innovation).[4] In this book, the scope will not be limited to national boundaries. The process of globalisation has increased the degree of inter-sectoral specialisation. Countries, and even similar sectors in different countries, specialise around their national knowledge base and comparative advantages. As a consequence, internationalisation has contributed to an increase in inter-country specialisation and a deepening of the international division of labour. Nation-specific factors of innovation-based competition have become more important, as has the international dimension of networks of production (OECD, 1999). Differences in historical institutional settings explain country-specific paths of development and innovation climates. However, innovation systems and networks operate at various geographical levels (supranational, regional, local) or are specific to certain industries, technologies or clusters. Most innovation systems cannot be fully understood if their scope of analysis is limited to national characteristics.

This book reviews the cluster perspective on innovation. Clusters can be seen as reduced-scale innovation systems. The dynamics, system characteristics and interdependencies are similar to those for national innovation systems. The cluster approach provides a number of advantages over the traditional sectoral perspective in analysing innovation and innovation networks. These advantages are not limited to the analysis of innovation processes, but extend to innovation policy making as well. Cluster-based policy aims to remove the imperfections of innovation systems (systemic imperfections) by facilitating the efficient functioning of these systems.

3. What is new in cluster analysis?

The cluster concept focuses on the linkages and interdependencies among actors in the value chain in producing products and services and innovating. Clusters differ from other forms of co-operation and networks in that the actors involved in a cluster are linked in a value chain. The cluster concept goes beyond "simple" horizontal networks in which firms, operating on the same end-product market and belonging to the same industry group, co-operate on aspects such as R&D, demonstration programmes, collective marketing or purchasing policy). Clusters are often cross-sectoral (vertical and/or lateral) networks, made up of dissimilar and complementary firms specialising around a specific link or knowledge base in the value chain.[5]

By specifying strict boundaries for industries or sectors (mostly based on some statistical convention), the traditional research approach fails to take into account the importance of interconnections and knowledge flows within a network of production (Rouvinen and Ylä-Anttila, this volume). Compared to the traditional sectoral approach which focuses on strategic groups of similar firms with similar network positions, the cluster concept offers a new way of looking at the economy and is more in line with modern and interaction-based innovation theory, with new market developments and with the changing character of market-based capitalism (Dunning, 1997; Roelandt *et al.*, this volume). The cluster concept offers an alternative to the traditional sectoral approach. To simplify, the sectoral approach focuses on horizontal relations and competitive interdependence (relations between direct competitors with similar activities operating in the same product markets), while the cluster approach also focuses on the importance of vertical relationships between dissimilar firms and symbiotic interdependence based on synergism. Although innovation is stimulated by horizontal struggles between competitors operating on the same product markets, vertical relations between suppliers, main producers and users are equally important for creating innovations.

Table 1 summarises the principal differences between the traditional sectoral approach and the cluster-based approach.

Table 1. Traditional sectoral approach *vs.* cluster-based approach

Sectoral approach	Cluster-based approach
▪ Groups with similar network positions	▪ Strategic groups with mostly complementary and dissimilar network positions
▪ Focus on end-product industries	▪ Include customers, suppliers, service providers and specialised institutions
▪ Focus on direct and indirect competitors	▪ Incorporates the array of interrelated industries sharing common technology, skills, information, inputs, customers and channels
▪ Hesitancy to co-operate with rivals	▪ Most participants are not direct competitors but share common needs and constraints
▪ Dialogue with government often gravitates towards subsidies, protection and limiting rivalry	▪ Wide scope for improvements in areas of common concern that will improve productivity and increase competition ▪ A forum for more constructive and efficient business-government dialogue
▪ Search for diversity in existing trajectories	▪ Search for synergies and new combinations

Source: Adapted from Porter, 1997.

The aim of the research programme of the OECD Focus Group on Cluster Analysis and Cluster-based Policy was to gain a better understanding of successful innovative behaviour in clusters. Box 1 summarises the common starting-points of the Focus Group work, which were theoretically and empirically embedded in the literature on innovation systems. The empirical work carried out in this field seems to support these central starting-points.[6] Most of the research on innovation reported in this publication focuses on mutual interdependency and interaction among actors in the value chain. These works have in common their focus on networks of strongly interdependent firms and linkages between business and the knowledge infrastructure (universities, research institutes). This interdependency can be based on trade linkages, innovation linkages, knowledge flow linkages or on a common knowledge base or common factor conditions.

Box 1. Starting-points adopted by the members of OECD Focus Group on Cluster Analysis and Cluster-based Policies

Firms rarely innovate in isolation, but rather in networks of production. Most innovative activities involve multiple actors and stem from the combination of various actors' complementary and specialised competencies and knowledge.

The synergy arising from the combination of complementary knowledge of dissimilar firms and knowledge organisations and the need for firms to cope with their increasing dependency on their environment are the driving force behind the emergence of innovative collaborative agreements and the formation of clusters.

The common theoretical starting point lies in interaction-based innovation theory and the innovation systems approach which basically defines innovation as an interactive learning process requiring knowledge exchange, interaction and co-operation among various actors in a production network or value chain.

Important innovations stem from "new" combinations of complementary and dissimilar knowledge and competencies.

Different types of networks and markets require different innovation styles.

Cluster initiatives originate in the trend towards forms of governance based on networks and partnerships. This coincides with a trend in policy making away from direct intervention towards creating mechanisms and incentives to indirectly facilitate the networking process. The role of the government as a facilitator of networking and as a catalyst, broker and institution builder, needs to be redefined.

The contributions to this book *analyse linkages and interdependencies* among the actors in a value chain or innovation system *at different levels of analysis* and with *different techniques,* depending on the questions to be answered.

Table 2 presents the levels of analysis, using variations on the cluster concept and a different focus of the analysis. Some studies focus on the firm level and analyse the competitiveness of a *network of suppliers around a core enterprise*. This type of analysis is used to study the firm in order to identify missing links or strategic partners in innovation projects that encompass the whole chain of production. In this case, cluster analysis is directly linked to action and strategic business development (for example, the Ottawa cluster in this book). Other contributions concentrate on the meso level, typically conducting some kind of SWOT or benchmark analysis at the level of the interrelated branches in a value chain. Most of the Porter studies carried out for various countries (Denmark, Finland, the Netherlands, Sweden and the United States) use this level of analysis. Finally, some country contributions focus on linkages within and between *industry groups* (mega-clusters, such as those in Finland or the Netherlands, for instance), mapping specialisation patterns of a country or region across the economy as a whole (macro level).

Table 2. Cluster analysis at different levels of analysis

Level of analysis	Cluster concept	Focus of analysis
National level (macro)	Industry group linkages in the economy as a whole	□ Specialisation patterns of a national/regional economy □ Need for innovation and upgrading of products and processes in mega-clusters
Branch or industry level (meso)	Inter- and intra-industry-linkages in the different stages of the production chain of similar end product(s)	□ SWOT and benchmark analysis of industries □ Exploring innovation needs
Firm level (micro)	Specialised suppliers around one or more core enterprises (inter-firm linkages)	□ Strategic business development □ Chain analysis and chain management □ Development of collaborative innovation projects

Cluster methodologies also differ in their use of techniques. This report illustrates the various research techniques used in the literature, namely:

- *Input-output analysis,* focusing on trade linkages between industry groups in the value chains of the economy (Hauknes, this volume; Roelandt *et al.*, this volume; Bergman *et al.*, this volume).

- *Graph analysis,* founded in graph theory, identifying cliques and other network linkages between firms or industry groups (DeBresson and Hu, this volume),

- The third category is *correspondence analysis* (for instance, factor analysis, principal components analysis, multi-dimensional scaling and canonical correlation). These techniques aim to identify groups or categories of firms or industries with *similar innovation styles* (Vock, 1997; Arvanitis and Hollenstein, 1997; Spielkamp and Vopel, this volume).

- The qualitative case study approach along the lines of the Porter studies conducted in the various countries (Rouvinen *et al.*, this volume; Drejer *et al.*, this volume; Stenberg *et al.*, 1997; Roelandt *et al.*, this volume).

A clear distinction should be made between approaches focusing on linkages between (dissimilar) actors in networks or value chains (approaches 1 and 2), and general quantitative cluster techniques as such (approach 3) to detect objects with similar characteristics (Meeuwsen and Dumont, 1997). The first group of techniques can be used to identify network linkages of production or innovation (using input-output tables or innovation interaction matrices), while the latter group of techniques can be used to identify different styles of innovation and division of labour in innovation.[7] Although the latter approach differs fundamentally from the value chain approach, the methodologies can be combined. Finally, it should be noted that the case study approach (approach 4) has been utilised in a number of countries (including Denmark, Finland, Italy, the Netherlands, Sweden and the United States), mainly using Porter's diamond network approaches as a framework for analysing the competitiveness of the local production structure. In most countries, these *monographic cluster case studies* were complemented by statistical analysis. Case study material can provide more in-depth information and can be used to interpret the results of statistical analysis. Table 3 summarises the major advantages and disadvantages of monographic case studies.

Table 3. Advantages and drawbacks of monographic case studies

Advantages	Drawbacks
□ Increase knowledge about the "real economy"	□ Strength in exports is often the main indicator used to identify the most competitive clusters
□ Contribute to the recognition that strong innovative networks cut across different industries	□ The usability of cluster charts is therefore limited
□ Stimulate the debate about the strengths and weaknesses of the national economy	□ Porter's diamond is basically a heuristic device
□ Focus on the importance of knowledge (pooling) and upgrading for all industries	□ Porter analysis is a tool for mapping competitiveness and system dynamics at the meso level. For strategy formulation, it needs to be combined with other tools and methods for use at both the macro and micro levels
□ Illustrate the variety in the geographic scope of clusters	
□ Highlight the role of institutions in supporting and facilitating innovation	□ Basically qualitative
	□ Results are hard to compare across clusters
□ Pave the way for cluster-specific policy making	

Table 4 summarises how various analysis and cluster techniques as well cluster concepts have been used in a number of countries. Most of the countries included in the table combine various techniques to overcome the limitations of using a single technique or approach.

Table 4. Level of analysis, cluster technique and cluster concept adopted in various countries

	Level of analysis			Cluster technique					Cluster concept
	Micro	Meso	Macro	I/O	Graph	Corres.	Case	Other	
AUS		x	x	x			x		Networks of production, networks of innovation, networks of interaction
AUT		x	x			x	x	Patent data and trade performance	Marshallian industrial districts
BEL	X				x			Scientometrics	Networks or chains of production, innovation and co-operation
CAN		x	x	x			x		Systems of innovation
DK	x	x		x	x		x		Resource areas
FNL	x	x					x		Clusters as unique combinations of firms linked together by knowledge
GER	x	x		x		x			Similar firms and innovation styles
IT		x		x					Inter-industry knowledge flows
MEX		x	x				x		Systems of innovation
NL		x	x	x			x		Value chains and networks of production
NOR			X	X					Value chains and networks of production
SP		x		x			x		Systems of innovation
SWE		x					x		Systems of interdependent firms in different industries
SWI	x	x				x	x	Patent data	Networks of innovation
UK	x	x					X		Regional systems of innovation
USA		x		x		x	X		Chains and networks of production

4. The policy dimension

As indicated above, the notion of innovation systems has gained in popularity among policy makers, especially since the mid-1980s. Although the systemic role of government is still a matter of debate, an increasing number of countries are focusing on removing systemic imperfections and on improving the efficient functioning of their systems of innovation. The contributions to this report clearly show that many countries use networks of production or clusters as their starting point for action.[8]

The ability of firms to innovate successfully strongly depends on their capacity to organise complementary knowledge by participating in strategic production networks. Firms wishing to innovate must interact and exchange knowledge with customers, competitors and specialised suppliers of machinery, services and inputs. The emergence of clusters tends to be a market-induced process with little governmental interference. Why then should governments have a role to play in strengthening or facilitating the emergence of strategic and innovative clusters? In practice, four rationales for government action became apparent from the research. The first pair are rather classical rationales, namely creating favourable framework conditions for the smooth functioning of markets and the externalities associated with investment in R&D and, more generally, knowledge creation. The third lies in the fact that government itself is an important player in some parts of the economy. The final argument is directly related to the innovation systems approach: the goal of governments is to remove systemic imperfections in their innovation systems. Before discussing the changing role of the state in industrial policy making, these four rationales for government interference are briefly explained.[9]

The changing role of government

It could be argued that the establishment of alliances and the combination of various skills in production chains take place in the market. Following this line of reasoning, the primary task of government would be to facilitate the dynamic functioning of markets and to ensure that co-operation does not lead to collusive behaviour which restricts competition. This fairly classical rationale can best be summarised as "creating favourable framework conditions" to facilitate the smooth and dynamic functioning of markets, *e.g.* through vigorous competition policy, smooth macroeconomic policy or regulatory reform.

The second well-known rationale for government intervention are the externalities associated with investment in R&D and, more generally, knowledge creation.[10] The argument is that, as long as the social rate of return on investment in R&D and knowledge creation is larger than the pure private rate of return on investment, then investment should be facilitated. If this is not the case, under-investment in R&D will result. This is true in fields such as energy, the environment, infrastructure or large-scale innovation projects on the electronic highway. This is the argument used in discussions on publicly funded research in universities and public research institutes. Stimulating co-operation between firms, on the one hand, and the public R&D infrastructure (universities, research institutes), on the other, might increase the social return on publicly funded R&D. More firms will be able to profit from public R&D efforts, potentially increasing in the diffusion of knowledge, particularly towards small and medium-sized enterprises (SMEs). In most countries, SMEs have not yet been able to fully capture the benefits of increased external linkages and knowledge sharing. Many SMEs are unaware of the opportunities offered by co-operation with other firms and knowledge institutes.

The third rationale for government action in the field of innovation and industrial organisation is, in one sense, a classical one, namely the fact that government itself is an important player in some markets. This face provides an opportunity for governments to put pressure on the various market players to come up with innovative solutions to societal problems (building bridges and roads, public transport, traffic congestion, pollution, health care, and so on). In some countries, technology procurement policies have apparently been transformed into public procurement policies with a view to enhancing innovation; with the idea that a demanding customer might be able to "pull through" innovations that otherwise would not have occurred. The aim of these new-style government procurement policies is not to support national industries, but rather to challenge firms and groups of firms (including firms from abroad) to come up with innovative solutions. For government, they often mean new ways of procuring projects, *e.g.* using different forms of contracting out or using functional instead of detailed technical specifications.

The competitiveness of a country's innovation system depends upon the synergies that arise from the interaction between actors involved in the innovation process. A rationale for economic policy, directly deduced from the innovation systems approach, refers to removing systemic imperfections[11] which hinder the realisation of these synergies: informational and organisational failures and externalities. These systemic imperfections can, for instance, result from a lack of strategic information (on market developments as well as on public needs), bottlenecks in dialogue and co-operation between the various actors, or environmental and knowledge externalities. Policy responses to systemic imperfections encompass, for example:

- Establishing a stable and predictable economic and political climate.
- Creating favourable framework conditions for the efficient and dynamic functioning of free markets and removing market imperfections.
- Stimulating interactions and knowledge exchange among the various actors in systems of innovation.
- Removing informational failures by providing strategic information.
- Removing institutional mismatches and organisational failures within systems of innovation, such as mismatches between the (public) knowledge infrastructure and private needs in the market or a missing customer in the value chain.
- Removing government failures and government regulations that hinder the process of clustering and innovation.

This list clearly illustrates that the rationale of systemic imperfection is broader than the old market imperfection argument since the point of departure is the functioning of the innovation systems as a whole.

In the contributions to this book, clustering is portrayed as a *bottom-up and basically market-induced and market-led process*. Nevertheless, it has also revealed the need to redefine the role of the government as a facilitator of networking, as a catalyst of dynamic comparative advantage and as an institution builder (Morgan, 1996; Roelandt *et al.*, this volume; Lagendijk *et al.*, this volume). The notion of systemic imperfections as well as the growing importance of clustering has resulted in a redefining of the role of governments in industrial and innovation policy making in a number of countries. In most countries, this changed perspective has resulted in the creation of support structures, such as broker and network agencies and schemes, and the provision of platforms for constructive dialogue and knowledge exchange. In some countries (Denmark, Finland, the Netherlands, the United Kingdom and the United States, for instance) cluster-based policy initiatives originated in a trend towards new forms of governance and incentive structures designed to reduce systemic imperfections within innovation systems. In the majority of countries, the *changing role of governments in industrial and innovation policy making* coincides with a shift away from direct intervention towards indirect inducement. The main task of the public policy maker has become one of facilitating the clustering process and creating an institutional setting which provides incentives for market-induced cluster formation (Morgan, 1996).

5. A brief guide to the book

The research programme of the OECD Focus Group on Cluster Analysis and Cluster-based Policy Making addresses the following research questions:

1. Which clusters can be identified economy-wide?
2. How do clusters innovate? Which innovation styles are most successful in which clusters?
3. How do the same clusters in different countries vary in their economic and innovation performance and how can the differences in performance be explained?

4. What are the lessons to be learnt from the above for policy making?

5. Which policy instruments have been used in the various countries and what is the role of cluster analysis?

6. What are the key instruments and pitfalls of cluster-based policy making?

This publication reviews the various cluster methodologies and policy instruments adopted in the participating countries.

Part I focuses on the *methodological dimension of the cluster approach* and answers the question of how clusters can be identified. *Chris DeBresson and Xiaoping Hu* (Chapter 2) review the methods available to identify innovative clusters. *Johan Hauknes* reports in Chapter 3 on the Norwegian attempt to identify clusters at the meso level using input-output analysis. *Alfred Spielkamp and Katrin Vopel* in Chapter 4 aim to identify categories of firms with similar innovation styles, using firm-level micro data.

Part II presents *empirical work* and illustrates how the cluster approach can be used as a working method and a market-led economic development tool. It starts with two contributions with a clear regional focus. *Arnoud Lagendijk and David Charles* (Chapter 5) show how the cluster approach is used in UK regional policy making. *Jane Marceau* in Chapter 6 provides an overview of the various clusters that can be identified in Australia, ranging from the wine and equine cluster to the bio-medical cluster. In Chapter 7, *Roger Heath* explains the emergence and growth of high-tech clusters in the Ottawa region. The next two contributions look at the degree to which information and telecommunications have developed as key technologies, and have given rise to an economic cluster. *Pim den Hertog and Sven Maltha* in Chapter 8 describe the emerging information and communications cluster in the Netherlands, while *Christina Chaminade* (Chapter 9) provides a picture of the electronics and telecommunication cluster in Spain. The final contributions to Part II describe the use of the cluster concept as an economic development tool in developed and developing countries. In Chapter 10, *Ed Bergman and Ed Feser* discuss the US case, and in Chapter 11 *Giovanni Ceglie* reports on experiences in developing countries such as Mexico, Nicaragua, Jamaica and Honduras.

Part III elaborates on the *policy dimension*. Although cluster policies have been in existence for only a short while, some countries, for example Denmark, already have a brief history of cluster policy, and in Chapter 12, *Ina Drejer, Frank Skov Kristensen and Keld Laursen* describe how cluster policy has developed in that country. *Theo Roelandt, Pim den Hertog, Jarig van Sinderen and Norbert van den Hove* provide an insight into the Dutch case in Chapter 13. *Michael Peneder* reports on Austrian experience with cluster analysis and cluster-based policy making in Chapter 14. The impressive and wide-scale cluster analysis carried out in Finland is reported in Chapter 15 *by Petri Rouvinen and Pekka Ylä-Anttila*. An overview of cluster policy models is presented in Chapter 16 in which *Patries Boekholt and Ben Thuriaux* provide a typology of cluster-based policies in OECD countries.

The final chapter provides a synthesis of the contributions on cluster analysis and cluster-based policy making in this publication and summarises the answers to the research questions.

NOTES

1. This research programme is part of the project on National Innovation Systems co-ordinated by the OECD.

2. Given these various definitions, it is not surprising to find key elements of NIS such as "innovation", "system", "national" and "institution" interpreted quite differently by the various scholars. It can even be said that the notion of NIS itself is "conceptually diffuse" (Edquist, 1997, p. 27). Major contributions include Freeman (1987), Lundvall (1992) and Nelson (1993). An overview of the main characteristics of, as well as differences in, the various systems of innovation approaches is provided by Edquist (1997).

3. Edquist (1997, p. 26) – and others – have pointed to the conceptual ambiguity surrounding the term "institution".

4. Edquist (1997, pp. 3-15).

5. Clusters, defined in this way, can be seen as reduced-scale innovation systems. This implies that dynamics, system characteristics and interdependencies similar to those described for national innovation systems can be said to exist for individual clusters. It also implies that the idea of systemic imperfections can be used as a starting-point for developing cluster-based innovation policies (see Section 1.4).

6. See, for instance, DeBresson (1996), who also provides a good state-of-the-art methodological overview in Chapter 2 of this volume; European Commission (1997); Hagendoorn and Schakenraad (1990).

7. Some countries, such as Germany and Switzerland, focus on this element, assessing categories of firms with the same types of innovation styles, knowledge sources and knowledge transfer mechanisms, and identifying success factors for innovation (see, for example, Vock, 1997; Arvanitis and Hollenstein, 1997; Spielkamp and Vopel, this volume).

8. See Boekholt and Thuriaux (1998) for a review.

9. These four rationales are not specific to cluster policy making and can be adopted in other fields. In this book, we limit ourselves to industrial policy making aimed at stimulating the emergence of innovative clusters.

10. Research has shown that in the majority of OECD countries, the social rate of return on investments in R&D and human capital largely exceeds the private rate of return (Mohnen, 1996).

11. See also Boekholt and Thuriaux, this volume; Porter (1997); Roelandt *et al.,* this volume; Rouvinen and Ylä-Anttila, this volume; Dunning (1997).

REFERENCES

Arvanitis, S. and H. Hollenstein (1997), "Innovative Activity and Firms' Characteristics: An Exploration of Clustering at Firm Level in Swiss Manufacturing", paper presented at the OECD Workshop on Cluster Analysis and Cluster-based Policy, Amsterdam, 10-11 October.

Boekholt, P. and B. Thuriaux (1998), *Overview of Cluster Policies In International Perspective*, report prepared for the Dutch Ministry of Economic Affairs, Technopolis, Amsterdam.

Boekholt, P. and B. Thuriaux (1999), "Public Policies to Facilitate Clusters. Background, Rationale and Policy Practices in International Perspective", this volume.

Chaminade, C. (1999), "Innovation Processes and Knowledge Flows in the Information and Communication Technologies (ICT) Cluster in Spain", this volume.

Cimoli, M. (1997), "Methodologies for the study of NIS: A Cluster-based Approach for the Mexican Case", paper presented at the OECD Workshop on Cluster Analysis and Cluster-based Policy, Amsterdam, 10-11 October.

Debackere, K. (1997), "Cluster-based Innovation Policies: A Reflection on Definitions and Methods", paper presented at the OECD Workshop on Cluster Analysis and Cluster-based Policy, Amsterdam, 10-11 October.

DeBresson, Ch. (ed.) (1996), *Economic Interdependence and Innovative Activity,* Edward Elgar.

DeBresson, Ch. and Xioping Hu (1999), "Techniques to Identify Clusters of Innovative Activity: A New Approach and a Toolbox", this volume.

Drejer, I., F.S. Kristensen and K. Laursen (1999), "Studies of Clusters as a Basis for Industrial and Technology Policy in the Danish Economy", this volume.

Dunning, J.H. (1997), *Alliance Capitalism and Global Business.* Routledge, London.

European Commission (1997), *Second European Report on S&T Indicators, Part IV,* Brussels/Luxembourg.

Feser, E.J. and E.M. Bergman (1997), *National Industry Clusters: Frameworks for State and Regional Development Policy,* University of North Carolina, 1997 (prepared for *Regional Studies).*

Hagendoorn, J. and J. Schakenraad (1990), "Interfirm Partnerships and Co-operative Strategies in Core Technologies", in C. Freeman and L. Soete (eds.), *Information Technology and Employment: An Assessment,* Science Policy Research Unit, University of Sussex.

Heath, R. (1998), *The Policy Utility of Systemic Analysis* (draft version), internal report.

Held, J.R. (1996), "Clusters as an Economic Development Tool: Beyond the Pitfalls", in *Economic Development Quarterly*, Vol. 10, No. 3, August, pp. 249-261.

Hove, N. van der, Th. Roelandt and T. Grosfeld (1998), *Clusters Specialisation Patterns and Innovation Styles,* Dutch Ministry of Economic Affairs, The Hague.

Jacobs, D. and A.-P. de Man (1996), "Clusters, Industrial Policy and Firm Strategy: A Menu Approach", in *Technology Analysis & Strategic Management,* Vol. 8, No. 4, pp. 425-437.

Lagendijk, A. and D. Charles (1999), "Clustering as a New Growth Strategy for Regional Economies? A Discussion of New Forms of Regional Industrial Policy in the United Kingdom", this volume.

Marceau, J. (1999), "The Disappearing Trick: Clusters in the Australian Economy", this volume.

Meeuwsen, W. and M. Dumont (1997), "Some Results on the Graph-theoretical Identification of Micro-clusters in the Belgian National Innovation System", paper presented at the OECD Workshop on Cluster Analysis and Cluster-based Policy, Amsterdam, 10-11 October.

Mohnen, P. (1996), "R&D Externalities and Productivity Growth", *STI Review*, No. 18, Special Issue on Technology, Productivity and Employment, OECD, Paris, pp. 39-66.

Morgan, K. (1997), "Learning by Interacting. Inter-firm Networks and Enterprise Support", *Local Systems of Small Firms and Job Creation*, OECD, Paris.

OECD (1996), *The Knowledge-based Economy*, free brochure, Paris.

OECD (1999), *Managing National Innovation Systems*, Paris.

Ormala, E. (1997), "New Approaches in Technology Policy – The Finnish Example", *STI Review*, No. 22, Special Issue on "New Rationale and Approaches in Technology and Innovation Policy", OECD, Paris, pp. 277-283.

Peneder, M. (1999), "Creating a Coherent Design for Cluster Analysis and Related Policies: The Austrian 'TIP' Experience", this volume.

Porter, M.E. (1997), *Knowledge-based Clusters and National Competitive Advantage*, presentation to Technopolis 97, 12 September, Ottawa.

Poti, B. (1997), "The Interindustrial Distribution of Knowledge: The Example of Italy", paper presented at the OECD Workshop on Cluster Analysis and Cluster-based Policy, Amsterdam, 10-11 October.

Roelandt, T.J.A., P. den Hertog, J. van Sinderen and N. van den Hove (1999), "Cluster Analysis and Cluster-based Policy in the Netherlands", this volume.

Rouvinen, P. (ed.) (1996), *Advantage Finland. The Future of Finnish Industries,* The Research Institute of the Finnish Economy, ETLA, Helsinki.

Rouvinen, P. and P. Ylä-Antilla (1999), "Finnish Cluster Studies and New Industrial Policy Making", this volume.

Spielkamp, A. and K. Vopel (1999), "Mapping Innovative Clusters in National Innovation Systems", this volume.

Stenberg, L. and A.-C. Strandell (1997), "An Overview of Cluster-related Studies and Policies in Sweden", paper presented at the OECD Workshop on Cluster Analysis and Cluster-based Policy, Amsterdam, 10-11 October.

Sweeney, S.H. and E.J. Feser (1997), "Plant Size and Clustering of Manufacturing Activity", University of North Carolina (forthcoming in *Geographical Analysis)*.

Vock, P. (1997), "Swiss Position Paper for the Focus Group on Mapping Innovative Clusters of the OECD NIS Project", paper presented at the OECD Workshop on Cluster Analysis and Cluster-based Policy, Amsterdam, 10-11 October.

Vuori, S. (1995), *Technology Sources in Finnish Manufacturing,* Series B 108, The Research Institute of the Finnish Economy, ETLA, Helsinki.

Vuori, S. (1997), *Technology Sources and Competitiveness – An Analysis of Finnish Industry,* The Research Institute of the Finnish Economy, ETLA, Helsinki.

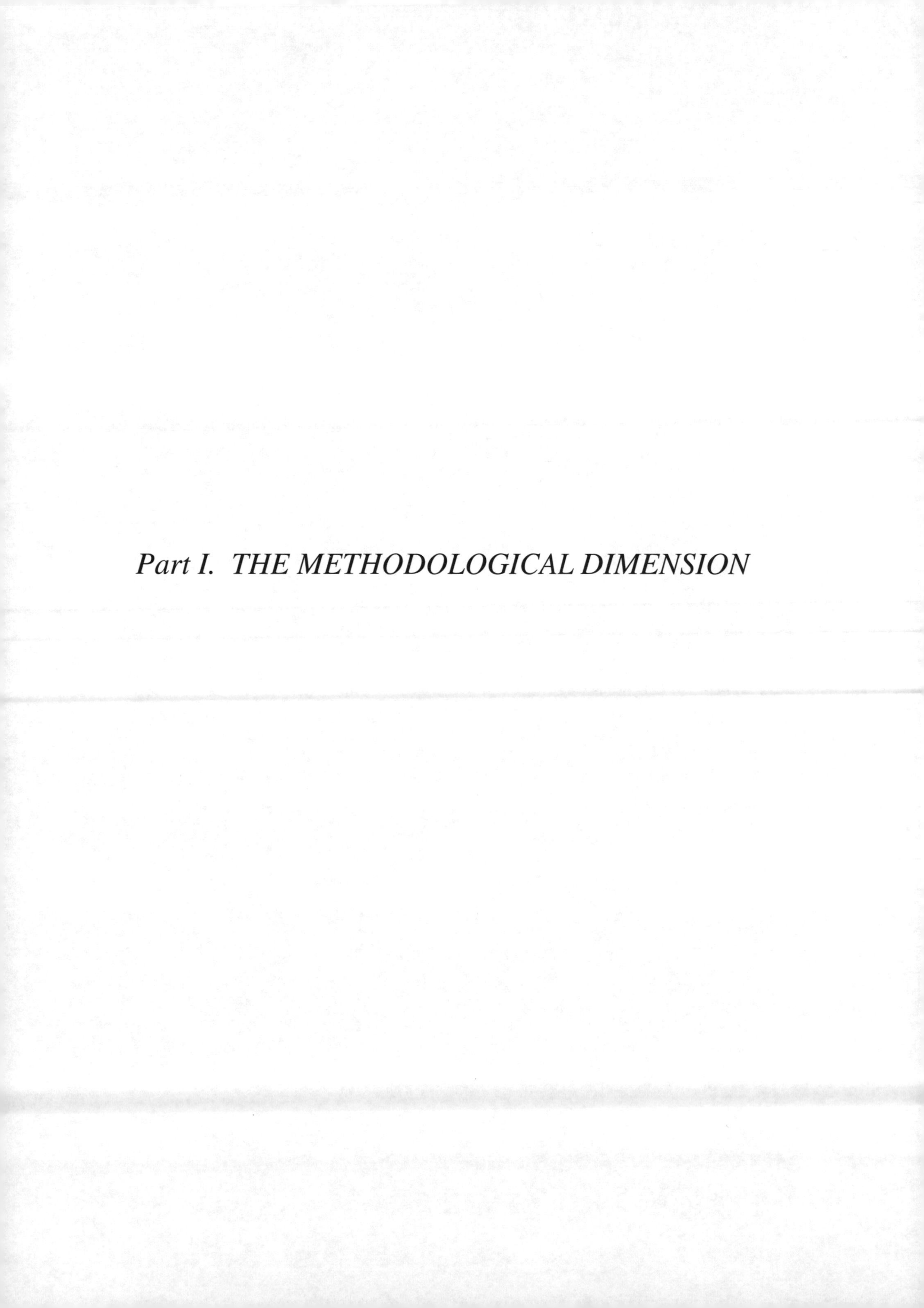

Part I. THE METHODOLOGICAL DIMENSION

Chapter 2

IDENTIFYING CLUSTERS OF INNOVATIVE ACTIVITY: A NEW APPROACH AND A TOOLBOX

by

Chris DeBresson
University of Quebec in Montreal, Canada, and Tsinghua University, Beijing, China

Xiaoping Hu
Consultant, Montreal

1. Introduction

This book is certainly not the first to address this difficult issue. At the end of the 19th century, Alfred Marshall with "industrial districts", Joseph Schumpeter with "innovation clusters", Eric Dahmen with "development blocks", François Perroux with "development and growth poles", economic geographers with industrial and "high-technology" agglomerations, all repeatedly called attention to the fact that most economic phenomena – and innovation in particular – are polarised in space. They also proposed an explanation for this; some even proposed analytical tools and policy instruments.

We will not dwell here on the intellectual history underlying these attempts, nor on the difficulties of making the above concepts analytically operational for public policy. No one has ever denied the central importance of identifying the national or regional loci that are the main springs of growth, increased welfare and well-being in our economies. If these approaches were not fully successful, it was not because policy makers did not consider them important: the source of growth and the origins of disparities have remained central to the preoccupations of policy makers and analysts. The unresolved issues which have hindered wider policy use of the "cluster" concept are related to the following questions: Why do activities cluster? Why is clustering important? Can, and should, one do something about it? If so, what and how? And how can one measure clusters? This chapter only addresses the last issue.

Over the last decades, science policy has shifted towards the policies designed to stimulate innovation (OECD, 1978). In its general review, *Technology and the Economy: The Key Relationships*, the OECD paid renewed attention to structural and systemic elements (OECD, 1992). Although renewed interest in systems has begun to focus on "national innovation systems" (Lundvall, 1992; Nelson, 1993), approaches focusing on reduced-scale systems, *i.e.* inter-industrial innovative clusters and networks, have been more successful. New opportunities to make headway exist, largely hinging on the recent availability of wide-scale representative, comparable surveys of innovative activity as a result of the *Oslo Manual* and the co-ordinated Community Innovation Surveys (CIS). However, these large rich data sets can only provide opportunities; these then have to be exploited using appropriate analytical instruments. That is the topic of this chapter.

Like an orchestra director inviting his orchestra to play its symphony one more time, "noch ein mahl mit ahnung!", the OECD's Technology and Innovation Policy Committee (TIP) and the Dutch Ministry of Economic Affairs in particular, are again inviting policy analysts to attempt to locate the nodes of growth and increased welfare in the economy, with a view to developing policies which take them into account, but to attempt this time to do so with a new focus and a new set of data – innovation and innovative activity data. The following is intended to get us started on this route.

Preliminary methodological caution

Clustering, polarisation or concentration (of firms, networks of firms, technologies, innovations, adoptions) have to be determined in relation to specified spatial dimensions. Space can have many dimensions: attributes – such as size, internal organisation, etc.; physical transport, communication, cultural, technological, functional distances, etc. Distance or proximity must be measured along these dimensions using some established technique. All the above have to be specified and the choice justified.

Much of the richness – and the confusion – in this field is that each researcher privately chooses different objects, different referent dimension of space, proximity and distance measurement tools, often with no specification, let alone justification or comparison with other approaches. Obviously, the results will differ. They will rarely be comparable; each shedding light on a particular aspect. The policy maker will still have to make a subjective judgement on which are the most relevant for policy.

In the following, we review an array of different instruments, most of them developed by others, a few developed by ourselves. We then propose an original and new approach (the proof of concept of this approach is found in DeBresson and Townsend, 1978). This new approach is based on the establishment of *innovative interaction matrices* compiled (or estimated, when necessary) from the recently available innovative activity surveys.

The basic approach is described in Section 2: Mapping innovative activity in economic space. Section 2.1 describes the compilation of innovative interaction matrices from surveys; or in the absence of surveys allowing such compilation, Section 2.2 demonstrates the estimation of innovative matrices. Section 3 reviews some useful analytical techniques which can be applied to this data and mapping. Section 3.1 provides a comparison with economic matrices. Section 3.2 provides a comparison with patent matrices. Section 3.3 describes concentration indices to measure the degree of innovative activity clustering in economic space. Section 3.4 identifies the member industries and shapes of the clusters using directed graph analysis. Section 3.5 maps the relationship between the input-output (I/O) structure and the location of innovative clusters with triangulated I/O matrices. Section 3.6 looks at overlaying innovative clusters and industrial complexes. Finally, Section 3.7 illustrates how different types of agents (firms or networks) are classified within the innovative matrix.

Because of the policy focus of this book, we will not dwell on the technical details, which can found in DeBresson (1996). Each technique is described briefly, with a general evaluation and summary of the advantages and disadvantages of each method.

2. Mapping innovative activity in the economy with innovative interaction matrices

What do innovation interaction matrices tell us? They reveal the breeding grounds of change in the economy. As innovation-induced fixed investment and increased demand are key factors underlying economic growth, the innovative matrices reveal the key locations in which growth originates and then diffuses to the rest of the economy. Since innovation and its adoption always reveal creation, acquisition of new knowledge or use of existing technological competences, the mapping of innovative activity highlights the segments of the learning economy in which firms are most dynamic.

What is new? Innovative activity matrices yield information which is not available from I/O matrices, R&D and patents data. R&D does not account for non-formal forms of learning, and therefore accounts for only a fraction (around 10%) of innovative knowledge. Patented inventions represent technological *opportunities*, only a fraction of which (between 10 and 40%) are ever used commercially. Furthermore, R&D and patents only provide the source of origin, *i.e.* the supplier. Potential users of research results and inventions have to be conjectured.

Why use matrices to map innovation? Both users and suppliers are crucial to the genesis and development of all innovations. In addition, in order to make the mapping of innovative activity comparable to a detailed mapping of the economy and patented invention (Evenson, 1997), we adopt Leontief I/O tables as a reference template.

What is an innovative matrix? An innovative matrix is a square matrix, with the suppliers of the innovative output classified by (preferably three or four-digit Standard Industrial Classification) industry in the rows, with the most frequent user industry of innovative outputs identified in the columns. The measurements in the cells can be varied: absolute or relative; frequencies or propensities (or percentages); simple occurrences or innovative sales or investments. These measurements can be modulated according to sub-set matrices by degree of novelty (new to the world, new to the country or simply new to the firm); incremental or radical; the degree of new learning required, etc.

How are the matrices built? Some innovative activity surveys enable such matrices to be *compiled* directly from survey replies; others do not. In the latter case, they have to be *estimated.* The following section reviews these alternative methods and provides examples of each.

2.1. Method 1: Compiling innovative interaction matrices from survey information

Compiling innovation matrices from survey data is relatively easy provided that a question is included related to the typical users of the firm's innovative output. This has been done for Canada (DeBresson *et al.*, 1984, 1986, and 1996, Chapters 3 and 13), for Italy, for China (DeBresson, 1996, Chapters 3 and 10), for Denmark, for Finland and for Switzerland.

An example: The 1981-85 Italian innovative-interaction matrix

How to read the figure. The lower left-hand corner of Figure 1 (or the rows in a matrix) represents the supplier industries of the innovating business unit, while the lower right-hand corner (or columns in a matrix) represents typical user industries. The value of each cell – or the height of the stalagmite of each cell in Figure 1 – represents the share of supplier industry "i"'s innovative sales (for that business unit's economically most important innovation) to industry "j".

Figure 1. Innovative-output coefficient in Italy, 1981-85

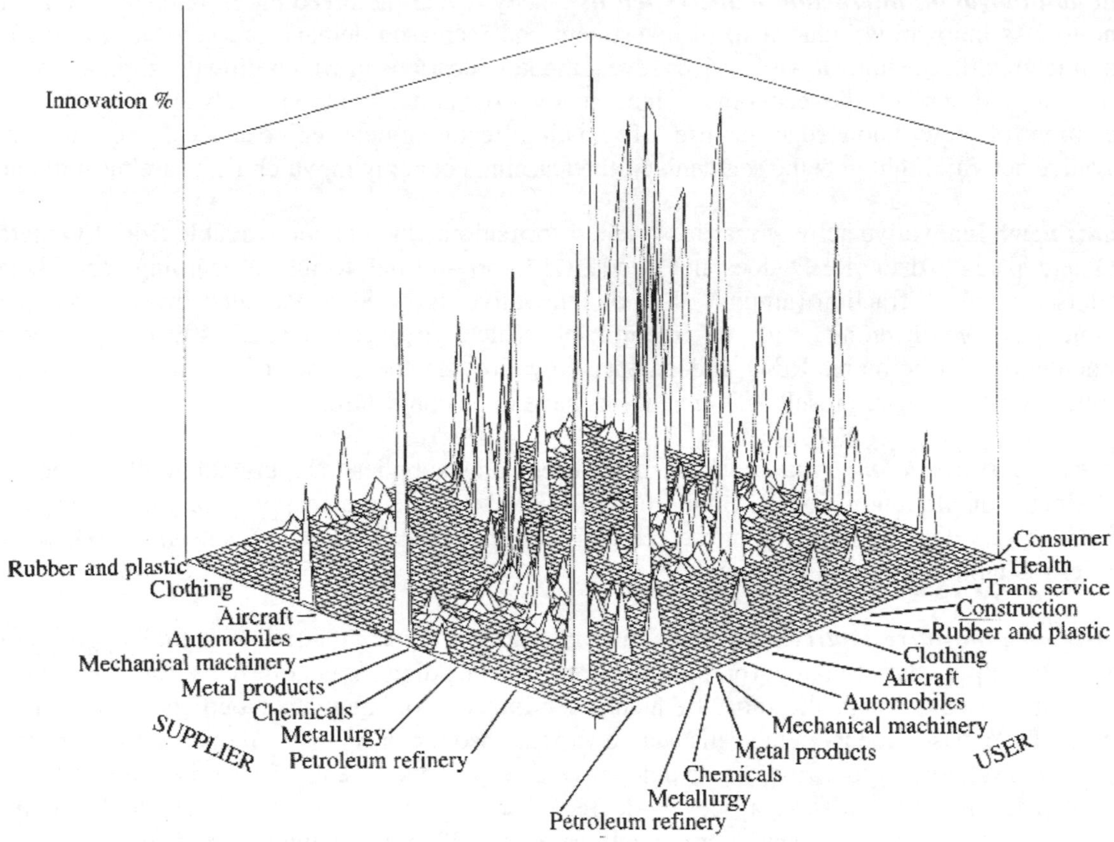

Source: DeBresson, 1996, p. 107.

Key findings. In absolute terms, during the 1981-85 period in Italy, the mechanical machinery and metal products, chemical and petroleum products industries were the most important and pervasive suppliers of innovative outputs to the economy. The final consumer and construction industries were key users.

How can the innovative matrices be interpreted? The words "supply" and "use" should not be taken literally when referring to innovative activities. In the generation of an innovation, information flows both ways between the supplier and the first user. Suppliers are known to innovate *with* as much as *for* the initial or typical user. Some of these two-way interactions are summarised in Table 1.

This is why we refer to innovative *interaction* matrices rather than to flow[1] matrices. The phenomenon these matrices describe, measure and enable to analyse is that of bi-directional (two-way) exchanges and interactions. These interactions activate or increase the *fund* of techno-economic knowledge. Neither can they be seen as *capital goods* matrices because the continuous use of knowledge does not, as is the case for machinery, always imply its eventual exhaustion, but very often, on the contrary, its sharpening, adaptation and updating.

Table 1. Important interactions between suppliers and users in the innovation process

Supplier: the product or service	*Client*: use of the product or service in a process
Search for a commercial application for supplier's technical competence	Search for a technical solution to an identified problem and need
	Specification of performance requirements
Conceptual design of solution	Assessment of design and further specifications
Value engineering	Assessment of technical choices
Prototype model to demonstrate technical feasibility	Testing and assessment of technical performance
	Specification of operating and fixed-cost requirements
Value analysis, new prototype	Testing, adjustment requirements, and firm orders
Pre-production engineering	Quality control, and operation cost control
De-bugging, quality and production cost control	
Production and launch	Economic use

Source: DeBresson, 1996, p. 68.

Limitations. The first measurement problem is how to fit the "new" innovation into the categories of the "old" economic structures represented in the I/O matrices. We must limit our goals to mapping where, within the *old* economic space, innovative activities arise. It is useful to know, for instance, that the new technology of semiconductor electronic digital computers first found financial support and a commercial breeding ground in business machines built for government, defence, higher education and research, and the insurance industry. Innovations do not emerge *ex nihilo,* but from *somewhere* in economic space. This is the object of our observations and pattern analysis.

Second, business units' technological assets are often even more diversified than their product output. Innovative activity surveys, however, only provide general information about the *innovative activities* of the business unit (such as its typical user-industry for all its innovative output). It is therefore wise to restrict oneself to (easier to manipulate) square industry-by-industry matrices.

2.2. *Method 2: Estimation of an innovative activity matrix in the absence of a complete survey*

When is estimation necessary? In cases where the innovative activity surveys are incomplete and have failed to request the crucial economic variable of the recipient market, *i.e. who uses the innovative output*, the matrix cannot be compiled. However, a rough outline of the innovative activity matrix can sometimes be estimated.

When is estimation possible? The matrix can be estimated for a large or medium-sized country, if reliable and representative innovation survey data exist with the suppliers disaggregated at the three- or four-digit Standard Industrial Classification level and a recent, disaggregated I/O table.

How? The recommended method of estimation is based on a now well-established recurrent correlation between the supply vectors of compiled – *i.e.* non estimated – innovative activity matrices and the corresponding I/O matrix. Four times in large economies we found .80 R-squares between the two sets of vectors; twice for Italy and twice for China. Even in the case of Denmark, we found a .50 R-square for intermediate goods and .30 when services and domestic demand were included, although this leads us to reject the applicability of this estimation technique for a small, open economy. As we have not yet found stronger correlations between innovative activity matrices and other economic data, we chose this correspondance as the foundation for ballpark estimations.

An example: France[2]

Using information from the French enterprise survey on the share of business contributed by each respondent in each industry, we can use the French 1990 domestic requirement output coefficient matrix to estimate which sectors are most *likely* to use the innovative outputs indicated in the supplement to the enterprise survey. The estimation of use and matrices of innovative activities required that we use the 1990 French matrix for intermediate exchange. We then added the use vector of final consumption of families to this matrix. Next, we calculated the new total of the supply vector.[3] Then we divided the values in every cell of the supply vector by their corresponding supply totals. This gave us the conditional probability that if an innovation is supplied by industry it will be used by industry. All the supply vectors in this form compose the *transition* matrix, which we used to input the probable use of the innovation by one industry or another. All vectors of total supply were used to estimate corresponding matrices.

Using a control population. The assumptions constrain our interpretation of the results. We will display the estimated location of innovative and non-innovative activities in the economic tables. We therefore limit ourselves to comparing estimates of innovative activity-related sales with sales by the control population of non-innovative enterprises. This produces the following two basic estimates, one for the matrix of innovative activities (Figure 2), the other for non-innovative activities (Figure 3) in the French economy.

Figure 2. Innovative-activity matrix in France, 1986-90
FRF 100 000 million

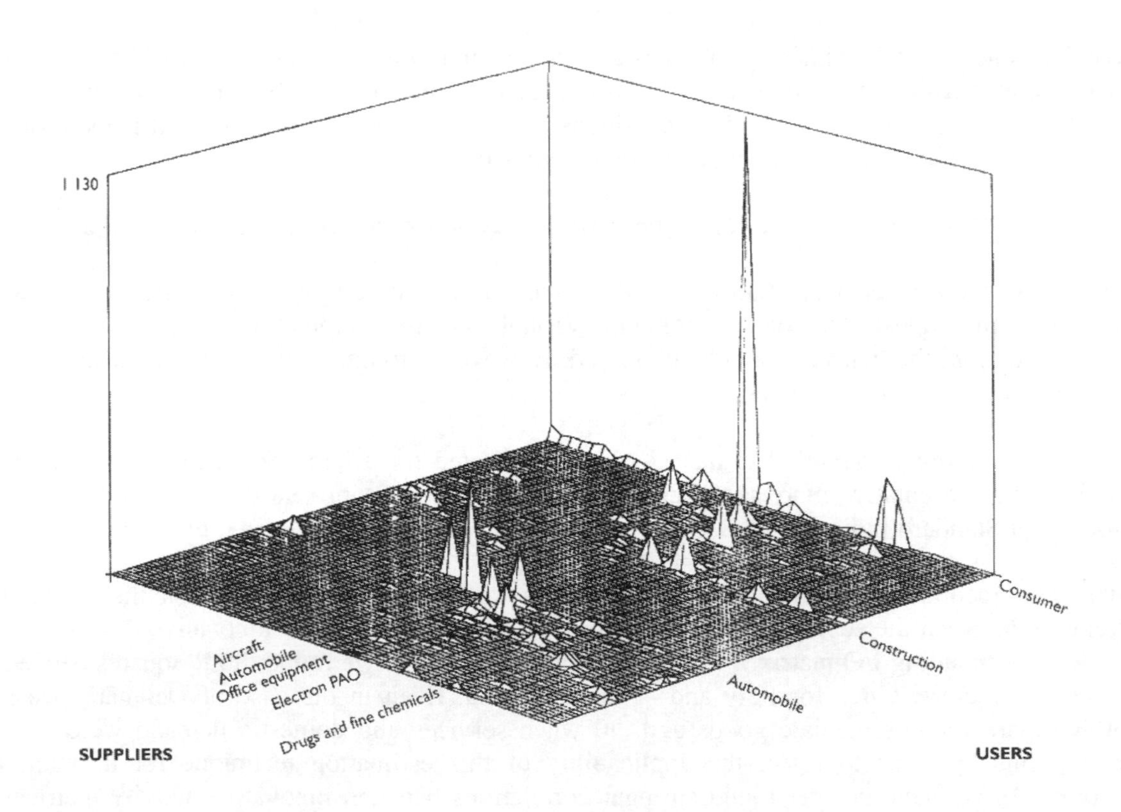

Source: DeBresson, 1996, p. 182.

Key findings. The locations of innovating activities are clearly distinct from those of non-innovating activities. This provides indirect evidence that innovative activities cluster only in parts of French economic activity. Although the innovative matrix was estimated, this approximate result can be teased out.

Figure 3. Non-innovative-activity matrix in France, 1986-90
FRF 100 000 million

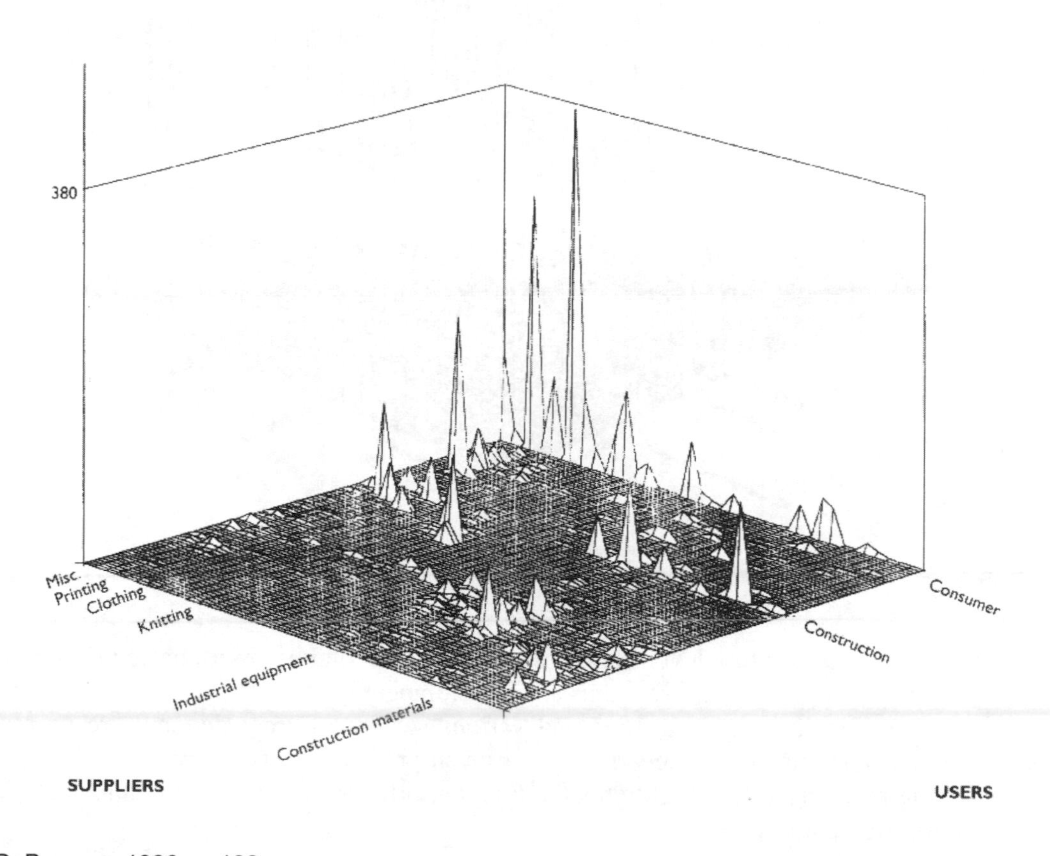

Source: DeBresson, 1996, p. 183.

Turning to the industries that are the core of innovative activity in France, the automobile, pharmaceuticals, fine-chemicals, electrical-equipment, electronic-domestic-appliance, and aeronautics industries dominate the innovative activity matrix. These are also the sectors that export most of the innovative outputs, indicating the contribution of innovative efforts to France's position in international competition. In addition, these industries are also the pioneering innovators – in the Schumpeterian sense. It should be noted that these are not the same sectors as those in which innovative activity in Italy was concentrated. In contrast, the non-innovative economy is also located in different areas: the rubber, wood-products (paper, furniture, etc.), clothing, construction-equipment, and materials industries.

A third, in-between, pattern provides some insights, although it refers to a grey zone. We know from innovation studies that some innovative activity is geared towards local demand. Much innovative activity is adaptive. Local inputs (labour, materials, energy) or demand (regulatory requirements, specific needs of clients) can lead to adaptive activities – even by late adopters – although these may not result in exports. We find that late adopters and those innovative activities that export little show a different structure. These innovative activities are far less clustered.

Figure 4. Late-adopter innovative-activity matrix in France, 1986-90
FRF 100 000 million

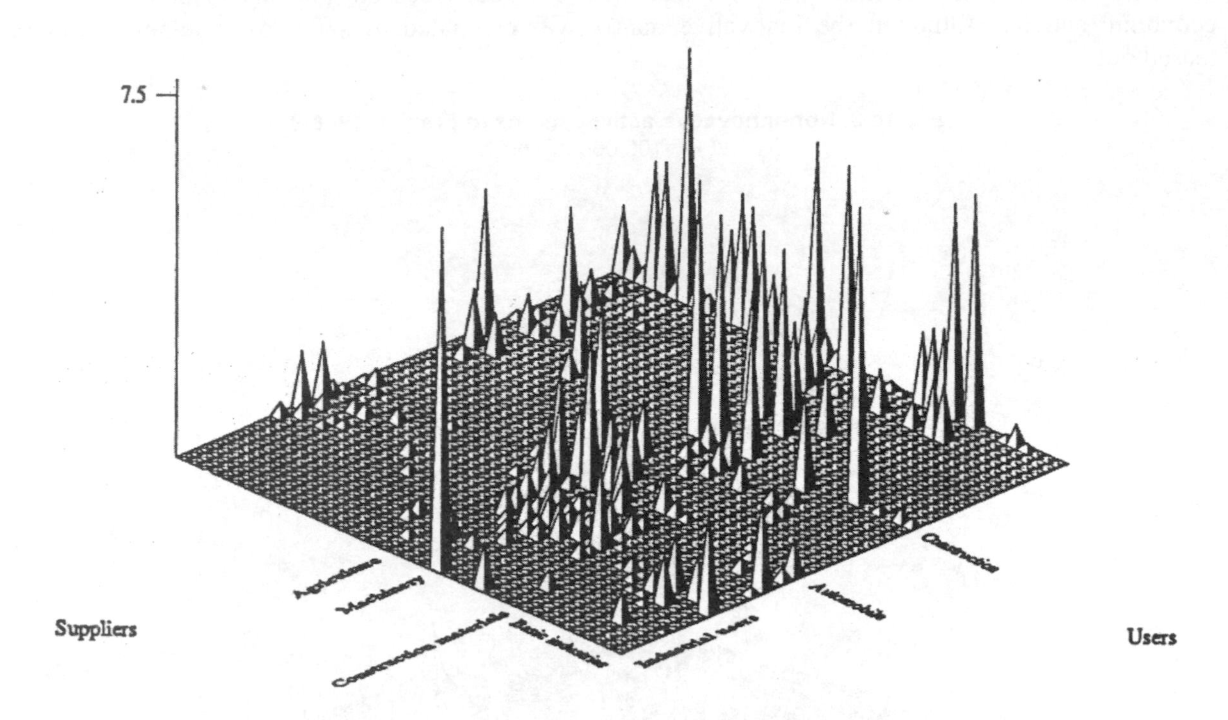

Source: DeBresson, 1996, p. 184.

These dispersed business and low-level innovative activities encompass more basic primary industries and industries with greater levels of research and development (not only chemicals, aeronautics, and automobiles). As a result, many industries and various sectors of final demand are users of these innovative outputs. The consumer appears to use a much greater variety of innovative inputs than in the two preceding mappings. This highlights the fact that different types of innovative activities may have different location patterns.

Contribution of the estimation method. Philippe Kaminski made a major contribution to our capacity of identifying the location of innovative clusters, in particular by estimating both the innovative and the non-innovative sales matrix, and then comparing the patterns in both. We subsequently applied this method to five countries in the CIS survey: Denmark, France, Germany, Italy and the Netherlands. Each time, the innovative and non-innovative sales matrices are markedly different.

Limits in information yield. Unfortunately, we used the OECD's STAN fairly aggregated I/O matrices which yield little detail. As recommended by Theo Roelandt, such cluster analysis must be done at a lower level of aggregation, at least equivalent to three-digit SIC.

Limits in reliability. In two cases, Denmark and Italy, we were able to compile innovative interaction matrices from survey responses about innovative firms' user industries, and we were therefore able to control whether the estimations for these two countries were good approximations. We found the method was much less reliable for small, open economies. We believe that in these cases, only the "total requirement matrix", should be used; that is, including imports.

Absolute prerequisite. An innovation survey which is representative of the whole economy and reliable is a pre-requisite for estimating supply vectors. To the question of whether one can, in the absence of even an incomplete innovative activity survey, estimate where innovative activity clusters could be, is a resounding "no". We have found no correlation between the country specificities in the ranking of supplier industries of innovative activity and country-specific economic data. And, as Zvi Griliches (1994) has pointed out, it is a dubious intellectual enterprise to attempt to *predict* innovation. Perhaps the best we can do is try to predict its most likely location.

As we shall see in the last section, we found in four countries out of five that innovative activity tends to clusters where the domestic economy has most backwards and forward linkages. This should enable us to predict roughly in which *corner of the economy* innovative activity is most likely. But this is far from an estimate: there are many industries in that corner, and some – like the paper and publishing industry in France – were not innovative at all. This would be equivalent to telling a miner who is asking where gold is most likely to lie: "Go North-West!" Not very helpful! I/O tables alone cannot offer a surrogate.[4]

Final diagnostic on the estimation method. Is it possible to make any useful estimation of an innovative matrix if innovative activity surveys have failed to ask for the crucial economic variable concerning *who makes economic use of the innovative output*? with the answer to this question is a qualified "yes". In the unfortunate[5] case where direct information about the users of the innovative output is not available from surveys, for large or medium-sized countries with representative and reliable innovation surveys and recent I/O tables, their estimation can provide a ballpark location of innovative clusters – which is certainly better than no knowledge at all.

Which of the two methods is superior? The first method – direct compilation – from surveys is far superior. Even in large countries, the estimations provide us with only 80% of the picture; what may be of critical policy importance is the 20% which is not captured. In addition, the most precise information comes from a *relative* measure of innovativeness, *i.e.* relative to economic activity, which requires an independently established innovation matrix. For small, open economies (such as Denmark), the estimations are probably always less than half correct. The analytical limitations stem from the assumptions which are necessary in order to make estimations. As will be shown below, many analytical tools yield rather vague findings with estimated innovation matrices and can only be, or are better, used with compiled innovative matrices. Obviously, the more sophisticated the analytical tools, the greater the precision required in the measurements.

3. Analytical tools

3.1. Comparing the innovative matrix with I/O matrices[6]

What does the comparison reveal? Comparing the innovative matrix with I/O matrices enables the policy analyst to answer the following questions: How similar or different is the location of innovative and general economic activity? Which economically active channels sustain innovative activity? Which do not? Where is innovative activity occurring without or in spite of very low levels of economic activity, perhaps indicating economic bottlenecks and the birth of new economic channels? It also provides a relative measure of innovative activity, *i.e.* relative to economic activity.

How is the comparison performed? The method is very simple. All that is required is that the format and size of matrices, as well as the measuring rod, be comparable. Innovative domestic sales and/or exports, current and/or fixed innovative investments and the corresponding matrices (domestic, import or total requirement, or capital goods matrices) can be used. The values of the two are normalised using supply (or demand) coefficient – or percentage – matrices or vectors.

An example: Italy, 1981-85

The innovative activity matrix can be compared with the total, domestic, import and capital requirement matrices of Italy in 1982 and 1985. To highlight the patterns and compare qualitatively different phenomena, transition (or conditional probability) matrices are calculated, where the size of the supply vectors is normalised by dividing each cell by the line total of each supply vector. This results in an output coefficient matrix, sometimes called the "marketing profile". In Figure 1 above, the percentage of the supply of innovative interactions by one industry to another is shown on the vertical axis; in Figure 5, the vertical axis displays the percentage of goods and services from a supplying industry which goes to a specific supplier, or, in technical terms, the transition supply vectors of the total requirement matrix.

Figure 5. Total requirement output coefficient in Italy, 1982

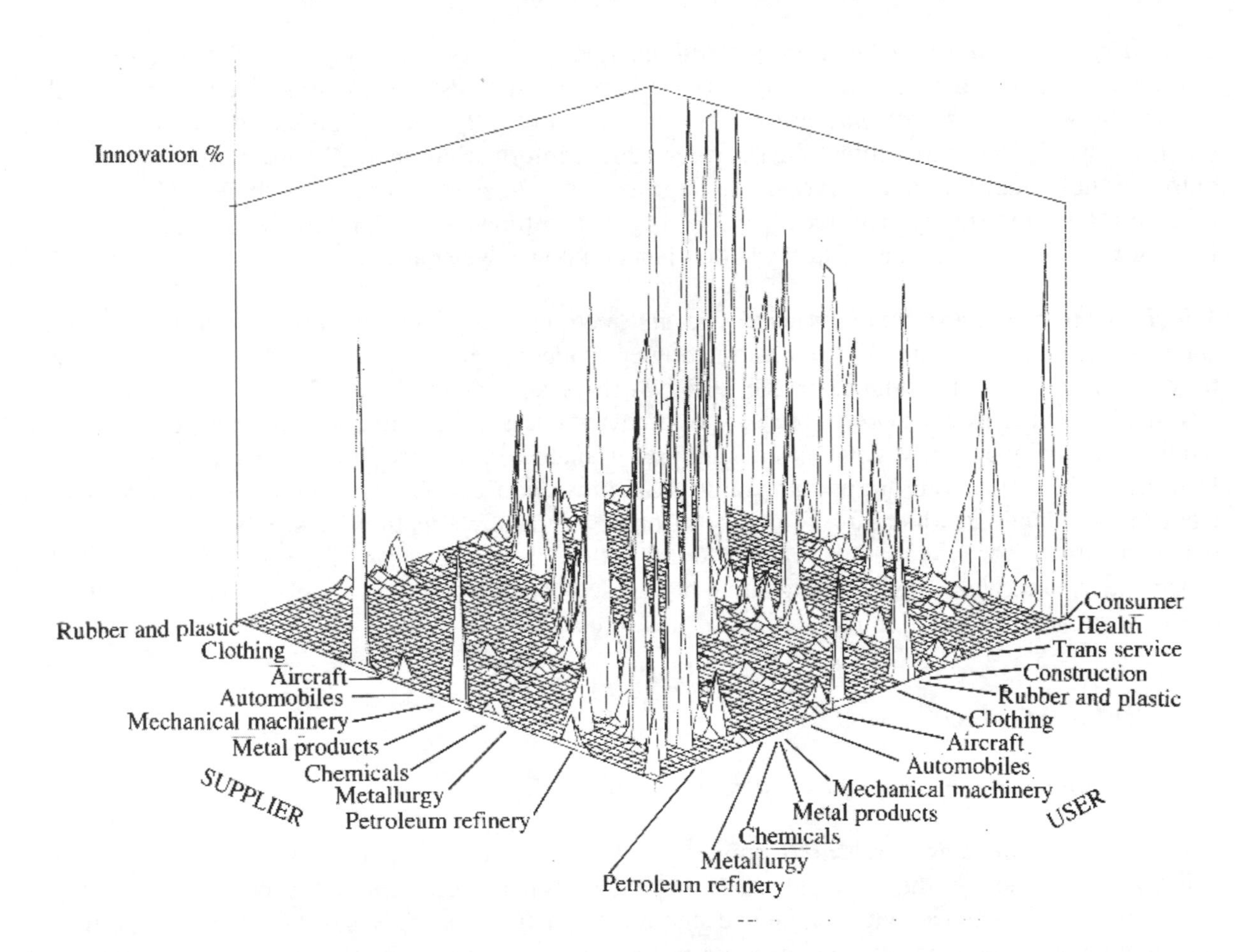

Source: DeBresson, 1996, p. 108.

In most cases, one would not expect high levels of innovative activity in the absence of economic activity. The radical innovations that create new economic markets from scratch are likely to be exceptions. In only four cells (out of 274) were there some innovative activities without flows of goods or equipment; at the disaggregated level (43 by 66), only 30 cells (out of 1 518) show that some innovative activity exists when goods and services flows did not reach the 100 million lire level, *i.e.* less than 2% of cases.

Understanding graphics, and in particular these complex forests of stalagmites, can be difficult and treacherous. More precision can be gained by simple quantitative analysis, for instance correlating the two vectors and isolating the outliers and tail-ends, *i.e.* those cells of the matrix which represent economic channels where the levels of economic activity and innovative activity diverge. This indicates more precisely areas in which high levels of economic activity exist but where there is little innovative activity, or vice versa. Overlaying innovative interaction matrices with domestic requirements of goods and services matrices enables us to identify more precisely, in relative terms, where firms in the learning economy are innovative.[7]

Warning concerning the applicability of this method. Obviously this method cannot be used with estimated innovative matrices. Because the former are estimated with economic I/O matrices, they cannot be compared with the latter. The only possibility is to compare the innovative matrix with the non-innovative matrix (as in Section 2.2).

3.2. A summary mapping using a triangular matrix

What does it reveal? In comparison to the method described above, this technique reduces the information to the essential. It can quickly show the heart of the national economy, its most integrated part, and which very innovative interactions are sustained by strong economic channels, which are not, and which strong economic channels have low levels of innovative activity.

How to triangulate the innovative matrix? Triangulation is a standard I/O method of reordering a matrix in order to identify the most integrated sectors. In any matrix, many cells are empty. This is particularly the case for small and medium-sized countries' I/O tables; it is even more so for innovative interaction matrices because they are even sparser due to the fact that innovative activity is clustered, as noted above. Triangulation reorders and summarises the information of the matrix in a meaningful way. Industrial classifications made by statistical offices are usually not economically meaningful for analysing interdependence because industry numbers are attributed by administrative fiat as new activities emerge – or, more precisely, many years after they emerge. Leontief developed a simple reordering technique that locates industries with the most backward and forward linkages in the upper left-hand corner.

An example: Innovative activity in Italy 1981-85

We used this simple technique on the Italian domestic requirement matrix and then added the relative figures for innovative business active in each cell (Table 2) (see DeBresson, 1996, Chapter 11 for its use for France, Chapter 12 for China, and Chapter 13 for Canada). The relative measure of innovative activity used in the cells is the following: taking two corresponding supply (or output) coefficient vectors for the same industry in the innovative-interaction and the intermediate-goods matrices, we calculated the difference between the coefficients in corresponding cells. These differences have a unimodal distribution. We then split this distribution into five groups with different shades of grey. Cells that are white had neither innovative activity nor exchanges of goods. We thus have a measure of innovative activity relative to economic activity.

Economic activity is concentrated in the upper left-hand corner. Innovative activity represents only a fraction (approximately one fifth of the cells) of this activity. In general, innovative activity is clustered in the most active economic core in those industries which have the highest degrees of first-order backward and forward linkages (this was checked econometrically to be the case with Italy and with China and France; DeBresson and Hu in OECD, 1996; DeBresson, 1996, Chapter 21).

Information yield. The method yields rapid analytical value added and provides decision makers with detailed information, without revealing any confidential information. However, there are preconditions for its use.

Table 2. Triangulated domestic requirement matrix for Italy (1982) with relative innovative activity (1981-85)

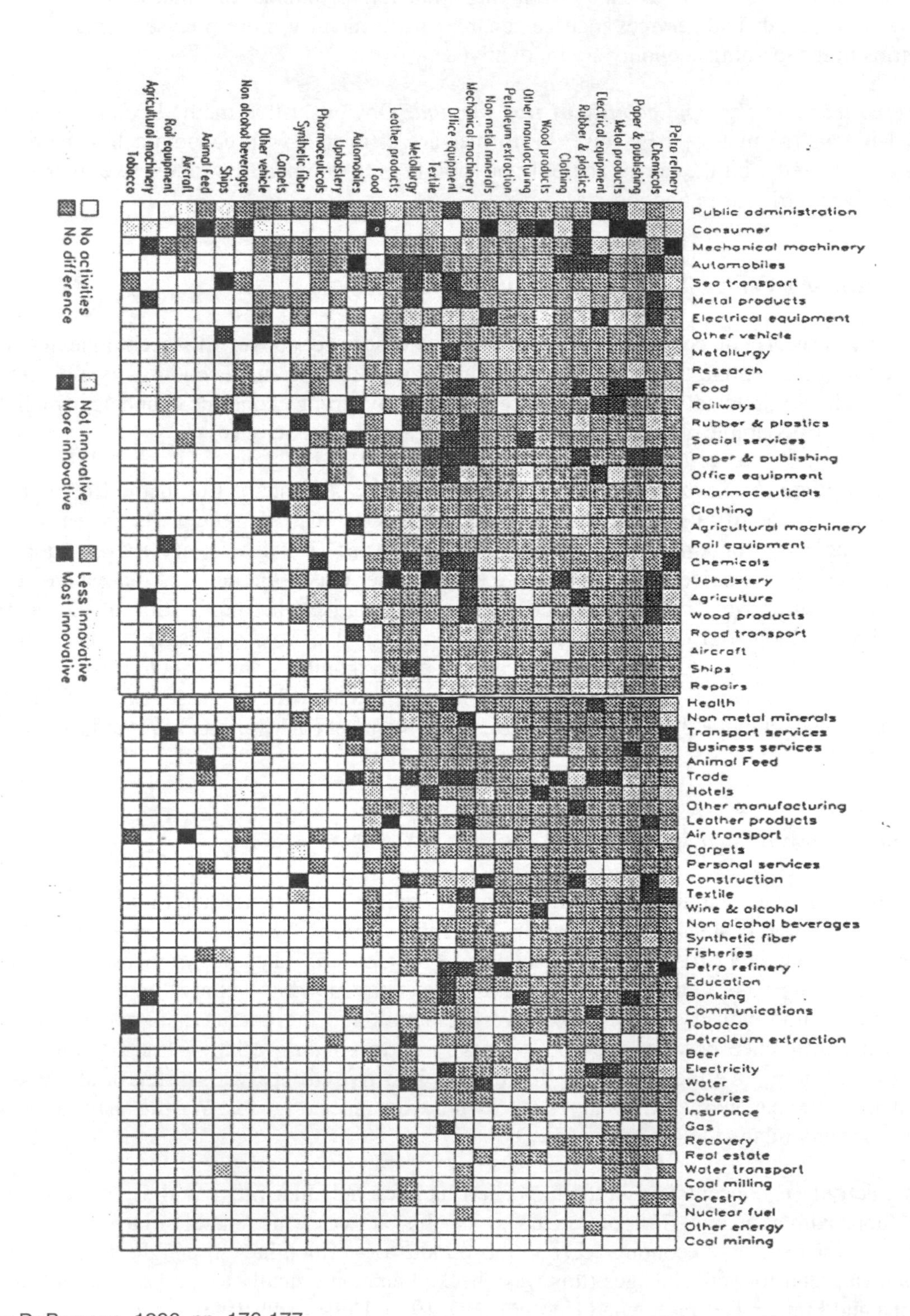

Source: DeBresson, 1996, pp. 176-177.

Preconditions. First, the matrices should to be fairly disaggregated. As is the case for the directed graph techniques described in Section 4, this type of analysis provides very few extra insights unless one has disaggregation at levels of three digits or more. Secondly, again the results are much sharper with innovative matrices compiled from direct survey information than with estimated matrices. Compiled innovative matrices allow a *relative* measure of innovative activity to be calculated. Thirdly, similar measures should be used for both domestic goods and service requirement matrices and innovative interaction matrices. Straight frequency counts were used in Table 2, but innovative and non-innovative sales would be better. This would allow one to correctly measure innovative activity clustering *relative* to non-innovative activity and general economic activity.

Communication problems. This technique, however, has a serious drawback. Triangulated matrices are useful for the analyst because he or she can take the time to understand them, but they are too cumbersome for rapid exposition of results to harried policy makers. They require that the reader carefully follow a line and a column and, while this is the normal routine of the quantitative analysts, policy makers are trained to focus their attention differently. As triangulated matrices have to be "explained" at some length by the analyst-presenter, it is just as easy to use words or numbers. Whereas complex three-dimensional graphics (as in Figures 1, 2, 4, 5, and 6) catch the attention of the audience, two-dimensional triangulated matrices do not.[8]

A necessary complement. The data can be summarised using another standard technique, consisting of a regression curve with an index of economic linkage on the horizontal axis and a measurement of innovative performance on the vertical axis, with observations – in particular outliers – labelled with the name of member industries (OECD, 1996).

Figure 6. Correlation between innovative activities (1981-85) and economic linkages (1985), Italy

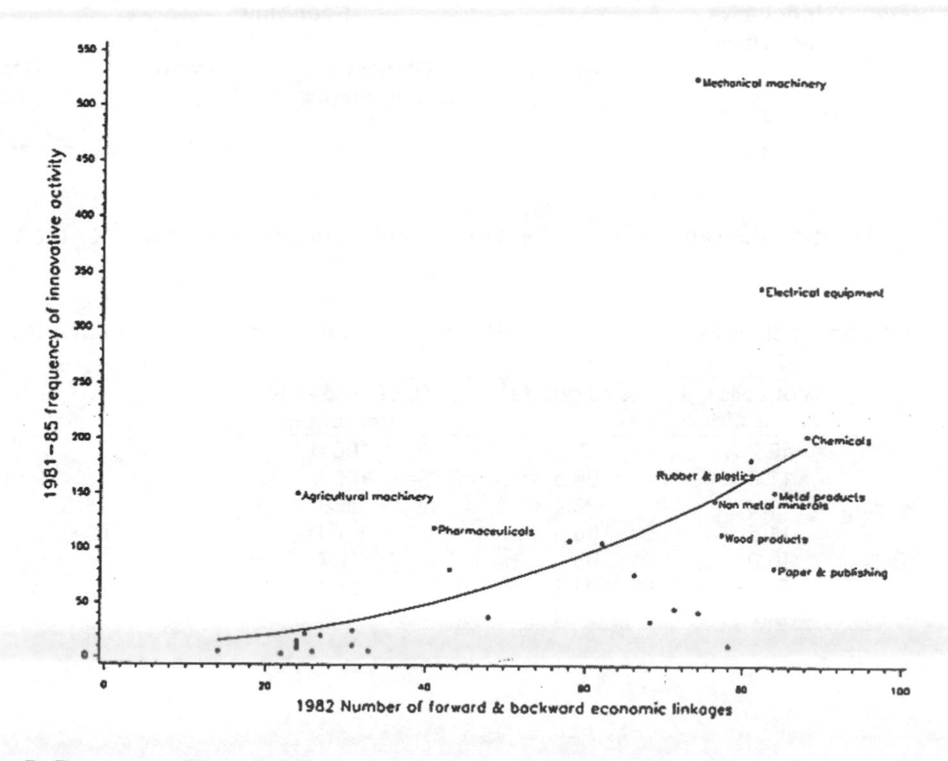

Source: DeBresson, 1996, p. 318.

3.3. *Degree of clustering of innovative activity in economic space: simple concentration ratios*

What does this method reveal? To what degree is innovative activity in the national economy clustered? To find out, we need more relative measurements, such as the above. One also needs to know how diffused or concentrated the innovative stimulus is through the economy.

How can one find out? One can use simple density and concentration indices, such as:

$$Density = \frac{SUM \quad SUMz_{ij}}{N_2}$$

where z_{ij} is the linkage between the supply sector i and the user sector j in the respective matrices and N is the total number of possible cells between the different matrices. All the values of cells are normalised to either one or zero to represent the existence or non-existence of an economic or innovative linkage. To standardise each comparison, all matrices are aggregated to the same size. This simple index borrowed from network analysis (Knobe and Kuklinski, 1982) will give us a first gross estimate of the relative concentration in the innovative and economic tables. We can then take into consideration the value of the cells with a simple concentration ratio for the basis of comparison of different matrices.

An example: Italy

In all cases, the density of the innovative activity matrix is lower than the other matrices. Innovative activity is clearly clustered in only part of the economic space.

Table 3. Density indices for different matrices

Matrix size	World firsts	Innovative activities	Economy			
			Total	Domestic requirements	Imports	Capital flow
43x66	.072	.11	.48	.46		
12x23		.43				.70

Taking now into consideration the value of the cells with a simple concentration ratio for the basis of comparison:

Table 4. Concentration indices for innovative activity and total economic requirement matrices

% of cells	% of goods	% of innovative interactions
48.2	100.0	100.0
11.8	96.5	100.0
11.6	96.4	96.8
11.5	96.3	95.9
10.0	95.3	71.7
7.5	92.7	23.0
4.3	86.2	4.5

The above tables both use percentages. Table 3 calculates the percentages based on the existence or non-existence (a dichotomous measure) of the linkage, while Table 4 calculates them based on frequencies within cells. These percentages can also be calculated on innovative and non-innovative sales (DeBresson *et al.*, 1997). The advantage of Table 4 is that it provides a measuring rod with greater variance and more immediate relation to economic impacts.

Figure 7. Correlation between innovative activities (1981-85) and economic linkages (1985), Italy

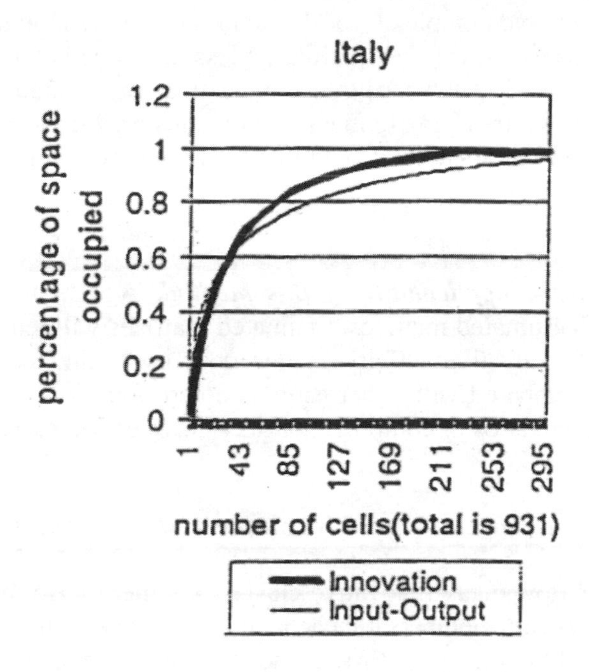

Table 4 can be represented graphically (Figure 7). If the innovative activity curve is at some time above that of goods, then clustering of innovative activity is greater than that of economic activity. Innovative and economic activity can be plotted by taking the number of theoretically possible cells on the horizontal axis and the percentage of occupied cells on the vertical axis.[9]

3.4. *Comparing the degree of innovative activity concentration in different countries*

What does a comparison of innovative activity concentration reveal? Nations and regions like to compare performances. The fact that innovative activity is more concentrated than the economy can be compared in relation to other countries' experiences. Is innovative activity more clustered in one country than another? One would expect that a very innovative country would have more innovative industries, while a marginally innovative country would experience strong innovative polarisation.

If one country's innovative activity is highly concentrated while the other's is less so, this may indicate that the first country should focus more on the diffusion of innovative impulses through the economy, on the spread and distributive effects of innovation, domestic or foreign, on the economy; the second country does not have this problem (David and Foray, 1995).

For instance, from past observations and preliminary calculations, it would appear that the German economy has innovative activity which diffuses throughout the economy (DeBresson *et al.*, forthcoming), whereas in France, innovative activity is more polarised in certain industries and locations (Kaminski *et al.* in DeBresson, 1996).

How can this be measured? As in Figure 7, we define N as the *total effective economic space* of a country. By *total effective economic space* we mean only those cells in the I/O table which have figures, *i.e.* not zeros. This, of course, depends on the minimum level of a cell value allowed in the country's I/O table.

Normalising national IO tables for comparison. Just as the size of the matrices must be the same between countries, the minimum considered value of a cell has to be equivalent in order to allow inter-country comparison. One solution would be to take the same threshold value for each country in monetary equivalents. But the size and complexity of different national economies must also be taken into account. Obviously, if we take identical size matrices, a less developed country such as Greece or Portugal will have much less of its I/O table cells occupied than the United States, Japan, China, Germany or France. Therefore, if we are trying to measure how clustered innovative activity is in their respective economic spaces, then we must normalise to identical reference points. N, the *total effective economic space* seems to be the most appropriate way to normalise this comparison (see DeBresson *et al.*, forthcoming, for examples).

Cautionary warning regarding the applicability of this method. Again, this method is far more reliable with compiled than with estimated matrices. Estimated matrices will tend to underestimate the clustering and concentration of innovative activity. If one does use estimated matrices to compare countries, they should only be compared with other estimated concentration figures; conversely, one should only compare concentration ratios of compiled matrices between themselves.

3.5. Identifying members and shape of innovative clusters: directed graph analysis

What does the analysis reveal? How many innovative clusters are there? Are they inter-connected or not? Which firms and industries are members of each cluster? What is the degree of integration between the different members of these clusters? Are some members more central than others in these clusters? The answers to these questions will allow policy makers to better understand innovation dynamics and how to sustain the growth or influence of these clusters in certain directions.

How can one find out? Directed graphs provide a very simple and unambiguous way to answer such questions with an innovative matrix. In this section, we will present simple dot-and-arrow graphics and the corresponding matrices. These graphics are immediately accessible to policy makers and analysts, yet they correspond to rigorous mathematical definitions and reproducible results (an example of the corresponding matrix form is given in the text and a mathematical definition of a di-graph is given in the notes; others can be found in DeBresson, 1986, Chapter 10; the analytical algorithm is obtainable from the authors).

What is the logic? At a very elementary level, one can differentiate supplier-user interactions using simple *set theory* concepts. We can distinguish whether they are: *i)* unidirectional; *ii)* symmetrical (if a-b, then b-a); or *iii)* transitive (if a-b, and b-c, then a-c). These simple notions enable us to distinguish extreme types of clusters. A first extreme type of cluster occurs when all industries are innovating with each other; we will call these clusters "cliques". Another type of cluster is when linkages are unidirectional; we will call these "trees". In between, we can define "agglomerations" or "complexes" as innovation clusters that have some symmetrical or transitive linkages. Of course, there are innovative industries that are independently innovative and do not interact with any others; we call these "points of development" or "enclaves". The rank order of clusters (see list below) must be seen as an indicator of synergy and integration; the higher the level of cluster, the higher the suspected level of positive (or negative) synergy.

Definitions. Six elementary structures can be distinguished:

1) **Development point.**[10] An isolated enclave industry with innovative activities totally enclosed within itself[11] and not linked with any other industries.

Directed graph:

This, for example, will correspond to the following, and only to the following, matrix configuration:

Supplier/User	Z	A	B
Z			
A		INTERACTIONS	
B			

(Due to space constraints, we will no longer display the corresponding matrix: see DeBresson 1996, Chapter 10.)

2) **Innovative couple.** When two industries supply innovative outputs to each other symmetrically, we use the word "couple".

Directed graph:

3) **Standard tree.** When one industry supplies innovative outputs to some industries that do not supply it in turn, we have the (upside-down) image of a (hierarchical) tree.

Directed graph:

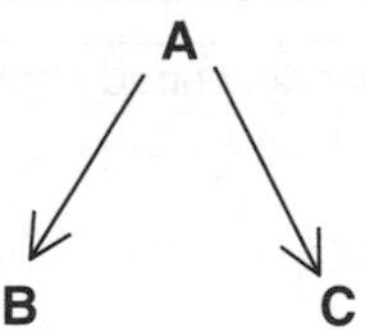

4) **Non-standard tree.** Some hierarchical relationships look rather like the roots of trees. These are known as non-standard trees.

Directed graph:

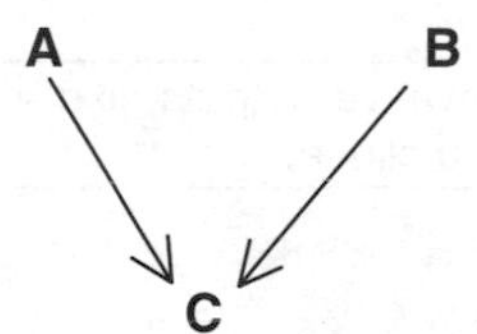

5) **Standard cycle.** When one industry supplies another with innovative outputs, and that industry in turn supplies another, and eventually supplies innovative outputs back to the original industry, we have a standard cycle, with circularity.

Directed graph:

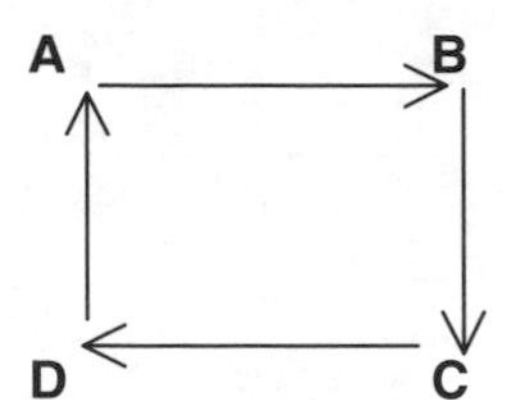

6) **Non-standard cycle.** When, however, complete circularity of supply relationships between the industries does not occur, we have what is known as a non-standard cycle.

Directed graph:

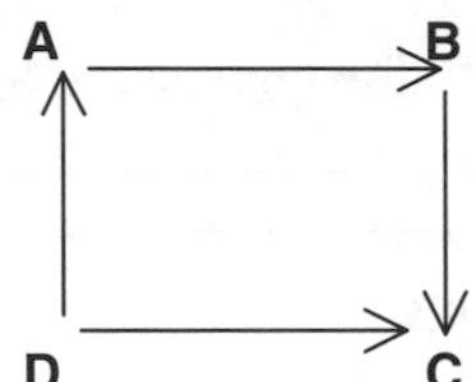

From the six elementary structures presented above, it is possible to define three *composite* structures, thus covering *all* possible clusters:

1) **Simple agglomeration.** This is a composite cluster with even less integration.

Directed graph:

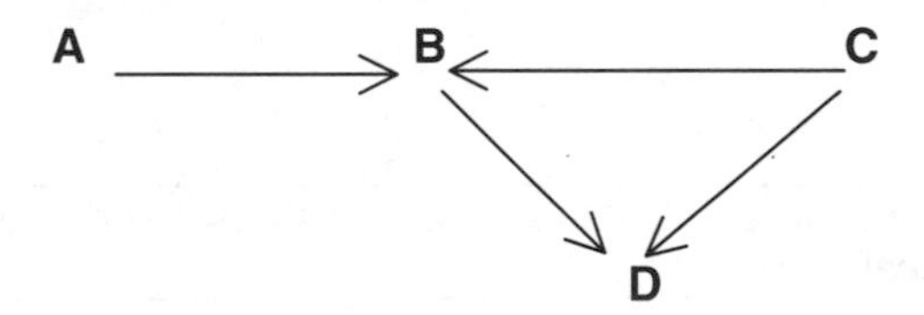

2) **Technological complex.** A composite cluster with less integration than that described above.

Directed graph:

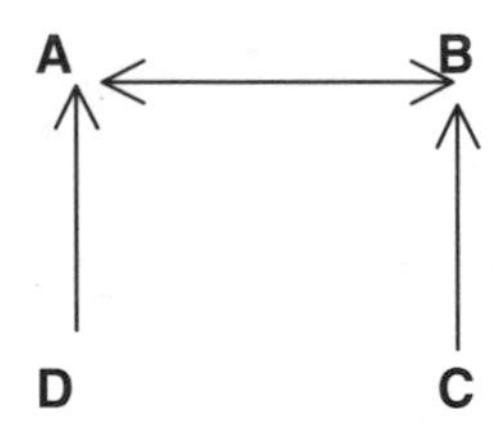

3) **Clique.** When all participating industries supply each other with innovative outputs, like a tightly knit group of friends who all interact with each other, this relationship is referred to as a clique.

Directed graph:

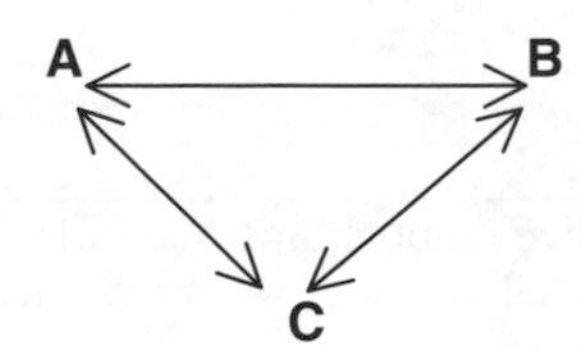

Robustness. Although this method of analysis is simple and lends itself to straightforward visual interpretation, it is also logically (mathematically) and empirically (statistically) rigorous. There exists only one corresponding matrix to each graph and vice versa. Mathematicians say there is a *bijection*

between the set of matrices and the set of directed graphs.[12] Given a specification of minimum linkage or interaction between two members,[13] a computer algorithm is capable of rigorously identifying a digraph in a controlled and reproducible form from a matrix.

Sensitivity. Graph analysis is very sensitive to the minimum level of linkages between members. As there must be some static noise and imprecision in the process of surveying, low levels should not be considered. What criteria should be used to chose a minimum level of linkage? Such a choice must not be arbitrary and must be consistent with the analytical goal. As our goal is to identify closeness and integration in the clustering, we chose the following criterion: a minimum level of innovative interaction without, however, losing a higher-order graph. Concretely, this means that the algorithm is rerun with higher and higher threshold levels until one loses a higher level of graph; one then takes one step back.

Interpretation. These different directed graphs represent different levels of integration in different types of clusters of innovative interactions. The highest order cluster, or di-graph form, is the clique, then the technological complex, simple agglomeration, standard cycle, non-standard cycle, tree, couple and point. Systems theory tells us that higher-level graphs will absorb external shocks better, without changing form, whereas lower-level graphs with less integration and feedback loops will be more likely to change structure.

An example of an application with Italian data

Let us take the example of Italy's 1981-85 innovative activity. The Italian system of innovative interactions appears to consist of two separate sub-systems: a simple innovative agglomeration around consumer goods and a non-standard tree in producer goods, with weak links between them.

How can the graphs be interpreted? In Figures 8 and 9, the arrow indicates the direction of economic exchange activity that sustains the innovative interaction – that is, the direction of the supply of goods and services to a user. Between suppliers and users of innovation, however, information flows both ways. Often the user gives more information to the supplier than vice versa. More importantly, the relationship must be seen as an interaction from which new knowledge is created for both. The arrows in Figures 8 and 9 refer not to information exchange but to the normal economic activity in the exchange of goods that sustains the economic activity.

Main findings. The *primary agglomeration of innovative interactions* is formed by innovation clustering around *final demand goods* (Figure 8). Many industries supply innovation to two main users. This cluster accounts for the largest share of innovative linkages in the Italian system.

The second cluster of innovative interactions, a non-standard tree (second also in number of industries and innovative interactions) is in producer goods around the chemicals, metal products, mechanical-machinery, and automobile industries (Figure 9). This cluster of innovative industries accounts for a much smaller share of innovative linkages in Italy.

In 1981–85 the producer-goods cluster of innovative activities had a weaker level of internal integration than that in the consumer-goods cluster. The two clusters are only weakly related by a few industries: machinery, metal products, electrical, electronic, and chemical innovations. These are the only potentially *nodal* industries.

Figure 8. Final consumer innovation agglomeration
16 industries, 17 interactions

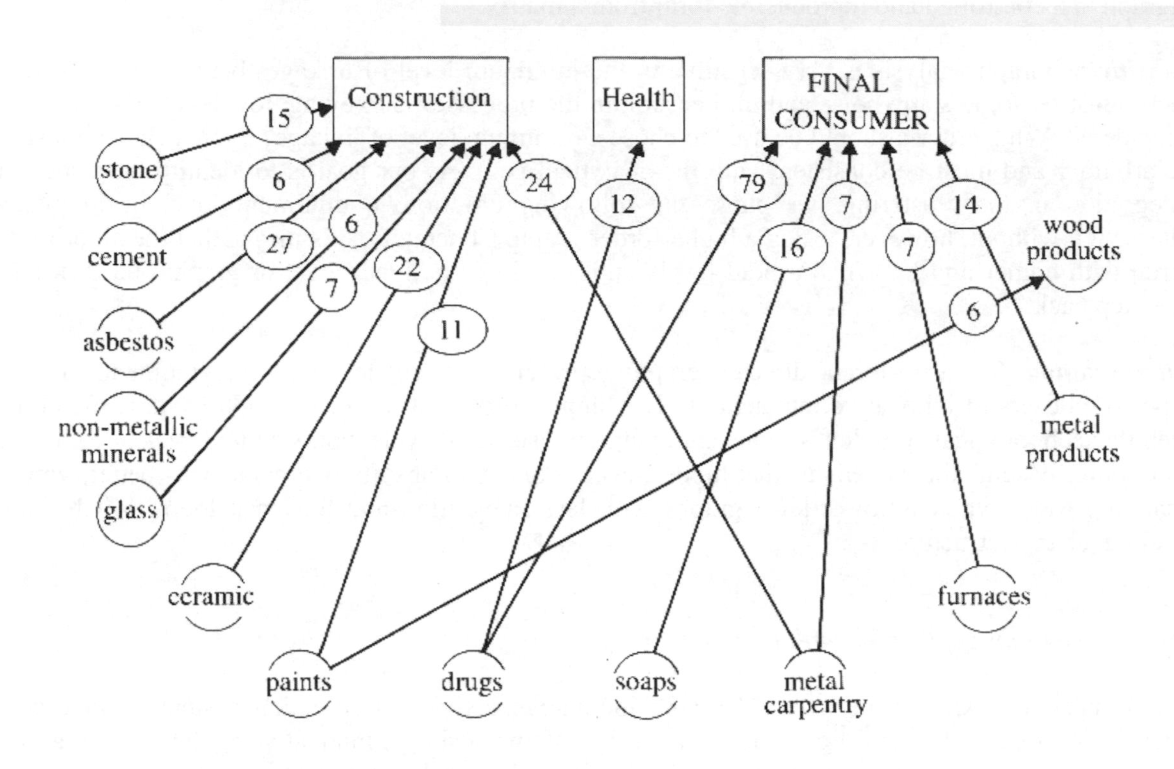

Figure 9. Innovation agglomeration in producer goods
14 industries, 15 interactions

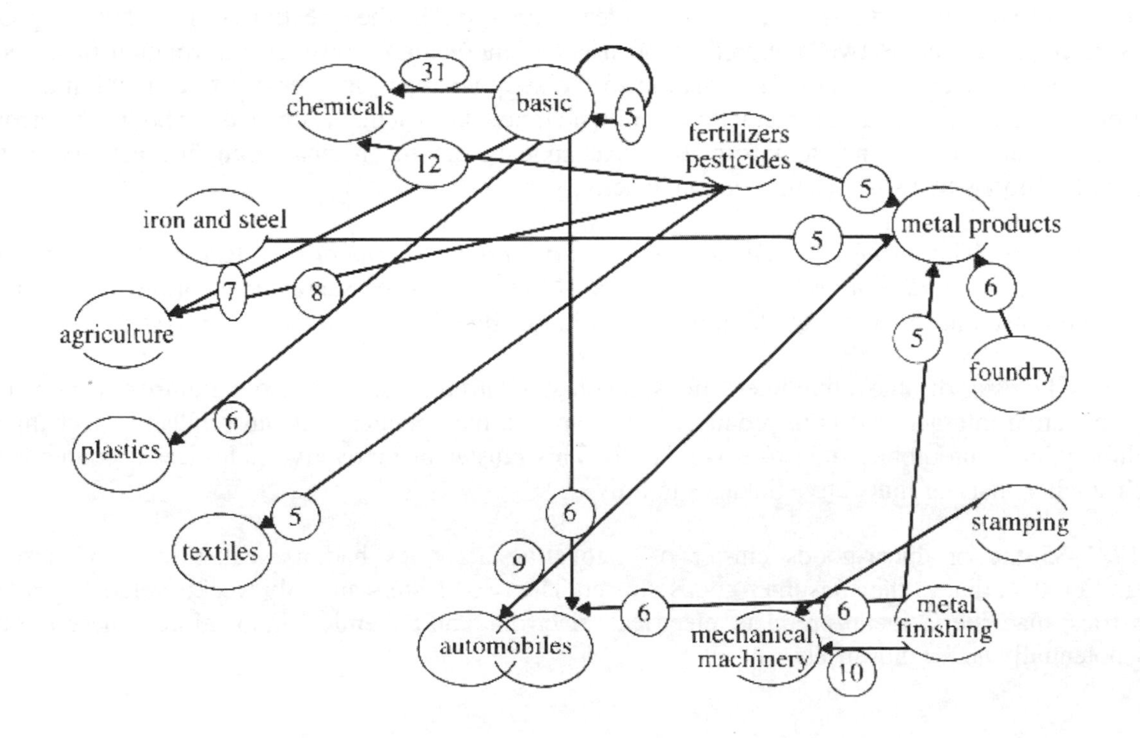

The system of innovative activities in Italy is more heavily oriented toward final consumer goods than towards capital goods or other producer goods (see DeBresson, 1996, Chapter 7). This could explain why, in the early 1980s, a modest amount of well-focused technological effort, much of it with imported inputs, contributed considerably to economic growth and benefits in Italy. As well, the lack of integration might be associated with the flexibility of the Italian industrial system.

Preconditions regarding the use of directed graphs. In order to yield interesting information, this type of analysis requires a disaggregated data set of at least three- or four-digit industries and compiled innovative interaction matrices from direct observations lifted from surveys. If such analysis is carried out on more aggregated data sets or from estimated innovative activity matrices, it will not yield interesting or useful information.

Advantages. The first essential value added is the ability to visualise graphically *which* industries innovate *together*. Second, they are able to pin-point the *weakest links* in a system, precious information for long-term technological policy and the prospect of stimulating new linkages, and the emergence of new clusters and, over time, helping the structure and dynamics of the economy to evolve. Thirdly, these graphics are descriptive and *easy to understand* by policy makers. They focus the attention of the policy maker and convey the essential information with minimum effort from the reader or audience. Policy makers like such graphics because essential information can be efficiently communicated to their superiors. They can become instruments in rhetorical arguments for policy advocacy. However, at this stage, not too much can be expected as to analytical value added.

Data limitations. Even this simple mathematical calculus is too sophisticated compared with the state of the art in surveys and measurement of innovative interactions. To clarify, in a research system, like any system, the efficiency of the system is limited by its weakest component. Research requires the combination of theory, observations, measurement and analytical tools. The weakest points in this particular analytical application are the methods of observation and the means of measurement of innovative interactions between sectors. Until these have been refined, our personal evaluation is that the intellectual purchase of the use of directed graph is limited to the three advantages stated above. In particular, until better measurement of the learning and acquisition of new technological capability and more reliable measurement of innovative investment and costs in innovation surveys are obtained, the actual values of the linkages between members (the arcs of the graph theorist or conditional probabilities in the corresponding Markov chain) should not be given too much importance.

Some improvements are underway. The Danish survey made some important improvements in the measurement of interactions (by asking questions about the importance of the user's contribution), and it may be worth using this technique on a Danish disaggregated matrix. One could also draw from the CIS survey the index on the importance of contact with clients in order to weigh the measure of these arcs and run the directed graph analysis only on the sub-set of innovative outputs "new to the industry" (as opposed to "new to the firm") and "significant" (as opposed to "incremental"). One could also attribute weights to arcs according to the costs and investments of innovative activity. Further work is required.

3.6. Relationship between innovation clusters and industrial complexes

Analytic opportunity. One would expect that innovative clusters would be located within inter-industrial complexes (Czamanski, 1973), that innovative poles be located within growth poles. As there is a fairly established tradition of I/O analysis for identification of industrial complexes (although policies related to these have been abandoned), and now that we know how to identify innovative clusters, we should obviously seize the opportunity to combine the two techniques.

What does this relationship reveal? Inter-industrial complexes are precise and easy to manage reference templates with which to pinpoint clusters of innovative activity. Are innovative interactions always embedded within these complexes? While this seems to be generally the case, on the other hand, one would expect that dynamic development and structural change would come from the innovative interaction of two sectors which had no prior economic relations. Identifying this relationship would be useful for regional planners.

How one can compare industrial complexes and innovative clusters? Graphical comparisons are easy to make. A graph of economic flows between industrial sectors from the inter-industrial complexes identified in the matrix and the corresponding innovative interactions is drawn. Careful visual analysis will do the rest. There are so many data problems with identifying complexes, that it makes no sense to use correlations in this case.

An example: Lombardy[14]

By comparing goods and service flows with innovative interactions between industries in Lombardy (Figure 10), we can identify when these relations overlap and when they do not.

Figure 10. Innovative interactions and exchange of goods within Lombardy

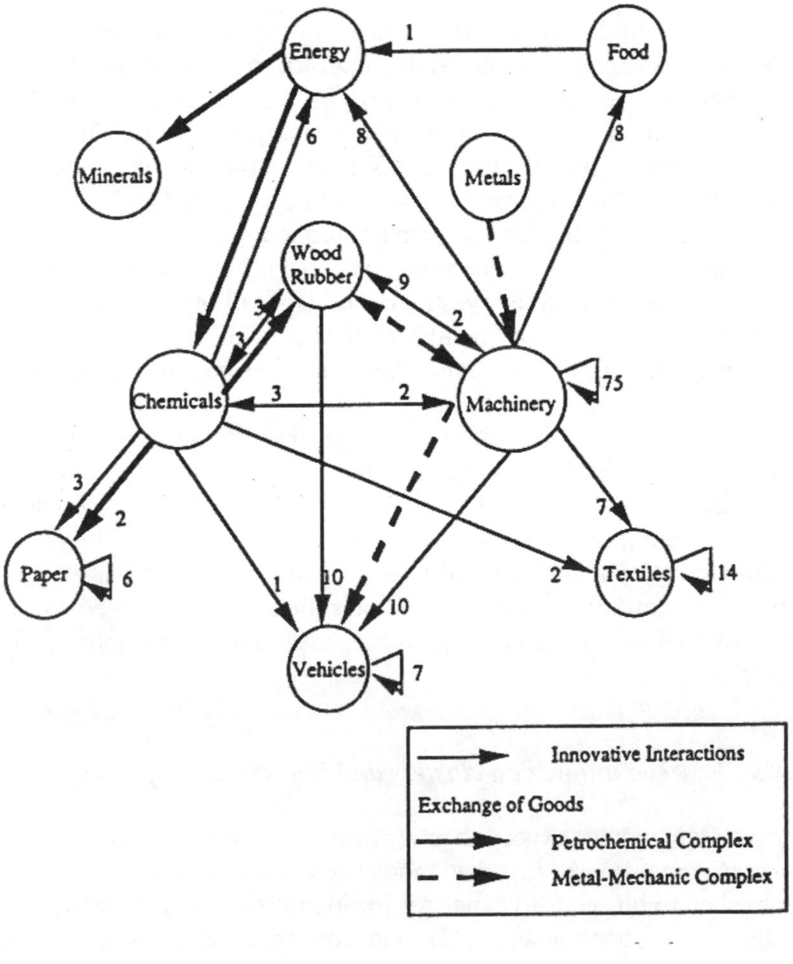

Source: DeBresson, 1996, p. 266.

Main findings. This comparison shows where there is an economically sustainable base for innovative activities: from chemicals to paper to textiles; between chemicals and the wood/rubber sector (in both directions); and between machinery and the wood/rubber sector (in both directions). However, there are also a number of industrial sectors that exchange goods and whose economic relationships could *a priori* sustain innovative interactions but do not (metal with machinery, metals with transport equipment, and the food industry with itself).

More interesting are cases where innovative interactions exist between two industries even though they exchange few goods and services. One would expect important innovations eventually to create new economic linkages between sectors. The machinery industry innovates in Lombardy with the transport-equipment, food, paper, and textiles industries; the chemicals industry innovates with the transport-equipment and machinery industries; the wood/rubber industry innovates with the transport-equipment industry. In all these cases, it is estimated that innovative interactions can occur locally without very strong local economic exchanges.[15]

Data limitations and prerequisites. Because of the nature of the data, we personally are not too confident with the above results for three reasons: *i)* the closed Lombardy matrix was estimated; *ii)* as was the closed Lombardy innovation matrix; and *iii*) the matrices are too aggregated. But we did go along with Italians in their judgement that this analysis did offer some reliable new information. In general, however, one should follow some very demanding data requirements: good disaggregated matrices which are closed, not estimated, and which are constructed from a recent industrial census.

3.7. Typologies or taxonomies of innovative agents: firms or networks?

Accounting for variety. Whatever the detail of information at the level of industrial exchanges gained using the preceding methods, one is not able to satisfactorily capture the richness and variety of different innovative agents. The economic agents classified in an industry can themselves be classified according to agents' attributes, and then mapped in the economy (Cesaratto *et al.*, Chapter 8, in DeBresson, 1996). Industrial classifications are largely arbitrary, and therefore it may be more useful to classify innovative production units, enterprises, business lines, firms or even networks of innovative firms by other characteristics and then, in a second stage, find their location in the economic environment.

What can one learn? Many different types of innovative agents contribute to innovative clusters. Do some play more important roles than others? Do some innovative clusters rely more on some types of firms than others? How can public policy efficiently support this variety of agents? Or, should it support some types more than others?

Review of precedents. There have been half a dozen typological and taxonometric exercises to date concerning innovative agents (Pavitt, 1982; Miller and Blais, 1992; Cesaratto *et al.*, 1993, 1996; DeBresson *et al.*, 1998), but only one as yet has attempted to locate these different types of firms in the economic environment (Cesaratto *et al.*, 1996).[16] Convergent findings confirm that there is a variety of types of innovative organisation, that no one type is more important and that all seem to be interdependent. Findings also confirm that one can establish a typology or taxonomy which limits itself to a manageable set of half a dozen or so well-identified types, indicating that there are probably some organisation homeostatic congruences at play. Cesaratto *et al.* (1993, 1996) have also shown, contrary to Pavitt's (1982) assumption, that these different types are not sector-specific but exist in various proportions in all sectors. As yet, it has been impossible to determine whether some types of firms are more efficient than others, suggesting that different types of innovative firms maximise different objective functions. However, most of the questions enumerated in the above paragraph have yet to be answered.

How can agents be classified? Three approaches can be taken: *i)* establish an *a priori* common sense typology, informed by theory and observed patterns at hand; *ii)* use taxonometric methods, such as principle component and quick cluster analysis, more or less allowing the data correlations and distances to classify themselves; *iii)* use the second method to substantiate intuition, but then use the first approach.

What is being classified? A key choice has to be made as to the unit – or taxon – to be classified. Firms, networks, industries, regions, etc., can be chosen, but it is important to choose only variables which exclusively refer to the taxon itself – not its environment.

How can classification be successfully achieved? It is not obvious at the outset that efforts at classification will be successful. Most units in the population should be easily classifiable as belonging to one class; principle component analysis should explain most of the variance. The units should be much closer to each other within that class than with units of other classes; or the standard deviation should be a fraction of the mean of variable values within that class. Trial and error is the only way to find out.

Example: Classifying ten European countries' innovative firms and networks

1) Classification of innovative firms

Taking only the firm-specific variables of the CIS1[17] (to the exclusion of their sectors or network relationships), a principal component analysis was run. This procedure basically collapses together all into one component those variables which are highly correlated with each other. In our case, this resulted in three components: weight of innovation versus adoption, market orientation of innovative sales and type of innovative costs. However, this analysis only explained approximately 60% of the variance. In other words, much of the variety of firms cannot be classified using the present survey variables in CIS.

The second step is to perform a cluster analysis with three variables, *i.e.* the variable in each component that has the highest "loading" in the component, provided it is not so skewed in its distribution that it does not enable the population to be split into classes. Many of our variables (size, cost) were skewed and thus had to be transformed. Whatever the number of classes we chose, however, the variance within the class was high, and the standard deviation was a high multiple of the mean within the class. This indicates that, while firms were closer to other firms in the same class, they remained very different from one another.

The number of groups has to be arbitrarily chosen on the basis of judgement so that the results are sensible. This is where we come back to Method 1, *i.e.* an *a priori* typology, but one which has been informed by prior metric analysis.

The resulting classification of innovative firms separates the adopters – the majority – from those who introduce new products and processes to the industry. The two populations can then be broken down according to whether they are principally domestic market or export-oriented in their innovative sales, and whether their innovative expenditures are mainly current or constant investments. This results in the following classification (Table 5).

Table 5. A tentative typology of innovative firms in Europe

Class	Features	Share of population
Innovators	Introduce products and processes which are new to the industry; some are exporters, others invest heavily in equipment, and some do both.	12%
Adopters only through fixed investments	Adopt innovations almost exclusively through acquisition of machinery	34%
Adopters, exporters and investors	Adopt innovations for the export market and make fixed investments	13.5%
Exporting adopters	Adopt and export, but make few fixed investments	~13%
Marginal adopters	Even adopted innovations are only a fraction of their sales; few fixed investments; no exports	A large tail-end of the population ~27%

2) Classification of innovative networks

Using the same hybrid method, we chose another taxon: networks of firms. With an even smaller number of variables relating to networks, it made no sense to perform a principle component analysis. A quick cluster analysis with a different number of classes provided some information about the metric split, which suggested the classes or types of innovative networks portrayed in Figure 11.[18]

Retrospective assessment. Although taxonomies and typologies may be useful descriptive exercises they are not always successful nor do they always yield results useful for public policy. Innovative surveys do not yet enable the majority of firms to be classified in a satisfactory way: efforts made to date do not explain a large enough proportion of the variance, nor are firms close enough to each other on any dimension within their class to convince the analyst that clear cut classes exist. However, we have been more successful with a very simple classification of innovative networks.

What is the correct unit of analysis? It may be that the unit chosen for the classification is not the right one. Other analytical results from the CIS show that the majority of innovative efforts require networks of firms, usually involving at least two firms and very often involving quite complex networks (DeBresson *et al.*, 1998; OECD, 1999). If such results are confirmed, then not firms, but networks of firms are the appropriate classification unit.

So what? The most interesting question, however, remains to be answered: Are certain types of innovative networks more prevalent in certain industries or inter-industrial innovative clusters? This question was suggested by Theo Roelandt. The answer could be found relatively easily, by mapping each innovative network by the industry of the respondent firm. However, further research is required.

Figure 11. Types of innovative networks

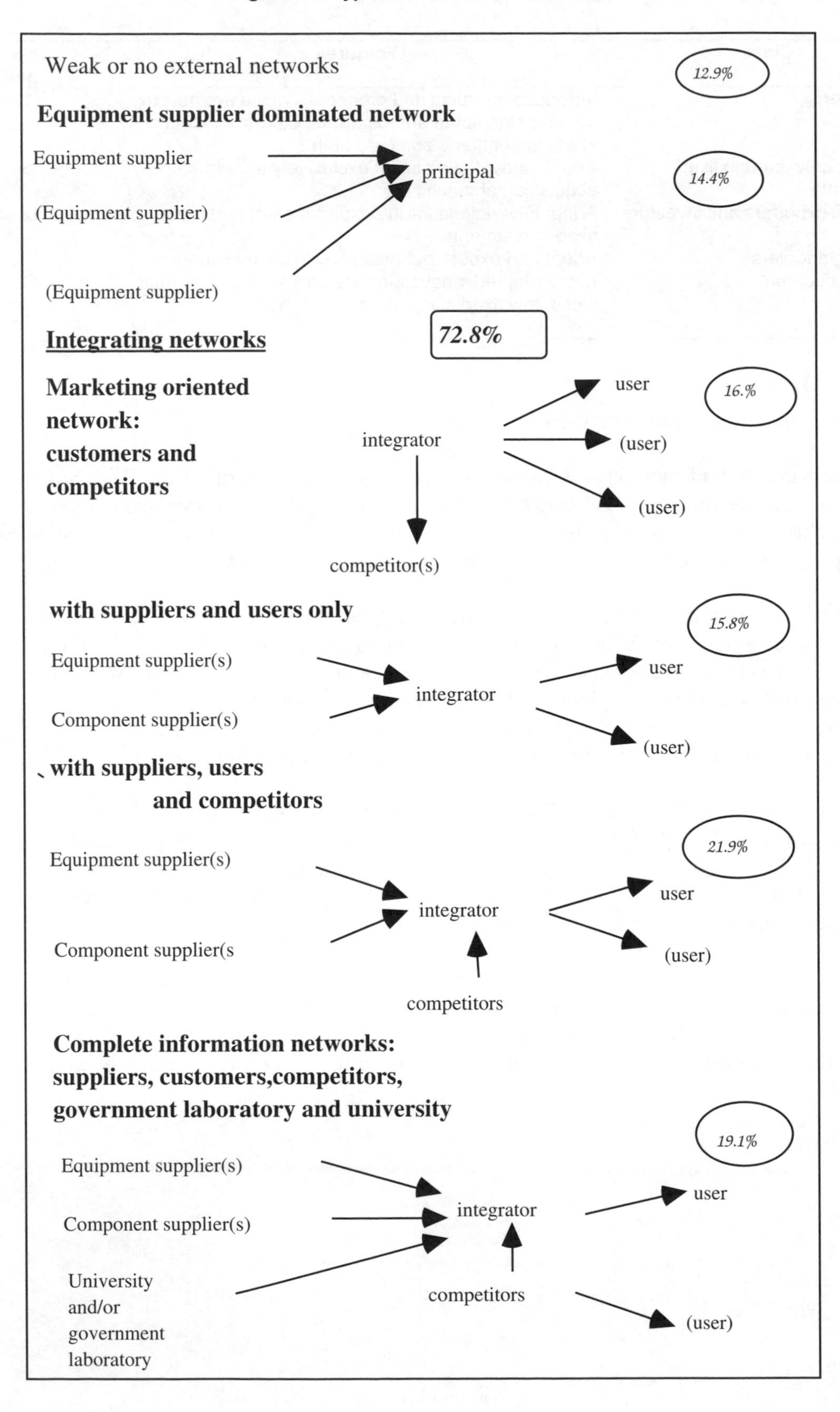

Conclusion: Progress to date

This paper presents a new approach, using innovative matrices to identify clusters and a preliminary toolkit of analytical instruments. The evaluation of the toolkit is summarised in Table 6 below.

Table 6. Summary of instruments, reliability, information yield and ease of communication

Analytical instrument	Maturity and reliability of technique	New information yield	Ease of communication to policy makers
Comparison with IO	OK	*Most innovative vectors in economy*	Needs highlighting and commentary
Triangulation and regressions	OK	*A great deal of detailed information and causal analysis*	Table too busy; regression graphics are better
Concentration ratios	OK	*Degree of clustering*	Simple and straightforward
Comparing concentrations	OK	*Relative innovativeness of the economy*	Simple and straightforward
Graph analysis	Requires refined survey measurements	*How many clusters are members of each integration. Missing links*	Tells a story, easy to communicate, little commentary required
Industrial complexes	Data problems	*Strong and weak links*	Easy to communicate
Typologies and taxonomies of agents	Probably not yet	*Sometimes. Different types of innovative agents*	Meaning not yet clear

Some progress has been made in identifying clusters of *innovative activity*. Using this new approach and simple tools, we were able to objectively establish beyond doubt that innovative activity always clusters, but to different degrees in different countries, and in different industrial areas in different countries. These different economic locations will have different economic impacts. Even when clusters in different countries encompass similar industries, they differ in scope and shape. Inasmuch as Perroux's development poles and Dahmen's development blocks both imply innovative clusters, we have contributed to identifying one of their key components.

The progress that has been made, however, is mainly due to new data, rather than to analytical sophistication. Simple analytical techniques sufficed for the job. More sophisticated techniques are likely to yield further insights and either make the initial ones more robust or qualify them.

There is one exception. With directed graph analysis we may have almost reached the limits of the present data. Our measures of the *interactions* between innovative partners are too sketchy to give any credence to the quantitative weights of the linkages within the clusters (Section 3.5).

However, this is just the beginning of a research programme which has been given a new life with the advent of representative and comparable innovation surveys. Our understanding of different levels of innovative performance in similar clusters remains limited. We know very little about the extent to which inter-firm innovative networks are contained within and/or straddle different inter-industrial clusters, regions and nations. Much can be learnt about differences between innovative clusters in different countries.

Some directions for future research are obvious: *i)* more varied measurements of innovative activity should be used, in particular those relating to costs, current costs and fixed investment, and economic benefits of innovative activity (revenues minus costs) in a way that can be linked with economic analysis; *ii)* more relative measurements, such as propensities, need to be developed to aid in identifying clusters; *iii)* efforts should be made to identify patent clusters – representing technological opportunity – and compare them with innovative clusters; *iv)* with CIS2, an effort should be made to carry out inter-temporal comparisons and predict the location of clusters; and *v*) analysis should be deepened as to the factors influencing innovative performance, in particular with regard to four factors: degree of networking; embeddedness in clusters or interindustrial linkages; exposure to foreign technology; and firms' own efforts and investment.

NOTES

1. One could interpret these matrices as flows of information – as neo-classical economists studying *externalities* and *spillovers* do. What matters here, however, is not the information itself but what is done with it, *i.e.* innovation. This denotes a capability to access, select, interpret, use, assimilate and add value to the information – so that it becomes much more than simple information.

2. This method was developed in collaboration with Philippe Kaminski; for further details see DeBresson *et al.* (1996), Chapter 11 and the related Appendix 4 for the technical details of the procedure.

3. This total will not correspond to any equation in standard Leontief or Stone balances, but is *ad hoc* for our exercise in locating the innovation in the economic space. As we do not later perform any balance analysis, this *ad hoc* procedure has no negative consequences.

4. One favourable reviewer of our work rather hastily suggested that correlations between innovation surveys and I/O tables afforded a way to economise the expensive outlay of innovative surveys (Schnable). While these correlations may allow us to more or less correct for the absence of direct information on users, and this only in the restricted case of large economies, they do not enable us to dispense with innovation surveys representative of the whole economy.

5. It is unfortunate that countries go to considerable expense to perform innovative activity surveys while omitting to ask, for only a small marginal cost, the key question of the economic use of the innovative output, when it has been proven eight times out of nine that respondents lose little time in replying, quite accurately checking a list of user industries, and that the cost of the extra question is incremental.

6. This draws on many chapters from DeBresson (1996) and DeBresson *et al.* (1994).

7. The other important dimensions of the learning economy are the inventive, R&D, formal education and capital goods matrices. However, the first three of the above dimensions are only *potential* or *capabilities,* in the sense that these inputs may not be ever actively used.

8. One solution may be to make these triangulated matrices three-dimensional as in Figures 1 and 2.

9. Figure 3 was obtained from the 1990-92 Italian data, whereas Table 4 was calculated using 1981-85 data.

10. As an example, this directed graph has the following mathematical definition: a summit is called a development point if and only if (if) (i,l) is an element of D and for each J=i, (i,j) and (j,i) are not elements of A.

11. Whether or not innovative activity is totally contained within an industry obviously depends on the level of aggregation. It is therefore desirable to take the highest level of disaggregation possible; however, respondents are incapable of identifying users in greater detail than three- or four-digit SIC.

12. This *qualitative* topological analysis can be extended to become *quantitative*. By attributing the frequency of the cell to each arc in the directed graph, and considering this quantity as the conditional probability p_{ab} (if industry *a* is supplying innovative outputs, it will do so to industry *b*), we can model the process as a Markov chain where the output coefficient matrix becomes the transition matrix of the Markov chain. Here again, there are bijections between graphs, matrices and Markov chains. We do not, however, use quantification in a Markov chain here because we believe it would be stretching the reliability of the data information (see the end of this section). But the opportunity

for such an extension does open the door to dynamic analysis (DeBresson, 1996, Chapter 22; DeBresson, 1991).

13. Graph theorists call the value of relations the "arc value".

14. This technique was developed at the initiation of Alberto Silvani, senior researcher at the Institute for Studies on Scientific Research and Documentation, National Research Council in Rome and present science policy advisor to the Italian government (DeBresson, 1986, Chapters 17-18).

15. Three remarks should be made on this interpretation of Figure 10. First, intra-industrial exchanges are notoriously poorly represented within I/O matrices, and therefore we cannot assume that important economic intra-industrial feedback does not exist for most sectors in this graph. Second, it may be that interindustrial economic exchange among some sectors does exists, straddling Lombardy and other regions, yet the innovative interactions are in Lombardy. Third, we have neglected the energy category because of a lack of correlation in previous studies.

16. Pavitt's (1982) seminal paper was both a classification of sectors and firms, and was used subsequently by scholars as a reference to classify firms and industries.

17. As yet, this analysis has been performed only on Eurostat's micro-aggregated data, which reduces the variance; this implies that results with real data would be weaker.

18. As the CIS surveys only supply the information linkages of the respondant, but not the linkages of its partners among themselves, it is not possible to use directed graph analysis with CIS to establish a rigorous taxonomy.

REFERENCES

Cesaratto S. and S. Mangano (1993), "Technological Profiles and Economic Performance in the Italian Manufacturing Sector", *Economics of Innovative Output and New Technology*, 2, pp. 237-256.

Cesaratto, S., S. Mangano and S. Massini (1996) in C. DeBressson (ed.), *Economic Interdependence and Innovative Activity: An I/O Analysis,* Edward Elgar, Cheltenham.

Dahmen, E. (1950), *Entrepreneurial Activity and the Development of Swedish Industry 1919-1939*, American Economic Association Translation Series, Homewood, 1970.

Dahmen, E. (1988), "Development Blocks in Industrial Economics", *Scandinavian Economic History Review,* 36, pp. 3-14.

David, P.A. and D. Foray (1995), "Accessing and Expanding the Science and Technology Knowledge Base", *STI Review*, Special Issue on Innovation and Standards, OECD, Paris, pp. 13-68.

Dahmen, E. (1991), *Development Blocks and Industrial Transformation: The Dahmenian to Economic Development,* B. Carlsson and R.G.H. Henriksson (eds.), Almquist & Wiksell, Stockholm.

DeBresson, C. (1980), "The Direct Measurement of Innovation", paper presented at the OECD Conference on Science and Technology Indicators, Paris, 15-19 September.

DeBresson, C. (1986a), "Conceptual Notes on the Measurement of Innovation", paper presented at the OECD Workshop on the Measurement of Innovation, Paris.

DeBresson, C. (1986b) "A l'ombre de la dynamo technologique : technologie, innovation et pôle de développement", *Politique*, No. 10, pp. 55-90.

DeBresson, C. (1987), "Poli tecnologici di sviluppo: verso un concetto operativo", *L'Industria: revista de economia et de politica industriale.*

DeBresson, C. (1989a), "Les pôles technologiques du développement : vers un concept opérationnel" *Revue Tiers Mondes*, XXX(118), pp. 245-270.

DeBresson, C. (1989b), "Breeding Innovation Clusters: A Source of Dynamic Development", *World Development* 17(1), pp. 1-16.

DeBresson, C. (1991), "Technological Innovation and Long Wave Theory: Two Pieces of the Puzzle", *Journal of Evolutionary Economics,* 1, pp. 241-272. Reprinted in C. Freeman (ed.) (1996), *Long Wave Theory*, Edward Elgar, Cheltenham, pp. 392-423.

DeBresson, C. (1996), *Economic Interdependence and Innovative Activity: An I/O Analysis,* Edward Elgar, Cheltenham.

DeBresson, C. and F. Amesse (1991), "Networks of Innovation: A Review and Introduction to the Issue", in C. DeBresson, R. Walker and J. Utterback (1991), pp. 363-379.

DeBresson, C., X. Hu, I. Drejer and B.-Å. Lundvall (1997), "Innovative Activity in Some OECD Countries: A Comparison of Systems", report to the OECD Secretariat.

DeBresson, C. and B. Murray (1984), *Innovation in Canada: A Report to the Science Council of Canada,* 2 volumes, CRUST reprint, New Westminster.

DeBresson, C., B. Murray and L. Brodeur (1986), *L'Innovation au Québec*, Les Publications du Québec, Quebec.

DeBresson, C., G. Sirilli, X. Hu and F.K. Luk (1994), "Structure and Location of Innovative Activity in the Italian Economy, 1981-85", *Economic Systems Research,* 6, pp. 135-158.

DeBresson, C. and J. Townsend (1978), "Note on the Interindustrial Flow of Technology in Postwar Britain" *Research Policy*, 7(1), pp. 48-60.

DeBresson, C., R. Walker and J. Utterback (1991), "Networks of Innovators", special issue of *Research Policy,* 20(5).

Knobe, D. and J.H. Kuklinski (1982), "Network Analysis", *Quantitative Applications in Social Sciences*, No. 28, Sage, London.

Malerba, F. (1993), "The National System of Innovation: Italy", in R.R.Nelson (ed), *National Innovation Systems: A Comparative Analysis*, Oxford University Press, New York.

Miller, R. and R.A. Blais (1992), "Configuration of Innovation: Predictable and Maverick Modes", *Technology Analysis and Strategic Management*, 4(4), pp. 363-386.

OECD (1978), *Policies to Stimulate Industrial Innovation*, 3 volumes (G. Bell, J.-E. Aubert, C. DeBresson, P. Dubarle), Paris.

OECD (1992), *Technology and the Economy: The Key Relationships*, Paris.

OECD (1996), *Innovation, Patents and Technological Strategies*, Paris.

OECD (1999), *Managing National Innovation Systems*, Paris.

Pavitt, K. (1984), "Sectoral Patterns of Technical Change: Towards a Taxonomy and a Theory", *Research Policy,* 13, pp. 343-373.

Perroux, F. (1931), *Les relations économiques franco-allemandes*, M. Giard, Paris.

Perroux, F. (1948), *Le Plan Marshall et l'Europe nécessaire au Monde,* Librairie de Médicis, Paris.

Perroux, F. (1948) "Esquisse d'une théorie de l'économie dominante", *Économie Appliquée*, No. 1, pp. 243 300.

Perroux, F. (1950), "Le Pool du charbon et de l'acier, illusions et réalités", *Nouvelle Revue d'Économie Contemporaine* , No. 16-17, November.

Perroux, F. (1954), *L'Europe sans rivage*, Presses universitaires de France, Paris.

Perroux, F. (1955), "Note sur la notion de pole de croissance", *Economie Appliquée,* 8, pp. 307-320 (summarised in Perroux, 1964, pp. 178-190).

Perroux, F. (1986), "Les concepts d'espace et de temps dans la théorie des unités actives", paper presented at the Lösch Colloquium, Warsaw, Institute for Space Economics, May.

Schumpeter, J.A. (1912), *Theorie des wirtschalisches Entwicklung,* Humblot, Berlin (this edition includes an important Chapter 7 which was deleted from subsequent German editions and translations). English translation (1934), *The Theory of Economic Development: An Inquiry into Profits, Capital, Credit, Interest and the Business Cycle*, Oxford University Press, London.

Chapter 3

NORWEGIAN INPUT-OUTPUT CLUSTERS AND INNOVATION PATTERNS

by

Johan Hauknes*
STEP Group

1. Introduction

Modern approaches to innovation and technical change place increasing emphasis on the systemic analysis of innovation processes and the determinants of innovation performance. Questions about the relationship between firm growth and evolution, industry development and differentiation and its structural macroeconomic effects are at the core of modern innovation analysis and theorising. Resource-based theorising of the firm, finding many of its core ideas and assumptions in Edith Penrose's seminal work (1959), is central to learning-based approaches to innovation systems: learning processes are seen as instrumental for the ability of the firm to shape its own development and environment so as to foster firm growth and industrial development.

This suggests that clusters of firms and industries are (increasingly) important aspects of firm and industrial development. With learning and capability formation as a main basis for innovation performance, and innovation as the main driver of industrial development and structural change, a resource-based approach must be fundamental to the understanding of innovation processes. The importance of local learning environments, or innovation systems, suggests that a natural unit of analysis in aggregate approaches to industrial development may not be the single industry or product market; these learning processes bridge industrial divisions and product groups. Learning is structured through an extended network, and any analysis of industrial development must of necessity consider inter-industrial and inter-organisational interaction.

Mapping such relations, which entails mapping the topology of the innovation system (Hauknes, 1998a), yields an appreciation of the connectivity and separability of systems of innovation. The innovation clusters that emerge from such an analysis form a natural focus for studies of innovation dynamics, just as they comprise the essentially behavioural related variables for such an analysis. A mapping like this obviously would have to include diverse sets of interrelations between firms and their environment in order to cover modes essential for the development of capabilities and innovative performance. It follows from this that such a mapping would quickly bring us beyond relations mediated through market relations.

* The author gratefully acknowledges the comments and coercion of the editors of this volume.

Lundvall's learning-based approach to innovation systems stresses the important role of user-producer relations, and more generally, the relations that are constituted in or through the market place (see, for instance, Lundvall, 1992, and Edquist, 1997). At the firm level, it should be evident that for most firms, their relations with customers, competitors and suppliers are the most significant links to their environment, to the extent that these agents constitute the major dimensions of this environment. It is not unlikely that these immediate relations shape the major learning modes for a majority of firms. This supposition finds strong support in the many innovation surveys based on the OECD's *Oslo Manual* of innovation indicators (OECD/EUROSTAT, 1997), where one very consistent result is the importance attached by the respondents to customers and suppliers as sources of information for innovation.

This paper will focus on economic transactions at a sector level to identify bounded cluster structures in the Norwegian economy. Though an approximation, these structures will form part of a more general mapping exercise. From the importance of customer and supplier relations as channels for acquiring informational inputs to innovation processes, it is also a reasonable assumption that these commercially based relations will form a significant part of a more completely mapped innovation landscape.

Hence, we suggest a general mapping, starting with the identification of dominant characteristics of innovation systems or clusters, starting from the centrality of user-producer links. With transaction data we internalise user-producer links by distinguishing clusters of strongly interacting industries or firms. Using the argument outlined above, the internal structure of these clusters describes dominant channels of learning. In addition, the composition of clusters may be interpreted as a signature of the overall degree of functional diversity of learning by user-producer interactions. Internalising user-producer linkages, we may look for residual modes or fingerprints of innovation at the cluster level. To the extent that specific characteristics of use of innovation inputs and composition of innovation performance survive at cluster level, they are indicators of specific innovation patterns at cluster level, with rather obvious implications for policy formulation. In this paper we will consider the sectoral use of three supplementary input factors to innovation: R&D; outsourcing of knowledge-intensive business services (KIBS), such as management consultancy, various IT services, technical design and engineering services; and use of professional personnel, as proxied by personnel with tertiary education.

On this basis, we will first discuss how clusters may be identified in economic transaction data. The second part of the paper will briefly outline relevant aspects of the six clusters identified. Lastly, we will distinguish innovation patterns of these clusters by the use of supplementary data.

2. Identification of clusters

2.1. The data

National accounts data provide a structural decomposition of the Norwegian economy at the sectoral level that is useful for structural mappings. Input-output tables, an integrated part of the system of national accounts, are the only data that describe economy-wide structural relations. For this exercise, we used Norwegian input-output data for 1993.

The input-output data used is a modified sector-sector table, based on make & use tables at the most disaggregated level available. Norwegian national accounts, based on the ESA 1995, contain nearly 150 industrial production sectors, convertible to NACE classification, distributed over five institutional sectors (Annex 1). At the most disaggregated sectoral level, this gives a resolution into

179 industrial-institutional sectors altogether. For the sake of this analysis, we have aggregated into 161 industrial sectors, with 18 "public" SNA production sectors and 143 "private" production sectors. The resulting input-output table of the Norwegian economy describes flows of intermediate manufactured and service goods between these sectors.

The resulting input-output table has nearly 21 000 sector-sector links,[1] with 151 of the 161 sectors having intermediate deliveries to other sectors.[2] Sectors that do not participate in intermediation networks are all sectors with no or negligible output. Given this denseness of linkages, even a one-step identification of interacting sectors would link almost all sectors, producing one giant, economy-wide cluster. Obviously, there is no true way of identifying clusters. Any algorithm must be developed to balance several factors, the resulting clusters in an ideal sense being structurally stable. We will identify clusters by mapping paths described by sector-sector intermediate flows, looking for weakly interacting substructures of the complete input-output matrix.[3] Thus, the essence of the approach is to look for block-diagonal structures in the input-output table (Leontieff, 1986), where blocks of interdependent sectors form input-output clusters with inter-cluster trade flows being small compared to intra-cluster flows.

The analysis of innovation patterns is supplemented with two further data sources. R&D-based variables are taken from the 1995 national survey of R&D in the business sector, performed by Statistics Norway in 1996. This survey includes questions on innovative performance of firm units,[4] asking firms at unit level about the introduction of new or significantly improved processes or products over the period 1993-95, as well as on the unit's participation in technological co-operation in the survey year. There are thus two main reasons for using 1995 data, rather than 1993 data corresponding to the vintage of input-output data. In contrast to the 1993 survey, the 1995 survey sample was a statistical sample allowing data scaling and cross-sectoral comparisons. Secondly, only the 1995 survey included questions on innovation performance. A further data set prepared by Statistics Norway on the basis of administrative registers links all employees in Norway with their main employers for the period 1986-96. Here we use these data for information on sectoral shares of higher educated personnel in 1993.

2.2. *Cluster definition*

Most of the inter-sectoral links are weak in the sense that:

- links may involve sectors with negligible economic activity;
- the flow along a single emitting link from a given sector is small compared to the other emitters from the same sector; or
- a receiving link is weak in the same sense from the perspective of the receiving sector.

For the Norwegian 1993 input-output table this conjecture is indeed confirmed (see Table 1 for the case of forward linkages). The table describes some reduced input-output networks and compares them to the full network, given in the first row. As suggested by the conjecture, the table describes the reduction in terms of two cut-offs. The first cut-off restricts *link strength* between sectors, restricting attention to links carrying flows above a certain fraction of total intermediate deliveries from each sector. The second cut-off restricts the network to *significant sectors*, sectors representing at least a minimum fraction of total intermediate deliveries.

Table 1. Input-output networks

	Links	Sectors	Delivering sectors	Receiving sectors
Off-diagonal flows	20 985	156	151	156

Cut-off schemes	Cut-off 1	Cut-off 2	Links	Sectors	Delivering sectors	Receiving sectors
Prime links	0%	--	151	151	151	54
	15%	--	107	119	107	45
All links	15%	--	163	125	107	60
	10%	--	249	137	127	73
		1‰	185	108	92	65

The table identifies variant cut-off schemes to illustrate both the reduction effect of cut-offs and the sensitivity of network structure on variations in the suggested cluster-identifying algorithm. The first scheme identifies the maximal forward linkage from any sector, measured in value terms, irrespective of both the relative link strength and sector weight in the total inter-sectoral flows. By definition, this scheme identifies 151 sectors as delivering sectors, and hence 151 links. Restricting to maximal links that represent at least 15% of intermediate deliveries from any sector reduces the number of sectors to 119, with 107 as delivering sectors, and hence 107 links. Retaining this cut-off of link strength while including all links adds just a few new sectors, while considerably expanding the number of links. A further reduction of the cut-off to 10% increases the number of links by another 50%. Thus most links are indeed weak, and with a fairly strong dependence of the density of resulting links on the cut-off. With this cut-off going down, clusters tend to merge. To exclude merging of clusters through inclusion of negligible sectors as bridges, the effect of the second cut-off is illustrated in the last row. A combined cut-off on link strengths of 10% and sector size of 0.5-1‰ produces a network of comparable sectoral coverage to the two 15% schemes, but with a richer network structure.[5] Interpreting significant links as major channels for interactive learning thus requires a consideration of both types of cut-offs.

This suggests a process of identification of clusters in a reduced input-output network, the full input-output table is reduced by neglecting weak links and small sectors. Identification of structural input-output clusters is suggested as a variational principle on reduced networks. We attempt to decompose the input-output matrix of transaction flows to minimise inter-cluster flows, with clusters that are robust towards variations in the cut-offs. Robust clusters remain unchanged in their main structural features over ranges of cut-off parameter values. In addition, we require that the resulting clusters should make economic sense, calling for a qualitative assessment of the sectoral content of the clusters and the inter-sectoral relations. Hence, we should not expect to be able to make a complete decomposition but allow for residual sectors without any clear cluster formation.

Attempts to identify input-output clusters face the problem of overlap between clusters and heterogeneities within sectors. Overlap, inter-cluster trade flows, is minimised by the cluster identifying schemes. Heterogeneities of sectors are closely related to the detail of sectoral classification.[6] The degree of homogeneity of any classification is broadly measured by the number of sectors and the size distribution of sectors within the chosen classification scheme. A high aggregation of any sectoral area may lead to the possible misidentification of clusters that would be resolved into separate parts "belonging" to other clusters in a more disaggregated analysis. This calls, at any level, for a judicious assessment of the cluster structures that emerge from a given classification scheme.

3. Clusters in the Norwegian economy

3.1. Decomposing the Norwegian economy

On the basis of the Norwegian 1993 input-output data, we have identified five reasonably well-defined clusters in the Norwegian economy, as well as a network of information intensive activities. These clusters are:

- *agrofood* industries;
- a cluster representing the main supply network and refining activities related to *oil and gas extraction*;
- a cluster of activities related to *construction*;
- a paper and graphical cluster;
- an inter-related set of *transport* activities; and
- a cluster of information-intensive activities.

The sectoral content of the six cluster networks is described in Annex 2. Including four trade sectors, we have been unable to relate the remaining 55 sectors with any specific cluster, mainly due to their small size. A few large sectors remain outside the identified clusters and networks, the main part is accounted for by public administration, education and health and social services, representing two-thirds of sectoral product outside the clusters and networks.

These six networks accounted for 62% of GDP in 1993, with more than 50% residing in the five clusters (Tables 2 and 3). The share of labour costs is substantially lower, reflecting the dominance of various labour-intensive public services in the residual outside the clusters. The 104 sectors account for nearly 45% of domestic consumption, while their share of exports is more than 75%. Through these six clusters, we have covered a major share of the Norwegian economy along several economic dimensions.

Table 2. Decomposition of the Norwegian economy, by cluster

	Number of sectors	Share of GDP	Labour costs	Consumption	Exports
Six clusters	104	62.3%	46.1%	43.9%	76.8%
Public and social services	15	19.2%	32.0%	38.7%	0.4%
Trade	4	10.1%	12.5%	10.1%	6.4%
Other, excluding trade	35	8.4%	9.4%	7.4%	16.4%
Total	158	100.0%	100.0%	100.0%	100.0%

The clusters vary from highly capital-intensive to relatively labour-intensive. The oil and gas cluster is the most capital-intensive of the clusters, accounting for nearly one-quarter of the national capital stock. Table 3 gives the average capital-labour ratios of the clusters, relative to the national average. The agrofood and construction clusters are medium to high labour-intensive, while the transport cluster is on the higher end of capital intensity. The oil and gas cluster has a capital-labour ratio nearly six times larger than the national total, while the capital-labour ratio of the paper and graphics cluster and the information network is about half the average for the economy as a whole, it is also lower than the capital-labour ratio of public and social services.

Table 3. Clusters in the Norwegian economy

	GDP[1]	Employment	Capital stock[1]	Relative capital-labour ratio
Agrofood	8.9%	11.0%	9.6%	0.870
Oil and gas	17.7%	4.2%	23.4%	5.609
Construction	7.7%	7.9%	5.7%	0.719
Transport	9.2%	7.9%	11.5%	1.456
Paper and graphics	2.9%	2.9%	1.6%	0.569
Information intensive	11.8%	8.8%	4.5%	0.516
Total	58.2%	42.6%	56.3%	1.321

1. Relative to national capital stock, excluding household-owned dwellings.

Following identification of clusters, the full input-output table may be decomposed, an aggregated table describing intermediate flows at cluster level and a series of cluster maps outlining the trade flows within the individual cluster networks. While the latter is considered to be a map of internalised user-producer links in the next section, Table 4 outlines some structures of the resultant input-output table at cluster level. The four penultimate columns distinguish, respectively, intermediate trade flows within each cluster network, between or into clusters, into public administration and social services and into industrial sectors that are not linked to clusters, as shares of total intermediate deliveries from the respective category. The last column gives the share of total intermediate deliveries in total production for each category.

Table 4. Production structure in the Norwegian economy
Percentages

	Intra-cluster[1]	Inter-cluster[1]	Public and social services[1]	Extra-cluster[1]	Intermediate deliveries[2]
Agrofood	79.9	12.2	3.8	4.1	45.1
Oil and gas	55.1	28.4	3.6	12.9	27.5
Construction	50.3	20.4	17.4	11.8	33.6
Transport	50.2	24.8	8.3	16.8	37.8
Paper and graphics	32.2	25.5	13.3	29.0	64.5
Information intensive	27.8	32.9	13.4	25.8	51.1
Six clusters	51.3	23.5	9.9	15.3	39.0
Public and social services	--	62.7	14.4	22.9	6.3
Private	40.0	31.4	10.3	18.3	38.3
Total	38.8	32.3	10.4	18.5	33.4

1. As a share of intermediate deliveries from category.
2. As a share of total production.

For the six clusters, 39% of total production is allotted to domestic intermediate use. The high export share of some clusters indicates that the "real" technological intermediation rate of these clusters is higher. The cluster with the largest share of intermediate demand is paper and graphics, nearly two-thirds of its output feeds into domestic intermediate inputs. Trade with clusters account for 75% of intermediate trade flows from the clusters, most of which is within each cluster. About 10% of intermediate deliveries from clusters are to public and social services, with another 15% to other industrial sectors. For all clusters, intermediate deliveries are dominated by intra- and inter-cluster trade flows, with intra-cluster flows greater than inter-cluster trade for all except the network of information-intensive activities. The lower intra- and inter-cluster shares of the paper and graphics and information networks is to a large extent explained by trade with wholesale trade. Note that the agrofood cluster appears to be almost completely isolated from the other clusters in this table. The two general "networks" of trade and information-intensive activities are more heavily disposed towards inter-cluster trade, reflecting their wider ranging links to many sectors.

We conclude that we have been able to define delimited clusters that cover a major part of the Norwegian economy, with aggregate features that in most measures reflect the overall structure of the Norwegian business sector. On the other hand, there is considerable inter-cluster variety in several economic characteristics, suggesting that the structure of economic change may vary significantly between clusters, supporting the assumption that we should be able to identify cluster-specific modes of innovation.

3.2. *The individual clusters*

3.2.1. *Economic dimensions and inputs to cluster*

This section describes intra-cluster structure and functional content of the individual clusters. Table 5 outlines some relevant economic dimensions of the six cluster networks. The table identifies gross product of each cluster (measured at factor cost) in billion NOK, total employment and employment of higher educated personnel (in thousands), total capital stock. The three last variables indicate forms of intangible investments in the cluster, R&D expenditures, intermediate use of knowledge intensive business services and frequency of use of public innovation policy initiatives by firms in the cluster.

Table 5. Economic dimensions and inputs to clusters

	Agrofood	Oil and gas	Construction	Transport	Paper and graphics	Information-intensive
Gross product (GNOK)	62.0	123.7	53.6	64.0	20.2	84.7
Total employment ('000)	223.7	84.9	161.5	160.6	58.6	193.0
Higher educated personnel (HEP)[1]	4.3	14.6	7.1	4.3	4.8	25.6
Capital stock (GNOK)	171.8	420.4	101.8	206.4	29.4	82.1
R&D[2] (MNOK)	519	2 326	698	251	304	2 864
KIBS inputs[3] (MNOK)	3 571	11 167	4 527	3 468	1 835	12 974
Public assistance[4]	22.5%	29.7%	25.1%	17.1%	19.8%	16.1%

1. Higher educated personnel is defined as personnel with formal education at levels comparable to ISCED 6 or higher.
2. R&D expenditures in 1995.
3. KIBS is defined in the table as the sum of intermediate inputs from sectors 720, 730, 74X, 642, 112, P74X, excluding 747, 800, P800.
4. Share of firms taking advantage of public innovation policy initiatives in the period 1993-95.

The cluster maps in Figures C1 to C6 (Annex 3) outline the main internal trade flows of the clusters, here interpreted as main channels for interactive user-producer learning. We will use this information and the six cluster maps to describe the economic characteristics and internal structure of these clusters, as a way of identifying aspects of processes of economic and technical change. In the next section this will be the basis for an attempt to describe cluster-wide innovation modes.

3.2.2. The agrofood cluster

The gross product of the agrofood cluster amounted to NOK 62 billion in 1993. The capital share of gross product is about 0.55, with a capital labour ratio of NOK 750 000, or ECU 100 000, per employed person. This capital share is fairly high. About 5% of total business sector R&D expenditure is at the low end of the scale, however. One in six firm units in the agrofood cluster performed R&D in 1995, each unit spending, on average, some NOK 4.5 million. One in three firm units are innovators. With less than 3% of employment, the cluster is the least intensive user of higher educated personnel of all the clusters and networks, reflecting the food industry's traditionally high reliance on low- and unskilled labour.

Overall, we may characterise the agrofood cluster as a medium-to-low-intensive user of KIBS inputs. However, inputs from public technology infrastructures (PTI) – intermediate inputs from contract R&D institutes (NACE 73), and educational institutions (NACE 80),[7] – represent about 6%. This may be compared to the share for the total private sector of 4.3%. One reason for this relatively high share of PTI inputs may be the presence of the agriculture sector in the cluster; the agrofood cluster has a substantially higher relative rate of interaction with PTIs.

The agrofood cluster has its main emphasis in the agricultural sector. In addition, we have included the production of fertilisers and other by-products of the chemical industry, a major input factor into the agricultural sector. Similarly, fisheries appear as a major user of the output of Sector 351 – shipbuilding. The agricultural sector plays a key role in the cluster, generating key inputs into the major parts of the food industry, meat products, dairy products and milled products, as well as the residual Sector 158, dominated by bakery products. The interaction with oil and gas extraction reflects the catering activities at oil and gas installations in the North Sea.

With the inclusion of recipient sector specification of investment data, we would expect some interaction with investment goods producing sectors to show up in the data, reflecting an expected significant role for capital-embodied technological change in these industries, as shown by the relatively high remuneration of capital inputs indicated by the capital share.

Besides its agricultural focus, the cluster is weakly coupled to fishing, fish farming and related manufacturing. The only input to fish farming that survives the various cut-offs is fish feed, which again is used with manufactured fish products as the basis for feed production. The "fishy" sub-cluster is unrelated to the more general agrofood cluster other than through this indirect route. The dominant mode of trade flows from these sectors is directly to final demand sectors, either through domestic consumption, or through the substantial exports of fish and fish products. This seems to be a basic characteristic of the agrofood cluster; with few exceptions, it consists of industries that might best be characterised as vertically integrated industries fed by the agricultural sector.

3.2.3. The oil and gas cluster

The oil and gas cluster is dominated by its prime functionality, the extraction, refinement and distribution of petroleum products. Total product in 1993 accounted for NOK 124 billion, or about 16% of total GDP. The dominant feature of this cluster is its capital intensity. Total employment in the cluster, about 85 000, amounted to just 4% of total employment in Norway in the same year, while the capital share in gross product was more than 75%. The capital labour ratio is by far the largest in Norway. The forward linkages are mainly restricted to its production chain, with little indication of significant multiplier effects. The input structure is more diverse, but its structure and possible backward multiplier effects are closely related to its capital intensity. The wider technology intensity, through KIBS inputs, higher educated personnel and R&D expenditures, is similarly related to its intensity in physical capital. These tangible and intangible aspects of the cluster's capital intensity, together with its size and network structure, suggest that the main impacts of the cluster on the overall technological performance of the Norwegian economy are at the macroeconomic level; on development of total investments, effects related to changes in the offshore sector's investment activity on supplying industries and factor markets, and the effects on specialised labour markets.

The oil and gas cluster is a prominent feature of the Norwegian economy along almost all dimensions. Considering that about 50% of the total labour stock with education at ISCED level 6 or above is employed in education and social welfare systems, the cluster's share of the higher educated labour stock outside these sectors is more than 13%; a share comparable to that of business sector R&D expenditure and KIBS inputs. In total, this cluster plays a decisive role in Norwegian technological performance, and hence gives a considerable impetus to the development of "intangible capital" providing sectors.

From the cluster map, it might reasonably be asked whether these sectors should be identified as a cluster. The cluster structure describes oil and gas extraction, its supplying industries and two major "user" industries; the petrochemical industry and the sector of pipeline transportation of oil and gas. The main interactions with other clusters are with the transport cluster on the right, concentrated on marine transport and shipping (Sectors 611 and 613).

We note three features. First the "supply chain" structure of the cluster suggests that it is open-ended: the network may be expanded in particular through the supply industries of the oil and gas sector, an expansion that would rapidly cover a diffuse and functionally diverse cross-section of major parts of the Norwegian economy. In this sense, this cluster, in contrast to those described above, is not insensitive to the application of cut-offs. Through a qualitative assessment of the linkages, their strengths and the functionalities of the sectors involved, we have limited the cluster to the core structure.

Secondly, we note the inclusion of pipeline transport in the oil and gas cluster, rather than in the transport cluster. Further, Sector 608 is economically separated from the sectors included in the transport cluster.

Thirdly, the sector of engineering and technical consultancy and related activities has been included in the cluster. The size of the petroleum-extracting sector and the technologically complex nature of offshore petroleum extraction and processing, makes it no surprise that the sector is the major user sector of engineering and technical consultancy. In fact, Sector 742 (NACE 74.2 and 74.3) represents more than 70% of the KIBS inputs into oil and gas extraction, or 16% of its total domestic intermediate inputs. The oil and gas cluster is the recipient of nearly 50% of the intermediate output of Sector 742. The total KIBS inputs have three equally sized components:

- trade flows within Sector 742;
- services from Sector 742 to oil and gas companies;
- KIBS inputs from other KIBS sectors.

As discussed above, an extension from the intermediate flows to include investment flows will probably further increase the strength of the interaction between the two sectors of technical and engineering consultancy and oil and gas extraction. Hence we have included Sector 742 in this cluster.

3.2.4. The construction cluster

The total gross product of the construction cluster is about 8% of national GDP. The cluster's share of the total capital stock of about 4% indicates a fairly low capital intensity; it also has the lowest capital share of these clusters. The capital labour ratio of the cluster is estimated at NOK 0.63 million. This figure is comparable to similar ratios for the agriculture and paper and graphics clusters. The lower capital share in GDP suggests, however, that the remuneration of capital is lower in the construction cluster. R&D measures suggest that the overall R&D expenditure in the cluster amounted to some NOK 700 million in 1995, corresponding to about 1.3% of value added, comparable to the oil and gas cluster. There is, however, a considerably skewed distribution across sectors in R&D performance. When Sectors 243 (chemicals, paints and varnishes), 246 (other chemicals) and 311-314 (electrical machinery) are included in the cluster, about two-thirds of total R&D expenditures are covered.

The structure of KIBS inputs shows a different pattern; these industries represent less than 9% of the cluster's KIBS inputs, while the construction sector NACE 45 accounts for more than half the KIBS inputs. The largest contribution of KIBS inputs into these core sectors is from technical and engineering consultancy, Sector 742. The cluster's share of HEP in employment tends towards the lower end of the scale.

The construction cluster is a network with a constituent part located in the inter-relationships between Sectors 452, 453 and 454, corresponding to NACE 45.2 and 45.3. The core role played by these three sectors is evident in the cluster map; the triangle of these sectors forms the backbone of the cluster. The left-hand side of the map, concentrated around Sectors 452 and 454, concerns erection and completion of buildings. The three wood-product industries (Sectors 201, 202 and 203) are linked to this sub-network, highlighting the position of wood as a construction material. Electrical and plumbing installation, painting and similar activities are located in Sector 454, reflected in this sector's input structure towards the lower left-hand side of the map. The map suggests that production of furniture, Sector 361, is included in the cluster. Rather than reflecting the general structure of the Norwegian furniture industry, this relates primarily to a specific part of this sector, the production of complete kitchen installations. As the kitchen furniture industry (corresponding to NACE 36.13) represents only a minor part (about 15% in value-added terms) of the NOK 2 billion Norwegian furniture industry, we have chosen to exclude this sector from the cluster.

The map includes a group of industries producing mineral products, such as glass products, ceramics, bricks and tiles and cement and plaster. This group of industries serves all three construction industries, including the civil engineering dominated activities of Sector 453. While the building-related sectors are suggested in the map as being more closely related to building property through "user" Sectors 700 and 704, the engineering-based activities of Sector 453 are closer to public administration and utilities as major user sectors, probably to a large extent reflecting major infrastructure projects, such as roads, or the construction of the new Gardermoen airport, etc. Road construction is a significant part of the explanation of the presence of the public sector P453, including the construction activities of the National Road Authority (*Statens Vegvesen*). The three core sectors are characterised by a share of intermediate deliveries in total production of 25-50%, and by final demand almost exclusively being oriented towards investment. A sectoral resolution of these investment flows will probably further increase the linkage between these sectors and their major "user" sectors.

3.2.5. The transport cluster

The 9% share of the transport cluster in national GDP in 1993 is similar to its employment share. The capital-labour ratio is significantly larger than the ratios for all the other clusters, with the exception of the oil and gas cluster. The cluster scores low on all intangible dimensions compared to other clusters. R&D expenditure in 1995 of NOK 250 million implies an R&D intensity of 0.4% relative to GDP, the lowest intensity of the six clusters and networks we have described here. Furthermore, about three-quarters of these R&D expenditures are accounted for by the sector's manufacturing transport equipment, NACE 34 and 35. However, with HEP personnel, the two manufacturing industries represented less than 6% of the cluster's employment of HEP personnel; while nearly 60% is accounted for by land and marine transport. The most intensive user category of HEP is air transport, with a share of nearly 7%. All in all, this suggests a differentiated pattern of innovation and technological change in the various segments of the transport cluster.[8]

The transport cluster is an inter-related network of transport functions and associated services. We may broadly identify three sub-structures in the cluster map, with land transportation located in the upper right of the diagram, marine transportation in the middle left with related support services, air transportation towards the bottom, while other support and auxiliary services are located in the middle. In contrast to the paper and graphics and oil and gas clusters, there is no immediate suggestion of an overall "filière" structure. Rather, the transport cluster as identified here is a loose network of interrelated activities.

3.2.6. The paper and graphics cluster

The paper and graphics cluster is the smallest cluster. Its 1993 capital share is the lowest of the clusters considered here, of the same order as the capital share of the construction cluster. The intensities of intangible investments are somewhat larger than the average. This is to a large extent due to the presence of the pulp and paper industry; this industry dominated R&D expenditures in 1995.

This cluster regroups activities mainly concentrated on pulp, paper and graphical production. As shown in the cluster map, the cluster has an open structure that is readily understandable in terms of flows of pulp- and paper-based production activities. The sectors included in this cluster are listed in Annex 2. The central chain of this cluster consists of the pulp and paper industries, NACE 21, and the graphical, printing and publishing industries, NACE 22. In view of the fact that the publishing industry includes both book and newspaper publishing, the linkages to sectors surrounding the cluster are almost intuitive.

One noteworthy feature of the cluster map is the presence of the advertising sector as a major user of graphical products. It is part of a rather strong link between the printing and publishing industry and KIBS, around technical, accounting and administrative services, as well as advertising services.[9] Note that this cluster is also a relatively intensive user of KIBS inputs; the ratio of KIBS inputs to GDP is the highest of the five clusters, and also higher than the relevant ratio for the oil and gas cluster.

3.2.7. Information-intensive activities

The information-intensive network represents a gross product of about NOK 84 billion, or about 12% of GDP, and employs 193 000 workers. The overall capital output ratio is below unity, with a capital-labour ratio, or average (tangible) capital behind each employee, of just NOK 425 000. This is in striking contrast to a capital share of 48%. The imbalance between these measures contrasts with the role of intangible investments and assets in these industries, almost 30% of business sector R&D expenditures are linked to the industries in this network. It follows from this that the R&D/GDP ratio is high compared to most other industries, and significantly higher than the similar ratios for the clusters described above.

Compared to the other clusters, the average R&D expenditures of R&D performing firm units is high. In 1995, the average R&D performer spent nearly NOK 23 million on R&D, considerably more than even the oil and gas cluster. There is, however, a lower propensity for firms to perform R&D; in 1995, 40% of firm units in the oil and gas cluster were R&D performers compared to only 22% of firm units in information-intensive activities.[10] The share of firm units involved in technological co-operation is second only to the oil and gas cluster. In both cases, units belonging to KIBS sectors represent major shares of co-operating firm units.

R&D tends to be mainly performed in house, only about 27% of total R&D expenditures were contracted out. This share is somewhat lower than the shares for other parts of the economy, and considerably lower than for the relatively R&D-intensive oil and gas cluster. A more substantial difference emerges, however, when this share is compared with that of R&D that is contracted to national R&D institute and institutes of higher education. Less than 8% of externally contracted R&D is contracted to such institutions, suggesting a considerably weaker overall linkage between these sectors and PTIs.

The KIBS inputs into this sector are high, corresponding to 16% of value added. Although this reflects the inclusion in the network itself of several KIBS sectors, intra-sectoral flows among KIBS sectors cover less than 20% of total KIBS inputs. Even after adjusting for this, the network still remains an intensive user of KIBS. Similarly, it should be noted that the network is a heavy user of personnel with higher education levels. Nearly 15% of employees have an educational background level corresponding to ISCED 6 or higher. Nearly one in four employees with tertiary education outside the social service systems is employed in these sectors.

The network is concentrated in financial, business and communication services, with a broad set of linkages to other sectors and clusters across the economy. As shown in the annex tables, this network is, with the exception of the construction cluster, the largest or second largest originator of inter-cluster intermediate inputs of all those we have looked at. Note that in addition, public and social services are large recipient sectors of the output from this cluster. The sectors in the upper left corner of the cluster map are only weakly connected to the rest of the network; advertising, miscellaneous business services and labour recruitment are clearly more closely related to major users of the network than to the network itself. There are two sectors in particular that tie the network together, computer services (Sector 720) and the variegated category of business services (Sector 741).

4. Innovation patterns and intangible investments

We have identified six clusters and networks through transaction flows of goods and services representing more than 60% of Norwegian GDP and more than 75% of intermediate inputs and exports. Although there is considerable ambiguity in this approach, the fact that we have managed to identify clusters with relatively clear economic and functional characteristics lends support to our approach. The rationale for this approach was an attempt to identify some main features of systemic innovation, on the hypothesis that user-producer links, *i.e.* traded flows, form a substantial basis for the topology of systems of innovation. In this last section we will outline features of investments in innovative activities at cluster level, concentrating on four gross dimensions, capital intensity, R&D measures, employment and inputs from knowledge-intensive business services.[11] Without time series presentation of variables, we will treat level variables as equilibrium values or nearly so, *i.e.* we assume that the documented levels reflect underlying technological and economic characteristics without significant expectation gaps. Table 6 outlines these features of innovation activities at cluster level. The share of innovators – firms that have introduced product or process innovations over the three-year period 1993-95 – is used as an output measure. While 43% of firms in the oil and gas cluster are innovators, the share is 28% for the transport cluster. Although variable, variation in inputs is generally more pronounced than variation in this innovation output measure.

Capital variables – capital share, measuring the gross return to capital inputs, and capital-labour ratio as a measure of capital intensity – point to the outlying character of the oil and gas cluster. With a capital-labour ratio of NOK 5 million and capital share of 76%, this sector is strongly characterised by technology intensity. The capital share of income of the agrofood sector is more than 50%, indicating the significant role played by capital in this cluster. Its capital-labour ratio is slightly higher than the all-economy ratio, excluding the oil and gas sector and household-owned property. The small capital base of information-intensive activities and comparatively high capital share of income, suggests that considerable returns on intangible assets are included. This situation is reversed for the transport cluster, suggesting a more dominant role for capital-embodied investments. This also applies to the construction cluster, taking into consideration its capital intensity, while the paper and graphics cluster appears to have either a higher gross return on capital investments, or a stronger reliance on intangible investments.

Table 6. Patterns of tangible and intangible investment

Percentages

	Agrofood	Oil and gas	Construction	Paper and graphics	Transport	Information-intensive
Innovators	35.6	42.8	34.4	42.2	27.8	35.8
Capital share	0.55	0.76	0.37	0.38	0.48	0.48
Capital-labour ratio (thousand NOK per employee)	768	4 952	630	502	1 285	425
R&D performers	16.4	39.2	23.8	11.5	12.6	22.0
Technological co-operation	12.4	34.6	18.4	19.2	16.6	22.0
R&D expenditures (million NOK)	519	2 326	698	304	251	2 864
KIBS inputs (million NOK)	3 571	11 167	4 527	1 835	3 468	12 974
R&D/GDP	0.8	1.9	1.3	1.5	0.4	3.4
KIBS*/GDP	5.8	9.0	8.3	9.1	5.4	15.7
Share of HEP in employment	2.9	19.1	4.7	8.8	3.7	14.6
R&D/firm unit (thousand NOK)	4 513	16 151	4 282	8 440	5 576	22 731
External R&D/Total R&D	32.1	41.5	21.5	34.7	37.6	26.8
PTI/External contract R&D	34.9	44.3	21.1	8.4	19.4	7.7

The intangible variables in Table 6 expand on these hypotheses. In contrast to the oil and gas cluster, the share of R&D performers and R&D intensity implies a more substantial role for non-R&D innovation patterns in the other clusters, while the information network, although skewed in terms of R&D performers, is highly R&D-intensive. KIBS intensities show a similar pattern, but also serve to strengthen the impression of the paper and graphics and construction clusters as dependent on "softer" modes of innovation. Somewhat surprising is the weak role of technological co-operation in all clusters except oil and gas. The share of higher educated personnel in employment varies considerably, from a low 3% in the agrofood cluster to a high 19% in oil and gas. Again, the paper and graphics cluster is shown to be medium-to-high intensive.

The last two indicators in Table 6 describe interactive aspects of R&D investments. The first measures the share of total R&D expenditures performed outside the R&D investing firm, in other companies and in R&D institutions. The second variable gives the share of external R&D contracted to public technology, or knowledge, infrastructures. The most prominent feature is the weak integration of the "soft" innovators of information activities and paper and graphics with such PTIs. The specialisation of PTIs towards the oil and gas cluster is also pronounced.

Figures 1a and 1b compare five measures of input requirements of the clusters. The plots measure five kinds of inputs to production: labour inputs, measured in terms of total employment, (physical) capital stock, as shown in the national accounts data described above, KIBS inputs from input-output data, the number of employees with tertiary education, and R&D expenditures in the cluster or network. In each case, it shows the ratio of the share of the cluster in the relevant variable and the share of the cluster in total GDP, or what is identical, namely the input to gross product ratio in the cluster relative to the input to GDP ratio for the whole economy. A cluster with a ratio larger than unity thus requires more of the relevant input per unit of output in terms of value added. The HEP variable has been somewhat modified. Due to the large share of HEP employees in education and health services, the ratio for the HEP variable was constructed from the parallel data restricted to the economy outside these services. The only effect of this modification for the clusters and networks considered here is to scale the axis; it does not affect the relative values of the clusters.

The axes of total employment and capital stock are measures of traditional physical inputs into production. An industry or group of industries with a value along the labour axis above unity thus requires more labour inputs in production than its share of GDP and the average share of labour in GDP would suggest. There are similar considerations for the other axes. A sector with a high (low) capital-labour ratio, being relatively capital- (labour-) intensive, would score high (low) on the capital axis and low (high) on the labour axis. The HEP axis, which measures the relative intensity of employment of personnel with tertiary education above ISCED 6 level, is suggested to be a combined measure of: *i)* the structural composition of the labour force within the industry, and hence of the limitations of the homogeneity assumptions of labour inputs; and *ii)* a measure of the requirements for intangible inputs that are present in these labour inputs as a consequence of their educational background and the specific experiences that this allows.

Figure 1a. Input requirements in the oil and gas, paper and graphics and information-intensive activities clusters

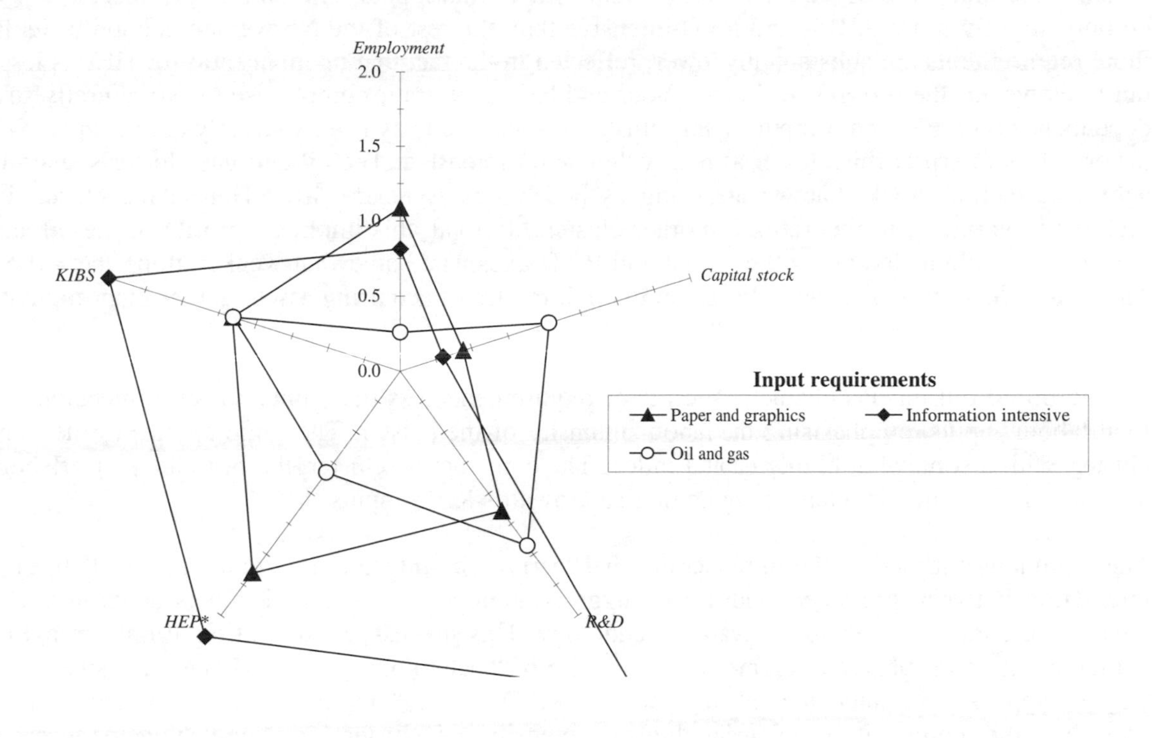

Figure 1b. Input requirements in the agrofood, construction and transport clusters

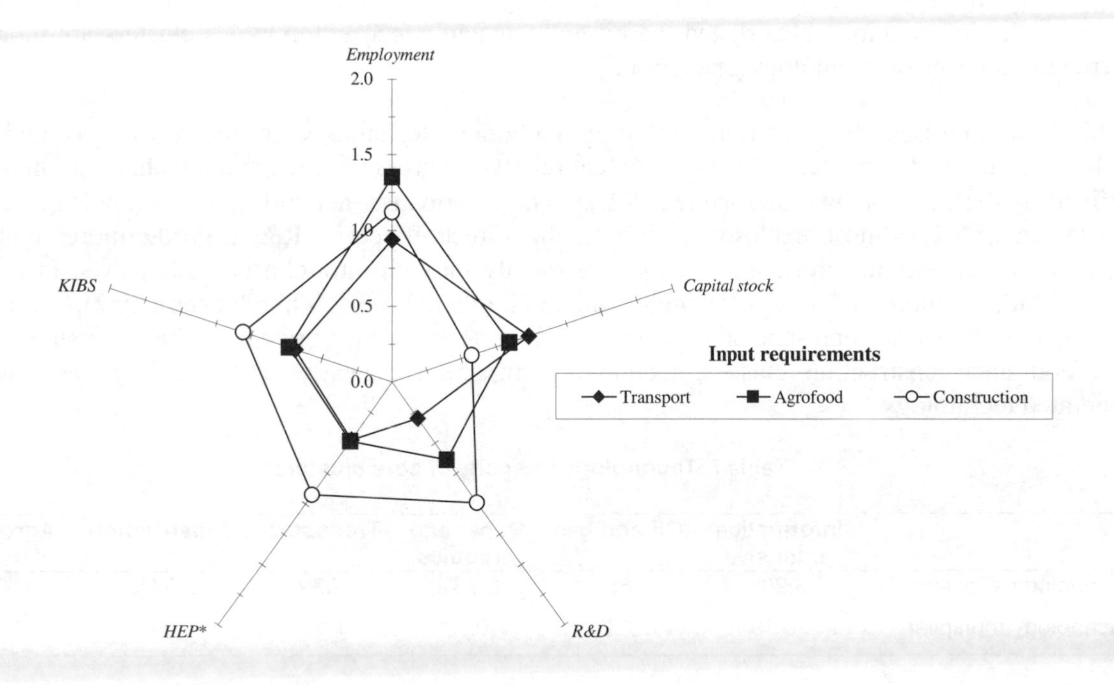

Figure 1a identifies the three networks: oil and gas, paper and graphics, and information-intensive activities. Comparing the plots, the latter two networks are extensively dominated by intangible inputs, which score high on these axes, with low values for capital inputs. The oil and gas cluster appears to be only slightly more KIBS- and R&D-intensive than the rest of the Norwegian economy. Its labour input requirements are substantially lower, reflected in the fact that its input ratio for HEP is less than unity. However, the differences in the labour and HEP input ratios emphasise the structurally stronger dependence on HEP labour inputs in this cluster. Capital intensity is only slightly above unity. We will nevertheless interpret this cluster as relatively capital-intensive. The oil and gas cluster's share in the aggregate capital stock that we are using as benchmark is nearly 40%. Hence, the cluster is less inclined to variation in this ratio than other clusters. In total, this implies a profile of the oil and gas cluster that is elongated along the capital and R&D axes, and somewhat weaker along the KIBS axis. This is a reflection of a strong reliance on explicit modes of acquiring assets, rather than implicit ones such as through KIBS inputs.

The agrofood cluster has characteristic factor requirements less than unity for all dimensions except total labour inputs, emphasising the labour intensity of the cluster. The same applies to the transport cluster, with a somewhat higher capital index. The construction cluster, the birthplace of offshore (oil and gas) engineering, is relatively more inclined towards KIBS inputs.

The argument outlined in the introduction, that learning-by-interacting is related to challenges in the interaction between producers and users, suggests attempts to grade trade flows according to some measure of content of new or innovative technology. This suggests a redefinition of link strengths, to a combined measure of trade volume and a measure of "technology" or knowledge intensity. A simple way to approximate such a method is to use R&D adjusted input-output coefficients.[12] Figure 2 describes the composition of technological intensity of clusters, distinguishing between R&D performed in each sector, *i.e.* direct, and indirect (domestic) inter-sectoral embodied technology flows.[13] For the three clusters, transport, agrofood and construction, imported embodied technology is larger than the aggregate intra-sectoral technology generation within each sector of the clusters. In contrast, the information network and the oil and gas and paper and graphics clusters are more self-sufficient in terms of technology generation.

Table 7 summarises the effect of including embodied technology in terms of two indicators; technology intensity as total technology content relative to gross product, and a technology multiplier, defined as the ratio of own and indirect R&D. The information network has a technology intensity greater than 5%, almost exclusively due to the direct effect of R&D. Furthermore, embodied technology deposits in information sectors are mainly through intra-cluster trade flows. In contrast, intra-cluster technology flows play a minor role in all other clusters, while technology inputs from the information network represent at least one-third of technology imports to these clusters. In the agrofood and construction clusters, technology imports are nearly twice as large as internally generated technology.

Table 7. Technology intensity of core clusters

	Information-intensive	Oil and gas	Paper and graphics	Transport	Construction	Agrofood
Technology intensity	5.20%	1.65%	2.73%	1.83%	2.50%	1.77%
Technology multiplier	0.19	0.39	0.70	1.15	1.68	1.81

Figure 2. Technology content of clusters

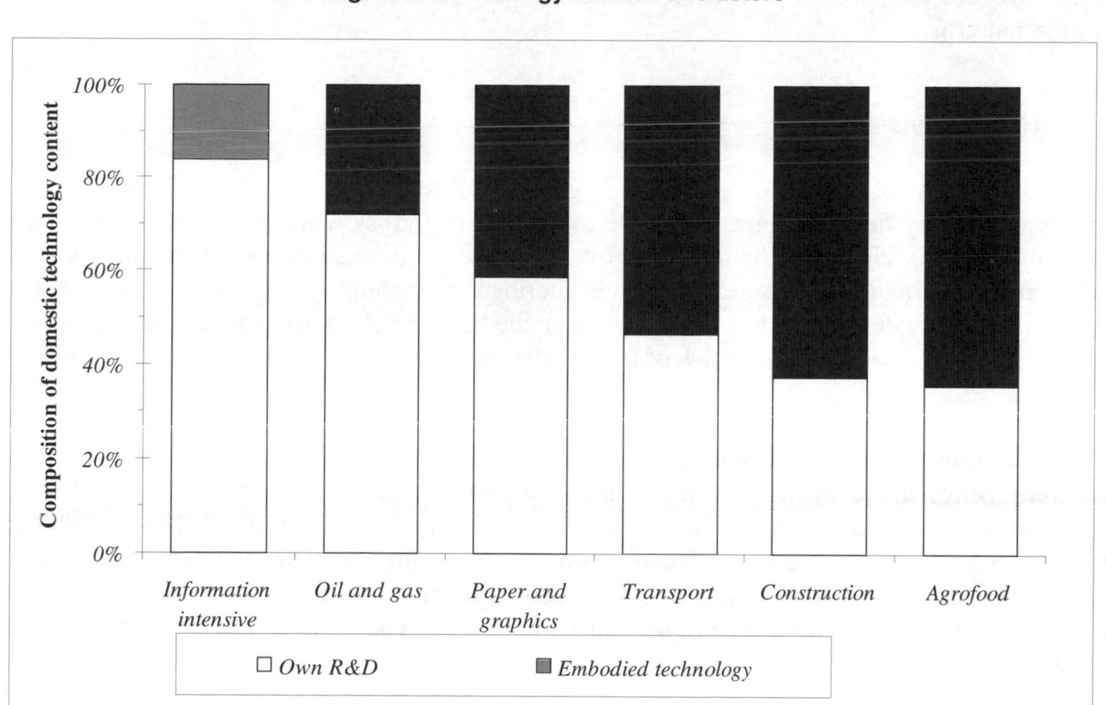

In summary, we may distinguish broad patterns for each cluster according to the relative dominance of factor inputs and embodied technology in each profile. These patterns are interpreted as fingerprints of gross modes of innovation and technical change at cluster level.

The three clusters, agrofood, construction and transport, are dominated by "implicit" modes of technical change, primarily through embodied use of capital goods and intermediate inputs. Overall, these clusters rely only weakly on "explicit" modes such as use of higher educated personnel, R&D performance and KIBS inputs. The relative weighting of capital intensity and intermediate technology flows suggests that technical change in the transport cluster is closely linked to technical change embodied in capital goods, while technical change in the construction cluster relates more to non-capital intermediate inputs, including softer KIBS inputs.

	Embodied technology	Capital	R&D	KIBS	HEP
Agrofood	***	**	0	--	---
Construction	***	-	0	*	-
Transport	*	***	---	--	---
Oil and gas	---	0	***	**	***
Paper and graphics	0	---	*	**	***
Information intensive	---	---	***	***	***

This situation is reversed for the three clusters, paper and graphics, oil and gas and information-intensive activities. Given the respective high and low capital intensities of the oil and gas cluster and the other two clusters, we may broadly distinguish between hard and soft modes of these explicit innovation patterns.

5. Conclusions

We have outlined a procedure for identifying aggregate clusters on the basis of input-output tables. Using supplementary data, we have been able to characterise gross features of innovation patterns at the level of these six clusters. The indicated overall mode of innovation and technical change in the agrofood and construction clusters are through intermediate technology imports, while the transport cluster is more dependent on technical change embodied in capital equipment. We may refer to these patterns as implicit, or embodied, modes of technical change and innovation.

The structure of technology flows described above is based on intermediate trade flows. By excluding inter-sectoral transactions of capital investments, we avoid double-counting capital goods in the innovation profiles above. However, this choice may imply an under-estimation of technology flows, in particular for the three clusters where embodied flows are most important. The underlying data show that technology flows into these clusters from relatively high-tech sectors such as chemicals and machinery production are significant. Together with information and KIBS inputs, they account for more than two-thirds of technology imports into the clusters. The inclusion of investment flows will further enhance this share.

We noted above that the agrofood cluster appeared to be isolated from trade flows with other sectors and clusters which, given the fairly low intensities of explicit dimensions of innovation, raised a question relating to the generation of technical change. Induced technology flows change this picture; the high-technology multiplier value for the cluster implies that it is far more heavily involved in inter-cluster technology flows. For both the agrofood and transport clusters, we expect the inclusion of investment flows to further enhance technology flows, reflecting capital-embodied technical change.

Although capital-intensive, technical change in the oil and gas cluster is more directly related to extensive use of firm-based R&D and professional personnel. KIBS inputs, mostly related to sub-contracted technical engineering, are also significant in the oil and gas cluster. The low technology intensity to a large extent reflects the high capital intensity and the importance of ground rent in this cluster.

The low capital and high technology intensities of the paper and graphics and information cluster implies the predominance of softer, or more intangible, modes of innovation. Overall, the paper and graphics cluster relies on its stock of professional personnel and KIBS inputs, while the aggregate fingerprint of the information cluster is heavily dependent on all three explicit dimensions of innovation inputs.

Our approach has two basic limitations. Firstly, we have focused solely on traded transactions as the mode of interaction between firms. Through these transactions we have internalised the user-producer links which are a vital part of the environment of individual firms, while we have excluded other forms of interaction from the analytical framework. Secondly, we have identified innovation modes at the level of clusters, which of necessity treats intra-cluster diversity and complementary divisions of innovation modes weakly. The consistent, and perhaps surprising, result of this work is that, even at the aggregate level of input-output-based clusters, it is possible to discern clear patterns of differentiation of innovation modes.

As a consequence, two lessons for innovation and technology policies may be drawn. One is negative, the fact that there is a considerable differentiation in innovation modes at this level directly implies that the strong emphasis on industry-neutral, or non-selective, innovation-oriented economic policies is misguided. These policies are not selective in implementation at firm level, the variation in innovation modes at cluster level suggests that the impacts of supposedly neutral policies will vary even at this level. An example may be the use of R&D tax credits or KIBS-based policies of technology diffusion.

The second lesson is that the existence of innovation patterns at the level of clusters points towards cluster-based innovation policies, where policy formulation would, from the outset, integrate central dimensions of the systemic interaction underlying innovation and technical change. Such inter-sectoral approaches to innovation policy may inform policy formulation on intra-cluster complementary relations of sectors and functions, within a framework of general patterns of innovation at the cluster level.

NOTES

1. Of the resulting 161 sectors, three had no economic activity in 1993. With 158 sectors, the maximal possible number of links is nearly 25 000. According to Table 1, the number of actual links covers 85% of this set of potential links; the actual sectoral network almost completely exhausts the potential set of links.

2. In constructing subsequent tables, figures and clusters we have subtracted intra-sector intermediate deliveries. In line with the approximation discussed below, of treating the sectors as (technologically) homogenous, we interpret intra-sector deliveries as expressing the effects of a functional differentiation within a homogenous technical production activity.

3 The data describe domestic flows, a full analysis of "technological" links would involve a sectoral decomposition of the sources of import flows, as well as a two-way sectoral decomposition of the flows of investments. Similarly, with the "innovation cluster" emphasis of this approach, a sectoral decomposition of exports would be necessary for a full analysis of such "technological" links. With our data we are implicitly making an assumption that the domestic intermediate flows are representative of the full technological flows. An assumption for the foreign trade dimensions, in line with the often-made assumption of input-output analysis about sectoral "import market shares" (see *e.g.* Miller and Blair, 1985), is clearly less reasonable for smaller economies than for larger ones; see Archibughi and Pianta (1992).

4. The survey unit in the Norwegian R&D surveys is the "firm" or "industry" unit, defined as all plants within a single enterprise that are classified in the same industry category at a detailed level.

5. The second cut-off is, contrary to the link strength cut-off, inversely related to the resolution of the sectoral classification. As the resolution power of the classification goes up, *i.e.* as the classification becomes more detailed, sectors are increasingly likely to be smaller than a given cut-off on sector size. In the same limit the specialisation of intermediate outputs would increase, more links would be stronger than a given cut-off on link strength. Similarly, as sectoral resolution gets low, specificities in sectoral trade flows are "washed out", while sectors grow larger. In this sense, the link strength cut-off is a low-end cut-off, while the sector size cut-off may be considered a high-end cut-off.

6. Among the largest sectors measured in intermediate trade flows, we would expect to find sectors that serve as important nodes in clusters. We would also expect a higher probability of finding heterogeneous sectors, especially if the industrial classification is coarse. The most prominent of these sectors is wholesale trade, which takes part in nearly 12% of total intermediate trade flows; as either a delivering or a receiving sector, it links up with a large cross-section of the economy. Following Roelandt *et al.*, we could identify trade as part of an economy-wide service cluster. Significant inter-sectoral trading between the underlying components of the wholesale trade aggregate is required to support this interpretation. If this is not the case, we must conclude that the generic nature of wholesale trade is an artefact of the statistical classification; it is an agglomeration of separated sub-sectors, each interacting with bounded sets of clusters. The data we have do not allow us to resolve this question. However, observation of the actual organisation of wholesale trade sectors suggests that the position of trade sectors in these networks is an artefact of the industrial classification. Hence, we will not include trade sectors as part of any cluster here.

7. The input-output table does not allow us to differentiate between different parts of the education system.

8. For a study of freight land transport in this cluster, related to Sectors 604 and 631, see Ørstavik (1998).

9. So much for the paperless office!

10. 572 of the 4 394 firm units surveyed in the 1995 R&D survey belong to the network of information intensive activities. Of these, 126 were R&D performers.

11. For KIBS inputs as inputs to innovation, see Bilderbeek *et al.* (1998) and Hauknes (1998b).

12. The methodology is outlined in Hauknes (1997) and (1998c).

13. The computation is based on sector-sector coefficients, rather than sector to final output coefficients. The embodied technology flows are distributed over the whole output of the recipient sector, giving a modified technology intensity of the sector, rather than of its final output.

Annex 1

PRODUCTION SECTORS IN THE NORWEGIAN NATIONAL ACCOUNTS

The account structure of the Norwegian system of national accounts is organised around five mutually exclusive institutional sectors describing the total economy:

22 Households' production for own consumption (the household sector).

23 Market activities (the market sector).

24 Central government and administration (the government sector).

25 Municipal government and administration (the local authority sector).

26 Non-profit institutions serving households (the PNP sector).

To each institutional sector corresponds a set of up to 149 industries or production sectors, with a total of 179 combined institutional-production sectors. Data for the institutional sectors of central and local government are based on annual public state and municipalities' accounts (*Stats- og Kommuneregnskapene*), while data for the remaining non-public sectors are based on industrial and other surveys organised and performed by Statistics Norway.

The five institutional sectors have been aggregated to two categories: a "private" sector, integrating the household, the market and the public non-profit sectors; and a "public" sector, covering the government and the local authority sectors. In the cluster descriptions, production sectors are distinguished by a 3-digit code, roughly corresponding to the related NACE classification, with the "public" production sectors identified with a preceding *P*. The resulting input-output table is described in terms of 161 production sectors.

Annex 2

CLUSTERS AND THEIR SECTORAL CONTENT

The agrofood cluster

SNA production sector	NACE classification	Description
10	011 + 012 + 013	Agriculture
14	014	Agriculture and husbandry service activities
51	0501	Fishing
52	0502	Fish hatcheries and fish farming
151	151	Meat and meat products
152	152	Fish products
153	153	Fruit and vegetables
154	154	Manufacture of vegetable and animal oils and fats
155	155	Dairy products
156	156	Grain mill products, starches
157	157	Prepared animal feed
158	158	Other food products
159	159	Beverages
242	2415 + 242	Fertilisers, nitrogen compounds and pesticides
351	35111 + 35112 + 35113 + 35116 + 35117 + 3512	Building and repair of ships
551	551 + 552	Hotels and accommodation
553	553 + 554 + 555	Restaurants and catering

The oil and gas cluster

SNA production sector	NACE classification	Description
P730	73	Public sector research and development
P742	742 + 743	Architectural, engineering activities and related technical consultancy, technical testing and analysis
P745	745 + 746	Labour recruitment, investigation and security
111	111	Extraction of crude oil and gas
112	112	Service activities incident to 111
232	232	Refined oil products
287	284 + 285 + 287	Other fabricated metal products
300	30	Office machinery and computers
334	334 + 335	Optical instruments, watches and clocks
352	35114 + 35115	Building and rep. oil platforms and modules
608	603	Pipeline transport
742	742 + 743	Architectural, engineering activities and related technical consultancy, technical testing and analysis

The construction cluster

SNA production sector	NACE classification	Description
P410	41	Water utilities
P453	45212 + 4523 + 4524 + 4525	Construction and civil engineering
100	10	Mining of coal and lignite
140	14	Other mining
201	201	Saw milling, planing of wood
202	202	Panels and boards
203	203	Builders' carpentry and joinery
204	204 + 205	Other wood products
243	243	Paints, varnishes
246	246	Other chemical products
261	261	Glass and glass products
262	262 + 263 + 264	Ceramic products
265	265	Cement, lime and plaster
266	266 + 267 + 268	Articles of concrete, lime and plaster
281	281 + 282 + 283	Structural metal products, tanks, containers, steam generators, a.o.
297	297	Domestic appliances
311	311 + 312	General electrical machinery
313	313	Insulated cable and wires
314	314 + 315 + 316	Other electrical equipment
451	451	Site preparation
452	45211 + 4522	Erection of buildings and frames
453	45212 + 4523 + 4524 + 4525	Construction and civil engineering
454	453 + 454	Building installation and completion
455	455	Renting of construction equipment
700	701 + 70202 + 703	Real estate activities
704	70201	Services from own property – households + calculated production own housing

The transport cluster

SNA production sector	NACE classification	Description
P601	601	Public railway transport
P631	63111 + 63113 + 6312 + 6321 + 6323 + 63401 + 63403 + 63409	Support and auxiliary services to road and air transport
P632	63112 + 6322 + 63402	Public support and auxiliary services to marine transport
340	34	Motor vehicles
353	352	Railway vehicles
354	353	Aircraft
355	354 + 355	Other transport equipment
502	502 + 50403	Maintenance of motor vehicles
601	601	Railway transport
602	60211	Regular bus transport
603	6022	Taxi operation
604	6023 + 6024	Other passenger and freight transport by road
605	60212	Tramway transport
611	61101 + 61102	International marine transport
613	61103 + 61104 + 61105 + 61106 + 61109 + 612	Domestic marine transport
620	62	Air transport
631	63111 + 63113 + 6312 + 6321 + 6323 + 63401 + 63403 + 63409	Support and auxiliary services to road and air transport
632	63112 + 6322 + 63402	Support and auxiliary services to marine transport
633	633	Travel agencies and tour operators
711	711 + 712	Renting transport equipment

The paper and graphics cluster

SNA production sector	NACE classification	Description
211	2111	Pulp
212	2112	Paper and paperboard
213	212	Articles of paper and paperboard
221	221	Publishing
222	222	Graphical services and printing
223	223	Recorded media
372	372	Recycling non-metal waste
921	921 + 923 + 924 + 925	Picture and video, library, archives and museums, news agencies

Information-intensive activities

SNA production sector	NACE classification	Description
P670	67	Auxiliary activities to financial intermediation
321	321 + 322	Electronic communication equipment
641	641	Post and courier
642	642	Telecommunications
651	6511	Central banking
652	6512	Other monetary intermediation
655	652	Other financial intermediation
661	6601	Life insurance
662	6602	Pension funding
663	6603	Non-life insurance
670	67	Auxiliary activities to financial intermediation
720	72	Computer and related activities
730	73	Research and development
741	741	Legal, accounting and auditing services
744	744	Advertising
745	745 + 746	Labour recruitment, investigation and security
748	748	Miscellaneous business activities, nec
922	922	Radio and television

Annex 3

CLUSTER MAPS

Figure C1. The agrofood cluster

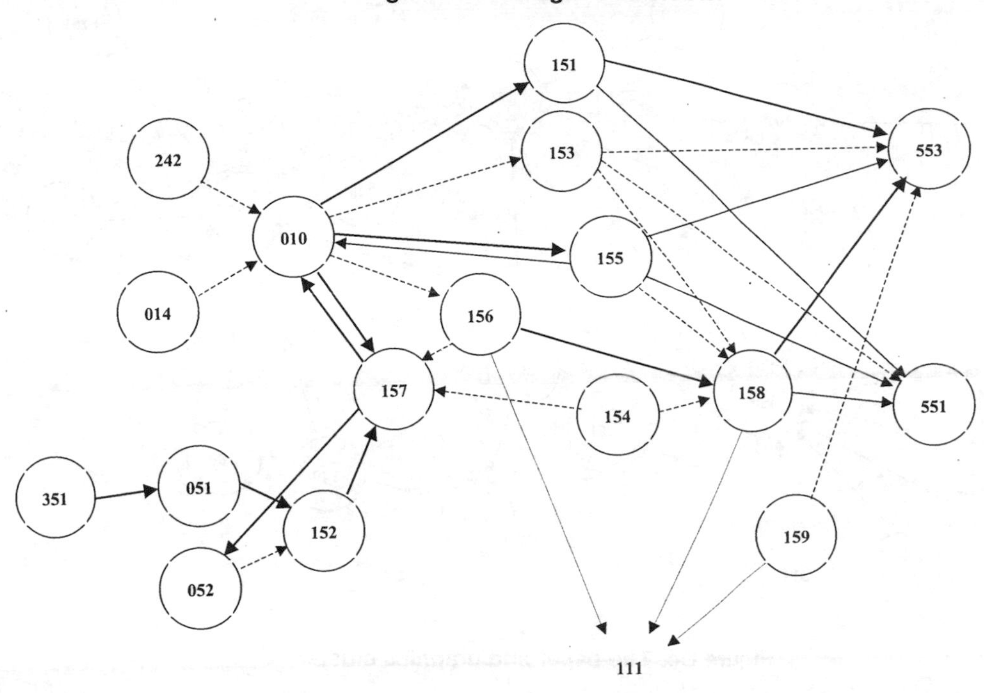

Note: Maximal forward or backward link > 15% with partner sector > 2.5‰ of total intermediate trade

Forward/backward links > 10% of deliveries/inputs, partner sector > 2.5‰

Forward/backward links > 10% of deliveries/inputs nec, partner sector > 0.5‰

Figure C2. The oil and gas cluster

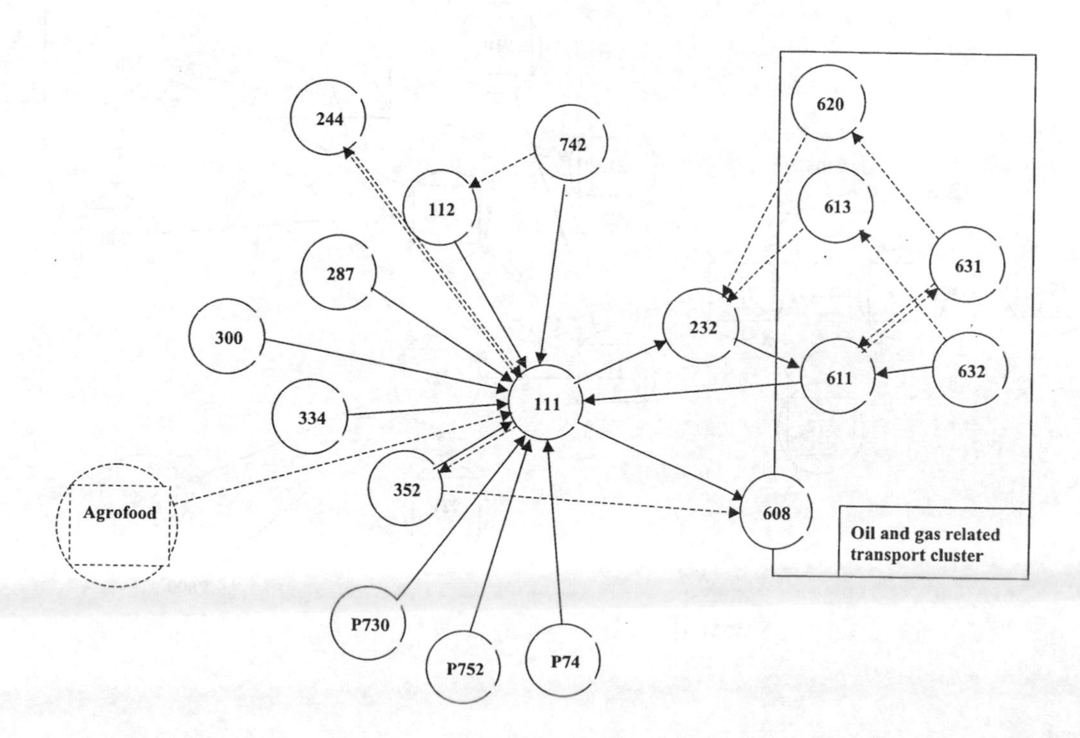

Figure C3. The construction cluster

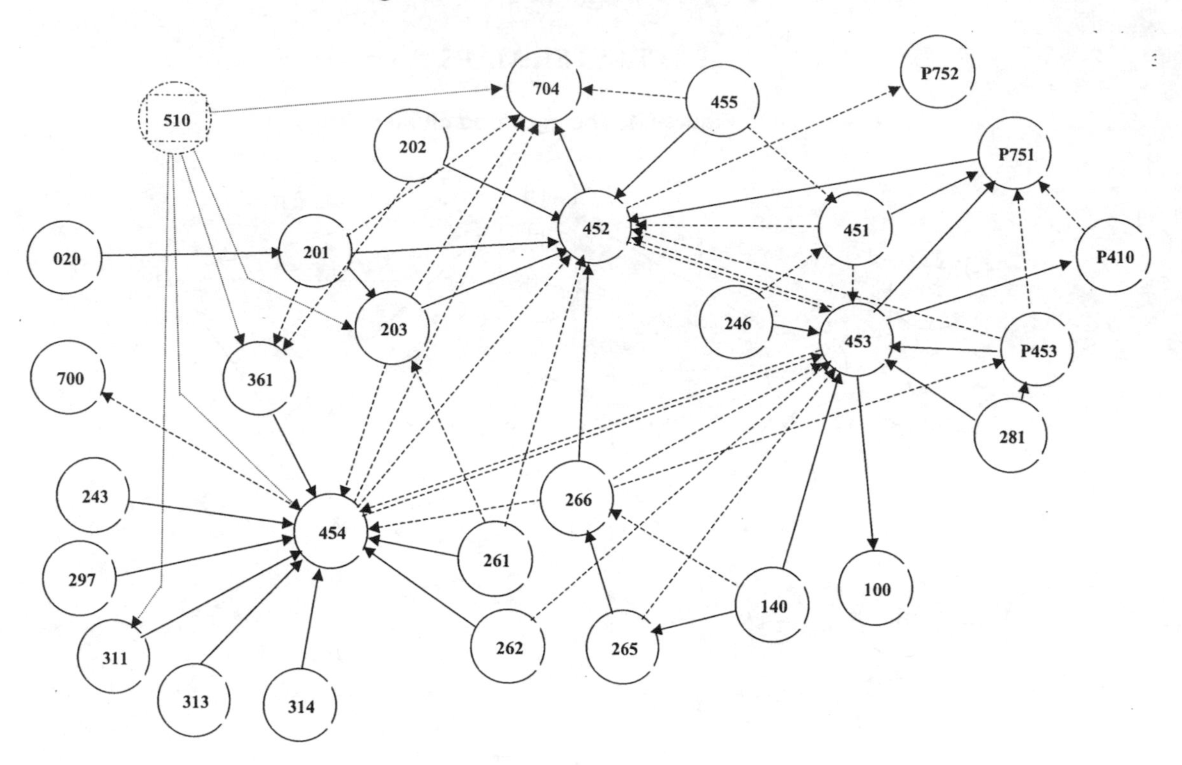

Figure C4. The paper and graphics cluster

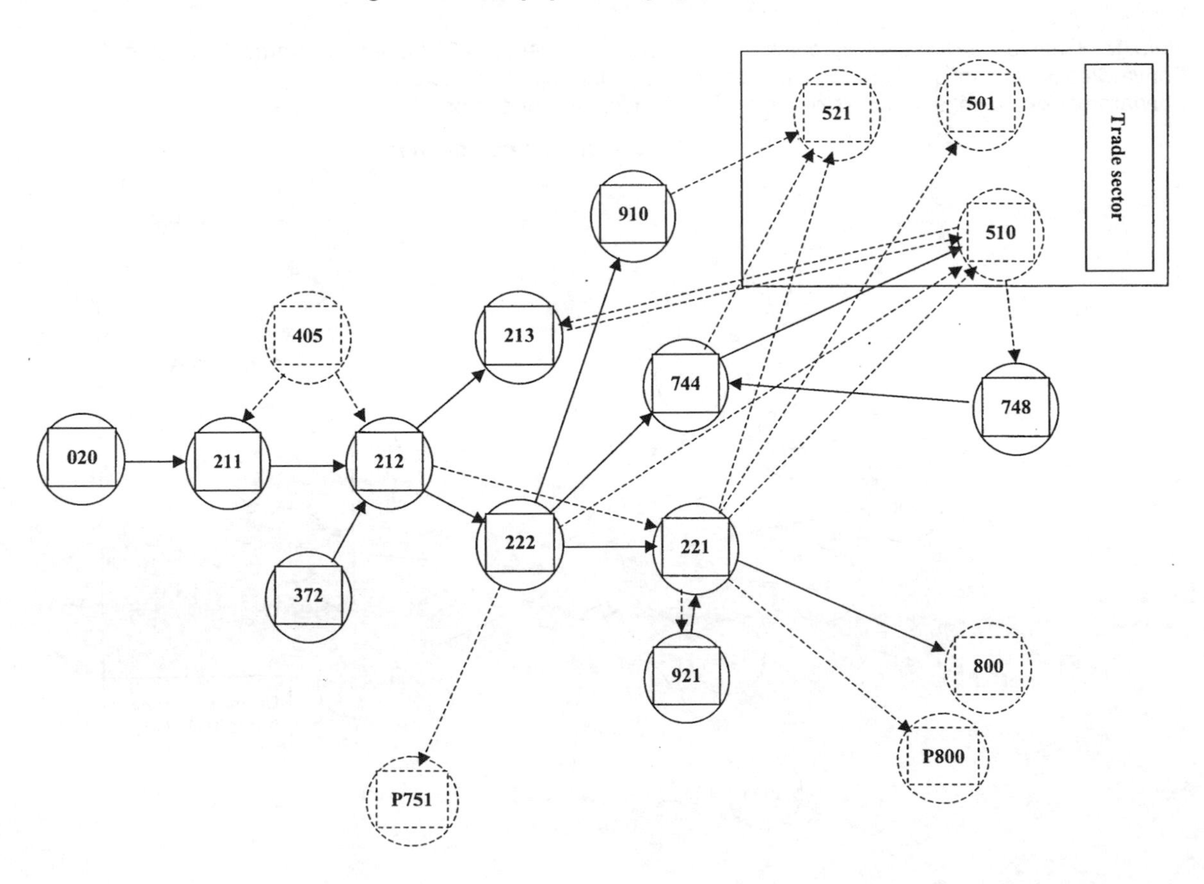

Figure C5. The transport cluster

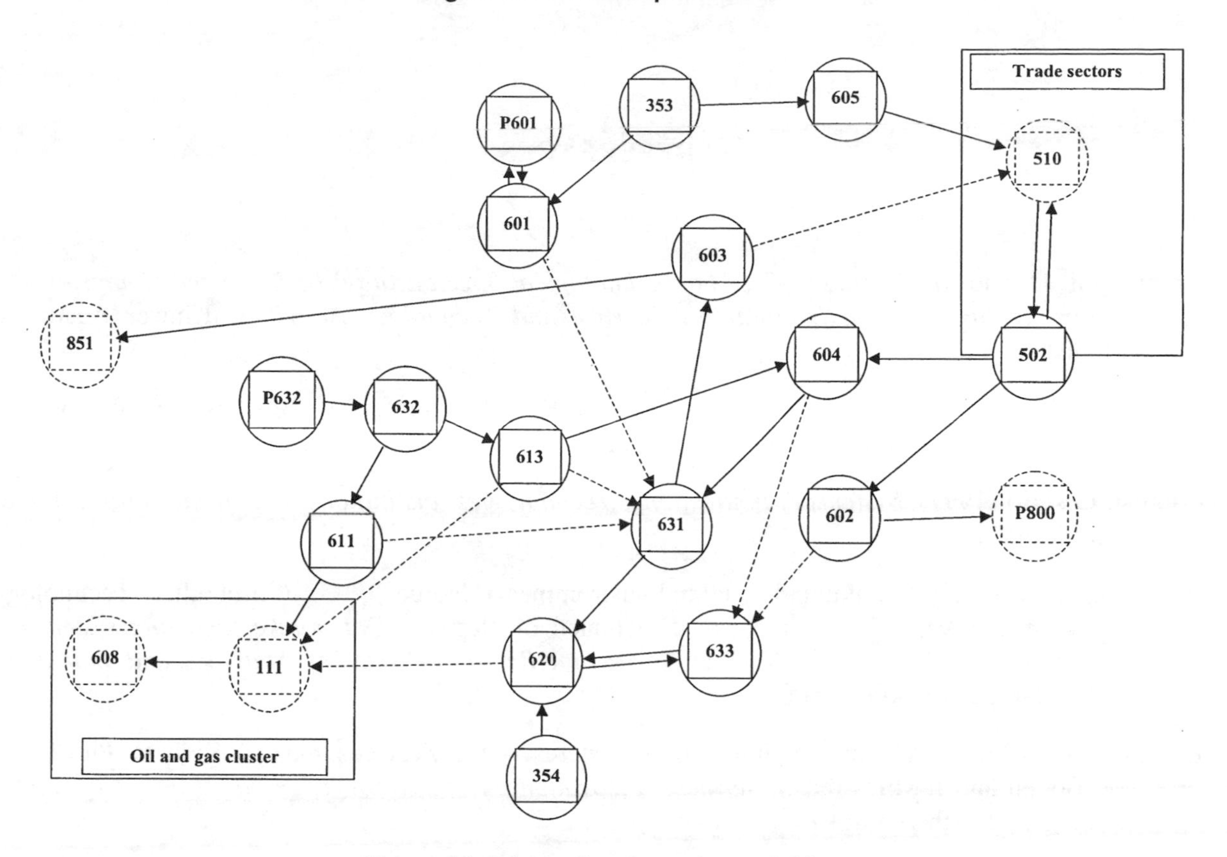

Figure C6. Information-intensive activities

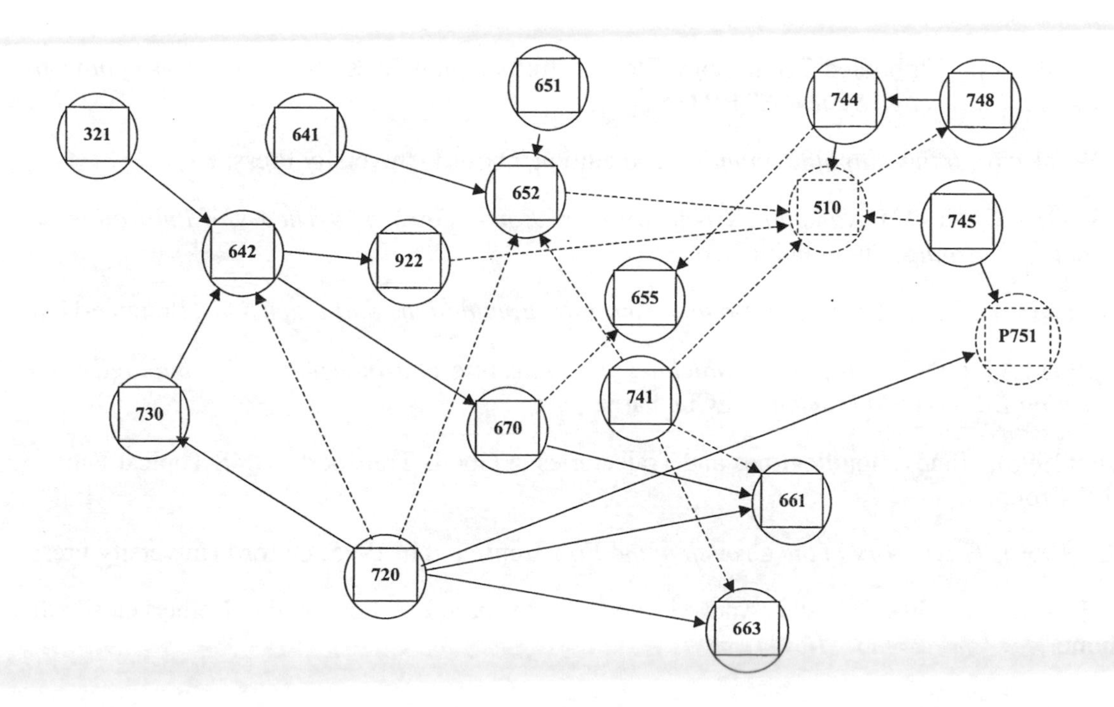

REFERENCES

Archibughi, D. and M. Pianta (1992), *The Technological Specialization of Advanced Countries – A Report to the EEC on International Science and Technology Activities*, Kluwer Academic Publishers.

Bilderbeek, R. *et al.* (1998), "Services in Innovation: Knowledge-intensive Business Services (KIBS) as Co-producers of Innovation", SI4S Synthesis Report 3, STEP Group.

Edquist, C. (ed.) (1997), *Systems of Innovation – Technologies, Institutions and Organizations*, Pinter Publishers.

Hauknes, J. (1997), "Produktbundne teknologistrømmer i Norge 1995" ("Embodied Technology Flows in Norway 1995"), in Research Council of Norway, *Det norske forskningssystemet – statistikk og indikatorer 1997* (*The Norwegian Research System – Statistics and Indicators 1997*), Norges forskningsråd.

Hauknes, J. (1998a), "Dynamical Innovation Systems – Do Services have a Role to Play?", in M. Boden and I. Miles (eds.), *Services, Innovation and the Knowledge-based Economy*, Pinter Publishers (forthcoming).

Hauknes, J. (1998b), "Innovation in Services – Services in Innovation", SI4S Synthesis Report 1, STEP Group.

Hauknes, J. (1998c), "Embodied Technology Flows", forthcoming in K. Smith (ed.), *Mapping the Norwegian Knowledge System*, STEP Group.

Leontieff, W. (1986), *Input-output Economics*, 2nd edition, Oxford University Press.

Lundvall, B.-Å. (ed.) (1992), *National Systems of Innovation – Towards a Theory of Innovation and Interactive Learning*, Pinter Publishers.

Miller, R.E. and P.D. Blair (1985), *Input-output Analysis: Foundations and Extensions*, Prentice-Hall.

OECD/EUROSTAT (1997), *Proposed Guidelines for Collecting and Interpreting Technological and Innovation Data – Oslo Manual*, OECD, Paris.

Ørstavik, F. (1998), "Innovation Regimes and Trajectories in Goods Transport", SI4S Topical Paper 5, STEP Group.

Penrose, E. (1959), *The Theory of the Growth of the Firm*, reprinted in 1995, Oxford University Press.

Roelandt, T. *et al.* (1999), "Cluster Analysis and Cluster-based Policy in the Netherlands", this volume.

Chapter 4

MAPPING INNOVATIVE CLUSTERS IN NATIONAL INNOVATION SYSTEMS

by

Alfred Spielkamp and Katrin Vopel
Zentrum für Europäische Wirtschaftsforschung, Mannheim (ZEW)
Centre for European Economic Research

1. Political terrain and objectives of innovation policy in Germany

New technologies and globalisation are changing the world and economic competition is taking on new forms. Information is the driver of knowledge and social development. In general, the changes taking place reflect the generation, transfer and adaptation of knowledge. Knowledge is recognised as the basis of productivity and economic growth. The "diffusion process" or "distribution power" of knowledge describes both an economic and a social process, for in an economy and society the transfer and use of information play an important role for the effectiveness of the innovative system and its potential to improve economic performance.

Compared to other countries, Germany still has a clear competitive edge when it comes to technology and know-how. Together with the United States and Japan, Germany is one of the leading technology producers in the world, and is the top supplier of technology in Europe. In addition to an excellent training system and a well-developed research infrastructure, important technological advantages include the innovative power and flexibility of firms and industries, and an efficient small and medium-sized business sector.

During the last decade, R&D-intensive sectors in Germany have increased their share of innovative products.[1] Between 1993 and 1995, the R&D-intensive industries contributed nearly half of the added value produced by German manufacturing industries. The major part of these contributions stems from electrical engineering, machine construction, the chemical industry, the automobile industry and data processing devices and equipment. On average, the "high technologies" accounted for 3.5% of the value added, and the "advanced technologies" 8.7%. Among the "advanced technologies", Germany ranks in first position, whereas in the "high or cutting-edge technologies" it comes third, behind Japan and the United States. The sectors' employment shares show a similar ranking. Nearly 13% of all jobs in Germany are generated by the R&D-intensive sectors. In contrast, strongly service-oriented countries such as the United States and Japan report only 6 or 9% (BMBF, 1997, p. 4).

Table 1. Domestic production, domestic demand and employment in R&D-intensive industries in selected OECD countries, 1993-1994/95

Industry	West Germany	United States	Japan	France	Italy	United Kingdom
Portion of gross value added in % 1993-95						
R&D intensive industries	12.2	8.5	11.5	7.7	6.4	8.0
• High technology	3.5	3.6	3.9	2.6	1.9	2.9
• Advanced technology	8.7	4.9	7.7	5.1	4.5	5.2
Non R&D-intensive industries	13.7	9.5	13.5	11.7	13.9	10.2
Manufacturing industry	25.9	18.0	25.0	19.3	20.4	18.3
Share of domestic demand[1] in % 1993-94						
R&D intensive industries	7.2	9.0	6.6	6.7	5.6	8.6
• High technology	4.0	3.7	2.2	2.6	2.5	3.0
• Advanced technology	3.3	5.3	4.3	4.1	3.1	5.6
Non R&D-intensive industries	16.2	11.0	14.6	12.1	10.8	12.1
Manufacturing industry	23.4	19.9	21.2	18.8	16.4	20.8
Share of employment[2] in % 1993-95						
R&D intensive industries	12.8	5.9	9.3	7.5	5.8	8.2
• High technology	3.1	2.5	3.1	2.3	1.4	2.7
• Advanced technology	9.7	3.4	6.2	5.2	4.3	5.4
Non R&D-intensive industries	14.9	9.7	13.8	11.0	14.7	11.4
Manufacturing industry	27.7	15.5	23.2	18.5	20.4	19.6

1. Gross value added of the respective industry plus net imports and less net exports as a percentage of domestic demand (private and public consumption/spending as well as gross investment). Net exports and net imports have been estimated using their share in the real output of the respective domestic production.
2. United States: 1993-1994; United Kingdom: 1993.

Source: OECD, STAN Database, *OECD Economic Outlook*. Calculations and estimates from the DIW (see BMBF, 1998, p. 4, *Germany´s Technological Performance*).

Knowledge is created through innovative processes, with R&D a critical input into those processes. The most important objective of national and international governmental R&D and innovation policy can be defined as the promotion and preservation of future corporate competitiveness.[2] From the point of view of the German Government, it is necessary to stimulate private companies to undertake and promote research and knowledge transfer. In addition, the structure of promotion programmes and supporting institutions has to be able to adapt to changes in innovation projects, especially when it comes to "advanced" and "cutting-edge" technologies.

An increase in the technological assets of an economy is undeniably based on a knowledge-based society, particularly on economic capabilities at the company level. Based on this perception, the central objectives of governmental research policy are derived from (BMBF, 1996, pp. 9-11):

- promotion of high technology;
- an innovation-oriented research policy;
- safeguarding and improving scientific excellence;
- strengthening and interlinking of the research system;
- international openness and co-operation.

Although an entrepreneurial spirit and the willingness to take risks in the development of new technologies are business characteristics which cannot be substituted by any governmental action, governmental R&D and innovation policy has to contribute to a dynamic innovative system. Readiness to innovate could also be encouraged by legislation. Therefore, the provision of favourable basic conditions has to go hand in hand with the development of co-operative networks within the innovative system. The funding and promotion of research, the stimulation of the exchange of knowledge between science and industry, and the creation of an environment that fosters innovative activities are vital characteristics of a comprehensive innovation policy.

The structure of the German innovative system has generally evolved through path dependent evolution, meaning that traditions and historical factors create a basis from which the system can depart. The division of competencies and responsibilities between the federal government and the *Länder* is seen as a key factor of the German innovative system. According to a very comprehensive view,[3] a national institutional framework is a set of rules and understandings which regulate the labour market, the educational system, corporate governance and product markets, together creating the economic context in which companies or their subsidiaries are embedded (Soskice, 1996, p. 16). The framework confronting German companies can be exemplified as follows:

- With regards to work organisation, training, hiring and firing, and some influence on wage bargaining, the labour market in Germany is relatively strictly regulated. However, the principles of consensus-based decision making between unions and employer organisations fulfil the requirements of a dynamic innovative system. The balance of opposing powers ensures the participation of both workforce and management. The innovation process is based on a commitment between workforce and management, who have to agree on R&D opportunities, risks and rewards.

- The regulation of the German education and training system has to be seen in the context of a regulated labour market. The system works on the basis of links between business and universities/science. Qualification standards and future training strategies are discussed in an open dialogue among companies, employee associations and educational institutions. From the enterprises' point of view, company training is long-term investment in human capital. This concept is anticipated by students and apprentices from fields of engineering who are willing to invest in their vocational skills. Basically, this system is able to cater for the companies' needs for specific know-how and skills; however, in terms of reacting quickly to market demands, it has its limitations.

- Corporate governance in the German innovative system provides capacities for long-term R&D and innovation projects at a relatively low level of risk and, at the same time, stable shareholding. Monitored by banks and consultants with the expertise to provide information on technological opportunities, companies are encouraged to engage in innovation strategies if the risks and costs can be estimated. This more-or-less consensus-based risk assessment reduces the likelihood of companies pursuing short-term radical innovation strategies.

- Alliances and co-operative partnerships between companies are also taken into account in German competition policy: the German system allows companies to work together on a consensual basis, and in many cases long-term co-operation partnerships formalised by contracts as well as intensive informal alliances are accepted, where such alliances are based on dialogue between companies and business associations.

In conclusion, the German innovative system favours innovation in traditional technologies, especially in important industries such as machine construction, chemistry, electrical machine construction or automobiles. The advantages of this system are obvious: there is a clear structure of responsibilities and competencies that has emerged over many years of continuous development. This development is backed by an industrial and R&D tradition determined through consensus-based governance structures and founded on individual experience and personal relationships. These conditions help to stabilise the system and enable it to cope with external shocks.

In recent years, these traditional strengths have increasingly become the target of criticism, being regarded as risky in a radically changing world. The emergence of new technologies such as computers, telecommunications etc., on the one hand, and considerable social shocks such as German reunification, on the other, indicate that the system may not be flexible enough to cope with extreme challenges. This is also due to the fact that in a system that places the main emphasis on high-quality incremental innovations, radical adjustments and radical innovations are very difficult to realise. However, the question of how to combine radical new strategies with existing competencies remains to be answered.

2. The German science and research framework

The German innovative system currently faces the task of reforming or transforming the existing system into a new one that is more aware of the importance of radical innovations in emerging technologies, and that is able to transform highly developed scientific knowledge into marketable (tradeable) products. Theoretically, the actors of the German innovation system should be divided into producers and users of knowledge and human capital, on the one hand, and knowledge transfer and bridging institutions, on the other.[4] In addition to universities and industry R&D facilities, a variety of public and semi-public research institutions exists, for example the Max-Planck-Gesellschaft (MPG), Fraunhofer Gesellschaft (FhG), and Helmoltz Centres.[5]

Through direct or indirect transfer to business sector enterprises, these institutions account for a large share of technology transfer. In terms of research in education and training, for example, achievements by public-sector research institutions stimulate industrial innovative success in three ways: by increasing human capital; by generating and publishing new knowledge from research work; and by directly supporting companies in solving industrial problems through the provision of services. Knowledge is channelled to firms in the form of:

- R&D co-operation projects;
- direct demand for applied research by industry enterprises, contract research;
- consultancy projects;
- personnel mobility;
- teaching of practice-oriented capabilities in technical colleges.

In addition, a variety of institutions serve as intermediary facilities for transfer activities. The crucial role played by institutions and their interrelationships in the innovation process as an element in the diffusion of know-how has been demonstrated. Therefore, a sound technology infrastructure is one of the most important preconditions in preparing an economy to meet global challenges. At the same time, institutions (in the sense of institutional, administrative facilities) can be a source of mismatches and barriers.

Figure 1. The science and engineering framework in Germany

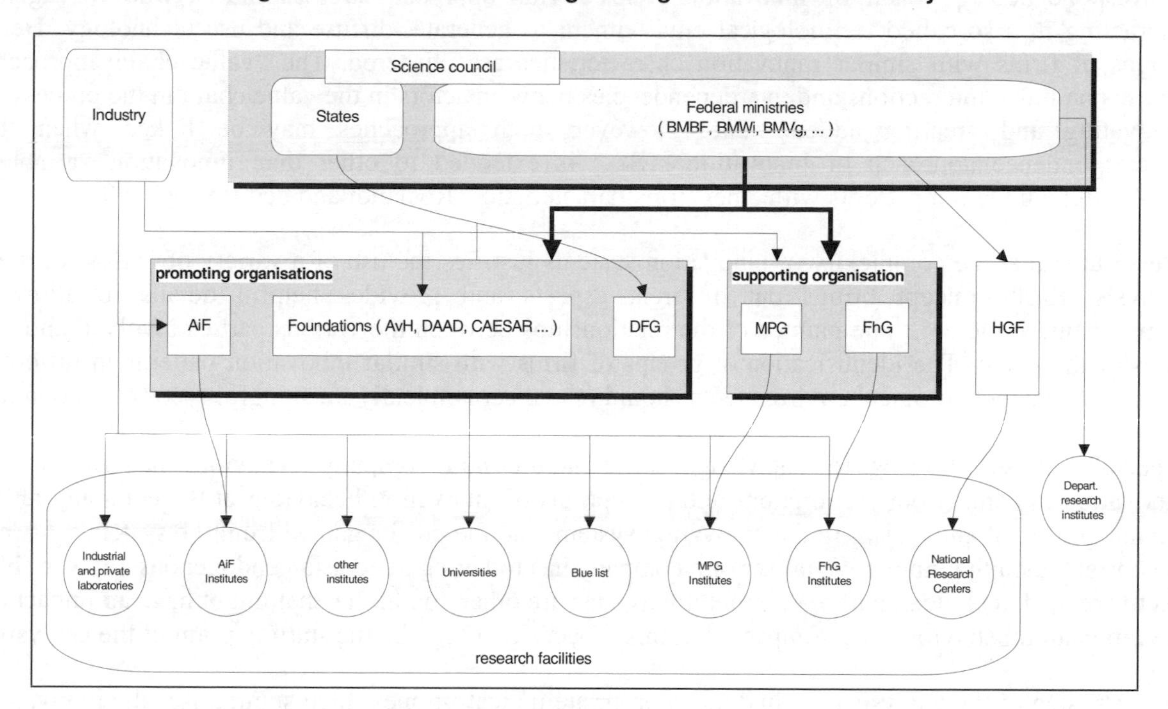

Source: Schmoch *et al.*, 1996.

From a theoretical point of view, it is possible to divide the innovation system into a technological, a structural, a regional or a national system of innovation; in practice, however, these elements work simultaneously, and focus of the analysis determines which categories are used. The active participants in the systems are companies, universities, academic research institutions, private and public sector educational facilities, political bodies and decision makers. In addition to structural preconditions, the crucial factors for success (in the sense of influencing the level and direction of technological change and contributing to the innovative power of an industry or a country) are primarily the behavioural patterns of industrial decision makers – the entrepreneurs and institutions involved – embedded in a proactive context of technical progress.

The efforts of the actors involved in the generation of new ideas, passing on knowledge, and translating it into marketable products (with all the surrounding circumstances of R&D, training and familiarisation, quality testing, market analyses, etc.), depend heavily on the extent to which the players profit from technological change and benefit personally from their innovative efforts. Key issues are the processes of interaction and the processes of competition and selection among firms with different capabilities and innovative performances.[6]

3. Innovative clusters at the company level

3.1. Correspondence approach of innovation styles

3.1.1. Conceptual framework

In analysing the innovation system, the emphasis is placed on the interdependencies between the corporate production system and the infrastructure of social and economic institutions, as well as other surrounding conditions.[7] For our purpose of cluster analysis, the methodology chosen was the "correspondence approach of innovation styles". This approach stresses the network of agents interacting in a so-called technological environment to generate, diffuse and use technology. Here, groups of firms with similar innovation characteristics are clustered. The "value chain approach" focuses on links, interactions and interdependencies between actors in the value chain in the process of innovating and creating added value. However, both approaches may be linked when the "correspondence approach of innovation styles" is extended to other than innovation variables, combining innovation patterns with other firm characteristics (Roelandt and den Hertog, 1998).

We assume that the complexity of innovation systems justifies the use of a variety of lenses for their analysis. Each concept brings out different aspects and provides helpful details to allow a comprehensive picture to be painted of the relationships between the various parties involved and the rules of the game. The identification of groups of firms with similar innovation patterns in different economic and technological environments should reveal economically meaningful types of innovation strategies. It may enable us to test the impact of factors necessary for the success of these strategies, especially those referring to knowledge infrastructure and company behaviour patterns in the innovation system. If our assumptions and perceptions of innovation behaviour at the company level, the integration of enterprises in the innovation system, and the profits gained from the system by firms are correct, then the innovative power of a company has to be regarded as an endogenous key variable. Sector-related, technical, partial or structural aspects are other conditions that might have an impact on the innovation behaviour of a company, but this aspect should not be the starting point of the analysis.

The first step of the analysis is to find the appropriate indicator; one which summarises the innovation and system behaviour of a number of firms. This indicator should reflect three factors, which together form innovative clusters:

$$(3.1) \qquad \text{Step 1: InnovativeCluster(IC)} = f\begin{Bmatrix} \text{Innovation (Inno)} \\ \text{Knowledge Transfer (Know)} \\ \text{Information (Info)} \end{Bmatrix}$$

- First, at the company level there are innovative capacities and potential which can be measured by the innovation intensity, or innovation and R&D performance, that occurs at different institutionalised company levels. This aspect refers to some conditions of appropriability, as well as to the accumulation of knowledge or the continuity of innovative activities. Therefore, we use the term innovation style or innovation status (*Inno*).

- Second, the company's capacities and readiness to interact are important. These can be measured by examining the networking activities or various knowledge transfer channels. The knowledge basis of a firm reflects the nature of know-how and the means of technology transfer and communication (*Know*).

- Third, the information and management behaviour of the firm is needed. For this, we use information sources important for the enterprise's innovation activities. These indicators capture both the sources of technological opportunities and the knowledge basis (*Info*).

The cluster defined using these variables is a pure innovative cluster. It is possible to examine whether such as cluster is sensitive to variables that are not directly linked to the innovation process. The variables used are size and industry characteristics:

(3.2) Step 2: Innovative Cluster (IC) = $f\left\{\begin{matrix}\text{Inno}\\\text{Know}\\\text{Info}\end{matrix}\middle|\begin{matrix}\text{Industry (Indu)}\\\text{Size (Size)}\end{matrix}\right\}$

Studies of innovative industries in Germany point to significant correlation between the innovative behaviour of a firm and the industry to which it belongs. Therefore, we suggest that *ex post* large firms have carried through their innovation processes more successfully than small firms.

3.1.2. Empirical findings

A very simple model of the above approach is an innovative cluster based only on innovation behaviour at the company level (Inno).

(3.3) IC_j = Innovative Cluster (IC) = $f\{\text{Innovation } (\text{Inno}_i)\}$ $j = 1...n, i = 1...4$

- Inno_1 = non innovative firm;
- Inno_2 = innovative firm without R&D;
- Inno_3 = innovative firm with R&D, but lacking an R&D department;
- Inno_4 = innovative firm with R&D and an R&D department.

The innovative power at the company level resulting from innovation and R&D activities is a first step to evaluating the technological competence of a firm, its market opportunities and conditions of appropriability. For this purpose, we differentiate four types of innovative firms, judging from the success of innovation activities at the company level.[8] The definition of these variables indicates an order of different degrees of innovative behaviour, or what we shall call an innovation style or innovation status. This order reflects the structural organisation of a company's innovation activities. It starts with no innovation, and ends with a highly organised and formalised system of innovative activities.

The objective of the cluster-oriented analyses based on this theoretical framework is to highlight the interdependencies between so-called innovation structures at the company level, and input behaviour and output results. We will answer the following questions:

- Does a company's innovative engagement and resource allocation depend on its innovation style?
- Is the performance of a company affected by its innovative power?

We chose this variable as a starting point because we have reasons to assume that the decision of being innovative, at which point in time and at which level of organisation, is one of the very first decisions made. The incentives to use a particular set of transfer and information channels ensue from the company's answers to these questions.

A first example for the alternative cluster approach, the "correspondence approach of innovative styles", refers to the innovation expenditure of small and medium-sized enterprises (SMEs).[9] We distinguish between different components of innovation expenditure, such as R&D, construction and design, training, etc., and three different types of innovative firms, the innovative clusters $Inno_2$, $Inno_3$, and $Inno_4$.

The innovation budget consists of different components, although the greater part is spent or invested in R&D, construction and design, pilot plants and projects, and investments in new products and processes. The distribution of these forms of expenditure does not vary much across the three groups of innovative enterprises; the above-mentioned components consuming nearly 85% of the budget. Among the different innovative clusters, it is obvious that R&D and investment, including the purchase of equipment, are substitutive actions. Firms without R&D invest nearly 40% in new equipment, whereas firms with R&D departments spend only 15% of their innovation expenditure on this item. On the other hand, they spend 30% on R&D, mainly for R&D staff. The results lead to the assumption that, for small and medium-sized firms, R&D activities are strategic, long-term business decisions. They are based on certain organisational forms, require highly qualified staff, and go beyond sporadic or operative work related to innovation.

Table 2. Innovation expenditure of SMEs, by innovative cluster
Percentages

Expenditure on:	Mean distribution of innovation expenditure of companies with 5-499 employees		
	Innovative without R&D	With R&D, but without an R&D department	With an R&D department
R&D	0	14	28
Pilot projects	19	22	20
Construction, product design	26	21	23
Gross investment for new products and processes	38	24	15
Market entry costs	3	5	5
Patent registration and renewal	1	3	3
Purchase of licences	3	2	1
Staff retraining	6	7	5
Other	4	2	1
Total	100	100	100

Source: ZEW (1995), Mannheim Innovation Panel.

Table 3 shows that firms with a more highly organised innovative behaviour (last column) also have more intensive export activities. On average, all firms that undertake R&D in a systematic way perform better than the average of the entire sample. The investment rates do not differ dramatically for firms that undertake innovative activities, no matter on which level. This points to a high potential for all groups of firms.

Table 3. Company performance in relation to innovation style
Percentages

Variables with a significant difference	Sample means	Non-innovative firms	Innovative firms without R&D	Firms with R&D, but without an R&D department	Firms with an R&D department
Share of turnover with new products	48.8	-	45.9	49.4	51.1
Share of firms with exports	56.8	28.9	42.5	63.4	83.7
Average export ratio of exporters	16.8	6.3	9.5	17.7	29.1
Average investment rate of investors	13.9	11.6	14.3	14.2	14.7

Source: ZEW (1995), Mannheim Innovation Panel.

It results that a company's performance and its innovative behaviour are comparable. It is also obvious that innovative firms without R&D behave differently from firms with R&D in the same industry. Of course, there are industry effects, for example we find more innovative firms in the chemical industry than in the textile industry, but innovative firms in these industries act similarly and show a similar firm profile or performance attitudes with respect to the different firm characteristics.[10] This gives rise to questions about whether and how enterprises are integrated in the national innovation system and how their input-output performance is influenced by their innovation cluster.

We define innovation styles not only by the formalised or institutionalised innovation or R&D activities, but also by alternative or complementary sources of innovation and channels of knowledge transfer. Taking these issues into account, the description of technology diffusion among firms in the national innovation system requires a more complex model. Therefore, we introduce an indicator for innovative clusters which covers the innovation structure, but also the knowledge and information aspects of companies' innovation activities:[11]

$$(3.4) \qquad IC_j = \text{Innovative Cluster (IC)} = f\left\{\begin{array}{c}\text{Innovation } (\text{Inno}_i)\\ \text{Knowledge Transfer } (\text{Know}_i)\\ \text{Information } (\text{Info}_i)\end{array}\right\} \qquad j = 1....n$$

Inno_i: i = 1...4; Know_i: i = 1...4; Info_i: i = 1...5

- Know1: firms stating that they had transferred or acquired knowledge through formal channels such as licences, or that they had purchased equipment, etc.
- Know2: firms preferring informal ways of acquiring information, *e.g.* through communication, hiring of skilled people, etc.
- Know3: firms transferring/acquiring know-how through international channels.
- Know4: firms using only national channels.
- Info1: firms using their own sources within the enterprise or within a group of enterprises.

- Info2: market-related external (vertical) sources of information such as suppliers or customers.

- Info3: market-related external (horizontal) sources of information such as competitors.

- Info4: science-based links to public R&D institutions and universities.[12]

- Info5: general information, such as patent disclosures, fairs, conferences.

We use the innovation structure in the same way as in the previous section, adding the knowledge and information variables. The characteristics of the knowledge variable will reflect formal and informal acquisition and transfer of technological know-how. A further distinction is made between national and international transfer mechanisms. Finally, we cluster innovative enterprises according to their sources of innovative ideas and according to the transfer channel used to direct information into new projects.

Equally as important as the transfer channels of new knowledge are the various sources of information that firms use to become acquainted with new ideas. The information attitude or management behaviour of a company comprises both the source of technological opportunities and the knowledge basis. Some sources, *e.g.* suppliers, customers or competitors, are closer to the market, while others are more related to science and research, *e.g.* universities or private or government R&D labs. It is interesting to examine to what extent information strategies in manufacturing industries are "pulled" by demand or "pushed" by technological development. In many cases, it is not technologies or products that are transferred but knowledge, enabling companies to develop market-driven innovations themselves and thus also expanding their innovative potential. In this context, the following questions need to be answered:

- What are the mechanisms that firms use to acquire and transfer technological know-how?

- What are the sources of information that can stimulate innovative ideas and activities?

The sources of information used in the creation of an internal knowledge base show an intuitive pattern: firms with rich and highly organised internal sources of innovation are more likely to use internal than external information sources (Table 4, $inno_4$). If a company carries out regular innovation activities and has annual R&D expenditure (Table 4, $inno_3$ and $inno_4$), the incentives to gather all the information available in the innovation system are high. However, for most firms, internal information plays an important role. More specific behaviour can be observed with companies with R&D departments ($inno_4$). In addition to the high importance attributed to internal knowledge diffusion, opportunities to access and exploit public information are highly developed. This is consistent with the argument that companies with higher innovation intensities experience more intensive demand-pull from customers and technology push incentives (public and other information) than do firms innovating at a lower level.

For innovative firms without R&D (Table 4, $inno_2$), indirect external and internal sources of information are the most important, while other external sources are used less frequently. It would seem logical that half of these firms try to gather information from sources that are easily exploitable. Within a company, information flows at almost no cost. Market-related external information cannot be obtained without resorting to costly resources. Besides the cost factor, another possible interpretation of the fact that only one-third of these firms uses public and direct external information could be the fear of loss of know-how; they may have little experience of the trustworthiness of external partners, especially in the case of public institutions. With regard to their information management, firms with own R&D activities differ from firms with less-formalised innovation activities. If a company is innovative and has regular R&D activities, the incentives to gather all the information available in the innovation system are high. However, internal information is very important for most firms. While internal sources of information are not of significant importance for companies innovating at a less-formalised level, external information resources are far more relevant. However, across all levels of innovation intensity, indirect external information is almost equally important (Table 4, $info_2$).

Table 4. Firms and their innovation styles and information behaviour
Percentages

Share of firms preferring certain sources of information	Non-innovative firms	Innovative firms without R&D	Firms with R&D, but without an R&D department	Firms with an R&D department
	$Inno_1$	$Inno_2$	$Inno_3$	$Inno_4$
Internal information $Info_1$		53	69.4	85.4
Vertical external information $Info_2$		51.1	54.2	59.7
Horizontal external information $Info_3$		38.4	46.1	57.6
Public information $Info_4$		37.1	53.1	70.4
General information $Info_5$		38.8	53.9	77

Note: Percentages refer to the whole sample of 2 859 firms.
Source: ZEW (1997), Mannheim Innovation Panel.

The use of transfer channels, whether formal or informal (Table 5, $know_1$ and $know_2$), is more likely in a more institutionalised innovation environment: the more formalised the innovation activities, the more elaborate and formalised the transfer channels used, *e.g.* joint ventures, co-operation with research institutes, etc. (Table 5, $know_1$). A logical conclusion is that firms with regular R&D expenditure and/or an R&D department have better opportunities to organise formal knowledge transfer. As mentioned above in the context of information strategies, the more formalised the innovation behaviour, the wider the range of transfer channels. The reason for this is better developed appropriability conditions. Looking at the last two columns of Table 5 ($inno_3$ and $inno_4$), it becomes clear that, in both groups of innovators with R&D, a slightly higher fraction uses formal channels of knowledge transfer. In firms without R&D, the opposite is the case. Enterprises with less formalised innovative activities depend on personal knowledge transfer, on a few highly skilled employees or on human capital in general.

Table 5. Firms and their innovation styles and knowledge transfers
Percentages

Share of firms preferring certain transfer channels	Non-innovative firms $Inno_1$	Innovative firms without R&D $Inno_2$	Firms with R&D, but without an R&D department $Inno_3$	Firms with an R&D department $Inno_4$
Formal channels $Know_1$		44.3	62.4	73.0
Informal channels $Know_2$		48.5	57.4	66.3

Note: Percentages refer to the whole sample, comprising 2 859 firms. A channel type was assigned to firms if they used that kind of channel in at least one form.
Source: ZEW (1997), Mannheim Innovation Panel.

Table 6. Firms and their information behaviour and knowledge transfers
Percentages

	Formal channels $Know_1$	Informal channels $Know_2$	International channels $Know_3$	Only national channels $Know_4$
Internal information $Info_1$	74.1	74.5	59	78.6
Vertical external information $Info_2$	59.3	58.8	50.4	60.8
Horizontal external information $Info_3$	54.1	53.7	41.7	53.8
Public information $Info_4$	63.4	61.6	48.3	60.5
General information $Info_5$	63.5	62.5	53.8	61.9

Note: Percentages give the share of firms preferring the respective information source; this percentage refers to the whole sample comprising 2 859 firms grouped into transfer channels. Channels and sources of information types were assigned to firms if they used each of the two characteristics in at least one form.
Source: ZEW (1997), Mannheim Innovation Panel.

The different sources of information are closely related to each other. If a firm uses at least one of these sources, this means in almost all cases that it uses, in fact, more than one, and sometimes a wide portfolio of sources. There is reason to assume that successful use of public sources of information, such as universities, goes hand in hand with the exploitation of indirect external information from suppliers and customers. Surprisingly, information strategies are little affected by the way in which the knowledge transfer takes place (Table 6, $know_1$ and $know_2$).

3.1.3 Characteristics of innovative clusters – size and industry effects

From the analysis, we assume that the decision to be innovative to a certain degree of organisation and the decision to use certain transfer or information channels are simultaneously determined variables.[13]

If the results obtained so far are placed in relationship with one another, the following three main clusters, or types, of innovative firms emerge:[14]

- (IC_1). Firms with an R&D department having a broad internal research and basis of knowledge: external information is, above all, used as a "technology push", even if other sources of information are also considered for global information management. Reflecting the specific company potential and human capital, formal transfer mechanisms with a high degree of organisation are preferred, although companies also maintain informal contacts at a high level.

- (IC_2). Firms with their own R&D activities which are less formalised, but nevertheless represent a very sound know-how and absorption basis for the company. Supported by this company potential, external sources of information are of considerable importance, showing a slight preference for the "demand pull" compared to the "technology push". Public research institutions are also important sources of stimuli. The degree of organisation of the knowledge transfer is still relatively high in this group. Formal transfer channels are of high importance.

- (IC_3). Innovative firms without R&D capacities which, besides their own qualities, strongly focus on market-related know-how from customers and suppliers: these companies rely on informal transfer channels and face-to-face exchange of knowledge. Due to lack of experience, they are little interested in formal collaboration, *e.g.* in the form of co-operative partnerships particularly with public research institutions, and rather react spontaneously to the current competitive situation. Innovation is seen more as every-day-business than as a continuous process.

In describing innovative clusters that are heterogeneous in terms of their innovative styles, and including in particular a heterogeneous use of the innovation system, the classification as clusters provides a sufficient background of information on companies' innovative behaviour and strategies. The substantial differences between innovative clusters that arise – even though decisions on how to use the innovation system are presumably made simultaneously within the company – are affected by differences in the degree of R&D formalisation, which can also be described as structural R&D.

Structural innovation behaviour is closely related to the size of the company. Thus, clusters following the definition IC_2 and IC_3 differ significantly in their size structure compared with IC_1 clusters. One implication of Schumpeter's hypothesis is that firm size and innovative behaviour are simultaneously determined. Concerning the choice of a transfer channel and source of information, firm size has some kind of threshold effect. Firms with less than 250 employees have a significantly lower propensity to use the full range of sources of information sources and transfer channels. In addition, most innovating firms that do not spend money on R&D systematically employ less than 250 people.

There are also industry effects. The level of organisation of innovative activities is considerably higher for R&D-intensive industries such as chemistry, machinery, the computer industry, electrical machinery, and the software equipment industry than for less R&D-intensive industries, such as mining, paper, wood or construction. However, knowledge transfer and information strategies are determined more by structural innovation behaviour than by the technological characteristics of the environment for which the industry is an indicator. There is no significant industry-related difference among firms with an R&D department which also use different transfer channels and different information sources. The large groups of innovative companies organise their knowledge transfer and information flows differently, using internal and external, formal and informal channels. However, it would appear that these different strategies complement each other.

Informal transfers are determined by an intensive exchange of experience and the hiring of qualified employees. Human capital therefore plays a key role in the innovation system. Even for firms with R&D departments, human capital makes a difference to how they organise the diffusion of knowledge. Formal channels, such as the purchase of licences, external know-how in the form of machines or co-operation within joint ventures, are easier to achieve. They may require a great deal of money and abstract knowledge, but they demand less creativity and personal effort. Informal channels depend heavily on the conditions and incentives provided by the firm. This tendency can also be observed in firms that innovate, not on the basis of systematic R&D expenditure but on different performance levels. Investment rates do not seem to be affected by the degree to which their innovation activities are organised. Moreover, firms that do not systematically innovate often have higher investment rates.[15]

With respect to firms' usage of the "national" innovation system, it is worthwhile mentioning that there is huge potential of co-operation with foreign firms and institutions that is not exploited and transformed into economic success. Judging from the cluster definition above, only a small fraction of firms innovate at a highly organised level and use international transfer channels. It is possible that only a small number of these firms, mainly the large ones, have the capacity to process the huge amount of knowledge entering the country from abroad – from foreign firms or institutions. In addition, small firms and firms with a low degree of formalised innovation in particular are confronted with a language problem. The propensity to use the national innovation system, especially the public sector, correlates with the size of a country. German enterprises, as well as firms in France and the United Kingdom, have closer relationships to national partners or institutions than companies in smaller countries such as Denmark or the Netherlands.[16]

3.2. Sources of innovation-related information

3.2.1. Effects of the German technological infrastructure

For many firms across all industries, it is true that universities or publicly financed research institutions are less important sources of information for their innovation activities than are internal and other external know-how. Depending on the organisational level of existing innovative activity, firms differ in the way they make use of the innovation system. The incentives to use external sources of information decrease with the degree of formalisation of a company's innovation activities. On the other hand, other sources of information can be better exploited when innovation activities are more formalised and when firms dispose of in-house capacities to use a wide range of information in their innovation process. This is especially true for public sources of information.

In international terms, the scientific infrastructure in Germany is highly developed. Nevertheless, the efficiency of the system has recently been called into question. It was claimed that there was inconsistency between the aims and motives of technological research institutions and the aims of innovating enterprises. Universities, senior technical colleges or polytechnics and research centres produce specialised know-how which, to be used by enterprises, requires a skilled workforce which has been trained accordingly. Therefore, it is not surprising that the importance of the technological infrastructure as a source of information increases with the level of structural R&D (IC_2, IC_3) and the size of the enterprise's R&D intensity.[17]

In this context, analysis has to be carried out to discover whether enterprises with high R&D intensities benefit most from information originating from universities, senior technical colleges and research centres. The 1996 Mannheim Innovation Survey showed that this is particularly true for the research centres (FhG, MPI), the majority of which are indeed aimed at industries with high R&D

intensities. In the case of universities and senior technical colleges, however, industries with medium R&D intensities benefit most from this source of information.

The degree to which academic research work is used by enterprises from "cutting-edge technologies", measured by the innovation expenditure of these enterprises, is 23%, corresponding to the share of this group of companies in the innovation expenditure of all innovating companies.[18] Enterprises from "advanced technologies" account for almost 50% of total expenditure, and thus exceed their share in innovating companies (less than 40%). The majority of research institutions other than universities are directed towards industries with high R&D intensities – "cutting-edge technologies" – while the representation of industries with medium R&D intensities – "advanced technologies" – is approximately proportional to their share in innovating companies. The non-R&D-intensive industries are under-represented, both with regard to their use of academic research work and their resort to research institutions other than universities (Licht and Stahl, 1997, pp. 38-39).

However, this does not account in any way for how the information is used in the enterprise's innovative process and how valuable the information is for industrial innovations. In the tradition of studies by Mansfield (1991), the aim was to identify the kind of public research that is utilised by private enterprises and that is in place for private research activities (Mansfield, 1991, 1998; Beise and Stahl, 1998). Hence, another question that arose was whether there had been any product or process innovations between 1993-96 that would not have been possible or would have been considerably delayed in the absence of the information gained from the above-mentioned external sources.

Weighted by the number of companies, the analysis comprised somewhat less than 38 000 companies, sub-divided in a ratio of approximately 25:75 into R&D-intensive "cutting-edge technologies" and "advanced technologies", on the one hand, and non-R&D-intensive enterprises, on the other. Only 10% of the enterprises surveyed (slightly less than 4% of the R&D-intensive, and slightly more than 5% of the non-R&D-intensive companies) stated that they could not have carried out the innovations in the absence of public research. Again, high R&D-intensity industries showed above-average representation (Table 7).

Table 7. The economic effects of academic research

	All innovators		*Of which:* companies which could not have carried out innovations in the absence of public research	
	Numbers		Total	
	Absolute	In %	Absolute	In %
All companies	37 780	100	3 430	9.1
Of which:				
• R&D-intensive industries	9 030	23.9	1 480	3.9
• Non-R&D-intensive industries	28 750	76.1	1 950	5.2
	Scientific support provided mainly by[1]			
Source	*[2]		2 480	6.6
Of which:	*			
• Universities	*		1 080	2.9
• Polytechnics	*		720	1.9
• Other publicly funded labs	*		1 470	3.9
Not classified	*		950	2.5

1. Figures are weighted by the number of companies involved.
2. Several answers were possible; in some cases the sum of the percentages therefore exceeds the total.
Source: ZEW (1997), Mannheim Innovation Panel (see Licht and Stahl, 1997, p. 39).

The scientific sources identified as the main supporters of public-research-based innovations are public research institutions (mainly FhG) and universities. Nevertheless, in view of their relatively low research budget, polytechnics also perform very well. As Beise and Stahl (1998) reveal, large companies tend to be biased towards universities, which are normally better equipped than polytechnics, while smaller firms obtain relatively more support from polytechnics. Compared to polytechnics, universities also play a role in innovation by small companies (Beise and Stahl, 1998, p. 11). These findings correspond to our empirical examinations and underline the definitions of innovative clusters used here, particularly the innovation styles of IC_2 and IC_3.

If these sample results are related to the economic success ensuing from public research support, almost 50% of the additional turnover is generated by companies in R&D-intensive industries which, as shown above, account for only one-quarter of all enterprises.

Table 8. Turnover on products which would not have been realised in the absence of research results of universities or publicly funded laboratories[1]

	Total		R&D-intensive industries		Non-R&D-intensive industries	
	In DEM million	In %	In DEM million	In %	In DEM million	In %
Total turnover	19 900	100	9 700	49	10 100	51
Of which:						
Universities	13 600	68	5 800	29	7 800	39
Polytechnics	8 000	40	2 500	13	5 500	28
Other publicly funded laboratories	10 800	54	3 800	19	7 000	36

1. Figures are weighted by the number of companies involved.
2. Several answers were possible; in some cases the sum of the percentages therefore exceeds the total.
Source: ZEW (1997), Mannheim Innovation Panel (see Licht and Stahl, 1997, p. 40).

Of these sales of research-based products, 68% can be attributed to universities. Research by polytechnics is used more for non-R&D-intensive products, suggesting that the high-tech sectors prefer research by universities and research laboratories for their main product innovations (Beise and Stahl, 1998, p. 12). It is therefore justified to say that R&D-intensive industries benefit most from the technological infrastructure.

3.2.2 Clusters towards innovation styles and production chain

The characteristics chosen to summarise innovative firms as clusters show a rough pattern of links of knowledge transfer and information exchange between companies and between companies and publicly financed research facilities. To highlight the former case, co-operation between single firms and whole industries, as well as the similarities of innovative behaviour of companies, we will combine the "correspondence approach of innovative styles"' with an analysis of input-output tables using the "value chain approach" to obtain a picture of industrial links within the economy.

In recent years, attention has increasingly focused on the flow of information along the value-added chain. Increased co-operation with customers and suppliers reduces risk and speeds up the whole process, thus increasing the quality of innovations. Since customers of advanced-tech enterprises are more likely to stem from the producing sector, while customers of R&D-intensive high-tech enterprises usually come from the service sector, it makes sense to differentiate between these two types of customers. Customers of the manufacturing industry represent an important source of information, especially for companies with medium R&D intensity. More than half of these companies think that information from these customers is vital for their innovative process (Licht and Stahl, 1997). The opposite view is held by companies with high R&D intensity. A minority of these firms view customers from manufacturing industry as an important source of information; on the whole, these firms prefer customers from the service sector. For non-R&D-intensive industries, the importance given to customers, from both the manufacturing and the service sectors, is relatively low. As far as suppliers are concerned, suppliers of intermediate products, materials and components are of major importance across the industry, while the importance of suppliers of machinery and equipment is considerably lower; the latter are only considered highly important by industries with medium R&D intensity.

As a first step, we drew a detailed picture of where information came from and where it went to, and in which direction technology transfer went. This information on the pattern of links between industries created by knowledge diffusion was completed by determining which companies were the suppliers, the customers or the competitors of firms within the innovation survey. This was obtained by input-output analysis which revealed a structural pattern of links and collaboration between industries and services. At first sight, differences in the use of transfer channels among industries are not that obvious. The service sector is one of the leading sectors with respect to industry linkages. In particular, consulting is used intensively to diffuse knowledge.

The structural analysis with input-output tables revealed an intuitive pattern of relationships between industries. Links, in the form of flows of intermediate inputs, serve to illustrate industry clusters and identify industries linked through the main products they deliver or use. This method illustrates the approach of clustering within or through the chain of values. Combining the aforementioned approach with the "correspondence approach of innovative styles", paves the way towards an integral analysis of flows of goods and knowledge.

4. Examining innovation policy – principles of an effective transfer of knowledge

The variety of institutions in Germany which are classified as "producers" of technical knowledge might be interpreted as a reflection of the heterogeneity of firms. Therefore, an efficient technology transfer or (German) national innovative system needs to be flexible and decentralised. At the company level, there are obviously different needs and wishes. Large companies use the innovative system more intensively than do SMEs, and all forms of knowledge transfer are taken into account. Technology transfer here is a strategic function and is part of the general information management of the company. In contrast, small and medium-sized firms prefer informal communication with local agents; they focus on personal contacts and prefer to be independent; they act spontaneously and on an operative level.

In general, any kind of technology transfer or collaboration among firms or between the business sector and research institutions is based on trust and experience. The socially interwoven nature of the academic and business communities is a crucial factor. When direct contacts and informal networks form the primary channels for transmitting scientific findings and technical knowledge in the innovative system, a flexible, decentralised and deregulated practice of technology transfer promotion can most effectively contribute to establishing and stabilising these informal networks.

Studies on technology transfer reveal that a substantial part of existing technology transfer in an innovative system is self-organised, taking place directly between scientists and corporate technicians and managers. Firms integrated in the national system of innovation and co-operating companies see a close correlation between knowledge exchange with universities and corporate success.[19] These firms differ from companies which are interested in starting up their first co-operative project and who wish to begin with a relatively small budget. The reason for this hesitation on the part of companies basically interested in co-operation is not a general liquidity problem; rather, during the first co-operative project, the company's risk is increased by uncertainty relating to both the procedures involved and the chances of success, combined with a yet-to-be established confidence in the partner. This creates an inhibitory threshold, which companies that have not yet co-operated with academic institutions have to overcome if they are to effectively utilise technology transfer from these institutions on a long-term basis.

Joint research projects are usually conditional upon ongoing R&D activities among the companies involved. The partner can contribute only complementary knowledge. In the case of co-operative projects with public-sector research institutions, in particular, companies cannot outsource the necessity to design market-driven product/process innovations. Universities and public-sector institutions, being remote from the marketplace, have only a limited ability to develop finished products for the real market.

In the final analysis, the potential role of politicians in improving technology transfer is somewhat unclear. Of course, there is no lack of suggestions on how technology transfer could be enhanced, but it should be borne in mind that not all companies participate equally in technology transfer:

- One important factor is size of firm. The strong relationship between size and the degree of formalised innovation activities, on the one hand, and of formalised innovation activities, on the other, suggests that once a certain size has been attained, it is far easier to establish systematic R&D activities and to successfully bring about innovation. It has also been demonstrated that the commencement of R&D activities constitutes the main inhibitory obstacle for small companies. What could be termed "threshold sponsoring", *i.e.* assistance to companies which do not yet have their own R&D activities, will increase the number of firms eligible for co-operative R&D projects with public-sector research institutions.

  An effective technology transfer policy does not replace R&D sponsoring for small and medium-sized companies, but rather complements it. "Threshold programmes" supporting small firms from all technological sectors could help to establish successful innovative institutions in these firms. There is a high potential of creative ideas in small firms that needs to be activated. The aim of these "threshold programmes" should be to stimulate and promote the innovative activities and capabilities of companies. Thus, it would be more effective to provide firms with information, such as on fairs presenting new technologies, seminars or other means of non-formalised technology transfer.

- To increase the efficiency of technology transfer or knowledge flows within the national innovative system, the usual recommendation is to set up specialised institutions, designed to act as an intermediary between research institutions and companies. Similar to the so-called employment exchanges in the labour market, the idea is to "place" the outsiders in technology transfer with the appropriate academics, and to compensate for any entrepreneurial shortfalls in project/innovation management. The crucial factor then will be a co-operative activity between these institutions that can draw optimal benefits from synergy effects, thus justifying the existence of an institutional network. This institutionalisation is aimed at involving the research institutions in business promotion, although its primary goal is to compensate for management/innovation shortfalls in companies. However, information and promotion are not sufficient. Information strategies imply that there is a correlation between innovative behaviour and the capacity to effectively use information and know-how. Therefore, it is necessary to create an innovative atmosphere and provide incentives for all dimensions of knowledge diffusion. Otherwise companies, especially those which lack experience with public information services, will hesitate to take up the opportunities. Market mechanisms, as they are traditionally conceived, would appear to have only limited use for co-ordinating the transfer of knowledge and technologies. In none of the environments inhabited by public-sector research can pure market conditions be found, neither in the labour market for researchers, nor in R&D contracts and engineering services. Assigning an *a priori* status to the market process thus bypasses the core of successful technology transfer in the research field. Promotion of technology transfer should not aim to compensate for market failures, but should help to overcome the problem of knowledge diffusion at the boundary between pure research and the development phase in the innovative process. Firms know more about their potential and their markets than any private or public institution intending to provide support for innovative activities can achieve through simply observing firms and their markets.

By comparison, concentration on the core question of how new research results can successfully find their way into companies (*i.e.* be actually applied), would not necessarily require institutionalised support. On the contrary, empirical feedback from successful technology transfers demonstrates the importance of self-organisation by innovative companies and technology-driven new businesses. The bottleneck of technology transfer in individual fields of technology is constituted by the small number of insiders that are actually involved in the national innovative system. Effective promotion of technology transfer should therefore aim at establishing long-term co-operative partnerships between companies and between companies and academics.

- For companies that do not innovate at all, it is not possible to dictate to them how to become "innovative". Institutionalised support does not provide a way out of the insider-outsider situation. Very few of these firms are willing and able to make use of mediating institutions. In addition to their weak in-house capabilities, they have little confidence in the experience of external partners, especially public institutions or mediators.

To the best of our knowledge, firms innovating at a certain level of organisation use a portfolio of information and knowledge transfer strategies that cannot simply be transferred to firms that are not (yet) innovative. While acknowledging that innovative in-house activities are necessary to keep abreast with international developments and competition, the creation of a general business environment which is highly supportive of innovative activities should be one of the main goals of innovation policy. Furthermore, firms need to have an absorptive capacity in order to transform knowledge into successful and profitable innovations.

Annex 1

THE SCIENCE AND ENGINEERING BASE

A summary of the financial sources, size and main research areas of the research institutions, universities and laboratories described above is provided below.

Financial resources and main support

Institution	Expenditure (in million DM)	Number of institutes	Employees	Public support (in DM million)	Share of support (federal/state)
AiF		107		170	100:0
MPG	1 533	98	11 901	1 429.9	50:50
FhG	1 261	49	6 099	578	90:10
Helmholtz Centres	4 171	16	22 501	2 900	90:10
Blue List	1 321	83	10 000	1 200	50:50
DFG	1.927			1 147	100:0
DAAD	372.6			354	90:10
AvH	87.7		78	87.7	100:0
Stifterverband	141.7			Assets of foundation	
Volkswagen Foundation	113		93	Assets of foundation; dividends	
CAESAR	750			685	
Federal institutions	2 867	57	18 682	2 867	100:0

Source: BMBF (1996).

- *Arbeitsgemeinschaft industrieller Forschungsvereinigungen* (AiF) comprises over 100 industry research organisations. AiF promotes applied research and development mainly in support of small and medium-sized firms. Since one of the goals of this organisation is to maintain a spirit of community and common interest, its activities focus on industries and branches.

- Max-Planck Gesellschaft (MPG) is a sponsoring organisation with 71 research institutions, active and represented throughout Germany. The MPG is primarily involved in pure research in selected areas of natural sciences, social sciences as well as the arts. It promotes new, promising research topics which are not yet adequately represented at universities. The MPG co-operates with universities, for example by allowing them to use their technical equipment. Expenditure amounted to DEM 1 616 million for 1994 and DEM 1 708 million for 1995. The MPG has a staff of 11 500, including 3 015 scientists.

- Fraunhofer Gesellschaft (FhG) is a non-profit sponsoring organisation with 47 institutes for applied research and two service facilities in 14 federal states. In addition, there are three offices in the United States. Carrying out contract research projects for the economy and the public sector, the FhG contributes to transferring pure research findings into practice. Institutional promotion by the federal government and the German states enables the FhG to deal with self-chosen research topics in order to secure their scientific potential and the development and continuous monitoring of new technologies. The FhG offers firms and public authorities its services in the area of microelectronics, information technology, production automation, production technology, materials and components, process engineering, energy and structural engineering, environment and health, technical-economic studies and professional information. Close relationships with universities are institutionalised through the joint appointment of Fraunhofer directors as regular university professors.

- National research centres, or Helmholtz Centres, are promoted jointly by the federal government and the state governments. Research using large-scale equipment and focusing on specific priority topics, primarily large accelerators, neutron and synchroton sources as well as observatories and telescopes, is the particular focus of the Helmholtz Centres, of which there are 16 in Germany. The Helmholtz Centres contribute significantly to long-term pure research in several fields through their own projects and as partners of universities and other research institutes.

- Institutions on the "Blue List" are characterised as one of the four cornerstones of the common promotion of research of the federal government and the German states. They cover all major fields of natural and social sciences, technology and applied technological research.

- DFG (German Research Council, *Deutsche Forschungsgemeinschaft*), is the major promoting and self-governing organisation for science and research activities in Germany. One of its main tasks is the financial support of research projects, supporting research co-operation and promoting young scientists. Secondly, it has an important role as adviser to policy makers on scientific questions. Thirdly, the DFG develops and entertains the relationship and co-operation with international research institutions.

- The German Academic Exchange Service (*Deutscher Akademischer Austauschdienst*) supports and organises exchange programmes for students, postgrad- and post-doctorate researchers with foreign universities and research institutes.

- Alexander von Humbold Stiftung (AvH) promotes foreign scientists and researchers. As an example, programmes launched by the AvH support co-operative research projects between Germany and other countries. In addition, scholarships and prizes are awarded to academics.

- Stifterverband is an association of firms, private non-profit organisations and private persons. It supports science and technology projects as well as institutes or other organisations which require financial or organisational help to perform R&D. The Stifterverband provides services such as statistics of economic indicators, seminars and an infrastructure for scientific activities.

- Volkswagen-Stiftung (VW-Stiftung) has three main tasks: *i)* promotion of pure research in special fields; *ii)* improvement of the infrastructure for research, teaching, and scientific communication; *iii)* promotion of research oriented towards co-operation with foreign countries.

- CAESAR (Centre of Advanced European Studies and Research). This foundation is oriented towards research and development of new technologies. A research centre in Bonn undertakes a combination of basic and applied research in the fields of technology and natural sciences.

Annex 2

SAMPLE DESCRIPTION AND VARIABLE DEFINITION

For the empirical cluster definition we used the German sample, comprising 2 859 firms that participated in the first part of the Mannheimer Innovation Survey (comparable with CIS data). The Community Innovation Survey (CIS) is an EU initiative and a joint survey of DG XIII/SPRINT/EIMS and Eurostat. The CIS was developed between 1991 and 1993, the year when the data was collected. In 1993, a common statistical survey questionnaire was distributed among the member states. The database contains almost 41 000 observations from the twelve states which were members at the time of the survey, and Norway.

Table 9. Innovation styles in relation to industries and size in Germany[1]
Percentages

	Non-innovative firms $Inno_1$	Innovative firms without R&D expenditure $Inno_2$	Firms with R&D, but without R&D department $Inno_3$	Firms with R&D department $Inno_4$
Utilities / mining	51.5	30.6	14.4	6.3
Food	26.3	43.2	17.0	13.6
Textile, leather	28.6	31.8	23.0	16.7
Wood, paper	29.4	49.7	14.7	6.3
Chemicals	10.3	11.9	20.0	57.8
Plastics / rubber	19.4	32.2	30.0	18.3
Glass, ceramics	18.9	25.6	34.4	21.1
Metal industry	14.7	25.3	34.7	25.3
Steel construction	21.3	31.4	31.4	16.0
Machine construction	9.1	16.9	29.9	44.2
Mechanical engineering	7.6	12.6	33.3	46.5
Other mechanical engineering	16.9	18.5	29.2	35.4
Data processing	13.6	12.7	18.2	55.5
Electrical equipment	16.8	16.0	28.6	38.7
Medical instruments	11.3	20.9	22.0	45.8
Automobile industry	10.8	20.4	31.2	37.6
Other vehicle construction	26.2	21.4	19.1	33.3
Furniture, musical instruments	20.8	23.8	26.7	28.7
Construction	43.0	36.9	14.8	5.4
Services	21.1	25.9	35.3	17.7
Aerospace	15.4	23.1	23.1	38.5
5-49	32.4	34.7	24.1	8.8
50-249	18.4	26.5	31.3	23.8
250-499	9.1	20.2	28.2	42.5
500-999	8.9	16.5	21.2	53.5
1 000+	9.2	8.0	18.4	64.4
Total	**20.2**	**25.5**	**26**	**28.3**

1. Figures are unweighted.
Source: ZEW (1997) Mannheim Innovation Panel.

The three canonical variables for the cluster definition are:

♦ Formalisation of innovative behaviour

$Inno_i$ describes the degree of formalisation or organisation of the innovative activities. This definition can be pictured as an ascending order of the degree of formalisation or organisation of R&D behaviour. Firms were asked whether they had introduced any technologically new or improved products or processes within the last two years.

♦ Formalisation of knowledge channels

$Know_i$ describes the transfer channels used by the firm. Firms were asked whether they really used the different transfer channels and with which countries they co-operate. Dummy variables were created for firms matching certain criteria. The aggregation of the information sources was carried out as follows:

Table 10. Formalisation of knowledge channels

	Obs	Mean	SD	Min	Max
$Know_1$	2 859	.48	.50	0	1
$Know_2$	2 859	.46	.49	0	1
$Know_3$	2 176	.39	.49	0	1

1. *Formal channels:* Know1 = 1 if firms purchased or sold licences or purchased or sold consulting services or purchased or sold research results of external research institutes or purchased or sold new technologies linked to investment goods or purchased or sold new plant that produces new technologies. Know1 = 0 else.
2. *Informal channels:* Know2 = 1 if firms exchange innovation experiences with other companies (or hire or exchange qualified personnel). Know2 = 0 else.
3. *International channels:* Know3 = 1 if firms transfer knowledge from or to at least one foreign country. Know3 = 0 using only national channels

♦ Formalisation of information sources

$Info_i$ describes the information sources used by the firm. Firms were asked how they evaluated different information sources. If the sum of the aggregated evaluations was higher than the mean of the sum for all companies, the dummy variable $Info_i$ was set at 1. The aggregation of the information sources was carried out as follows:

Table 11. Formalisation of information sources

	Obs	**Mean**	**SD**	**Min**	**Max**
$Info_1$	2 138	.75	.43	0	1
$Info_3$	2 199	.58	.49	0	1
$Info_2$	2 182	.50	.50	0	1
$Info_4$	2 167	.57	.49	0	1
$Info_5$	2 203	.59	.49	0	1

1. *Internal information:* Info1 = 1 if firms using information coming from different departments inside the company. Info1 = 0 else.
2. *Vertical external information:* Info2 = 1 if firms using information coming from competitors, consultants, private research institutes. Info2 = 0 else.
3. *Horizontal external information (market-related information):* Info3 = 1 if firms using information coming from suppliers, clients, customers. Info3 = 0 else.
4. *Public information:* Info4 = 1 if firms using information coming from universities, public research institutes, public technology transfer institutions. Info4 = 0 else.
5. *General information:* Info5 = 1 if firms using information coming from patents, fairs, conferences. Info5 = 0 else.

Table 12. Ordered Probit models of the degree of formalisation of R&D coefficients (standard errors)

Dependent variable Degree of formalisation of R&D	(1)	(3)
Internal information: $Info_1$	0.432 (0.060)	0.311 (0.623)
Direct external information: $Info_2$	-0.044 (0.057)	-0.051 (0.059)
Indirect external information: $Info_3$	-0.185 (0.054))	-0.093 (0.061)
Public information: $Info_4$	0.294 (0.054)	0.221 (0.061)
General information: $Info_5$	0.517 (0.055))	0.408 (0.058)
Formal transfer: $Know_1$	0.336 (0.055)	0.244 (0.057)
Informal transfer: $Know_2$	0.160 (0.053)	0.151 (0.056)
International transfer: $Know_3$	0.545 (0.053)	-0.236 (0.077))
Utilities/mining		-1.134 (0.212)
Food		-0.865 (0.169)
Textile, leather		-0.531 (0.163)
Wood, paper		-1.051 (0.168)
Chemicals		0.430 (0.149)
Plastics / rubber		-0.384 (0.145)
Glass, ceramics		-0.528 (0.176)
Metal industry		-0.567 (0.184)
Steel construction		-0.447 (0.132)
Machine construction industry		--
Mechanical engineering		0.131 (0.148)
Other mechanical engineering		0.011 (0.143)
Data processing equipment		0.550 (0.181)
Electrical equipment		-0.079 (0.163)
Medical instruments		-0.150 (0.147)
Automobile industry		-0.198 (0.170)
Other vehicle construction		-0.305 (0.253)
Furniture, musical instruments		-0.078 (0.176)

Table 13. Ordered Probit models of the degree of formalisation of R&D coefficients (standard errors)
(cont'd)

Dependent variable **Degree of formalisation of R&D**	**(1)**	**(3)**
Construction		-1.228 (0.176)
Services		-0.333 (0.150)
Aerospace		-0.399 (0.396)
5 – 49 employees		-1.33 (0.113)
50 – 249		-0.990 (0.098)
250 – 499		-0.681 (0.110)
500 – 999		-0.446 (0.120)
1 000+		-
Firm-size dummies Chi-squared (df) (p-value)		166.3 (4) (p<0.001)
Industry dummies Chi-squared (df) (p-value)		241.9 (20) (p<0.001)
N	2012	2012
Chi-squared (df)	461 (8)	887.3
log L	-2 044	-1 831
pseudo-R-squares	0.1013	0.1950

Note: The table displays the coefficients from an ordered Probit specification. They can be interpreted as the impact of the exogenous variable on the decision process in one of the classes of formalised innovative behaviour (innovative styles $inno_1$ - $inno_4$). A direct interpretation is not possible, only the sign and the level of significance contain information. For the ordered Probit analysis we used the assumption of an ascending status of innovative power and defined the innovation styles "inno" with values from 1-4. Inno = 1 if $Inno_1 = 1$; Inno = 2 if $Inno_2 = 1$; Inno = 3 if $Inno_3 = 1$; Inno = 4 if $Inno_4 = 1$.
Source: ZEW (1997), Mannheim Innovation Panel.

NOTES

1. R&D-intensive industries are classified as "high or cutting-edge" (pharmaceuticals, computer/office technology, radio/TV/telecommunications engineering, aircraft and spacecraft industry, precision instruments, optics/clocks) and "advanced" technologies (chemical industry, mechanical engineering, electrical engineering except radio/TV/telecommunications, railroad industry, automobile industry). The line is drawn according to the share of R&D in the turnover: "cutting-edge technologies" cover goods with an R&D intensity of more than 8.5%, while "advanced technologies" cover goods with an R&D share of more than 3.5% and less than 8.5% of turnover.

2. In economic terms, the external effects of corporate innovation activities have recently been presented as a justification for governmental intervention. The core argument, in this context, is that the national economic benefits of corporate research expenditure are higher than the benefits for the respective company. Thus, in terms of the national economy as a whole, companies are not investing enough in inventing and developing new products and processes. One explanation for this is that the knowledge-producing firm is unable to reap the full profits of its investment. For studies focusing on the situation in Germany, see Harhoff and König (1993), Klodt (1995), Meyer-Krahmer (1993) and Becher and Kuhlmann (1995).

3. In Section 2 we will discuss an institutional framework of the German innovation system that focuses on the distribution power of business and science. See also Annex 1.

4. Recent research on innovation has revealed that the innovation process, to a large degree, can no longer be depicted as a linear model of consecutive phases of innovation, where the research phase, the development of the product and the product design are implemented independently of each other in temporal succession. In the majority of industries or technologies, the phases and sub-processes involved are recursively interwoven with each other, they are mutually contingent, and recur in response to the increasing experience gained in the subsequent phases of innovation. The recursive model of the innovation process implies a long-term interaction between all the actors involved in the innovative activities concerned (Kline and Rosenberg, 1986).

5. See Annex 1 for the list of institutions.

6. Nelson and Winter (1982) single out four factors which are fundamental for the establishment of an innovation system and which emphasize the technological and/or sector-related focus: opportunity conditions, appropriability conditions, accumulation of technological knowledge, and the nature of the relevant knowledge basis.

7. For some years now, academics have been discussing systems of innovation in a multitude of theoretical and empirical studies. These discussions deal, above all, with the incidence and variety of collaborations between industry and the science sector and the impact on the innovative performance of the business sector. In trying to evaluate the innovative potential at the company level, alternative definitions of innovation systems as well as alternative innovation approaches might be useful. See Acs and Audretsch (1990), Kleinknecht (1987), Breschi and Malerba (1997), Carlsson (1994), Carlsson and Stankiewicz (1991), Cooke and Morgan (1994), DeBresson (1989), Edquist (1997), Harrision (1991), Marceau (1994), Smith (1995). See the following literature for works related to national systems of innovation: OECD (1996, 1997, 1999), David and Foray (1995), Freeman (1988, 1990, 1994), Lundvall (1992), Nelson (1988, 1993), Nelson and Rosenberg (1993), Porter (1990), McKelvey (1991), Patel and Pavitt (1994).

8. Alternatively, the innovation input could be a reasonable criterion for forming innovative clusters at the company level. The measure for a comparable group of enterprises would be expenditure on innovation. Innovation expenditure covers current expenditure (labour costs, materials, etc.) and

tangible assets (equipment, buildings, etc.). Another way of measuring innovation behaviour is to examine R&D intensity. In Section 3.3 we will cluster firms according to different levels of R&D intensity.

9. The following discussion on the basis of Table 2 is a brief summary of the in-depth analysis of innovation behaviour of small and medium-sized firms by Harhoff, Licht *et al.* (1996) on the basis of German CIS I data. See also Beise, Licht and Spielkamp (1995).

10. The analysis is based on regression models (Probits) where the industry dummy is controlled. This means that the results are significant among the clusters and that there is no industry bias. See also Annex 2.

11. At this stage, the defined clusters are still rather theoretical. No "natural" grouping of variables such as regional, branch or technology information are taken into account. Instead, groups of firms are described that can be categorised by a specific type of innovation structure with a characteristic transfer channel portfolio and typical information sources. For a description of the variables, see Annex 2.

12. See the description of "public R&D institutions" in Annex 1.

13. Based on the assumption that information and transfer strategies are independently chosen instruments (exogenous variables) to define the level of innovation, or innovation styles (endogenous variable), the above statements and the cluster definitions are additionally supported by an ordered Probit analysis as shown in Annex 2. We carried out this analysis for all combinations, taking simultaneity into account.

14. These results are based on a qualitative evaluation of the internal potential or the absorption capacity of a company's information management (*e.g.* "demand pull" or "technology push") They are derived by judging the degree of formalisation of the knowledge transfer channels.

15. Table 1 shows that expenditure on innovative activities is distributed between R&D and construction, pilot projects and investment in R&D. In addition, companies invest in physical capital such as machines, technology, etc. If the investment rate is defined as the ratio of investment to total expenditure, then firms that do not innovate or that have no systematic R&D expenditure but invest in new products or technologies have, by definition, a higher investment rate.

16. See MERIT (1995), PACE Report, Arundel *et al.* (1995). From the information available in this report it is not possible to judge whether the behaviour of German firms reflects a strong national German innovation system and technological leadership or whether it is a sign of weakness and a lack of internationalisation which might cause negative feedback in the future. See also Beise and Felder (1997), Harhoff, Licht *et al.* (1996), Licht (1994).

17. In the following, we use the R&D intensity as an indicator to distinguish between different styles of innovation. For simplification, we separate out R&D-intensive and non-R&D-intensive industry sectors. As pointed out in Section 1, R&D-intensive industries are grouped into "high or cutting-edge" (pharmaceuticals, computer/office technology, radio/TV/telecommunications, aircraft and spacecraft industry, precision instruments, optics/clocks) and "advanced" (chemical industry, mechanical engineering, electrical engineering, except radio/TV/telecommunications, railroad industry, automobile industry) technologies. The line is drawn according to the proportion of R&D in the company's turnover: "cutting-edge technologies" cover goods with an R&D intensity that lies above 8.5%, while "advanced technologies' cover goods whose R&D makes up more than 3.5% and less than 8.5% of turnover.

18. If total innovation expenditure for 1996 in DEM million is set at 100%, 23% of that expenditure was accounted for by "cutting-edge technologies", 47% by "advanced technologies" and 30% by non-R&D-intensive industries. The ratio of the three groups of enterprises according to their number is 23:36:41 (Licht and Stahl, 1997, p. 38).

19. For an in-depth analysis of the R&D co-operation behaviour of SMEs on the basis of German CIS data, see Beise *et al.* (1995). See also Harhoff and Licht (1996), König, Licht and Staat (1994), Licht (1994).

REFERENCES

Acs, Z.J. and D.B. Audretsch (1990), *Innovation and Small Firms*, Cambridge, Massachusetts and London.

Audretsch, D.B and Z.J. Acs (1991), "Innovation and Size at the Firm Level", *Southern Economic Journal*, Vol. 57, pp. 739-744.

Becher, G. and S. Kuhlmann (eds.), (1995), *Evaluation of Technology Policy Programmes in Germany*, Dordrecht.

Beise, M. and J. Felder (1997), "Innovationsverhalten der deutschen Wirtschaft im internationalen Vergleich", mimeo, Mannheim.

Beise, M., G. Licht and A. Spielkamp (1995), "Technologietransfer an kleine und mittlere Unternehmen", *Schriftenreihe des ZEW*, No. 3, Mannheim.

Beise, M. and H. Stahl (1998), "Public Research and Industrial Innovations in Germany", ZEW Discussion Paper, No. 98-37, Mannheim.

BMBF (1996), "Report of the Federal Government on Research 1996", Federal Ministry of Education, Science, Research, and Technology, Bonn.

BMBF (1998), "Germany's Technological Performance", Federal Ministry of Education, Science, Research, and Technology, Bonn.

Breschi, S. and F. Malerba (1997), "Sectoral Innovation Systems: Technological Regimes, Schumpeterian Dynamics, and Spatial Boundaries", in C. Edquist (ed.), *Systems of Innovation: Technologies, Institutions, and Organizations*, Pinter, London and Washington, pp. 130-156.

Carlsson, B. (1994), "Technological Systems and Economic Performance", in M. Dodgson and R. Rothwell (eds.), *The Handbook of Industrial Innovation*, Edward Elgar, Cheltenham, United Kingdom, pp. 13-24.

Carlsson, B. and R. Stankiewicz (1991), "On the Nature, Functions and Composition of Technological Systems", *Evolutionary Economics*, 1, pp. 93-118.

Christensen, J. (1994), "Asset Profiles for Technological Innovation", *Research Policy,* Vol. 24 (1995), pp. 727-745.

Cohen, W.M. and S. Klepper (1994), "The Anatomy of Industry R&D Intensity Distributions", *American Economic Review*, Vol. 84, pp. 773-799.

Cooke, P. and K. Morgan (1994), "The Creative Milieu: A Regional Perspective on Innovation", in M. Dodgson and R. Rothwell (eds.), *The Handbook of Industrial Innovation*, Edward Elgar, Cheltenham, United Kingdom, pp. 25-32.

David, P. and D. Foray (1995), "Accessing and Expanding the Science and Technology Knowledge Base", *STI Review*, Special Issue on Innovation and Standards, No. 16, OECD, Paris, pp. 13-68.

DeBresson, C. (1989), "Breeding Innovation Clusters: A Source of Dynamic Development", *World Development*, Vol. 17, No. 1, pp. 1-16.

Dodgson, M. and R. Rothwell (eds.) (1994), "The Handbook of Industrial Innovation", Edward Elgar, Cheltenham, United Kingdom.

Dosi, G. *et al.* (eds.) (1988), *Technical Change and Economic Theory*, Pinter, London.

Durlauf, S.N. and P.A. Johnson (1995), "Multiple Regimes and Cross-Country Growth Behaviour", *Journal of Applied Econometrics*, Vol. 10, pp. 365-384.

Edquist, C. (1997), "Systems of Innovation Approaches – Their Emergence and Characteristics", in C. Edquist (ed.), *Systems of Innovation: Technologies, Institutions and Organizations*, Pinter, London and Washington, pp. 1-40.

Edquist, C. (ed.) (1997), "Systems of Innovation: Technologies, Institutions, and Organizations", Pinter, London and Washington.

Ehrnberg, E. and S. Jacobsson (1997), "Technological Discontinuities and Incumbents' Performance: An Analytical Framework", in C. Edquist. (ed.), *Systems of Innovation: Technologies, Institutions and Organizations*, Pinter, London and Washington, pp. 318-341.

Freeman, C. (1988), "Japan: A New National System of Innovation?", in G. Dosi *et al.* (eds.), *Technical Change and Economic Theory*, pp. 330-348.

Freeman, C. (1990), *The Economics of Innovation*, Aldershot.

Freeman, C. (1994), "Innovation and Growth", in M. Dodgson and R. Rothwell (eds.), *The Handbook of Industrial Innovation*, Edward Elgar, Cheltenham, United Kingdom, pp. 78-93.

Harhoff, D., G. Licht *et al.* (1996), "Innovationsaktivitaeten kleiner und mittlerer Unternehmen", *Schriftenreihe des ZEW*, No. 8, Mannheim.

Harhoff, D. and H. König (1993), "Neuere Ansätze in der Industrieökonomik - Konsequenzen für eine Industrie- und Technologiepolitik", in F. Meyer-Krahmer (ed.), *Innovationsökonomie und Technologiepolitik,* Heidelberg.

Harrison, B. (1991), "Industrial Districts: Old Wine in New Bottles?", *Regional Studies*, Vol. 6, No. 5, pp. 469-483.

Kleinknecht, A. (1987), "Measuring R&D in Small Firms: How Much are we Missing?", *Journal of Industrial Economics*, Vol. 36, pp. 253-256.

Kleinknecht, A. and J. Reijnen (1992), "Why do Firms Co-operate on R&D? An Empirical Study", *Research Policy,* Vol. 21, pp. 347-360.

Kline, S.J. and N. Rosenberg (1986), "An Overview of Innovation", in N. Rosenberg and R. Landau (eds.), *The Positive Sum Strategy*. Washington, pp. 275-305.

Klodt, H. (1995), *Grundlagen der Forschungs- und Technologiepolitik*, München.

König, H., Licht, G. and M. Staat (1994), "FuE-Kooperationen und Innovationsaktivitaet", in B. Gahlen, H. Hesse and H.J. Ramser (eds.), *Europaeische Integrationsprobleme aus wirtschaftswissenschaftlicher Sicht,* Tübingen.

Licht, G. (1994), "Gemeinsam forschen – Motive und Verbreitung von Strategischen Alliancen in Europa". *ZEW-Wirtschaftsanalysen*, No. 4, Mannheim.

Licht, G. and H. Stahl (1997), "Ergebnisse der Innovationserhebung 1996" (Results of the 1996 German Innovation Survey), *ZEW-Dokumentation,* No. 97-07, Mannheim.

Lundvall, B.-Å. (eds.) (1992), *National Systems of Innovation: Towards a Theory of Innovation and Interactive Learning*, London.

Mansfield, E. (1991), "Academic Research and Industrial Innovation", *Research Policy*, Vol. 20, pp. 1-12.

Mansfield, E. (1998), "Academic Research and Industrial Innovation: An Update of Empirical Findings", *Research Policy*, Vol. 26, pp. 773-776.

Marceau, J. (1994), "Clusters, Chains and Complexes: Three Approaches to Innovation with a Public Policy Perspective", in M. Dodgson and R. Rothwell (eds.), *The Handbook of Industrial Innovation*, Edward Elgar, Cheltenham, United Kingdom, pp. 3-12.

McCann, P. (1995), "Rethinking the Economics of Location and Agglomeration", *Urban Studies*, Vol. 32, No. 3, pp. 563-577.

McKelvey, M. (1991), "How do National Systems of Innovation Differ? A Critical Analysis of Porter, Freeman, Lundval and Nelson", in G.M. Hodgson (ed.), *Rethinking Economics: Markets, Technology and Economic Evolution*, Aldershot, pp. 117-137.

MERIT (1995), *PACE Report: Innovation Strategy of Europe's Largest Industrial Firms*, Luxembourg.

Meyer-Krahmer, F. (ed.), (1993), *Innovationsökonomie und Technologiepolitik,* Heidelberg.

Nelson, R.R. (1988), "Institution Supporting Technical Change in the United States", in G. Dosi *et al.* (eds.), *Technical Change and Economic Theory*, Pinter, London, pp. 312-329.

Nelson, R.R. (ed.) (1993), *National Systems of Innovation: A Comparative Study*, Oxford.

Nelson, R.R. and N. Rosenberg (1993), "Technical Innovation and National Systems", in R.R. Nelson (ed.), *National Systems of Innovation: A Comparative Study*, Oxford, pp. 3-21.

Nelson, R.R. and S. Winter (1982), *An Evolutionary Theory of Economic Change*, Cambridge.

OECD (1996), *The Knowledge-based Economy*, free brochure, Paris.

OECD (1997), *National Innovation Systems*, free brochure, Paris.

OECD (1999), *Managing National Innovation Systems*, Paris.

Patel, P. and K. Pavitt (1994), "The Nature and Importance of National Innovation Systems", *STI Review*, No. 14, OECD, Paris, pp. 9-32.

Porter, M. (1990), *The Competitive Advantage of Nations*, New York.

Reger, G. and U. Schmoch (eds.) (1996), *Organisation of Science and Technology at the Watershed*, Heidelberg.

Roelandt, T. and P. den Hertog (1998), *Synthesis Report of the NIS Focus Group on Industrial Clusters*, The Hague.

Soskice, D. (1996), "German Technology Policy. Innovation and National Institutional Framework", *WZB Discussion Paper*, FSI 96-319, Berlin.

Scott, J.T. (1984), "Firm Versus Industry Variability in R&D Intensity", in Z. Griliches (ed.), *R&D, Patents and Productivity*, Chicago.

Schmoch, U. *et al.* (1996), "The Role of the Scientific Community in the Generation of Technology", in G. Reger and U. Schmoch (eds.), *Organisation of Science and Technology at the Watershed*, Heidelberg.

Smith, K. (1995), "Interactions in Knowledge Systems: Foundations, Policy Implications and Empirical Methods", *STI Review*, No. 16, Special Issue on Innovation and Standards, OECD, Paris, pp. 69-102.

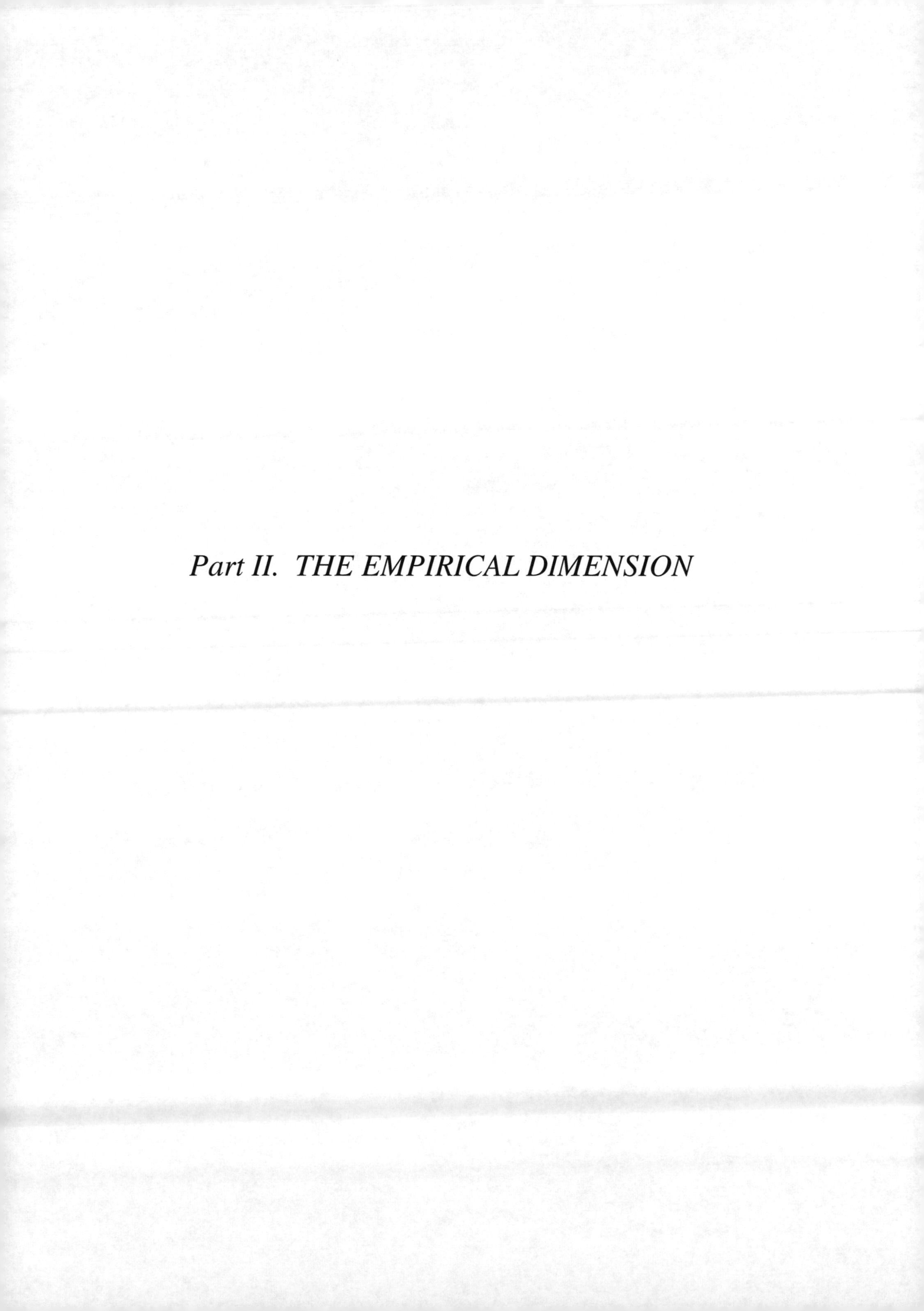

Part II. THE EMPIRICAL DIMENSION

Chapter 5

CLUSTERING AS A NEW GROWTH STRATEGY FOR REGIONAL ECONOMIES? A DISCUSSION OF NEW FORMS OF REGIONAL INDUSTRIAL POLICY IN THE UNITED KINGDOM

by

Arnoud Lagendijk and David Charles
CURDS, Newcastle upon Tyne

1. Introduction

Over the last decade, regions such as Scotland, Wales, Northern Ireland and the North East of England have provided examples of successful economic developments in sectors such as automotive manufacturing, electronics, chemicals, and some more traditional sectors such as food and drink, clothing and textiles. To a large extent, this success has been built on the attraction of foreign direct investors, combined with efforts to improve the local supply base and to foster linkages between local suppliers and foreign investors. In recent years, a trend can be observed towards more interest in skill development and the support of the "indigenous" sector, with more attention on non-manufacturing sectors such as multimedia and professional services. Against this background, cluster initiatives have emerged both as part of the desire to improve the benefits from foreign investments by supporting supply chains and other forms of inter-firm relationships, and to support networking among local firms, particularly SMEs.

This chapter will focus on the cluster strategies and initiatives developed in the different British regions, and will discuss how those institutions have modified regional industrial policy. Attention will be paid to the wider institutional context in which new policies have emerged, and the extent to which they reflect a move away from attracting foreign investment towards a more endogenous, innovation-oriented approach. Nevertheless, a key premise of the chapter is that the development of peripheral regions remains closely tied to their potential to capture foreign assets, whether in the form of production facilities or via direct technology transfer. In particular, the relationship between the cluster approach and the concept of systems of innovation needs to be addressed from the perspective of the position of regions in wider chains of production and knowledge transfer. The discussion is structured in five sections. Section 2 introduces the cluster approach and its relation with the innovation system concept, highlighting the political context in which cluster policies have emerged, followed by a discussion of the role of Regional Development Agencies. The third section will present the regional case studies, followed by the conclusion.

---

* The authors would like to thank the Welsh Development Agency, Scottish Enterprise, and various agencies in North East England for their assistance in this study. They would also like to thank Professor Kevin Morgan from the University of Wales for his good advice on the Welsh case.

2. From innovation systems to clustering: the development context for peripheral regions

2.1 Conceptual issues

Recent thinking and research on innovation has provided an important contribution to the understanding of the economic success of particular territories, such as nations or regions. The literature on innovation systems has underpinned two essential dimensions of innovation (Morgan, 1997): first, the role of systemic interaction between different agents in the innovation chain, particularly between producers and users of intermediate goods (Lundvall, 1992), and between business and the wider research community; and second, the fact that innovation processes are institutionally shaped. Different set-ups of institutions, with between them particular processes of networking, will develop different kinds of capabilities in advancing technological development and its commercialisation. Against this background, the concept of *systems of innovation* seems to present a framework of conceptual thinking and analytical research, based on a strong inter-disciplinary approach, rather than a full-blown theory (Edquist, 1997). One of the great challenges the *systems of innovations* approach is still facing, as shown through this volume, is how it can bring together the underlying dimensions of technological, institutional and economic change, and how this can be related to conventional economic concepts which justify policy intervention, notably systemic and market failures. Moreover, the role of the geographical dimension is far from resolved. While the national dimension provided the starting-point, inspired by the success of particular nations such as Germany and Japan, recent work on particular territories, as well as the role of transnational firms and systems, has shifted the attention to the level of regions, continental blocs and the global system (Edquist, 1997). Not only is there a question of which level is most pertinent, but also how different levels interact.

The latter issue is of particular interest in the case of more peripheral regions with a strong presence of foreign capital, and invokes the question of how externally owned capital may fit in a concept of innovation at a regional level. An intriguing debate has emerged around the role of multinational corporations in technology development. Porter's "home base" argument, which stated that it is the home base that determines most of the technological competence of multinational firms as well as receives most of the benefits, was challenged by authors like Dunning and Cantwell. The latter showed that in both host and home areas, various forms of interaction between firms and the environment determined the firm's technology potential as well as, depending on the specific strategies of the firms and the geographical context, its overall territorial impact. Cantwell (1991) provides evidence showing that the specific geographical structure of multi-plant firms is an important determinant of their innovative capabilities. In particular, although specific R&D functions may be concentrated in a few places, their presence in different areas allows them to tap into different areas of knowledge, or to apply their knowledge in different environments. Dunning (1992), elaborating on his seminal ideas relating the emergence of the multinational enterprise to the existence of imperfections especially in factor markets, suggests that multinational firms through various forms of common governance may play an important role in the development of particular industrial agglomerations. In his recent work, Dunning shows that vital resources and capabilities are increasingly controlled by multinational enterprises and that, particularly in recent decades, governments have been competing intensively for these resources (Dunning 1991, 1992). A different view is taken by Mowery (1995), who sees as the core component of innovation systems the "absorptive capacity" rather than the generation of new knowledge, and he emphasizes competition as an important factor force to increase this capacity.

Other scholars have emphasized the localised character of innovation, pointing at the revival of the phenomenon of industrial districts and the success stories of hi-tech regions such as Silicon Valley. Here a distinction can be made between authors who emphasize the role of networking in a particular socio-cultural context, as caught by the term of "innovative milieu" (Camagni, 1991), and authors who have adopted a more institutional perspective by developing the concept of "regional innovation systems" (Cooke, 1992). In the innovative milieu concept, the growth of a locally embedded innovation system is seen as essential in shaping the social routines and strategies of actors in the regional economy. The institutional approach pays more attention to the development of and interaction between specific technology-oriented organisations, such as universities, research centres, and training organisations, and business. The differences between these two approaches should however not be overstated. They share an emphasis on networking, the development of shared visions and the building of trust. Indeed, in practical research, both approaches are often seen as complementary.

While an author like Dunning articulates the link between local developments and global organisations, this aspect is missing in most of the local innovation approaches. Advocates of the "innovative milieu" and "industrial district" concept have been accused of presenting a particular phenomenon observed in a small number of highly successful regional economies as a general case. They thus seemed to undervalue the fact that there were organisational forms, at other geographical levels, which could also embody innovative capabilities. Some authors have even tried to trivialise the role of large firms by arguing that, with respect to innovation, they behave almost as indigenous firms (Malmberg and Maskell, 1996). Against this idea, authors like Gray and Markusen stress that the subsidiaries of multi-plant firms present a distinct organisational form, which, depending on the nature of their embedding in local as well as wider networks, may act as vital *hubs* in regional economic development (Gray *et al.*, 1996) (other hubs are also feasible, for instance regional universities). Moreover, critical analysts of localised innovation systems, including that of high-tech areas such as Silicon Valley, have argued that a large part of the knowledge underpinning the regions' economic success was obtained from elsewhere, and that success was based on the organisational capability to apply knowledge to a commercial environment (Dupuy and Gilly, 1994).

For most regions and nations, the most important asset is to have a market system that is able to capture new forms of knowledge and apply them within the context of the local production system. Multinational firms can act as important vehicles of technology transfer, even if a plant established in a host area starts with only routine production activities. Whether the establishment of a subsidiary by a multinational firm brings genuine benefits to a host economy depends on whether there are adequate mechanisms for information exchange and trading, whether the right incentives exist for co-operation and dialogue between local and foreign actors, and whether the foreign players fit into the existing institutional system of economic development and technology transfer. Dunning (1992), partly as a response to Porter's clustering concept, refers to two typical scenarios that can be followed by a foreign investor. One is the "easy pickings" scenario, which is accompanied by poaching local workers, driving out local competitors and a very low local-purchasing level. This will occur particularly when the conditions of exchange do not hold and systematic failures abound. Another is the "upgrading scenario", which includes the creation of partnerships with local firms and institutions, aimed at improving the local skill base and support to the development of SMEs.

The "upgrading scenario" should be seen as a two-edged sword. On the one hand, it provides the region with a vehicle to improve parts of the local economy. One way in which subsidiaries have been used is, for instance, as "model" plants for local firms, through their role as a demanding customer, as learning sites for best practice in business processes, skill development, etc. On the other hand, much of this upgrading has been directly beneficial for the foreign investor, serving its needs for better skills and suppliers. In many cases, the investments made by the subsidiary management and workers in the upgrading process have been matched by even higher investments (and subsidies) stemming from the local community and support channels in supporting "after-care" initiatives (Peck, 1996). To reach such a doubly beneficial scenario, however, it will be necessary to address informational and institutional failures within the regional system.

Several forms of "failure" may be mentioned that can be characterised as government failure. The last two decades have witnessed a proliferation of business support targeted on improving the innovative capacity of the regional economy. Some of these initiatives were explicitly spatial measures (technology parks, incubator centres, etc.), others involved non-spatial instruments covering a certain geographical area (technology advice and transfer centres, university knowledge extension centres, etc.). However, in recent years, it has been recognised that many initiatives in the area of innovation were too supply driven, failing to detect and adapt to the real needs of (potential) customers (Morgan, 1997). Missing was a proper articulation of the needs of firms, particularly SMEs, from the perspective of a longer-term development strategy. It has also been argued that many initiatives operated too much in isolation, that they lacked linkages with other local institutions. Such institutional mismatches have been aggravated by the short-term character of funding and frequent changes in technology policy driven by politics rather than business needs.

Table 1. Cluster-related systemic and market failure applicable to more peripheral regions

General categories of systemic and market failures	Specific applicability to peripheral regions	Contribution of cluster-based policies
Informational failures	Often severe, due to poor market and information infrastructure and segmentation of the regional economy (*e.g.* foreign *vs.* local firms)	Facilitating networks involving local and foreign firms; supporting information exchange systems and business links brokering services
Limited interaction between actors in innovation systems	Innovation systems poorly developed; in general, lacking co-ordination between key (public and private) actors in the regional economy	Improving regional facilitation of social networking and institutional links, notably through public-private partnerships targeting specific clusters
Institutional mismatches between (public) knowledge infrastructure and market needs	Proliferation of policy initiatives and support structures; co-ordination is, however, poor and lacks demand orientation	Rationalisation of support along cluster lines, *e.g.* through (real) service centres, university extension services.
Absence of demanding customers	Foreign plants may provide opportunities (also along non-trade lines), but they are often not sufficiently exploited	Creating strategies for the regional embedding of foreign plants.
Government failure	Affecting particularly the implementation of technology policy and provision of business support.	Improving the co-ordination of regional governance structure, *e.g.* through regional development policy with capacity to develop (top-down) cluster policies

Against this background, cluster-based policies can play two fundamental roles for the development of more peripheral areas. On the one hand, a cluster approach may be adopted to increase the "absorptive potential" of the regional economy, as well as to build a more strategic context in which local actors can work at the improvement of regional innovative capacities. A cluster policy, seen from this perspective, should provide an organisational framework that, through the notion of linking local businesses to central hubs, improves the embedding of firms in both local and global networks (Young, 1994). Such links can embody trade relations (local purchasing), but, perhaps more importantly, should also involve non-trade relations of information exchange encouraging a variety of inter-firm learning. In the long term, such interaction may also facilitate the alignment of activities and investments at the level of the regional economy, which may be the source of dynamic competitive advantages (Langlois and Robertson, 1995). On the other hand, a cluster policy, through structuring and integrating business support along sectoral and supply chain lines, may contribute to the effectiveness of business support. The essence here is to overcome substantial failures in government-based support provision and to address institutional mismatches in the policy and support system. Table 1 summarises the main justifications for cluster-based regional policy, and also includes specific policy contributions which will be discussed in more detail below.

2.2. Dimensions of regional cluster policy

While the concept of clusters has gained a prominent place in the discourse on economic development in policy at all spatial levels, the concept has been adopted primarily at the regional or sub-national level. At both the national and regional level, supply-side-oriented measures have been popular since the 1970s. However, it is at the regional level that a shift can be detected away from primarily infrastructure- and technology-oriented policies towards more comprehensive approaches based on clustering. Cluster strategies have been adopted, for instance, within several German *Länder* (primarily Nordrhein-Westphalen and Baden-Württemberg), many states in the United States, and many regions in Europe (Basque Country, Catalonia, Northern Ireland, Styria). Also at the local level, new policies have been developed with clustering objectives (Thomas and Shutt, 1996). While the last section has provided some general justifications for adopting cluster policies, this section will discuss in more detail what cluster policies may entail and how they differ from previous approaches. The main dimensions relate to the shifting overall approach to industrial policy, the role of sector policies and the attraction of FDI, and the trend towards networking and partnerships. The final part of the section will address a complex issue in which more theoretical and practical aspects of cluster-based policy are confronted with each other, that of industrial targeting.

Changing conditions of regional industrial policy making

One reason for the popularity of the cluster concept at the sub-national rather than the national level is the difference in economic policy perspectives adopted at both levels. The adoption of the cluster concept by sub-national authorities can be seen as an attempt to fill the gap left by the "hands-off" stance of central government. In particular, this seems to be case in Germany, the United Kingdom (Geddes, 1992) and the United States (Sternberg, 1991). In the case of the United Kingdom, it was the shift towards Conservative politics in 1979 that brought an end to planned industrial policy and led to the dismantling of the sectoral organisations of the Department of Industry and Trade. However, the absence of a national strategy and vision on industrial development did not mean that the state did not steer industrial development (Cowling and Sugden, 1993). The large-scale privatisation, with the emergence of new regulatory environment, and the support to foreign investments were two developments with a significant impact on the recent evolution of UK industry. What is important though is that the strongly ideological position of the Conservative government against "state intervention" also had a strong impact on local (more Labour-oriented) government. Increased control by central government over local authority finances led to the curtailing of resources for economic policy, thereby forcing local authorities to search for other sources of funding, such as Europe, and to engage in partnerships with local business.

In the development of local industrial policy and business support, cluster initiatives have generally emerged from sector- or technology-oriented policies. In Emilia-Romagna, a general shift can be observed from a policy largely organised along sectoral lines to a more horizontal, inter-sectoral focus (Gómez Uranga and Ozerin, 1997). In some cases, the emphasis on clusters has been triggered by a general feeling of disappointment with previous supply-side-oriented measures targeted on infrastructure and technology support. In particular, most of the technology parks and technology transfer centres which had been established on a large scale in the 1980s and early 1990s did not match the high expectations their role models had created (Hassink, 1996). The innovation strategies adopted in peripheral regions did not produce the kind of take-off that had happened with hi-tech sectors in core regions. Many of the initiatives were, to use the words of Cooke (1995): "too little, too late". Cluster strategies were also seen by some as preferable to the kind of place-marketing measures and subsidies used to attract foreign investment. While the latter may easily turn into wasteful negative-sum bidding games between rival authorities, cluster initiatives are regarded as growth creating. In the words of Sternberg (1991), cluster policies constitute a real alternative to the local economic developer's typical preoccupation with such zero-sum policies.

Several aspects of regional policy development can be mentioned which have promoted a clustering approach. Most significantly, clustering provides a link between policies focused on SMEs and those focused on inward investment, thus bridging the indigenous and exogenous components of economic growth. One common aim of cluster initiatives is to bring different types of firms and organisations together around one supply chain or one common resource or technology, thus improving economic integration as well as fostering communication and the transfer of knowledge between firms. Bridging the policies targeted on indigenous and externally owned firms is not only considered a step towards a more integrative development model that may overcome informational and government failures as well as reducing institutional mismatches, it is also seen as a way to overcome existing deficiencies in the two separate strands of regional development policy. In particular in the case of peripheral regions, SME-oriented policies have tended to be overshadowed, in interest as well as resources, by the efforts made to capture foreign investment. The latter, in turn, has been strongly oriented towards *attracting* investment, while paying less attention to the process of embedding established plants in the local economy. Because of its capacity to combine different sets of actors and bind different strands of policies and approaches, cluster approaches may support what Storper has called the emergence of more heterodox policy frameworks.

By encouraging a process of networking, clustering facilitates an environment in which firms can learn from each other rather than from support organisations. One of the crucial problems which small firms in particular face is that they lack the knowledge and information channels to identify their demands for business support. One of the problems of the business support sector has been that while they have generally been able to offer a standard package for upgrading business practices, they lacked the industrial and business-specific knowledge which could help firms to identify their more specific needs. For this reason, although they have played a vital role in assisting business *start-ups,* business support agencies have often lacked the credibility required for a stronger involvement of the established local business sector (Morgan, 1996). Clustering, by involving SMEs, larger firms and support organisations, is seen as a way to establish constant interaction between demand and supply. Clustering can thus both foster inter-firm learning, in which the more experienced firms can become the tutors of SMEs, and improve the interaction between business support agencies and their clients. Not surprisingly, many cluster initiatives have developed in regions with a background of industrial decline or crisis. Such a context has been instrumental in motivating both the public and business sector to search for novel ways of overcoming pressing problems, to create trust and to facilitate access to funding. In particular, regions benefiting from Objective Two funding under the EU Structural Funds have shown a high propensity to develop novel initiatives with financial assistance from the EU. The more interventionist measures are then also justified because they were aimed at preventing business failures and job losses.

An important issue is the extent to which clustering can be seen as part of a process of *institution building*. In general, clusters have been initiated by the establishment of various kinds of forums, as regular meetings of the firms and organisations related to a particular industry or value chain, mostly in the form of public-private partnerships. Over time, such forums can become more established organisations and even turn into a kind of association (Waits, 1992). The latter may be particularly important in regions which have generally lacked a strong "meso level" of economic governance between the state and the business sector. In regions with a strong presence of business associations, Chambers of Commerce, etc., clustering may contribute to the building of more strategic links between organisations and businesses. The latter motive has, for instance, been mentioned in the case of cluster initiatives in Germany (Rehfeld, 1995). Many of these initiatives have been supported through the promotion of private-public partnerships by national governments and the European Commission.

Industrial targeting: justified intervention?

The concepts of networking and the creation of forums around cluster building portray an image of clustering largely as a *bottom-up* process, in which the main task of the policy maker is to facilitate the networking process, to play, to use the words of Morgan, the "innovative interlocutor". From the perspective of business development and institution building, this is generally seen as the most suitable approach, which builds on the tradition of business-oriented policies. However, from a regional development perspective, there is also a structural dimension to clustering. Following Porter's (1990) analysis of "cluster maps", clusters present a way to depict the strengths and weakness of the regional economy, and have thus induced a revival of structural regional policy. Various regions have embarked on developing a cluster strategy, based on an analysis of the competitiveness of the existing economy and an identification of threats and opportunities in particular industries. To a large extent these approaches are focused on traditional sectors, although they might include an examination of cross-cut sectors from a cluster perspective. The analyses, often carried out by consultants such as Porter's company Monitor, are based on a combination of established statistical methods, such as employment and production data analysis, "shift and share", input-output analysis, and the use of technology indicators, combined with the capturing of qualitative information from industry representatives and experts about perceived strengths and weaknesses. They thus present a kind of cluster-oriented "SWOT" analysis of the region, which serves to build a strategy of industrial targeting.

As a response to the trend towards sector or cluster targeting, it should be noted that various academic observers as well as policy makers have expressed doubts about what is seen as a return to an interventionist, top-down approach in regional policy (Rosenfeld, 1997). A critical question seems to be to what extent targeting represents an *ongoing process* rather than a "one-off" definition of the preferred cluster map. A related issue is the extent to which formal cluster analysis is matched by a process of extensive consultation of local actors. Using the latter, cluster analysis may be used to create a collective vision of regional development which serves as a framework for initiating particular initiatives rather than as a step towards a top-down cluster policy. The most fundamental quandary, however, is to what extent governments can be seen to be capable of understanding future economic developments in sufficient detail to justify the prioritisation of certain activities. On the one hand, government failures in the domain of policy implementation may already have cast sufficient doubt over the endorsement of any form of more strategic government behaviour. On the other hand, the growth of new economic activities may be hampered by such intense systematic and market failures that certain forms of targeting may be warranted (as, for instance, documented by Langlois and Robertson, 1995). Industrial targeting may thus be seen as a way of addressing the structural lock-in of a regional economy in an unfavourable situation since market forces cannot be expected to provide a prompt solution.

2.3. The role of Regional Development Agencies in cluster approaches

Clustering initiatives, as shown so far, should be seen as an agenda item that has emerged around concepts of networking, institution building and industry targeting, rather than as a well-defined set of policy measures. Thus, rather than viewing intervention in terms of a specific form of mechanism, it is necessary to examine the process by which the role of state institutions has shifted to one of network facilitation and to examine the means by which they and other regional associations and actors interact.

On the regional scale, the cluster agenda has emerged from the contested nature of regional development policy since the early 1980s and the perceived weakness of traditional exogenous development strategies. With the slowdown in availability of foreign direct investment (FDI) in the early 1980s and the rise of new paradigms of endogenous development, many regions saw a fragmentation in policy between localised initiatives focused on SMEs, and the continued existence of inward investment agencies competing for a smaller share of mobile investments. The rationalisation of branch plants in periods of recession undermined the role of FDI and the perception of its stability and embeddedness, but also presented considerable challenges for local agencies to compensate for job losses (as shown, for instance, in the case of the withdrawal of the recently established plants of Siemens and Fujitsu in the North East of Britain).

Renewed investment opportunities in the late 1980s following recovery from recession and in the promise of the Single European Market, therefore presented regional and local agencies with opportunities to rethink the relationship between exogenous and endogenous strands of development. Fresh experience of closures sharpened thinking about the sustainability of inward investments and shifted the emphasis of regional strategies away from simply attraction towards retention, re-investment and maximising local spin-offs. Rather than competing simply on grants, agencies began to talk about building the "business case" for the long term, and an emphasis on locally specific untraded interdependencies as factors in the competitiveness of plants within their companies as well as within their industry.

These moves have paralleled a changing role for the state in industrial policy which has been described by Kevin Morgan as a shift from "direct intervention" to "indirect animation" (Morgan, 1996). In this transition, the essential role of the state is being redefined as that of an *animateur*, a facilitator of networking and institution building. Following this logic, the state should not try to take ownership of cluster initiatives, but should primarily work as a catalyst, a broker that brings actors together and supplies initial funding for research and the initiation of the networking process. Knowledge is an essential component of this role as catalyst. Not only do local state organisations need to gain insight into the strengths and weaknesses of the regional economy at the industry and business level, but they also need to acquire an in-depth knowledge of the local institutional structure. Moreover, such knowledge acquisition should be part of an ongoing process of reflection and monitoring. One of the most difficult demands for the state is that, while aiming to encourage collective learning within its constituency, it needs to become a learning organisation itself, following strict principles of how to act and when.

One way to organise this process is by assigning a strong role to an organisation that can act largely independently from the state bureaucracy, such as a regional development agency (RDA). While the state may retain responsibility for monitoring the overall process, such an agency can be in charge of commissioning processes of research and consultation, of providing the support frameworks for bottom-up clustering, and of translating long-term strategies into short-term actions. An essential task of an RDA is thus to invoke the support of other organisations, to act as a broker between business and actors such as research centres, education institutions, training providers, business associations, Chambers of Commerce, etc. This involves the complicated issue of how to create and monitor network-based forms of governance, which is essentially decentralised but requires some form of overall co-ordination. Batt (1994) advocates the creation of RDAs as the "institutional expression of regional political networks". As central moderators and facilitators, such agencies should act as a pivot in regional negotiation and in mobilising networks to establish a co-operative and consensus-based framework for industrial policy. RDAs should have the capacity to gather economic intelligence and create a platform for strategic thinking on regional development, engaging with the main partners involved in regional economic development through various forums, part of which could be cluster-based.

Analysing cluster-based policies: key points of evaluation

An important aspect of cluster approaches in regional industrial policy relates to the role of political context. While cluster initiatives aim at encouraging collaboration and at the creation of shared visions and strategies in the regional business sectors, they emerge from a political structure which shows a trend towards forms of governance based on networking and partnerships. In understanding the role and impact of cluster development, the interaction between networking processes in and between the political and industrial systems is of vital importance. Such political considerations can be seen in the various stages of cluster policy development outlined below.

Conception and cluster mapping: Cluster initiatives have generally been developed against a particular background in which there was a need for new approaches. One incentive has been the search for follow-up policies after attracting foreign investment; another the wish to promote the development and networking of SMEs. Cluster initiatives vary in the extent to which they are devised as top-down policies, generally based on a regional cluster analysis and mapping, or as bottom-up initiatives, linked to the support of particular groups of firms often by smaller business support organisations.

Objectives: Regional cluster policies generally pursue two sets of objectives. One is the support of business development through the creation of a favourable business environment, the tailoring and customising of pre-existing business support delivery, and, above all, the brokering of networks among businesses. The other is the improvement of the regional economic structure, through explicit or implicit forms of targeting. Whatever mix between these two levels is chosen, the cluster agenda should address systematic and market failures observed in the regional economy, and take account of government failures showing up in existing forms of business support and technology policy. Because cluster policies rely on public-private interaction and often evolve around certain hubs, a critical issue is the extent to which policies are manipulated by dominant economic or political players.

Methodology: Cluster policies generally involve two levels of organisation. The first level is that at which cluster initiatives are conceived, facilitated and monitored. The central question at this level is how these processes are embedded in regional policy networks to secure sufficient interaction between relevant actors (public agencies, business representatives – including from SMEs, technology centres, universities, etc.). The second level is that of particular cluster initiatives targeting a particular set of industries or businesses. At that level, the core issues are how networking processes are brokered, how inter-organisational learning is facilitated, and how the business support sector and regional technology infrastructure is involved. The methodology chosen will also influence the kind of businesses and organisations involved in cluster initiatives, and the extent to which these are representative for the wider value chains in which they operate.

Evaluation and monitoring: However important, evaluation poses great difficulties. The processes aimed for by regional cluster policies are not easy to measure, since they involve a gradual change in culture and routines of interaction rather than yielding concrete outputs. Strict monitoring systems may even be counter-productive, since they will force policy implementers to pursue certain targets (so many firms enlisted, so many contact hours, so many jobs "secured", so many cases of technology transfer) which may not reflect the optimal way of network brokering and tailoring of support. While certain quantitative indicators may be highly useful, evaluation should be based on a flexible and reflexive observation of the unfolding of policy initiatives. A critical question is to what extent there is a learning loop through which evaluation results feed back into the cluster strategy. Such loops should also ensure that policy measures are phased out once the justification for intervention disappears, that is, once systematic failures have been adequately addressed. Evaluation should also trace the final destination of expenditure and consequences of support assisted by the cluster measures. Since the initiatives generally only assist business development in an indirect way, the final impact of support may be concealed. Networking among firms may be geared to improve learning processes along the supply chain, for instance, to the benefit of local SMEs; it may also be used for brokering trade relations between selected local firms to the exclusion and disadvantage of other, potentially more efficient, businesses. This kind of assessment will generally be more useful than counting heads in network meetings.

A related issue is that of funding. Obviously, proper financial accountability is essential, also to prevent abuse through the redirection of resources via the networking processes. A problem is that many cluster initiatives are funded on the basis of short-term projects requiring regular applications for financial support and thus quick proofs of success. However, most cluster initiatives will only show real benefits in the long term, and may thus be more effective if they are funded on a stable financial basis which does not require frequent re-applications. It should also be noted that cluster initiatives, since they do not involve direct business subsidies or the establishment of technology parks or centres, are generally cheap forms of regional policy.

3. Clustering in UK regions: case studies

3.1. Introduction to cluster policies in the United Kingdom

Following on from the review of theoretical and political aspects of cluster policies, this section will examine the re-orientation of industrial policy in the United Kingdom, and specifically the emergence of cluster strategies at the regional level. Central to this process of change has been the erosion of national-level sectoral policies and a re-emergence around a cluster concept. Although the United Kingdom had pursued national sectoral strategies during the 1970s, during the 1980s the central government moved away from such initiatives, with the dismantling of sectoral teams within the Department of Trade and Industry, and a shift in R&D programmes towards collaboration in generic technologies. This left the regions and notably the smaller "nations"[1] of the United Kingdom (Scotland and Wales) as the main scale at which some sectoral strategies were continued. However, more recently, there has been in some ways a return to a sectoral or cluster perspective in technology policy through the Technology Foresight Programme ("Foresight"). Within a strong competitiveness perspective, Foresight brought together actors in certain industries or on certain themes, such as chemicals, IT or leisure and learning to identify structural weakness and create new development strategies (Table 2).

Table 2. Sector panels in the Technology Foresight Programme, 1995

Agriculture, natural resources and environment
Chemicals
Communications
Construction
Defence and aerospace
Energy
Financial services
Food and drink
Health and life sciences
Information technology and electronics
Leisure and learning
Manufacturing, production and business processes
Materials
Retail and distribution
Transport

Although some sectoral or cluster-oriented approaches have begun to develop nationally through Foresight, even within this programme some regional level organisations have played a key role, notably in electronics. This analysis will therefore focus on this regional scale. However, before examining the specific cases of the regions, some understanding of the nature of governance of the regions is required.

Regional governance in the United Kingdom

The governance of economic development at a regional level in the United Kingdom is very complex and varied, and in many cases does not approach the ideal of a decentralised yet co-ordinated, representative yet proactive governance system. The most powerful actors, apart from central government, are Regional Development Agencies (RDAs) in Scotland, Wales and Northern Ireland, but such organisations do not yet exist in England.[2] Various authors have sketched the history of RDAs in the United Kingdom, and the details of these histories will not be repeated here (Danson *et al.*, 1992). While the origin and range of responsibilities of the RDAs is rather different, they have tended to concentrate on foreign investment as a core element of their economic development strategies. However, their other responsibilities range from physical regeneration (including housing in some circumstances in Scotland), to community development, indigenous business development including venture capital, and training. In England, on the other hand, there is a more complex melange of Training and Enterprise Councils (TECs), local authority bodies and varying forms of publicly funded agencies with specific remits, such as property development, land regeneration and inward investment (Regional Development Organisations – RDOs). Even in Scotland, Wales and Northern Ireland, the nature and governance of development agencies varies greatly, and there remain other sub-regional actors such as local authorities, and, in the case of Wales, TECs.

In each territorial unit the RDA is accountable to the local representation of central government, respectively the Welsh Office, the Scottish Office and the Northern Ireland Office; in England, the more focused regional development organisations are accountable to the regional Government Offices. The relationship between central government and the RDAs has been a crucial determinant in the development and performance of the RDAs. Under the Conservatives in the 1980s, for instance, RDAs were compelled to become facilitating rather than interventionist, while they were also forced to seek funding from other sources.

Another result of changes in the organisation of local government is that RDAs/RDOs were confronted with the establishment of semi-independent Training and Enterprise Councils (TECs) formed from the break-up of a government training agency, and new one-stop shop business support agencies (Business Links), controlled by a board of local business representatives but funded by the central state. The exception is Scotland, where the Scottish Development Agency (SDA, forerunner of Scottish Enterprise) and government training agency were integrated in one network, Scottish Enterprise Network, with local delivery agencies, Local Enterprise Companies (LECs). In the Welsh case, the structure of the WDA has been radically changed by the shift from a theme-oriented to a geographical division (Figure 1). The decentralisation was seen as a way to bring the organisation closer to its customers, including a better integration with other local organisations, and to account for the territorial differences in the area.

Figure 1. Changes in the Welsh Development Agency

The Centralised Structure

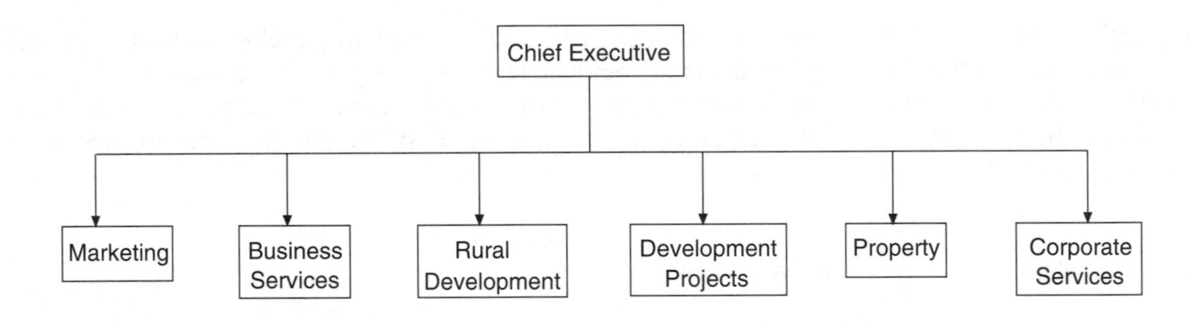

The Decentralised Structure

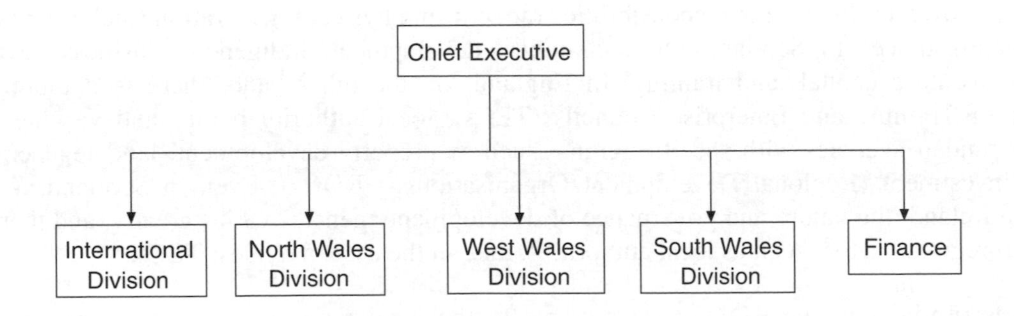

The British RDAs/RDOs, accordingly, are caught between the regional arm of central government, on the one hand, and a proliferation of local agencies, on the other, some of which are also controlled by central government. In this context, they have been struggling to find the appropriate organisational structure and linkages to support their position as core actors in the shaping of regional industrial policy. With a lack of local democratic structures and a tendency of the different organisations to compete on policy initiatives rather than to collaborate, however, it is not surprising that the RDAs/RDOs in Britain do not really match Batt's (1994) model of the "institutional expression of regional political networks". Local authorities have tended to become marginalised in the development, and even implementation, of regional industrial policies, which has encouraged some authorities to develop their own local initiatives with the help of European funding. Unions have also had little representation in the bodies of regional policy making. On the other hand, central government offices have considerable influence, as in the case of the Scottish and Welsh Offices which fund their respective agencies, or the government offices in the English regions which monitor RDOs.

One additional reason may be mentioned as to why RDAs have not reached the level of transparency, local openness and engagement which would make them more accountable and support their role as "local *animateurs*". Because they have developed largely through the game of winning foreign investment, an important hurdle that all RDAs need to overcome is that of managing their information flows. While some degree of secrecy is inevitable for organisations that deal with foreign investment and the creation of "regional competitiveness", it seems that, in the past, attitudes were such that they tended to build barriers rather than encourage debate and consensus building. This applied to WDA in the early 1990s (Morgan, 1994), and is still a problem in the case of an English RDO like the Northern Development Company (NDC). Part of the problem is that RDAs, rather than accepting a role as catalysts and brokers, are still tempted, as in the case of foreign investment acquisitions, to acquire full ownership and credit for the projects they are involved in. RDAs thus still face the challenge of becoming more transparent, and more accountable while retaining strategic power and effectiveness.

The following sections outline some of the cluster initiatives that have been developed in the regions discussed above, and which are summarised in Table 3.

Table 3. Sector/cluster orientation in UK regions

Region	Organisation	Cluster focus	Cluster methodology
Scotland	Scottish Enterprise	Information industries (electronics, software, multimedia), energy, food, textiles and tourism	Focus groups, co-ordinating policy initiatives along cluster lines (*e.g.* skills, technology)
Wales	Welsh Development Agency	Automotive sector, consumer electronic sector, medical devices and diagnostics sector, telecommunications equipment sector	Supply chain initiatives; links to centres of excellence
Northern Ireland	Northern Ireland Growth Challenge	Engineering, food processing, health technologies, software, textiles and apparel, tourism and leisure, tradable services	
North East England	Northern Development Company: Real Service Centre	Automotive, off-shore, food, electronics, business services	Varied top-down and bottom-up initiatives, but primarily agency-led

3.2. *Wales*

Emergence of cluster policies

As is the case for all regions, cluster initiatives in Wales have primarily emerged from policies geared to attracting foreign investments. An important incentive to developing cluster initiatives has been the programmes developed around the "after-care" for and embedding of foreign investors. The Welsh supplier programme "Source Wales" has been described as one of the most effective policies in this field (Morgan, 1994). Inspired by Japanese supplier development approaches, Source Wales included the creation of supplier clubs in core manufacturing sectors (particularly automotive, initiated by Calsonic, electronics, and aerospace). The success of Source Wales is shown by the fact that the programme has largely become self-financing. Another pillar of the Welsh approach is support for local firms to build joint ventures with foreign firms (Global Link, the successor of Eurolink), which involves close links with the four partner regions which make up the "Four Motors" group.

The core objective of clustering has been to establish forums with representatives from various interest groups, depending on the industries involved. Clusters primarily emerged out of the networking processes around supply-chain initiatives. Most cluster forums are assemblies of representatives of large firms, SMEs, research organisations, TECs, local authorities and local enterprise agencies. The main route to cluster development has been bottom up, in a highly customised fashion, with an important role assigned to more experienced agents, notably large firms. The motivations for clustering have varied widely (Table 4). Only in a few cases has the WDA, more in a top-down fashion, identified industries to initiate a new cluster. One such a case is the food sector, where the WDA saw a need for consolidation given the ongoing trends towards rationalisation and spatial concentration in the sector. Other industries have been targeted because they were perceived as presenting new growth opportunities, such as multimedia, call centres and financial services. In the area of attracting foreign investment, special teams have been formed with a sectoral focus. This includes two core sectors: automotive, and electronics, and various "emerging" activities: multimedia, medical, food, financial services and call centres.

Parallel to the bottom-up cluster development, the WDA has developed a country-wide sector development strategy, based on an analysis of the strengths and weaknesses of the Welsh sectoral economy. The Sectoral Initiative Programme was launched in 1990 (Table 4). This involved a selection of priority sectors where Wales was seen as relatively strong and a set of emergent sectors where the WDA could play a role in supporting growth. A more in-depth study was commissioned in 1996, which resulted in a more elaborate typology distinguishing between inward-investment-related targets, indigenous growth targets, industries in need of more defensive strategies, and long-term priority sectors. The latter were identified on the basis of supply-chain gaps, opportunities arising from technological developments and new infrastructures. While this type of economic intelligence has supported the identification of new clusters, the majority of established clusters are based on a "hands-on" approach, in which most information was obtained by direct and regular communication between WDA officials and firms.

Recently, cluster policies have become more oriented to indigenous development. Although less than in Scotland, the WDA has taken an interest in indigenous firm development but this developed separately from the foreign investment policies. There has been a shift to a more integrated approach, partly as a response to the understanding that, with the competitive bidding game for foreign investment becoming tougher, factors other than land and subsidies were required. Building supply chains was seen as one way to promote the region, as well as to tie in existing investors. In addition, with development agendas increasingly stressing the need to raise Welsh competitiveness by innovation and increased export capability, large firms were increasingly seen as helpful for upgrading indigenous sectors. A recent initiative in which bridging indigenous and foreign-owned sectors is seen as an important objective is the Regional Technology Plan (Morgan, 1997).

Table 4. Wales – cluster and sector strategies, 1990 initiative and actual state

1990 Welsh Sector Initiative: four priority categories
Cat. 1. Information technology, financial services, R&D – *sectors are fully resourced initiatives with a dedicated team and a full-time sector manager.*
Cat. 2. Automotive, aerospace, medical/health – *sectors where a range of activities have already developed and which have the potential to impact upon a substantial number of companies both in Wales and for inward investment targeting*
Cat. 3. Chemicals, garments, furniture, craft, packaging – *sectors where activities are already underway but the level of resources is likely to remain limited*
Cat. 4. Environment, energy, media – *sectors in which research is underway to identify the needs of a sectoral approach*

	Actual cluster forums and description of core targets/motivation
1.	**Automotive:** strong self-sustained cluster; targeted on supply-chain development, increasing exports, training.
2.	**Electronics:** core growth sector; supply-chain development, technological development, training.
3.	**Opto-electronics:** mainly in West Wales; new technology sector.
4.	**Medical:** more research-oriented.
5.	**Multimedia:** indigenous development, built around three TV companies in Cardiff.
6.	**Call centres:** short-term target sector.
7.	**Pharmaceuticals**
8.	**Aerospace:** creating global links.
9.	**Renewable energy:** possible emergent cluster where people have been introduced to each other.
10.	**Machine tooling:** industry which benefits from the growth of, and growing interaction between, the automotive and electronics sectors.

Source: WDA (documentation, interview).

Cases of cluster policies

The industry in which WDA has had the most success in creating new collaborative structures and encourage networking is the automotive sector. It is this industry which has played a major role in the building of the core strands of the WDA programme, such as Source Wales, notably the supplier clubs, the Eurolink/Global Link initiative and its involvement in the Four Motors group. The automotive forum has become an important institution in the region, from the perspective of both policy makers and businesses. Membership is generally seen as imperative by firms in the automotive sector, because of its emphasis on building locally integrated value chains and its impact on regional development strategies.

Another industry where clustering has been important is electronics. However, here two initiatives have been developed, one around foreign investors (particularly LG), and another which draws on smaller firms. The first initiative is particularly geared towards developing a training strategy for the region, motivated by the expectation that recent investments will enhance the problem of skill shortages in the region. The SME forum is also focused on skills. A separate group was set up because these firms fear that skills shortages will lead to poaching by foreign investors.

These main cluster initiatives are essentially sector-based although, especially in the automotive case, they work from a value-chain perspective. One of the few examples of interindustry linkages in Wales is that shown by the relationships between the automotive and electronics industries. One industry that has particularly benefited from this relationship is machine tooling, in which about 70 firms are operating, many of them SMEs. The demand for machine tools has been induced by the large foreign firms in electronics and automotive, one example being the growth of Valenite-Modco. This firm has become a major supplier to Ford-Bridgend. At present, the WDA is investigating further benefits from linking automotive and electronics; but this is the only cross-industry grouping which receives such attention.

One cluster which was established primarily on the basis of indigenous businesses and actors is the Welsh Medical Technology Forum. This forum is strongly geared to creating innovative networks in the region, led by indigenous industries, local universities, pubic organisations and representatives from the National Health Service (NHS). The total number of Forum members exceeds the 500 mark, and it has already led to the establishment of a centre for SMEs: the Medicentre in Cardiff (1994), which collaborates with the Welsh University College of Medicine and the University Hospital. Another "indigenous" initiative is the Garment Industry Initiative which has led to the establishment of a Garment Design Centre (1992). This Centre offers computer-based services to local firms in a "real service" style (Cooke, 1992). It is, however, one of the few cases where a strong involvement of other actors (non-business, non-support) can be observed.

Impact

Wales has received considerable attention from researchers for its recent attempts to build new forms of public-private partnerships and the range of activities undertaken in the field of attracting and embedding foreign investments, clustering and technology policies. What the impact of these initiatives has been on the socio-economic development of the region, however, remains a controversial issue. Evaluations have been primarily carried out on a case-by-case basis, without a strong feedback on following support measures. Some authors see most development from a positive angle. Cooke argues that Wales has struck the right balance between attracting foreign investments and the building supply chains and other forms of embedding, thus preventing the emergence of industrial *enclaves* around foreign investors. The impact of foreign investments has even been seen as vital for what is perceived as a "renaissance" of Welsh industry (Price *et al.*, 1994). In this context, the creations of forums and partnerships are presented as good cases of *institution building* which is engendering the "filtering down" of the benefits of foreign investment to other parts of the local economy.

This positive image has been refuted by Lovering (1996), who, on the basis of statistical evidence, observes little more than a "tiny regenerative contribution of foreign firms". In particular, he attacks the vision of Wales catching up with regions such as Baden-Württemberg. On the issue of embedding foreign investments, he comments: "Wales has a few tiny, compromised and basically top-down measures designed to attract foreign firms, to encourage businesses to talk to each other, and to support minuscule innovation efforts. These are worthy in their own right, and might just make a modest contribution to particular firms or sectors in the very long term. But they are minuscule, unco-ordinated, and lack economic resources and a political context" (p. 15). While it remains to be seen how "tiny" the impacts of the WDA measures are in the long term, at present it is obvious that most initiatives do not reach very "deep" into the Welsh economy and are still geared towards the larger, more successful firms in the economy.

3.3. *Scotland*

Emergence of cluster policies

Scotland has had a long tradition of sectoral support, notably in electronics where there has been a variety of institutions and strategies of various kinds since the 1950s. Foreign investments have driven most of these initiatives, although an early clustering strategy in the 1950s was based around Ferranti, a UK-owned defence contractor that collaborated with local firms in the development of new electronic technologies in a publicly funded, shared, laboratory complex. However, with the growth of US investment in the 1960s and 1970s, followed by Far Eastern firms in the 1980s and 1990s, the emphasis has shifted to defending and embedding foreign-owned firms within a Scottish electronics cluster.

Electronics is not the only sector to be the focus of support in Scotland, and there have been long-standing programmes for biotechnology, food and drink, textiles and clothing and others, but electronics has tended to be the most distinctive and perhaps most successful in terms of the external perception of Scotland as a world-class centre for electronics manufacturing.

Figure 2. Scottish Enterprise organisational structure

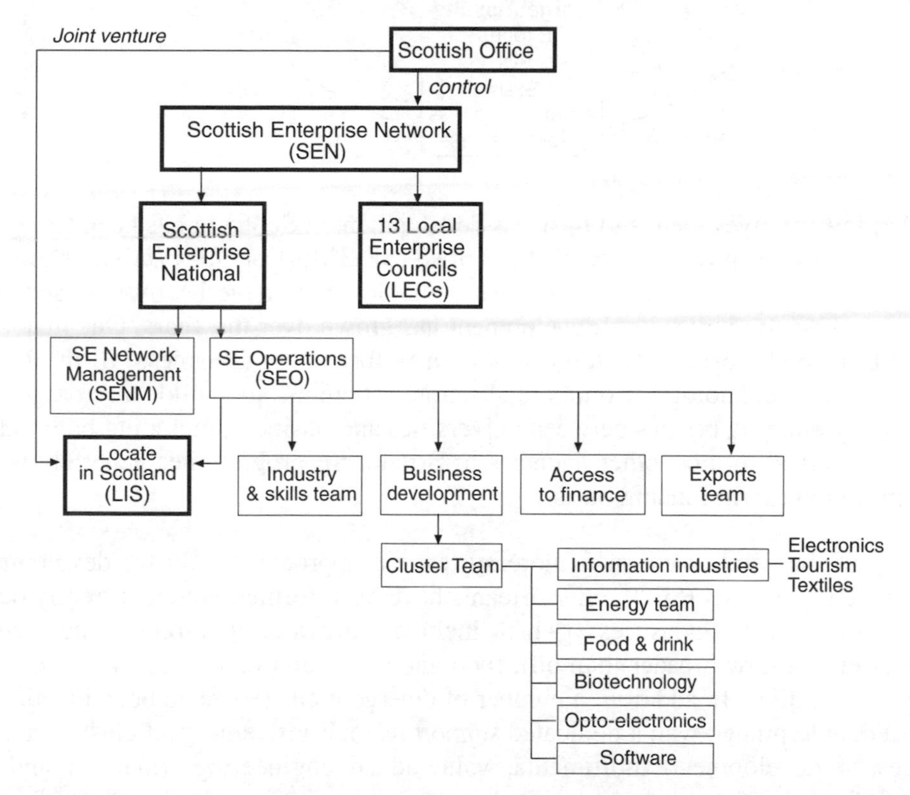

Note: Separate organisations with own governance structure are shown in **bold**; others = divisions dependent on Scottish Enterprise National.

Figure 3. Scottish Enterprise cluster approach

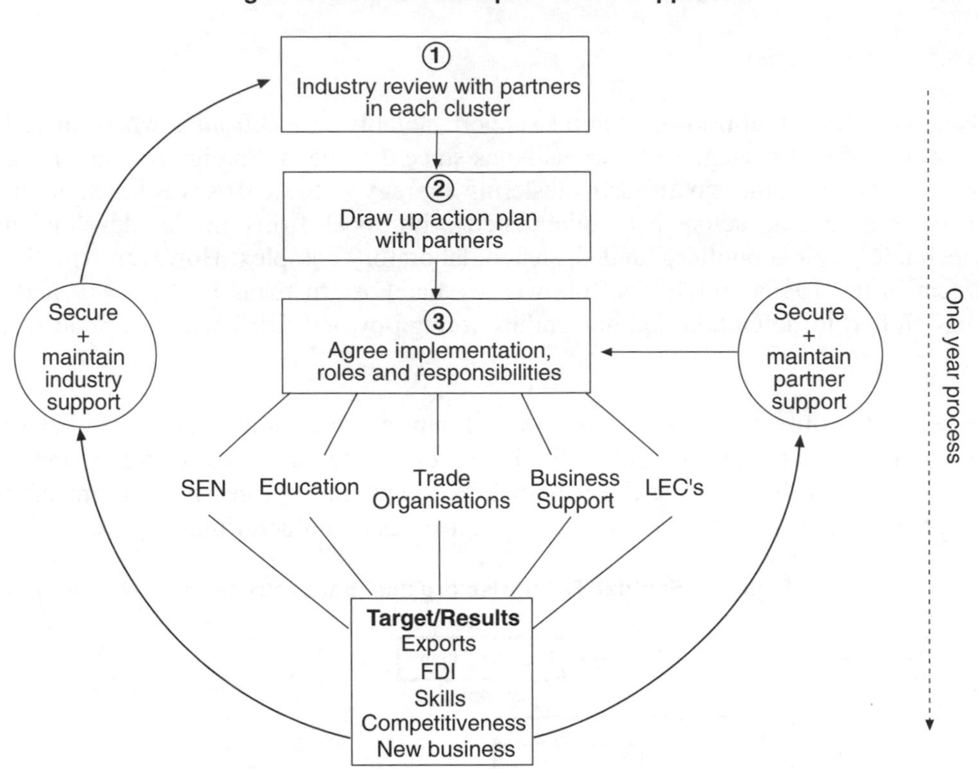

As noted earlier, inward investment and business development in Scotland has been led by the Scottish Development Agency, and now by Scottish Enterprise (SE). Most of SE's activities are organised in theme-oriented divisions: exports, skills, etc., but some activities have been organised within sector divisions. The emphasis on indigenous development has grown over the years. One major programme launched by SE recently especially targets new-firm formation (Business Birth Rate). Another programme, that of Technology Commercialisation, attempts to build a stronger indigenous technology base, by building bridges between universities and business, including both indigenous and foreign firms. In contrast to most other agencies in Britain, SE programmes generally benefit from a secure, medium- to long-term funding basis.

In recent years, SE has embarked on a more systematic approach to cluster development. Within Scottish Enterprise Operations (SEO), cluster teams have been formed which currently deal with four major clusters: "information industries" (which includes software, electronics, and manufacturing); energy (building on but now broader than oil); food and drink; and tourism; and two smaller clusters: biotechnology; and textiles. In addition, a number of emergent clusters have been identified, which so far have not been underpinned with a dedicated support team. These emergent clusters are presently in different stages of development: multimedia, value-added engineering, financial and educational services, chemicals and forest production. Finally, there is group of so-called "latent" industries, for which no initiatives exist at the SE network level, although cluster teams may emerge at the local level. This includes defence, financial services and chemicals. While different philosophies are applied in the initiatives, one common objective is to move from what are considered more interventionist sectoral policies to a position in which SE facilitates the creation of "knowledge-based networks". Involvement of other Scottish organisations is generally strong.

Cluster cases

A specific SE methodology has been developed to set up clusters, which is intended to lead to a cluster strategy over a period of approximately one year. As illustrated in Figure 3, the clustering initiatives are particularly targeted on identifying common needs in different fields of business support and economic development policy. One of the emerging clusters which was, at the time of writing, halfway through the process of consultation, is multimedia. The more systematic approach followed by SE can partly be attributed to the fact that it asked Monitor to undertake an in-depth analysis of the Scottish economy, which included a detailed comparison of key industries, notably in electronics, with core regions in the United States, on themes such as innovation, finance. The Monitor research was followed by a – perhaps more important – phase of consultation, resulting in a final cluster map and strategy which has been published very recently. Key criteria for the selection of clusters included: the weight and scale of the sectors, the possibility for SE to "make a difference", the potential for the Scottish industry to be a winner, the urgency of intervention, and the opportunities for global exports. Undertaking in-depth analysis is seen as important by SE. With the right identification of problems and possible solutions, and "smart" people on board, the design of a development strategy will be much more effective, both with respect to time and costs. It is too early to indicate the nature of the evaluation of cluster initiatives.

The electronics forums are most developed – with different forums for general electronics manufacturing: opto-electronics, software and, more recently, multimedia. These forums link into a rich network of initiatives, co-ordinated by the "information industries" group in Scottish Enterprise, some of which have spun off to become distinct companies. One example is SPEED, a logistics support and lobby organisation for the electronics sector which is supported by the major companies and SE, and has undertaken a number of initiatives to enhance the capability of firms to export from Scotland, to use electronic data interchange within the sector, and to build supply-chain linkages and adopt industry-wide solutions to logistics problems.

Over the last few years, clusters have become a core concept in SE's strategic and organisational development. The organisation is now striving to become a more cross-sectoral and inter-disciplinary organisation inspired by the cluster concept. This means that other policies, notably skill development and technology policies, have been linked to the cluster-targeting approach. One recent example of such initiatives is the "Alba" project, a skill development project developed around Cadence, a US-based semi-conductor design company. Since the Cadence plant, established in 1998, presents a high value-added activity with a high skill profile, it is seen as vital hub for cluster development.

3.4. Northern Ireland

The situation in Northern Ireland is slightly different from that in Wales and Scotland because of the way in which direct rule from London was imposed on a devolved parliament structure. Thus, the Northern Ireland Office has a Department of Economic Development (DED), with specialist agencies such as the Industrial Development Board, the Industrial Research and Technology Unit, the Local Enterprise Development Unit, etc. These units all undertake responsibilities equivalent to parts of SE/WDA, but co-ordinated by DED. Each has a private sector led board. DED has traditionally had much greater control over institutional development than the Scottish Office and its Industry Department, dating back to the 1970s when a Northern Ireland parliament had a Minister for Industry.

The Northern Ireland Growth Challenge (NIGC) is a private sector initiative by the Northern Ireland branch of the CBI, although it works in close operation with the Department of Economic Development (DED) of the Northern Ireland Office. The NIGC largely follows a top-down industry-targeting vision based on a cluster approach. It is the only development strategy with clusters as its main tool, associated with the ambitious mission of making "Northern Ireland *the fastest growing region in Western Europe*. Key to this growth is building more dynamic, competitive clusters to drive continuous innovation, up-grading and learning. In essence, it is a vision of a return to Northern Ireland's heritage of industrial leadership built on hard work, inventiveness and dynamic enterprise." (NIGC Interim Summary 2). Although the aim of the initiative is to focus not on detailed analysis but private sector action, the strategy is based on an overall analysis of the regional economy which follows the standard sequence of identifying strengths and weakness and developing a cluster strategy which includes a cross-cluster dimension. The upgrading strategy is built around a set of common themes from the management literature: achieving world-class standards, supply-base development, skills development, and place-marketing. In addition, the role of networking among businesses, government, universities and other groups is emphasized to build consensus and new forms of interaction, as is the need to improve infrastructure and the environment. These themes have been integrated and specialised in the "cross-cluster programme".

The strategy is firmly based on a Porterian approach, which is not unexpected since most of the work was commissioned from Monitor. Porter's influence can be seen in the way the NIGC repeatedly stresses the fact that the government, through its tradition of generously subsidising weak firms, has actually aggravated the core economic problems in a serious form of government failure, thus exacerbating the lack of competitiveness and a lack of an innovative, outward-looking culture.

The clusters targeted by the NIGC are: engineering; food processing; health technologies; software; textiles and apparel; tourism and leisure; tradeable services; and contracting. For each of these clusters, a range of initiatives has been developed in collaboration with private and public organisations

3.5. England

In England, strategic economic development policy at a regional scale, particularly for inward investment and sectoral development, has been left to the emergence of regional development organisations established by local authorities and other partners, with some funding from central government. These bodies tend to be non-profit companies established without statutory powers or formal relationship with government. In the North East, NDC emerged from a joint initiative of various local and national organisations (trade unions, business federation, local authorities and the Invest in Britain Bureau), building on an earlier organisation, NEDC, which fulfilled a similar function and was funded by Invest in Britain (IBB). While these organisations (others exist, such as INWARD in the North West) have had government support for their inward investment role (mainly for marketing activities), where they have been eligible for European funds or other non-FDI-related UK support, they have been able to expand into other forms of activity such as investor after-care, supply-chain work and cluster strategies. NDC is following its neighbouring agencies by moving towards a more sector-based structure.

In other English regions, the history of sector initiatives and the emergence of partnership-based governance networks have been important factors in the development of network-oriented local industrial policy. In the late 1970s and early 1980s, industrial decline in manufacturing sectors triggered the development of sectoral policies at the local level. Some of the larger councils, such as the later abolished Greater London Council, undertook comprehensive sectoral studies as a basis for local industrial policy. The political changes in the 1980s led to a less analytical, more instrumental approach focused on single sectors, primarily targeted at SMEs. In particular, organisations such as the TECs encouraged higher levels of user engagement and an elaboration of specific areas, such as training, supply-chain development and accreditation. Such initiatives were largely driven by funding conditions and were not part of larger strategic plans. In general, no link was established between local sector-oriented initiatives and the inward investment strategies followed by NDC and INWARD. The latter evolved as purely focused on inducing "a quick injection of jobs" by simple means of place-marketing and competitive bidding for investment opportunities.

Cluster initiatives in England largely emerged out of sectoral policies inspired by the "industrial district" model. They evolved either in conjunction with spatial policies, such as the cultural district in Sheffield, which is viewed as one of the few successful cases of "industrial district" creation. Or, they were inspired by models of inter-firm networking and resulted in initiatives to bring small numbers of firms together in business clusters (Shutt and Pellow, 1997). An example of the latter is the support of the North Tyneside Real Service Centre – its name being a clear reference to its Italian inspiration – to the development of five business clusters consisting of between five and ten firms, in sectors ranging from offshore and software production to consultancy and design activities. There are also examples of partnership models being applied to local sectors, such as in the case of Leeds. From the four sectoral organisations which have emerged to date, one, the financial cluster, has become self-sustained (Thomas and Shutt, 1996). Another example of a partnership model is the "World Class Supplier Base" Programme for automotive suppliers in the West Midlands, which has taken on a "hub and cluster" model. This initiative, which has grown out of supply chain programmes of the national Department for Trade and Industry (DTI), is backed by a coalition of Chambers of Commerce, universities, TECs, and Industry Centres.

What is lacking in these and similar cases, however, is the embedding of the local, bottom-up initiatives in a wider perspective on local or regional development. The initiatives primarily present isolated projects driven by particular agents and supported by specific sources of funding on a local scale. One of the problems in England is that, with the exception of London and the North East, there is no well-developed regional structure on which more comprehensive development strategies could be built.

An important factor in the emergence of cluster policies by a number of regional and local actors in North East England has been the European Structural Funds, and the insertion of an action line on clusters of competitive advantage in the Single Programming Document for 1994-96. This has stimulated a number of new projects and initiatives with a broad cluster approach, and led to greater debate within the region about the merits of a sectoral or cluster approach.

One sector where a cluster approach is currently emerging in the North East, following the examples of Scotland and Wales, is electronics. Over the years, the numbers of electronics firms in the region has ebbed and flowed, with NDC taking an opportunistic approach rather than the more targeted approach taken in Scotland. The core of the industry in the region has been consumer goods manufacturing (mainly Asian in origin in recent years), with a more traditional components industry which is of UK and mixed foreign ownership. A key departure, however, has been the arrival of first a Fujitsu, and then a Siemens, semi-conductor plant. In both cases, the local universities have been very active in developing new training courses but this has been only the first stage of a wider local response. The Siemens investment in North Shields coincided with the development of a research and training facility: the Centre for Advanced Industries. The character of this centre was transformed when Applied Materials, the supplier of the bulk of semi-conductor manufacturing equipment to Siemens, decided to establish a European training centre there. This was followed by an aspirational strategy to enhance and develop the electronics sector with the establishment of the North East Microelectronics Centre, a forum for the industry based on training needs, and involvement in the new National Microelectronics Institute which is based in Scotland. Over the last few years, a new sectoral strategy has developed, led by the NDC, which has responded to the region's aspirations by establishing an electronics "division". Obviously, the recent announcements that Siemens and Fujitsu will pull out from the region will have serious consequences for the continuation of this policy.

At the local level, bottom-up initiatives have emerged, aimed at promoting indigenous development, particularly of SMEs, by facilitating and supporting business clusters. This generally involves the joining of between five and ten firms under the umbrella of a formal cluster organisation, in which firms exchange experiences, share resources, and develop common strategies in areas such as product development, marketing and training. Besides the support for the development of local industrial networks, such business clusters may contribute in several ways to local economic development and policy. Business clusters present an environment through which SMEs can identify and express their needs for business support and can have a stronger voice in local industrial policy making. They may also present a more effective and sustainable way of building relationships between the business support sector and SMEs. Finally, at the local level, such clusters present a vehicle for the authorities to support or launch the development of a particular economic activity.

One case of support for business clustering is offered by the North Tyneside Real Service Centre (RSC), operating in the area east of Newcastle upon Tyne. The RSC is a spin-off from the Council's economic development department. With the assistance of ERDF funding, the RSC was set up in 1994 to develop and implement a local cluster strategy. The first two clusters developed were Argonautics (marine engineering) and Pegasus (consultancy to the pipeline industry). In both cases, the initiatives responded to a crisis situation in which the council was facing a loss of innovative capability in the area. Argonautics was set up to retain some of the marine engineering capacity after the bankruptcy of the last shipyard (Swan Hunters). Pegasus was developed to retain some of the expertise in pipeline fabrication and testing after the closure of the British Gas Engineering Research Station at Killingworth (which employed around 500 people). The Pegasus cluster has been particularly successful in creating new business by preserving some regional expertise in the area of pipeline design and maintenance. Over the last two years, its six member firms have been able to develop new expertise (notably in pipeline rehabilitation) and to access new markets abroad through an aggressive cluster marketing strategy. After Pegasus, the RSC has initiated cluster development in the areas of design communications, software and management systems. In all these cases, networks have been formed between previously isolated SMEs through which, collectively, new market positions have been acquired. The RSC has developed an advanced monitoring system and reporting structure which is especially geared to justify the continuation of its (generally short-term) funding.

4. Conclusion

Clustering has become a popular concept in the domain of regional policy making, underpinning new initiatives geared to facilitating networking processes along (inter)sectoral and value-chain lines. Cluster policies can be seen as an innovative step in regional policy making, not only because of the emphasis on networking, but also because they may build a crucial bridge between two levels of regional economic development:

- The business level, where cluster initiatives may promote inter-firm trading and inter-firm learning (in many cases, the latter may arguably be more important than the former) as well as improving links between business and the regional technology and business support infrastructure.

- The structural level of the regional economy, where cluster policies, through a strategy of targeting, may support the reorientation of regional economic development towards growth sectors.

A number of theoretical arguments can be put forward to explain why policy interventions targeting these two levels may be justified. First, intermediate markets in regional economies, notably small firms in more traditional sectors, may suffer from informational failures, and cluster policy may thus be geared to breaking business isolation and facilitating the co-ordination of modernisation and investment strategies among related industries. Second, the regional economy may be locked into an unfavourable economic structure and thus needs initial triggers to develop new sources of growth and employment creation. Growth may be triggered, for instance, around certain hubs, such as foreign investors or a university, through a cluster approach. Third, past experience with business support has shown a worrying level of mismatch between supply and demand. Through the development of new types of interaction between support agencies and their (potential) clients along cluster lines, occurrences of government failure may be reduced and the effectiveness of existing support mechanisms enhanced. Fourth, clustering may be used to improve the interaction between the regional institutional system at large, by offering a platform for discourse on business development as well as the long-term economic development of a region. This may be expected to reduce institutional mismatches.

The last dimension bears directly on the issue of regional innovation systems. Particularly in the case of more peripheral areas, it is essential that these systems be seen not as autonomous entities important for knowledge production. Rather, regional innovation systems should be associated with a coherent institutional structure facilitating and directing processes of learning in a region. In peripheral regions, virtually all sources of learning can be expected to be external, and a crucial role may be played by knowledge hubs represented by large externally owned firms, universities, training colleges, etc. The innovation system, therefore, should aim to steer the process of institutional linking and useful forms of knowledge transfer. In this perspective, clustering provides strategic direction and, in particular, connections to groups of actors already linked through supply chains or other inter-firm relations. It is in these particular contexts that an open-ended concept of regional innovation systems will benefit cluster developments and vice versa.

While the discussion has highlighted the potential benefits of cluster policies, this should be contrasted with the practical implementation of cluster policies. The case studies presented here show the difficulties in assessing the justifiability of initiatives, in measuring outputs, and in finding the right approach to funding. Learning at the levels of the policies themselves still seems to be poorly developed. Moreover, while targeting features in most approaches, this remains a controversial issue for public intervention. In all peripheral regions, the case for exploring and triggering new directions of development is easily made; however, how business opportunities that can be "unlocked" by policy initiatives should be identified and monitored remains an open question. One worry may be that, especially in regions dominated by foreign investors, clustering will improve the interaction between foreign firms and parts of the indigenous sector in a region, but through a process of "picking winners" rather than through a more widespread modernisation of local SMEs. There remain, accordingly, a number of critical issues that should receive further attention while advancing cluster approaches in regional policy making: for instance, who will be involved in strategy development (question of representation and dominance)? How will the strategy be implemented (relation between bottom-up and top-down) and funded (tension between the short-term nature of project funding and the long-term results)? Which industries will be targeted? And, how will the clustering process be governed and evaluated?

NOTES

1. The United Kingdom has a varied institutional composition, and so the term region has varied meanings. Scotland and Wales are classed as nations, united with England within the United Kingdom, and so are not usually referred to as regions, especially given current devolution trends. Northern Ireland also had special status with its own Parliament until the 1970s, now being reintroduced, although Northern Ireland is usually termed a province in reflection of a colonial status. The term region is usually reserved for the English regions, which are mainly of a similar population scale to Scotland. Further confusion arises from the existence until recently of a tier of local government in Scotland known as "Regional Councils".

2. Development agencies in Scotland, etc., are still termed RDAs here for convenience. In England, RDAs are currently being established and will be operational early in 1999. These will follow a different model to the existing RDAs and, in the first instance, will network together a number of existing regional-scale bodies concerned with inward investment, land and property development and probably skills development.

REFERENCES

Batt, Helge-L. (1994), *Kooperative regionale Industriepolitik. Prozessuales und institutionelles Regieren am Beispiel von fünf regionalen Entwicklungsgesellschaften in der Bundesrepubik Deutschland*, Peter Lang, Frankfurt amd Main.

Camagni, R. (1991), "Local 'Milieu', Uncertainty and Innovation Networks: Towards a New Dynamic Theory of Economic Space", in R. Camagni (ed.), *Innovation Networks: Spatial Perspectives*, Belhaven Press, London, pp. 121-144.

Cantwell, J.A. (1991), "The Theory of Technological Competence and its Application to International Production", Chapter 2, in D.G. McFetridge (ed.), *Foreign Investment, Technology and Growth*, The Investment Canada Research Series, University of Calgary Press, Calgary, pp. 33-70.

Cooke, P. (1992), "Regional Innovation Systems – Competitive Regulation in the New Europe, *Geoforum* 23(3), pp. 365-382.

Cooke, P. (1995), "Keeping to the High Road: Learning, Reflexivity and Associative Governance in Regional Economic Development", Chapter 13, in P. Cooke (ed.), *The Rise of the Rustbelt*, UCL Press, London, pp. 231-245.

Cowling, K. and R. Sugden (1993), "Industrial Strategy: A Missing Link in British Economic Policy", *Oxford Review of Economic Policy* 9, pp. 1-18.

Danson, M.W., M.G. Lloyd and D. Newlands (1992), "Regional Development Agencies in the United Kingdom", Chapter 11.4, in P. Townroe and R. Martin (eds.), *Regional Development in the 1990s – The British Isles in Transition*, Jessica Kingsley, London, pp. 297-303.

Dunning, J.H. (1991), "Governments, Economic Organisation and International Competitiveness", in L.G. Mattsson and B. Stymne (eds.), *Corporate and Industry Strategies for Europe. Adaptations to the European Single Market in a Global Industrial Environment*, Elsevier Science, Amsterdam, pp. 41-74.

Dunning, J.H. (1992), "The Competitive Advantage of Countries and the Activities of Transnational Corporations", *Transnational Corporations,* February, pp. 135-168.

Dupuy, C. and J.-P. Gilly (1994), *Collective Learning and Territorial Dynamics: A New Approach to the Relations between Industrial Groups and Territories*, Institut d'Économie Régionale de Toulouse.

Edquist, C. (ed.) (1997), *Systems of Innovation. Technologies, Institutions and Organizations*, Pinter, London.

Geddes, M. (1992), "The Sectoral Approach to Local Economic Policy", Chapter 1, in M. Geddes and J. Benington (eds.), *Restructuring the Local Economy*, Longman, Harlow, pp. 1-21.

Gómez Uranga, M. and L. Ozerin (1997), "Comparing Regional Innovation Systems in Transition. The Case of Three Regional Development Agencies", paper presented at the EUNIT Conference on Industry, Innovation and Territory, Lisbon, February.

Gray, M., E. Golob and A. Markusen (1996), "Big Firms, Long Arms, Wide Shoulders – The Hub-and-Spoke Industrial District in the Seattle Region", *Regional Studies* 30(7), pp. 651-666.

Hassink, R. (1996), "Technology Transfer Agencies and Regional Economic Development", *European Planning Studies* 4(2), pp. 167-184.

Langlois, R.N. and P.L. Robertson. (1995), *Firms, Markets and Economic Change: A Dynamic Theory of Business Institutions*, Routledge, London.

Lovering, J. (1996), "New Myths of the Welsh Economy", *Planet. The Welsh Internationalist* 116 (April/May), pp. 6-16.

Lundvall, B.-Å. (1992), "User-producer Relationships, National Systems of Innovation and Internationalisation", Chapter 3, in B.-Å. Lundvall (ed.), *National Systems of Innovation. Towards a Theory of Innovation and Interactive Learning*, Pinter, London, pp. 45-67.

Malmberg, A. and P. Maskell (1996), "Proximity, Institutions and Learning – Towards an Explanation of Regional Specialization and Industry Agglomeration", *European Planning Studies* 5(1), pp. 25-41.

Morgan, K (1994), "The Welsh Development Agency. The Fallible Servant", mimeo, University of Wales.

Morgan, K. (1996), "Learning-by-interacting. Inter-firm Networks and Enterprise Support", in OECD, *Local Systems of Small firms and Job Creation*, OECD, Paris,

Morgan, K. (1997), "The Learning Region: Institutions, Innovation and Regional Renewal", *Regional Studies* 31(5), pp. 491-503.

Mowery, D.C. and J.E. Oxley (1995), "Inward Technology Transfer and Competitiveness – The Role of National Innovation Systems", *Cambridge Journal of Economics* 19(1), pp. 67-93.

Peck, F.W. (1996), "Regional Development and the Production of Space – The Role of Infrastructure in the Attraction of New Inward Investment", *Environment and Planning* A 28(2), pp. 327-339.

Porter, M.E. (1990), *The Competitive Advantage of Nations*, Macmillan, London.

Price, A., K. Morgan and P. Cooke (1994), "The Welsh Renaissance: Inward Investment and Industrial Innovation", CSA Report No. 14.

Rehfeld, D. (1995), Disintegration and reintegration of production clusters in the Ruhr area. Chap. 6. In: The rise of the rustbelt. (Ed: Cooke, P.). UCL Press, London, 85-102.

Rosenfeld, S.A. (1997), "Bringing Business Clusters into the Mainstream of Economic Development", *European Planning Studies* 5(1), pp. 3-23.

Shutt, J. and N. Pellow (1997), "Industrial Districts and Business Support Strategies – Contrasting Evidence from the North of England", paper presented to the Conference on Networking and Small and Medium-sized Enterprises, Bologna University, 19-20 June.

Sternberg, E. (1991), "The Sectoral Cluster in Economic Development Policy: Lessons from Rochester and Buffalo, New York", *Economic Development Quarterly* 5(4), pp. 342-356.

Thomas, K. and J. Shutt (1996), "World-class Leeds: Sectoral Policy and Manufacturing Alliances in the 1990s", Chapter 5, in G. Haughton and C.C. Williams (eds.), *Corporate City? Partnership, Participation and Partition in Urban Development in Leeds*, Avebury, Aldershot, Hants, UK, pp. 81-102.

Waits, M.J. (1992), "Arizona – Preparing for Global Competition through Industrial Clusters", *Spectrum – The Journal of State Government* 65(3), pp. 34-37.

Young, S., N. Hood and E. Peters (1994), "Multinational Enterprises and Regional Economic Development", *Regional Studies* 28(7), pp. 657-677.

Chapter 6

THE DISAPPEARING TRICK: CLUSTERS IN THE AUSTRALIAN ECONOMY

by

Jane Marceau
University of Western Sydney Macarthur

1. Introduction

Much public attention has been focused recently on ways in which modern western economies can make the transition from earlier forms of industrial organisation to those characteristic of the "learning economy". In the "learning economy", innovation is seen as at the heart of economic development. In innovative industries, especially those whose products depend to a significant extent on scientific information, both products and the processes used for production tend to change relatively rapidly. Some countries, including Australia, are beginning to realise that the existing structure of their economies and the forms of their business organisations are not going to lead them smoothly to competitive success in the coming century. New policies are being devised which seek to create or sustain the development of clusters of innovative activity.

In order to develop such policies, governments need to have maps of the clusters of activity which currently operate in their economies. In some countries, a special effort has been made to find clusters and to develop effective assistance policies. Australia, however, is only just beginning to realise the importance of putting on new analytical "lenses" to seek and find clusters and map their activities. One thing which seems clear when looking at Australia is that there are few, if any, clusters of the "Third Italy" or German kind, or even of the kind analysed by Porter (1990). The country as a whole, so far, has even fewer active cluster development policies, although there are several promising sets of initiatives in the states, notably South Australia and some areas of Victoria.

Developing policies to improve that situation need to be built on a firm understanding of why these "natural" clusters have been so slow to develop.

This chapter is in four sections. Following this Introduction, Section 2 outlines some different definitions of clusters and networks so that the many aspects of "clusters" and networks appear and links them to views on the functioning of national innovation systems as an analytical approach to looking at what is happening in the economy. The section is intended to illuminate the field, not to provide a matrix for examining the Australian data.

The first section of Section 3 presents a picture of the background situation in Australia using input-output data as a measure of the indirect linkages in the economy and to indicate why there are problems in locating existing and encouraging new clusters, especially where one is looking for innovative activities. The second part of Section 3 presents available data on clusters of several different kinds in Australia. The third part indicates the results for cluster development of the particular characteristics of Australia's industrial structure and of past policies and current restructuring. Section 4 makes some policy suggestions for future cluster development

2. Clusters, networks and innovation systems

Concepts of clusters

There have been numerous definitions of "clusters". Some focus on the "horizontal" nature of relationships between small and medium-sized firms which both compete and collaborate with each other. This is what one may call the "democratic" view of how clusters differ from other inter-firm relationships. It is the one typified by the industrial districts of Marshall (1920) or the "Third Italy".

Others would see what I have elsewhere called the "webs" or relationships established between large firms and their core suppliers as also in many cases leading to clustering both among those firms and among smaller enterprises in the industry who supply one lead firm rather than the industry as a whole. This would perhaps have been particularly the case where there were many small specialist firms attached through supplier relationships to lead firms in the automotive industry, for example. Such webs are often viewed as essentially hierarchical relationships which just happen to involve inter-firm rather than intra-firm relationships. In these clusters the lead firm both pays the piper and calls the tune. This is what one could call the "vertical supply chain" notion of clusters. In these circumstances, purchasing power and other dominant management practices determine clustering.

Most observers emphasize the geographical co-location of firms involved in clusters of activity, seeing the daily interactions as part of the essential dynamics of the cluster which set it apart from collaborative activities which involve firms in different regions or countries. That definition is now coming up for greater scrutiny as what Harrington (1991) has described as "virtual clusters" are being discovered. In this chapter, however, geographical propinquity is seen as a key criterion of the existence of functioning clusters.

Geographical co-location, however, may not be enough. In some cases there are no real inter-firm linkages except sales to the same customer. In others, there may be what have been described as "emerging clusters". Firms in these use either a common resource base or have some common resource needs but few, if any, clear relationships have usually developed among the firms involved in innovation or production. These may be emerging because there is increasing recognition among the players that they do indeed have common interests or, more rarely, because they are the subject of policy attention which wants to raise the productivity of the firms concerned.

One way in which such intervention may occur is through the action of leading-edge customers which the firms may or may not realise they have in common. This may be most common in defence-related procurement, but could also be encouraged by actors such as hospitals who wish to reduce costs, improve quality or upgrade reliability of service. Clusters formed in this way may not share the characteristics of hierarchy or "democracy" of relations between firms in clusters which evolved more or less in isolation.

Emerging clusters may also gradually develop buyer-supplier links among firms who previously "traded out" and hence develop denser networks of linkages and greater complementarities of activity.

Finally, there is some uncertainty apparent in the literature as to whether clusters involve firms in the same or in related industries, whether intra- or interindustry links are essential. In the wool textile cluster in Prato, in Italy, for example, there are both textile companies and engineering firms which make textile equipment. Similarly, in the Finnish forest cluster, machinery manufacturers are a essential factor in the success of the cluster, and the cluster includes both paper manufacturers and the emerging firms which clean up after the paper processes. In that cluster the "forest" is the key link between the economic activities undertaken. In other work, the firms in clusters described are more strictly defined as being in the same segments of an industry, all making leather goods or ceramic tiles, or whatever.

Perhaps what all clusters have in common is that they are *potential* sites for networks of activity which are greater among these firms than "normal" market analysis would expect and which mean that as organisations the member firms agree to make their boundaries more "permeable" to partners and institute rules about collaborative and competitive behaviour which apply to members of the cluster but not to other enterprises either in the immediate environment or in the same industry in other places.

For the purposes of this chapter, I adhere to the definition of clusters put forward in the Focus group summary paper by Roelandt and den Hertog (Chapter 1 of this volume). Clusters are characterised as networks of production of strongly interdependent firms, knowledge-producing agents and customers linked to each other in a value-adding production chain.

Notions of networks

Networks of activity are the very essence of functioning clusters and as such are part of the definition. There are many other aspects of "networks", however, which it may be useful to outline here because they have both analytical power and influence in the formation of clusters on the ground.

"Network" is a concept much used by analysts struggling to make sense of organisational relationships between economic actors which do not fit either of the main traditional notions of what firms do since they are clearly based neither on market nor on hierarchy. In practice, networks are hard to define and may well be even harder to create or manipulate by policy action seeking a tool for economic development, at least in areas where there has been little prior connection between players (for a useful summary of work on networks, see Riemens, 1996).

Some writers consider investigation of network functioning as an analytical approach to understanding the dynamics of the productive structure. These writers see networks as filling "structural holes" or the interstices between formal organisations, and they see the dynamics of the networks as at least as important as the structures and formal organisations which they link (Burt, 1992). In some ways, the biomedical companies studied for this project also seem to be working in the interstices between the dominant players in the health system – the regulators, the hospitals and the large firms which import major equipment and consumables and medical doctors.

For other observers, networks are a newly prominent and empirically visible element of the emerging productive system. Even here, however, there are several interpretations of what is being seen and its importance. Networks are seen by some as the motors of the "Third Italy" or clusters model of economic development and as alternatives particularly to the large hierarchical productive forms of organisation (Piore and Sabel, 1984) which Chandler has so tellingly described (1990). For other observers, participation in networks is seen as a means for firms to deal with the burdens of financial, technological and information needs which are increasing in weight and complexity, especially for small and medium-sized firms, as leading-edge areas of the economy approach a knowledge-based production system. In this case, networks are often seen as having some of the characteristics of joint ventures, strategic alliances and other more formal and structural links. Development of such networks is increasingly seen as the way forward in the innovation race, especially for SMEs, and thus deserving of specific public policy encouragement (see, for example, writers such as Cooke *et al.*, 1990; Camagni, 1991; and others, who discuss the functioning of "innovative milieux").

Since networks in practice have different functions and serve different roles, they vary along several dimensions and contain several elements. The relationships generated by network interaction also vary. The most important include:

- Single or multi-stranded social relationships based on trust, including non-trading links that inform a business's strategic decisions.
- Production or industrial relationships (users-producers but perhaps also other players) based on commercial transactions and contracts.
- Exchange relationships based on linkages such as joint ventures specifically related to companies' innovation activities, such as joint pre-competitive R&D.

The major variables determining whether networks function smoothly or not seem to derive from the degree to which collaborating firms need to be or are geographically co- or closely-located; the nature of the co-operation (single or multi-stranded) between them, the period of linkage, the intensity or criticality of the relationship, the formality of the relationships and the degree of co-ordination that is involved in network activities. These may all need to be taken into account in developing policies for clusters. Finally, and related to this last issue, there is the question of the number of partners involved and the degree of equality in the relationships.

One final distinction among different types of networks is useful here. This is the distinction between networks of production and networks of innovation. Networks of production, which are seen in this book as the critical aspect of clusters, differ from networks of innovation. Networks of innovation may usefully be seen as the networks which enable knowledge to be gathered and generated for innovation. Networks of production, however, imply close links between suppliers and customers which may not be present to aid innovation by firms linked together in informal ways to generate the knowledge they need. One of the weaknesses of Australian industry, as suggested by this chapter, is that while we have good science and innovation links we have poor production links between firms due to the patchy nature of our industrial structure and the consequent dearth of high-quality suppliers from whom firms wanting to innovate can learn.

National systems of innovation

An important issue arises concerning the ways in which the networks which characterise clusters relate firms to the institutions and organisations of a country's system of innovation. The issue concerns the degree to which the success with which firms relate to the other players in their commercial environment depends on their own capacity for innovation and that to which their success depends on the innovative support given by the national economic system agencies. In policy terms, the question arises of the degree to which the capacity of networks to function and maintain themselves is linked to and dependent on the operation of other sets of institutions and policies specifically designed as "higher" levels of the national system of innovation. In particular, the creation and successful functioning of formal types of networks in industry may depend on the involvement not only of other industrial firms but also of different economic actors, including banks and other financial organisations, trade unions, research generators and governments which shape the rules governing network transactions, provide funds for networking activities and may diffuse resulting information.

The glue of the system, however, is formed by the firms whose operations bring together the different parts of the system. It is the relationships among firms which link them into clusters of activity which are the focus of this chapter.

3. Clusters in Australia

Input-output links

It is hard to measure directly the extent to which clustering occurs in an economy, although DeBresson and his colleagues are developing innovative ways of doing this (Chapter 2 of this volume). It is possible, however, to use indirect approaches, notably using input-output data. To indicate real clusters, input-output data should be at lower levels of aggregation than those which are available in Australia as a time series. While input-output data at the high level of aggregation available cannot strictly tell the observer anything very specific about collaborative relations in economic activity, they can nonetheless indicate areas of concentration of such linkages. Figures 1-3 trace the situation from 1975-93, the last year for which such data are available. Recent input-output data on Australia suggest that there are indeed relatively few interindustry and intra-industry links in the Australian economy. They show, moreover, that the links which do exist have been eroding or "hollowing out".

Figure 1. Australian input-output matrix (domestic), 1975

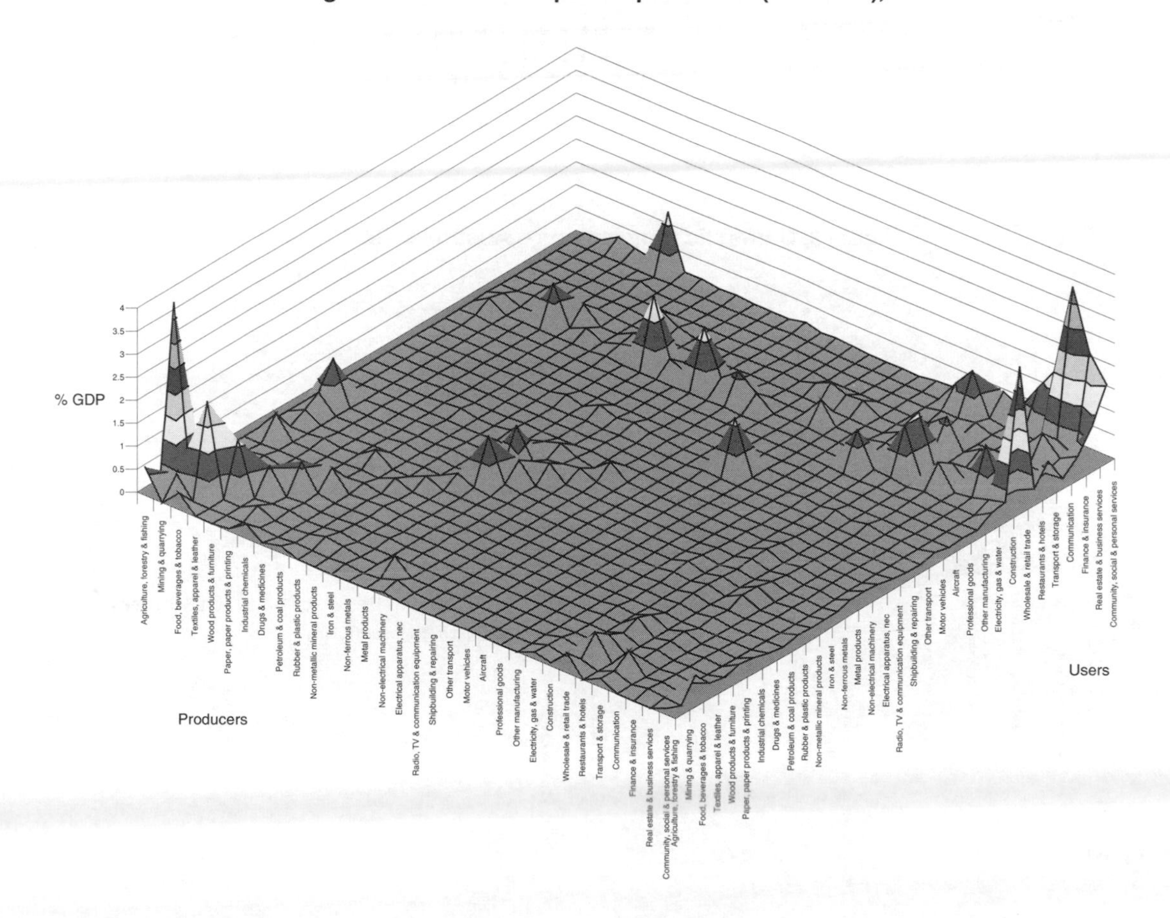

Figure 2. Australian input-output matrix (domestic), 1989

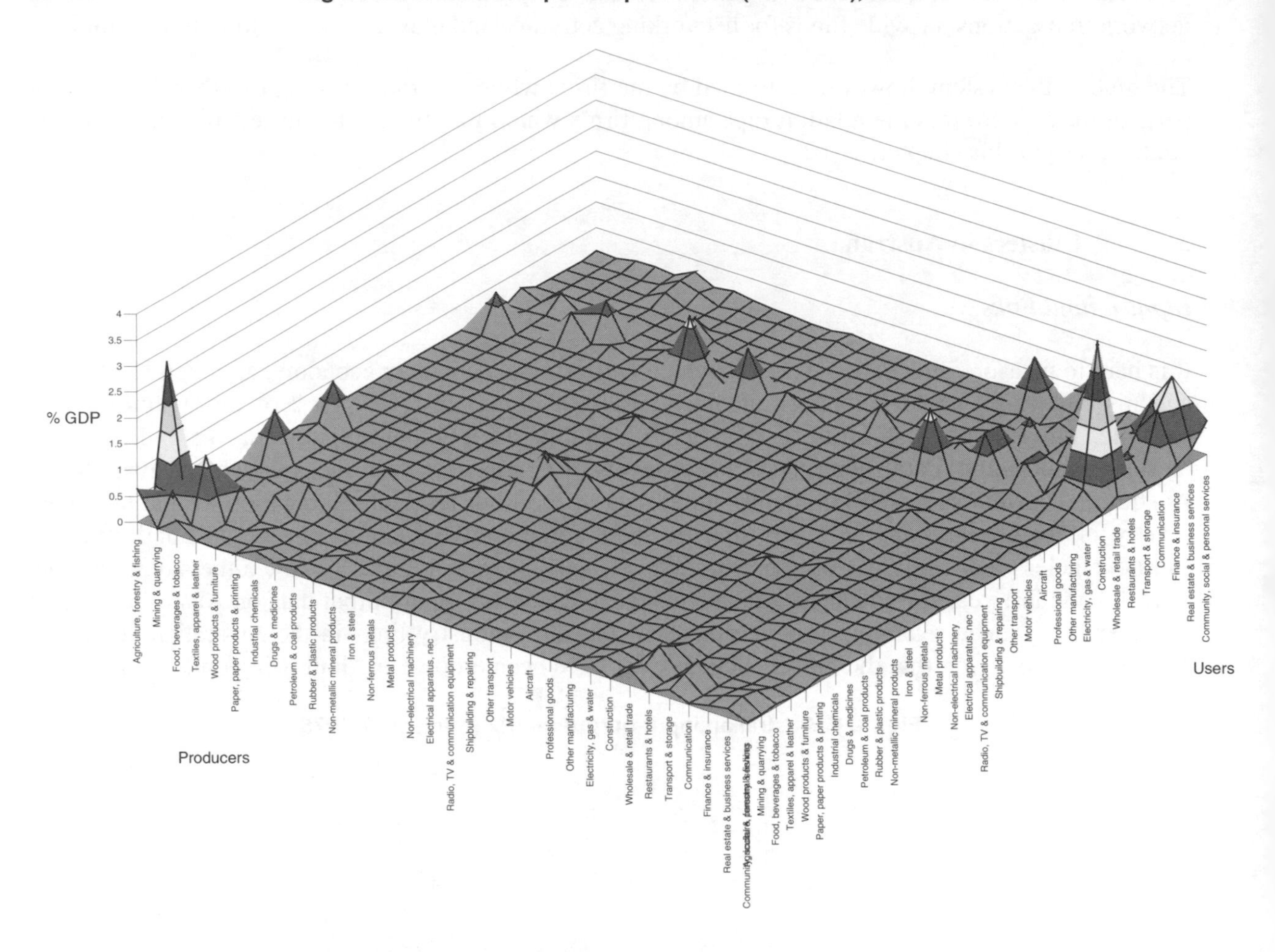

Figure 3. Domestic transactions – Australia, 1993

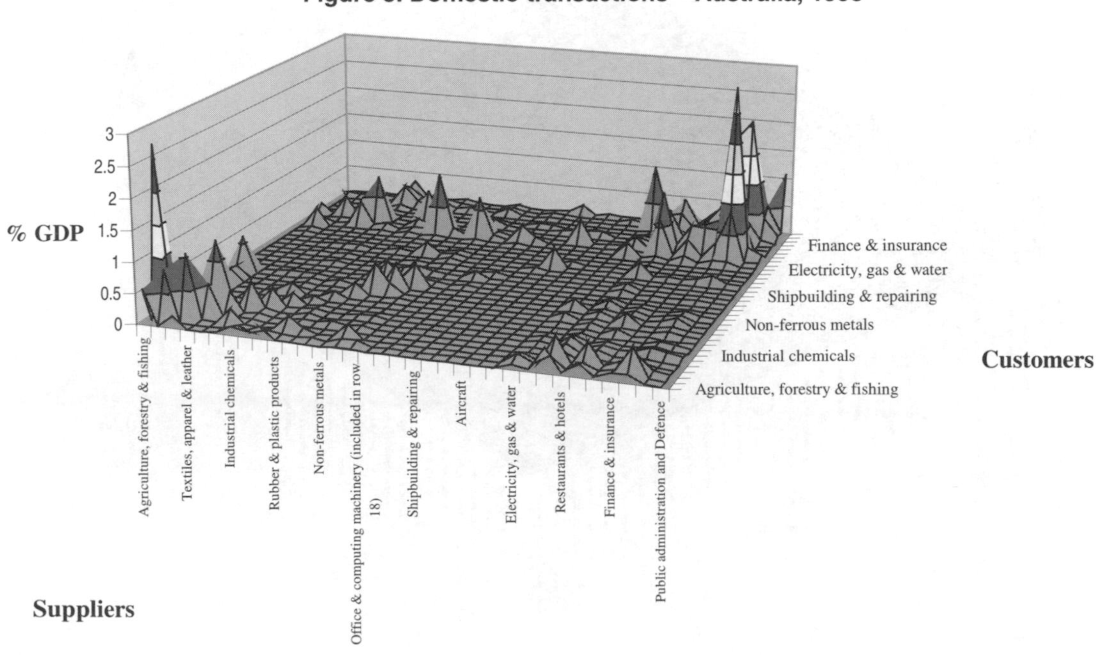

The linkages shown in the above charts are, of course, exchange-based transactions[1] rather than the co-operative linkages implied in the notion of clusters. It is likely, however, that the more intense the exchange-based links, the more likely it is that the economy can also foster co-operative links. In the total absence of exchange-based links, there is virtually no likelihood of other links forming.

Comparison of the charts at the beginning and end of the 15-year period 1975-89 shows that the average density of links across the entire domestic matrix fell to 0.063%, a drop of 11.3%, while the average backbone density became 0.373%, a fall of 15.6%. The data therefore show a reduction in the production and sale of intermediate goods within the economy as a proportion of domestic output and thus a reduction in the strength of the domestic linkages measured within the economy. This reduction was greater within industries than across industries.

Comparing the 1989 and the 1993 data indicates that the backbone over the period 1989-93 was stable, but the overall proportion of interactions in the economy fell a further 6%.

As Marceau, Manley and Sicklen say:

> "Overall the picture... is somewhat disappointing. Firstly, over the period under examination, Australia's export intensity increased substantially yet we see a decline in average linkage density. This suggests that the drive to exports through increasing overall trade intensity has not resulted in a net positive effect on domestic linkages. It also suggests that many of Australia's exporters may not have strong domestic inter- or intra-industry linkages. This may reflect the nature of our exports, being more resource-based with fewer pull-through effects...In terms of the importance of inter-firm connections...for innovation and learning this suggests that a key aspect of Australia's linkage potential may be weakening through a decline in the domestic industrial critical mass." (Marceau *et al.*, 1997)

Comparative data (Table 1) show that Australia has been doing worse in this respect than other resource-based economies such as Canada and Denmark and far worse than Japan in all areas except steel.[2]

Table 1. Changes in input-output densities, various countries

		1975	**1989**	**Change**
Australia	Average	0.071	0.063	-11.5%
	Backbone	0.442	0.373	-15.7%
		1971	**1990**	**Change**
Canada	Average	0.060	0.058	-4.3%
	Backbone	0.313	0.342	9.2%
		1972	**1990**	**Change**
Denmark	Average	0.061	0.060	-1.2%
	Backbone	0.332	0.322	-3.0%
		1975	**1990**	**Change**
Japan	Average	0.078	0.073	-6.5%
	Backbone	0.711	0.622	-12.5%
less steel	Average	0.072	0.070	-3.5%
	Backbone	0.531	0.555	4.5%

Source: Calculations from the OECD Input-Output Database.

Australian clusters

Australia seems to have very few functioning clusters, no matter which definitions are used. The input-output section above gives some indication of why this may be so, why so many potential areas of apparently common interest have not been translated by the players into common activities or even common policy demands. If clusters of firms and networks of inter-firm relations are the glue which holds national systems of innovation together and makes potential synergies actual, then it is important to realise that some features of the surrounding economic structures may militate against the development of true "cluster" collaborative and competitive relations. I return to this later in the chapter.

Geographical co-location

There have been several studies in Australia which have attempted to identify clusters. Almost all emphasize the aspect of geographical propinquity. Thus, for example, there is said to be a cluster on the North Shore of Sydney Harbour which brings together firms in the information technology (IT) sector. This co-location is eased by the presence of Macquarie University and a nearby industrial park. The North Shore is principally a residential area and the firms have largely grown up intermingled with the houses of their founders. The area is particularly beautiful and enjoys great amenity, being close to the CBD and to emerging office areas as well as to a university and sections of the Commonwealth Scientific and Industrial Research Organisation (CSIRO) and to several major medical facilities.

This is the area where Australia's major medical technology company, Nucleus, with its subsidiaries, Telectronics, now the world's third largest maker of heart pacemakers, and Cochlear, the world's largest maker of implantable hearing aids, is located. Nucleus is now heavily involved in a new form of organisation called a Co-operative Research Centre (CRC) which brings together public sector research (PSR) organisations and companies to develop new products in fields such as cardiac technology. This kind of industry depends greatly on easy access to appropriate IT products such as semi-conductors of the right size and reliability. This cluster seems to have been led by the availability of scientific knowledge, skilled labour and the presence of customers (surgeons) as well as the physical attractions of the place.

This arc has been described as follows:

> "The strip of industrial land between North Ryde and Lane Cove has...developed into an important industrial zone which houses the headquarters, commercial and warehousing space for a cluster of high-technology industries, particularly in electronics and computing. Macquarie University continues to expand its influence on local economic activity, linking in closer [sic] with high-technology developments." (*National Economics*, 1997)

How strongly the companies located there operate as a "cluster" in a stronger sense than simple co-location is not clear, however. Interviews conducted by the author concerning location decisions for some of the firms involved suggest that many are located there more for the physical amenity including the provision of right size property and the prestige which now attaches to location in Ryde, than to anything which relates to contacts among firms located there. This is not clearly a cluster in the sense of member firms either using a common resource base or being involved in local client-supplier relationships.

Other studies of New South Wales have located further sets of activities which have been categorised as clusters. These include some which involve local farmers reaching new arrangements with multinational food processors. Others have been identified in the older industrial areas around Sydney – Newcastle to the north and Wollongong to the south, both steelmaking towns dominated by the same firm, BHP.

In the service sector, analysts have identified finance and business service clusters in the central business district (CBD), freight and warehousing in the broader central city region and transport in Western Sydney. The second CBD, Parramatta, is also home to a cluster of legal and accounting firms, encouraged to re-locate there when some agencies of the New South Wales State Government moved to the Parramatta region some years ago and now servicing both government and private businesses in the retailing and hospitality fields as well as in transport and warehousing. Blacktown, also located in Sydney's west, has emerged as the centre of a cluster specialising in the distribution of food and consumer electronic products.

There are also thought to be some manufacturing clusters in Western Sydney, including food processing and some metals. In addition, there could be a small high-tech belt centred on telecommunications and electronics lead firms such as Alcatel, which makes optical fibre cables there, and Philips. There are also some small medical device firms located in purpose-built premises in an area adjacent to the western fringe of the inner city.

Given that there is now emerging an "Executive Belt" of prestige residential accommodation in the Baulkham and Pennant Hills area not far from Parramatta, it may be that these clusters will grow because they will find locally both the right premises and skilled personnel. Transport links to the area are good and improving fast since the zones are not far from the Olympic sites. They may also be encouraged by the construction of Sydney's second airport which may go ahead in the foreseeable future. At present, however, the same reservations concerning their functioning as true clusters apply to the groupings described above as to the North Shore sites discussed earlier. The companies may be co-located and their location may indicate some common resource needs. The density and intensity of the relationships between the firms operating there, however, seems to be limited.

My 1996-97 study of the biomedical industry firms located in the Hills District, for example, indicated that they were not there because they had any synergies with other firms and their decision to locate there had been taken quite independently of the presence of others (Marceau, 1998). Indeed, while the firms studied were all in the biomedical device industry, they were in such different segments that in most cases they had little in common except the hospitals which housed their major local clients. Similarly, there were few supply chain links. Location decisions had much more to do with proximity to the residences of the founders of the businesses or their successors than with anything else.

Similarly, recent work in Victoria has shown 120 multimedia companies co-located in an arc of Melbourne suburbs (Scott, 1997) but the degree of their interaction is not yet clear.

Because of its importance, geographical co-location is a theme which runs through all the following sections.

Natural resource-based clusters

Policy makers have often looked longingly at the rich resource base of the Australian economy and hoped to be able to develop strategies for adding value to natural riches. In most cases, they have had little success but it does seem that there are two resource-based clusters which may genuinely have enough commonality in their resource needs to function as clusters, at least for some time.

The first is the wine industry. Australia's success in the wine industry is too well known to labour here but the reasons why it has been a success have something to do with issues of interest to work on clusters. Until recently, most wine producers in Australia have been small and geographically concentrated in pockets. Although the wine districts have been widely separated in geographical terms, the producers have recognised their common interest in wine technology and oenology. Two education establishments have provided common training facilities, one in New South Wales, but near the Victorian border, and one in South Australia. These two institutions have developed world-class education and training for winemakers and, in conjunction with the horticultural research undertaken by the CSIRO, created the conditions for both a high-level skill base and a high-tech industry in which there could be many players.

In recent years, however, several developments have seen the dilution of the sense of common purpose of the industry. On the one hand, the number of wine growing areas has increased very considerably, now extending from Queensland in the north to Tasmania in the south, from New South Wales in the east to Western Australia, several thousand kilometres away, bringing in many new players who may have little in common with the "pioneers". At the same time, the industry has been the subject of a concentration movement which has seen several foreign multinational and local conglomerates, *i.e.* non-specialist firms, take over the biggest local winemakers and impose significant changes in the industry. In many areas it is no longer clear that the industry operates as a cluster anymore except that wineries are co-located and share the same resource base. Instead of interacting locally, many firms' activities take place outside the district, their markets are essentially both overseas and national, and they do their own marketing and training.

The second rural-based cluster deserving mention is the equine cluster which is emerging in the Lower Hunter Valley north of Sydney and next to the main New South Wales wine growing area. The equine cluster takes advantage of Australian interest and skill in anything to do with breeding and raising horses. The country lends itself to studs, provides excellent residential facilities and uses the local research base provided by the CSIRO and other agricultural research agencies. The co-location of several horse breeders has not remained simply that: the owners of the studs and tracks have come together to offer training to local and overseas clients in the industry and have co-operated to persuade the State authorities to provide a new TAFE (Technical and Further Education) college which offers training in all areas related to the equine industry and courses for foreign students.

Rural and industrial combinations

There is one example in Australia of the potential development of a new industry on the basis of a natural resource. This is the magnesium industry in Queensland.[3] Australia holds 70% of the world's deposits of the valuable cryptocrystalline magnesite, which can be transformed into magnesium metal for use in the automotive industry to reduce the weight of engines and hence the power needs of cars and hence again to lower pollution levels. It so happens that the major deposit is in Queensland, is close to the earth's surface and is located close to a deep-water port and to good rail and road links. The nearest city, Gladstone, is a major industrial centre, housing much of Australia's chemical industry, and hence a source of labour. It also so happens that the deposit is owned by one person who is determined that the value adding will be done in Australia rather than overseas as is so often the case. He has therefore worked closely with both state and federal governments to develop the site, and with the CSIRO to provide the necessary science and technology. He has also carefully planned the development of the magnesite into four different product lines. The most technologically advanced of these is magnesium metal. This is thus a very interesting case for Australia.

A number of lessons may be learned from this experience. It has become clear, for example, that in the case of the metal product, not enough work was done to link closely with users to work out what they actually needed and to have them join in development. Several times, the production plans have come close to fruition and several times problems have arisen. It now seems that the way may be clear for the creation of this industry. If the industry does happen, it will be on the basis on a highly unusual *innovation network* and, because the original company was able to take advantage of facilities provided for a cluster of other industrial activities in the same location, it will be the result of considerable government intervention in both research and finance arrangements.

A modern approach

South Australia has perhaps been the state most active in seeking to identify clusters and then to devise ways of further developing them. In an interesting example of international "technology" transfer, the Employers' Chamber of Commerce and Industry of South Australia, the state government and the Multi-Function Polis Development Corporation, have been employing an American firm called Collaborative Economics to develop a joint venture Silicon Valley approach to development in South Australia.

This partnership has introduced a new model for economic development in the state. The aim is to identify and develop existing and potential clusters. To date, cluster development initiatives have been launched in the defence and advanced electronics and multimedia fields and several more are in preparation. The latter include a Spatial Information Industry project and a Water Management cluster linking the big water and water quality players in the state. The group is now thinking of developing a "cluster incubator" to bring together more scattered enterprises (for details of these projects, see newsletters put out by the relevant business organisations and the state government).

A city network

Sometimes it is hard to distinguish networked firms which collaborate together from a cluster of related firms. The example which follows is a firm which is clearly a network and yet which also seems to be a cluster operating in the classic collaborate-and-compete manner.

Technical and Computer Graphics, TCG,[4] was founded by four computer specialists in 1971. As Mathews (1992) says, since then TCG has become one of Australia's most significant innovators in the fields of portable data terminals, computer graphics, simulators, barcoding systems, electronic data interchange, electronic identification tags and other applications of information and communications technology. Mathews goes on:

> "But TCG is no ordinary company. In fact, it is not a company at all. Instead, it has grown as a dynamic cluster of small firms, each specialising in certain operations, and each one feeding work to the others. Today it consists of an operating core of 24 companies, organised as a co-operative network or cluster but with a market coherence that gives the group profound advantages over its more conventional commercial rivals." (Matthews, 1992)

TCG operates as follows. One company in the group receives an order for data terminals and wants special features tailoring the machines to the supermarket's business. The contracting firm takes charge of the order as a whole and is responsible for delivery. It does not, however, conduct all the work but subcontracts it out to the different specialist forms of the group. In this way, one firm may do the software engineering and maintenance, one the manufacture of the terminals, subcontracting the manufacture of the components to companies outside the group as appropriate. Another company in the group may take care of any telecommunications areas of the order. The customer deals only with one firm which in turn contracts all aspects of the order to partner enterprises in the group and through them to others elsewhere in Australia.

Mathews says that the TCG group operates with minimal rules and maximum flexibility. Membership of the group is voluntary, and some companies have indeed left while others have joined. What makes the group of firms involved a cluster may be the fact that the companies are expected not to engage in head-to-head competition with each other but to specialise in their own market niches. This underwrites the co-operative relations and free exchange of information which characterise intra-group activities. Members make most profits from contracts with firms outside the group, relying on the group for a constant supply of work, thus removing some of the commercial risk each would otherwise face.

The overheads for the group as a whole are very low. Most business functions are carried out by the firms themselves and there is no central bureaucracy. Accounting for all is done by a member firm, R&D by another. Invoices, however, are issued centrally and payments made centrally by the remnant of the initial firm which has an equity stake in each member, providing linkage via ownership (Mathews, 1992).

As a result of its collaborative arrangements, the TCG group is:

> "a successful new product developer in the high-tech area...It skilfully leverages its development through the triangular networking method [linking client, partner firm and TCG], allying itself with larger partners and customers, and using the public policy framework as a catalyst." (Mathews, 1992)

Innovation networks

In many instances in Australia links between firms are stronger in the arena of innovation than in the arena of production. A recent study of the biomedical industry in Australia (Marceau, 1998) indicated much stronger relationships between firms, public sector research institutions and clients such as surgeons than between firms involved in the production process. These links were strengthened for a few firms by participation in CRCs, but most seemed to feel that the user-producer relations they established with clients were sufficient. They had few links with other firms in the same industry and thus even when they were co-located or closely located they did not know what the other firms were doing. Indeed, in one case where there are only two firms in the same segment in the country there was no contact at all: they simply competed. Links with supplier firms, moreover, were usually relatively weak and there were few partner firms and subcontractors who shared the design and production tasks. This lack of overlap between innovation and production networks is perhaps one reason why there are so few clusters of inter-related activity in most industries in Australia.[5]

The consequences for Australia of a patchy industrial structure

The discussion of the input-output picture above indicates one possible reason why Australia is not developing more clusters. The industrial structure is extremely patchy which means that many firms are unable to find the partners they need to develop innovative products and processes. In particular, too, the Australian industrial structure is biased towards the low-tech end of the spectrum so that companies in higher-tech fields find it even harder to find partners and develop complementary activities. Almost all machinery, for instance, has to be imported.

There are further factors which are likely causes of the lack of local clustering. One is the concentration of significant amounts of economic activity in the hands of only a few players. Industrial activity in many critical areas essentially takes place in vertically integrated large businesses. In Australia, only 0.4% of businesses (those employing more than 200 people or with turnovers of more than AUD 200 million) employ nearly 38.5% of all workers, make 64.3% of all capital expenditure and produce 49.7% of gross industry product. These firms earn 61.1% of all pre-tax profits and pay higher wages (they employ 38.5% of all workers, but pay 49.5% of wages). There has been a clear tendency for the smaller firms' futures to be linked not to each other's but to those of the giants in most sectors.

Most of these large firms are part of overseas-based multinational enterprises. There has been much controversy about whether dominance by overseas-based firms has adverse effects on local economic activity. Without buying directly into that argument, it should be noted that there is evidence that multinationals usually still undertake the greater part of their research and development activities at "home", although this proportion may be declining a little (Patel and Pavitt, 1998). This may help explain why in Australia business conducts relatively low levels of R&D (Marceau *et al.*, 1997) and in 1997 after changes to the tax concession rules was actually dropping. In the course of the author's biomedical study, she was told several times that any R&D ideas would be passed back to headquarters overseas and would usually be carried out there. Not carrying out R&D in Australia creates a situation where links between firms and education and training institutions are not deep or frequent, thus creating gaps in the innovation circuits on which innovative firms depend and making it harder for governments to find levers to improve national industrial technological capability.

The dependence of local firms on overseas headquarters may also explain why empirical studies in several fields indicate that, while firms may co-locate, they do not co-operate to any large degree. Industrial firms have a presence in Australia but have a rather more significant link directly back out of the country. The knowledge circuits follow the ownership links and the chains of production. I have already quoted evidence from my own study of the biomedical industry to indicate the lack of cp-operation between co-located companies. In some cases the issue is local policy blindness and real clustering of co-located firms has received little encouragement from local public authorities who make little effort to help them build links with local R&D institutions, for example. Mathews and Weiss (1991), in their study of the textile industry in Australia, similarly noted that there was some concentration of activity in small towns in Victoria, but that this did not imply any co-operative effort. Mathews and Weiss say that:

> "The irony of this situation is that the textile and clothing industries...have traditionally clustered together – but until recently have made little or no effort to achieve any competitive advantage through this clustering. In Victoria, wool producers in the Western District have fed woollen-based industry clusters in both Geelong and Warnambool. In Warnambool, for example, a proto-industrial district has existed for most of this century, with woollen mills, blanket mills and clothing firms all co-located. Yet Warnambool has no resonance in Australia as a 'centre of

excellence' in woollen textiles nor as an 'industrial district'. The Warnambool City Council has no policies to encourage networking between firms located in this vicinity...The Warnambool Institute of Advanced Education does not conduct any courses in textile technology or in textile-related areas." (Mathews and Weiss, 1991)

The same authors continue by looking at the clothing industry which has congregated in inner city areas of both Sydney and Melbourne. In Sydney, they say, "little has been done...to develop any kind of co-operative awareness of the advantages of sharing resources. In Melbourne, the city authorities are doing rather better at recognising the clothing cluster but for many years at least were still apparently unable to see the benefits of promoting clustering and networking between textile firms. This had the result that, although the now Melbourne Institute of Textiles and the National Spinning Skills Centre are co-located with firms in the City of Moreland, there was until recently little contact and no complementary activities such as encouragement for bringing dyestuffs and fabric printing into line with the needs of the textile industry" (Mathews and Weiss, 1991). Since that study, Moreland City Council has become more aware of the importance of its TCF firms and of encouraging links with R&D institutions. According to its submission to the 1997 national inquiry into TCF, the Melbourne Institute of Textiles has developed dynamic links to industry. If the Institute were to move, the Council claims, it would reduce the opportunities for local firms to have conducted the R&D needed to assist TCF competitiveness there. This suggests that the firms in Moreland may be indeed beginning to cluster and are no longer simply co-located.

The examples given above also indicate that industry attention has not been on the provision of collective infrastructure such as specialist training schools of the type which exists in the Nordic countries for a range of industries and which provide, for instance, design facilities. There has been little pressure to develop specific industry-oriented research facilities, or test laboratories such as that which the Danish government provided to the nascent wind turbine industry, enabling fast progress towards commercial viability of different turbine designs (see, for example, work by Peter Karnoe published in the early 1990s).

Instead, and perhaps fatally, too much attention has been paid both by firms and by industry associations to links with government which focus not on industry development but on the levels of protection which will be applied and the rate at which they will be reduced. This dependence on government action for an essential component of their profitability has developed a mentality among leading firms which does not encourage the creation and nurturing of horizontal links with partner firms. Only the more hierarchical linkages have been formed and these have been strengthened recently. Thus, for instance, in the auto industry, the belated shift to a "Toyotist" model of relationships between clients (core firms) and their suppliers has meant that there have been some developments in the creation of-long term relationships; these are recent and as yet not well established. Until the very last few years the focus remained on price rather than on design or other forms of collaboration while "partnerships" have really only been developed with larger firms.

Reliance by suppliers on operational policies developed within one large firm makes clusters based on hierarchical relationships very fragile. David Scott has recently indicated experience in the La Trobe Valley near Melbourne which shows that when the large electricity-generating plant operating there was stripped back to its core functions and privatised, the cluster of small firms with environmental clean-up and other expertise related to the operations of generators using brown coal declined. The small firms remaining were not able to continue to develop when the centrepiece of the cluster was no longer there (Scott, 1997). Similarly, in Gippsland, near Melbourne, a group of 20 small companies developed to supply a major oil enterprise serving in turn the offshore oil and gas industry, but their viability was threatened when the petroleum firm moved its headquarters to Melbourne and altered what it expected its suppliers to do. In the case of the La Trobe Valley, the central partner restructured in ways which removed business from the cluster, while in Gippsland the volume of business which came to be required from suppliers under the new company rules was beyond the capabilities of the firms in the cluster.

In short, in Australia, the economy was established in a way which discouraged horizontal relationships between firms who could collaborate and compete. Most national governments have always made the assumption that firms only compete and do not collaborate and have built a policy environment in which firms are encouraged to behave as though they were in isolation rather than engage in activities where collective behaviour to reduce the risks of innovation could make commercial sense. Only in 1994 was a policy adopted at federal level (some states had been earlier) which specifically encouraged the creation of networks – although this does not necessarily mean clusters. Borrowing from a Danish programme, Australian public authorities instigated a programme of "enterprise networks" through which groups of companies pooled resources and skills to lift their international marketing profiles by collective strategies which have an impact beyond what individual firms could have achieved alone. That policy was terminated in 1998.

Some positive results were achieved by the Business Network Programme but the programme did not have sufficient scale to have great impact. It may have begun, however, to alter firm expectations and to encourage the search for provision of collective infrastructure. An example taken from the pilot programme, for example, showed several innovative schemes for the provision of collective infrastructure and for bringing together players who would not normally "talk" to one another. Thus, for instance, an article in the *Business Review Weekly* (Rowbotham, 1991) reported that the pilot networking scheme had given assistance to a grouping of membrane and biotechnology companies. The network's research director is reported as saying:

> "By getting them to work jointly we will spread the risk of what will always be a high-risk exercise. An additional advantage is that it brings into a close working group a variety of talents and skills that would not otherwise talk to each other. Diagnosticians do not normally talk to microelectronics engineers." (Rowbotham, 1991)

The networks built through the state and federal business network and related programmes have also suffered from the fact that participating firms were often working in an arena dominated by a very few large firms. This is true, for example, of some Business Network Programme-created networks in the toolmaking field which I studied in 1998. These networks were trying to assist small firms to enter new markets. The reason such change was needed was that the firms were extremely dependent on decisions on new models made by the automotive firms. In the auto industry there have traditionally been many small suppliers co-located with their clients in one or two cities. It is clear, however, that while the Australian supplier firms may see their market as the local auto firms, the auto firms see their suppliers as located anywhere in the world. The clients thus now look over the heads of the smaller supplier clustered firms to areas further afield which are perceived to have greater capability to respond to new technological needs and demands for improved speed on delivery and quality. In this industry the creation of networks is very unlikely to ever be enough to counteract such client visions of their needs and their power to go elsewhere (Marceau, 1998).

4. Policies and policy practices

The case of Australia has indicated that some countries have not developed the clusters of activity which could make them strong international players or even strong local co-operative activity of a networked kind at all. While in the past some analysts may not have "seen" the clusters present in economies because they believed that the hierarchical organisation of the vertically integrated firm was the way in which all economies would develop (see the critique by Piore and Sabel, 1984), re-analysis has suggested in many countries that there have always been and are areas of the economy in which networks of small and medium-sized firms cluster together and operate in a mutually interdependent manner. Such clusters have been described in Denmark, Germany, Finland and elsewhere.

In Australia, in contrast, it seems as though there has indeed been very little development of networks and clusters. There are many areas which could be described as "potential" clusters, such as the networks which could develop between hospitals and biomedical firms and transform co-location into active cross-firm collaboration, but so far there has not been the catalyst to transform potential collaboration into the actual working together which makes co-located firms into networks of inter-related enterprises and the enterprises into clusters.

This chapter has attempted to describe those areas in which there is now some potential and some actualisation of clusters in Australia. It has also attempted to indicate some of the reasons behind the lack of spontaneous clustering in the Australian economy. Some of these are social and historical, notably the nature of Australia as an immigrant society and a rapidly developing urban fabric which has not engendered many longstanding communities.

Some of the explanation seems to lie in the structure of the economy and the ownership of major firms outside the indigenous business community which mean that lines of production and circuits of knowledge tend to be drawn vertically rather than horizontally. Some of it probably lies in the public policies developed to protect Australian industry and which have tended to lead firms to rely greatly on their relations with government and hence their ability to agree on levels of protection rather than to encourage them to turn towards other firms and hence to the collective development of an industry. Public policies developed for one set of industrial reasons, notably import replacement, have, given the set of players, thus tended to militate against the development of "alternative" organisational forms in the production arena.

Current public policies which have a more developmental focus may have some impact on the number and extent of the collaborative relationships which firms establish among themselves. It would seem desirable, for instance, to encourage firms to come together to discuss what the collective needs for the development of their industry are, and to assist governments to devise suitable approaches. In many fields there are potential clusters but firms are not in the habit of turning towards each other. In addition, educational institutions focus especially on individual students and their needs rather than on the needs of the enterprises around them. Many students will expect, once educated, to move to positions in firms located anywhere rather than to having any expectation that they may work in their local region. There is rarely any sense of the need to develop regional institutional links and few if any coherent policies which bring together the major institutions of the national system of innovation.

In other words, while in Australia most states have created regional development organisations of diverse kinds aimed at assisting local development, there is rarely any articulation between these and little attempt to adapt policies created at national level to particular regional or sectoral needs. There is a major problem in the minds of many policy makers, especially at national level, with making the distinctions often needed between what will assist one industry rather than another. Thus tax concessions are aimed at individual firms and the R&D they each undertake. There are no longer any incentives for sharing costs and benefits and hence for changing from co-location into real clustering and interaction on the lines of an industrial district.

Policies to encourage such a shift would nonetheless seem especially important in Australia because of the distances between markets, between clients and suppliers and between firms and the centres of educational, R&D, training, testing and other facilities. Special mechanisms could be developed to assist firms located in Adelaide to link with their counterparts in New South Wales, Victoria or Queensland, for instance. Perhaps with the advent of better access to IT communications systems, the needs for linkages apparent at state levels will permit real interaction across state lines and over the large distances concerned. Governments at all levels should be encouraging such interaction and the development of strong collective infrastructure. This is especially critical in Australia because the scatter of firms and their small size relative to the distance of their markets.

Marceau *et al.* concluded in 1997 that Australian policy makers need to find ways to ensure that three critical developments take place:

- First, given the structure of the economy, they have to ensure that the biggest private sector players are also the best – large corporations must not be allowed to get lazy because we depend too much on their activities.

- Second, they must find ways to ensure that new players can enter the arena. There must be specific policies to encourage the entry of the new and innovative firms needed to keep the economy moving forward in the new century.

- Third, they must develop policies which can help smaller players to achieve the scale they need not only to undertake the R&D and other innovative activities which will underpin their competitiveness but also to reach key markets and serve major customers both at home and overseas.

Developing these three sets of policies together could provide the impetus to cluster building which is currently lacking either in a systematic manner or on the scale needed. These policies require close collaboration between state and national policy makers and a willingness to bring together players who have not worked at all or not worked well together in the past. This is turn requires public sector co-ordination capacity and policy understanding skills beyond those needed simply for programme implementation. This in itself requires a major rethink of current Australian federal policy trends.

In short, policies need to be developed at different levels. Some must assist firms to find others, whether as suppliers or clients, quite a distance away so as to encourage "virtual" clusters. Others must teach firms to look for collaboration in their local region and encourage them to devise collective development ideas, as has been done with success in some areas of Canada (Quebec) and the United States (see relevant chapters in this volume). Only then will Australia have a real chance of realising the opportunity to capitalise on its good science and develop the institutions and links needed in the emerging learning economy.

NOTES

1. I owe these charts to the work of my doctoral student, Brian Wixted. The analysis is from Marceau, Manley and Sicklen, 1997, Chapter 7 and Brian Wixted's own work.

2. More recent comparative data are not immediately available.

3. This section is based on work carried out by my friend and student, Wendy Riemens, as part of the preparation of her doctoral thesis at the Australian National University. Further details may be obtained from her at the Department of Sociology, Flinders University in Adelaide, South Australia.

4. TCG was studied by John Mathews of the Industrial Relations Research Centre at the University of New South Wales in the early 1990s. The account given here depends entirely on that study.

5. Further details of this study can be obtained from the author.

REFERENCES

Burt, R. (1992), *Structural Holes: The Social Structure of Competition,* Harvard University Press, Cambridge, Massachusetts.

Camagni, R. (ed.) (1991), *Innovation Networks: Spatial Perspectives,* Belhaven, London.

Chandler, A. (1990), *Scale and Scope: The Dynamics of Industrial Capitalism,* Belknap Press of Harvard University Press, Cambridge, Massachusetts.

Cooke, P., F. Moulaert, E. Swyngedouw, O. Weinstein and P. Wells (1990), *Towards Global Localisation*, University of College of London Press, London.

DeBresson, C. and X. Hu (1999) "Identifying Clusters of Innovative Activity: A New Approach and a Toolbox", this volume.

Harrington, D. (1991), "Virtual Clusters: A Useful Paradigm for Australian Industry Policy", in M. Costa and M. Easson (eds.), *Australian Industry: What Policy?*, Pluto Press, Leichhardt, NSW, pp. 251-270.

Marceau, J. (1998), "The Triple Helix in National Context: Collaboration and Competition in the Biomedical Industry in Australia", *Industry and Higher Education*, 12(4) pp. 251-258.

Marceau, J., K. Manley and D. Sicklen (1997), *The High Road or the Low Road? Alternatives for Australia's Future*, report for the Australian Business Foundation Limited.

Marshall, A. (1920), *Principles of Economics* (8th ed.), London.

Mathews, J. (1992), "TCG: Sustainable Economic Organisation through Networking", Industrial Relations Research Centre, University of New South Wales.

Mathews, J. and L. Weiss (1991), "A Tale of Two Industries: Textiles in Italy and Australia", Industrial Relations Research Centre, University of New South Wales.

National Economics (1997), "Towards an Economic Development Framework for the Greater Metropolitan Region", draft Working Paper.

Patel, P. and K. Pavitt (1998), "National Systems of Innovation under Strain: The Internationalisation of Corporate R&D", SPRU Electronic Working Paper Series, Paper No. 22, http://www.sussex.ac.uk/spru/.

Piore, M. and C. Sabel (1984), *The Second Industrial Divide: Possibilities for Prosperity*, Basic Books, New York.

Porter, M. (1990), *The Competitive Advantage of Nations*, New York.

Riemens, W. (1996), "Innovation Networks and Public Policy", doctoral thesis, Australian National University, Canberra.

Roelandt, T.J.A. and P. den Hertog (1999), "Cluster Analysis and Cluster-based Policy Making in OECD Countries: An Introduction to the Theme", this volume.

Rowbotham, J. (1991), "Enterprise Networks Draw New Strength from Numbers", *Business Review Weekly*, 16 August, pp. 70-74, 76.

Scott, D. (1997), personal communication.

Chapter 7

THE OTTAWA HIGH-TECH CLUSTER: POLICY OR LUCK?

by

Roger Heath*

1. Introduction

The Ottawa region now hosts more than Canada's federal government: it is home to a growing cluster of high-technology firms active in telecommunications, microelectronics, software, and health and environmental products and services. Although widespread awareness of this emerging employment- and wealth-creating engine is recent, its roots go back at least 50 years to post-war Canada's early ventures in high-technology procurement. This chapter examines the emergence of Ottawa's high-technology cluster and tries to answer three related questions:

- What were the sources of the cluster's growth before it attracted local policy attention?
- Once started, how successful has local policy been in consolidating and accelerating the cluster?
- What are the general lessons from the Ottawa experience that might apply to other local clusters?

Given the scope of these questions, this note must paint a broad picture and depend mostly on secondary and informal material.[1] This is a common dilemma with local studies. Local surveys are relatively costly, and even when the cost is borne, confidentiality requirements limit response and analysis. Moreover, since many of our concerns are with earlier years, we must make do with whatever data was then collected. On the positive side, hindsight, coupled with the cluster analysis framework, adds clarity, but, overall, our opinions are only tentative, and it is hoped that readers will consider this essay an invitation for further discussion.

---

* Recently retired from Industry Canada, the author is now researching the Ottawa cluster under contract to the National Research Council. He would like to acknowledge the assistance of many people in the Ottawa region and, in particular, would like to thank Guy Steeds and Arvind Chathbar for their support.

2. Ottawa's high-technology cluster

In terms of geography, Ottawa is Canada's capital and a city in the province of Ontario. It lies north of the route linking Montreal and Toronto. By car, Ottawa is two hours from Montreal and five from Toronto. The Ottawa region contains several proximate municipalities and rural areas in both Quebec and Ontario[2] – all far enough away from other centres for Ottawa to be a distinct economic region.[3] The Ottawa Census Metropolitan Area (CMA), roughly the same as the region or cluster, compares as follows with other Canadian cities readers may recognise (Table 1).

Table 1. The Ottawa cluster – A comparison[1]

CMA	Population	Labour force	% University degrees	Per capita income	Commute time (minutes)
Ottawa-Hull	1 030 500	584 200	23	$21 800	20.1
Toronto	4 444 700	2 246 500	19	$22 200	20.7
Montreal	3 359 000	1 785 900	15	$18 500	19.1
Vancouver	1 891 400	940 900	16	$21 500	25.0
Edmonton	891 500	476 300	15	$18 900	22.1

Source: Data from Statistics Canada, mostly compiled by the Ottawa Economic Development Corporation (OED), March 1998.

In Canadian terms, Ottawa is a mid-sized area, relatively affluent and, at least until recently, economically stable because of the federal government. Education levels are relatively high, and its commute times, an indicator of density, appear average. In 1997, a local survey of high-technology companies found:

Table 2. The Ottawa cluster – High-tech quick facts, 1997[1]

Employment	48 312
Companies	815
Enterprise R&D	$1.7 billion
Revenue	$8 billion
Exports	$5 billion

Source: OED, March 1998.

Employment in the federal government, Ottawa's traditional main employer, was about 72 000 or about 1.5 times high-technology employment, but the gap has been closing fast and the two sectors may be equal employers by 2001.

What do we mean by high technology? The region's core competency lies in telecommunications, software and microelectronics. In addition, the region has growing expertise in health and environment/energy products and services. This is reflected in the following table of local high-technology firms that are currently listed on various stock exchanges (Table 3).

Table 3. The Ottawa cluster – Listed companies

Company	Product class	Local employment
Agiss Software Corp	Soft,it,tel	20
AIT Corp	Def,soft,tel	130
Calian Technology	Tel	400
Canadian Bank Note Ltd.	Printer	?
Canadian Marconi	Equip,tel,comp	250
Cognos Inc.	Soft	710
Corel Corp	Soft	950
Crosskeys	Tel	260
DY 4 Systems Inc	It,def	200
Haley Industries Ltd	Manufacturing	?
IDS Intelligent Detection Systems	Robotics	75
I-Stat Corporation	Automation, chemicals	270
International Datacasting Corp	Telecom, IT	40
JDS Fitel	Photonics, testing	1 600
Jetform	Software	475
Linmor	Software	?
Lumonics	Eqipment, IT	225
Microstar Software Ltd	Software	34
Milkyway Networks Corp	Software	50
Mitel Corp	Telecom	1 800
Mosaid Technologies	Telecom, IT	175
Newbridge Networks	Telecom	2 800
NewSys Solutions Inc	Software, telecom	150
Northern Telecom	Telecom, photonics	11 000
Nuvo Networks Management	Telecom	21
Plaintree Systems	IT	120
Res Int'l Inc	IT, software	15
Seprotech Systems	Energy, equipment	30
Simware Inc	Software	100
Thermal Energy International	Environment, energy	14
Timminco	Metals and alloys	?
World Heart	Medicine	70

Source: List published weekly by *Ottawa Citizen*, employment and product data from OED, 1998.

In sum, the Ottawa region has over 800 high-technology firms operating in five technology-led markets (telecommunications, information technologies, software, health, and medicine). The cluster's firms compete internationally.

3. The origins of Ottawa's high-technology cluster

Ottawa's first high-technology company, Computing Devices Canada, began by commercialising a military aircraft position and homing indicator in 1948. Similar ventures followed, *i.e.* Electronic Materials International, Northern Radio, and Mechron Engineering. These companies depended on government procurement, mostly military procurement, and benefited from proximity to the large government research infrastructure that emerged in Ottawa following the War.[4]

Since the 1960s, the Ottawa cluster has been dominated by Nortel, a major supplier of telecommunications equipment and Canada's largest R&D performer. Nortel came to Ottawa after US antitrust authorities severed Bell Canada's technological co-operation agreement with Western Electric in 1956. Since the Bell system then had an assured market in Canada, they subsequently undertook extensive research in Canada and located their laboratories in Ottawa (McFetridge, 1992).[5] Although Northern Telecom now has plants and research in several countries, its core research remains in Ottawa and Northern Telecom still employs almost a third of the total high-technology workers in the region. Northern Telecom rooted the "telecommunications" cluster.

By 1960, we see the emergence of another important cluster characteristic, the formation and growth of new local firms. The following table shows this:

Table 4. The Ottawa cluster – High-technology growth

Year	Number of firms	Employment	New companies
1960	20	2 600	10
1965	35	3 800	18
1970	140	9 400	120
1975	200	18 000	80
1980	300	22 000	120
1985	350	25 000	60
1990	350	25 000	20

1. Less than five years old and still in existence.
Source: Doyle (1991).

In 35 years, the region generated 418 high-technology companies and, even though many have failed, Ottawa's high-technology firms are still predominantly home grown – an unusual finding in Canada.

Much work has traced this fertility to spin-offs from existing firms. In 1988, an Ontario report estimated that the Bell-Northern group had spawned some 50 start-ups. McFetridge (1992) discusses an interesting table from Steed and Nichol (1995) that looks at the previous employment of high-technology entrepreneurs in Ottawa:

Table 5. Sources of high-technology entrepreneurship in Ottawa, 1982

Source company	Type	Percentage
Computing Devices	Foreign-owned since 1956	18.9
Leigh Instruments	Canadian	13.2
Northern Telecom	Canadian	28.3
Telsat	Crown corporation	3.8
	Other firms	7.5
	Federal government	28.3

Interestingly, three medium-sized firms, Computing Devices, Leigh Instruments, and Telsat, were unusually productive and, unlike some other clusters (Route 128), few entrepreneurs were originally university or National Research Council scientists. Ottawa's entrepreneurs tended instead to be engineers. The study showed that local conditions were clearly fostering start-up companies:

> "A final point concerns the emphasis of regional development approaches on promoting indigenous technical entrepreneurship. Ottawa's experience hardly seems to rate as indigenous when one-third of its technical entrepreneurs forming NTBFs (new technology-based firms) were actually foreign born and even went to secondary schools in their country of birth. However, it was the amalgam of local conditions that helped attract them and subsequently contributed to their forming NTBFs."[6]

The importance of local spin-offs has since been confirmed in two separate studies. In each, an extensive genealogy of Ottawa firms traces the previous employment of entrepreneurs (Doyle, 1995) or ownership and takeovers (Pricewaterhouse Coopers, 1998). Both confirm the vigour of small-company formation in the Ottawa cluster.

Until recently, the importance of these developments was largely unnoticed. The only reference to the cluster in a 1976 book on Canadian cities was:

> "Ottawa offers few advantages for heavy manufacturing; there is little prospect of attracting such firms and little effort is expended in doing so. Instead, large electronic, computer and research firms such as Bell-Northern Research are increasing in importance." (Nader, 1976)

At that time, the 10 000 high-technology jobs were almost invisible. Public administration was over 30% of the labour force and there was no thought that the region might lose government jobs (as has occurred through the 1990s). This means that the cluster's start-ups did not benefit from as wide an array of support measures; local, provincial, and national, as they do today. By default, they were spawned by national framework conditions and (unplanned) local circumstances.

The following paragraphs list the main factors behind Ottawa's high-technology firm creation:

- In the early days, the bulk of the federal government's R&D was undertaken in the Ottawa region, along with much of its high-technology procurement. This gave Ottawa entrepreneurs easy access to government technology and information on potential markets. Products that the government would have produced without hesitation during the War were now thought more appropriate for development by private firms. The federal technology establishment seeded Ottawa's high-technology business cluster.

- Starting-up a business in Canada and Ontario is relatively uncomplicated. Commercial and technical risks were great, and finance has always been difficult to arrange; but at least government rules and regulations were relatively simple. In a study that compares the United States with several European countries, Davis and Henrekson (1997) show that company size and (by implication) start-ups, relates inversely to the complexity of government regulation. Ottawa's company formation rates, essential to the cluster's style of growth, would be impossible in many countries.

These two factors were the major contributors to the cluster's start-up. The next development focused high-technology activities on telecommunications and related sectors:

- The location of Northern Telecom's Canadian R&D in Ottawa changed the cluster in many ways. First, it added significant mass, based on different concerns and cycles to the government. Moreover, telecommunications has a variety of special properties that have changed the sector. These include rigorous international competition, proliferating niche markets, and the need to practice alliance capitalism. Telecommunications exhibits natural clustering – it is one of those activities where the need to "be close to the action" so that research results still have a market when they eventually materialise, implies strong spillover benefits from locating in clusters.[7] Core competencies in telecommunications are closely associated with software and microelectronics. In the rest of this chapter, telecommunications, software, and microelectronics are called the "core" sector.

As momentum built, other factors began to impact:

- Through the 1960s and 1970s, the cluster provided novice entrepreneurs with mentors and exemplars. Starting a company was profitable and prestigious, and some people have done it several times. There have been many instances where laid-off employees have stayed in the region and founded new companies.

- As time went on, local institutions learned how to work with cluster firms. One or two major banks learned how to finance high-technology companies profitably. Appropriate business and technical advice began to appear and a variety of essential services like the construction of clean rooms, intellectual property lawyers, technical training, etc., began to appear.

- In a similar way, the region began to achieve a critical mass in its labour force. Some companies began to come here to access this labour force and others found their expansion relatively easy.

- For almost a hundred years the federal government's "National Capital Commission" has been active in increasing the beauty of the city (Nader, 1976). Consequently, Ottawa has a variety of parks, parkways and other features that impart an attractive ambience. These federal efforts have been complemented by local efforts to make the community "family-friendly". An attractive community helps anchor – even attract – the mobile high-technology workforce. When companies close, their workers try to stay even in the face of offers from elsewhere. Thus, many of Ottawa's high-technology firms are located in Kanata, an unusually well-planned and upscale satellite community. An attractive community is an important advantage.

- Ottawa's companies are research intensive and Canada has had R&D tax credits since 1949.

This consolidation stage, of course, continues and with increasing speed. In the next section, we will illustrate cluster synergies in a description of the cluster as it is stands today.

The medical and environmental/energy sectors as more recent cluster foci

These two sectors or recent cluster foci are scarcely a decade old and have a different history. The medical sector drew its technology from local university and hospital research, and from the core sectors. Many firms concentrate on telemedicine or the management of medical information, which make use of core sector skills. The style of company formation and even some of the key entrepreneurs also come from the core sector. The environmental/energy sector is also recent. Its technology base appears to rest in federal laboratories and university research. Again, companies often combine with core sector skills.

Because they are more recent, these sectors have benefited from the current range of local support agencies such as the stronger university-business ties, and the federal laboratories' current emphasis on business support in their start-up phase. They therefore serve to test whether this improved support has increased the cluster's effectiveness as an engine of growth.

4. Ottawa's high-technology cluster today

4.1. Local high-technology companies

These companies form the core of the cluster. Its culture, momentum and success mostly stem from their attitudes and actions. This culture did not arise from the companies alone; policy and local infrastructure played an important role. However, understanding the companies is the key to understanding the cluster.

In this respect, we are fortunate in having access to the Ottawa Economic Development Corporation (OED) database. This database is based on a biannual survey of local high-technology firms and collects information on employment, markets, alliances, activities and R&D. The data reflect the excellent knowledge of OED staff in, for instance, deciding which firms to survey. OED is not a national statistical office, and respondents need not reply. Nevertheless, the current database incorporates 796 firms and more than 49 000 local employees, reflecting the high propensity of local firms to support local development initiatives. We are using data that reflects updates as recent as August 1998.[8]

Local firms means any firm with activity in the Ottawa cluster – firms may be Ottawa only, Canadian, or the branch of foreign companies. Employee size ranges from 1 to 11 000 (Nortel) and mostly reflects only local employment. As expected, most companies are small.

Table 6. The Ottawa cluster – Company size

Size	Frequency
0 - 25	534
26 - 50	113
51 - 75	54
76 - 100	20
101 - 200	40
201 - 300	16
301 - 400	3
401 - 500	3
501 - 1000	6
1001 - 2000	3
2001 - 3000	1
> 3001	1

Only 15 100 (or 31%) of employment is in firms of 1 000 or more employees and 20 110 (or 42%) is in firms of 500 or more. On the other hand, 4 473 employees are in firms which employ less than 26 workers (10%). Clearly, size means that most Ottawa firms do not deliver stand-alone products, but are allies.

The survey does not use SIC codes, but MPG codes.[9] These codes define 20 sectors based on 264 sub-activities or products. Thus, the consultant sector has five service categories: software, hardware, communications/IT systems, consulting intellectual property, and other. Companies may indicate up to three categories. The data thus permit good estimates of company activity (Table 7).

Table 7. The Ottawa cluster – Firms in sectors

Sector	Firms	Employment	Employees / Firm	Relative size
Core sector				
Computer software	301	13 307	44.2	Small
Computer hardware	125	5 131	41.0	Small
Telecommunications	177	11 323	64.0	Large
Sub-assemblies & components	72	13 811	191.8	Large
Photonics	27	1 869	69.2	Large
Factory automation	33	751	22.8	Large
Test & measurement	44	2 075	47.2	Small
Manufacturing equipment	49	2 356	48.1	Small
Defence	63	4 018	63.8	Large
Sub-total	**702**	**4 5451**	**61**	
% total	65.0	79.6		
Medical				
Biotechnology	15	233	15.5	Small
Medical	25	1 279	51.2	Small
Pharmaceuticals	8	500	62.5	Large
Sub total	**48**	**2 012**	**42**	
% total	3.5	2.9		
Environment/Energy				
Environmental	72	1 319	18.3	Small
Energy	35	771	22.0	Small
Sub total	**107**	**2 090**	**20**	
% total	7.8	3.0		
Other				
Market research & development	28	212	7.6	Small
Transportation	30	1 337	44.6	Large
Chemicals	7	304	43.4	Large
Holding companies	7	131	18.7	Small
Advanced materials	13	309	23.8	Large
Consulting services	240	7 594	31.6	Small
Total	**1 371**	**68 630**	**50.1**	

Because firms can report more than one sub-activity in different sectors, the firms-by-sector total is 1 371, 72% more than the 800 firms that responded. This indicates that many firms engage in complex activities. The table groups the activities into the cluster's specialities. We have included Sub-assemblies & components, Photonics, Factory automation, Test & measurement, Manufacturing equipment, and defence in the core sector based on an analysis of individual companies. The bulk of Consulting services, Transportation, and Market research would also be part of the core sector if we began to split activities. Even without these parts, the core sector, with 65% of the companies and 80% of total employment, dominates the cluster. The medical sector has 48 firms and 2 012 employees; while environment and energy has 107 firms and 2 090 employees. About a quarter of the firms and 14% of employees engage in miscellaneous activities. The last two columns indicate the average size (employees) of firms in each sector. In the core sector, the large-firm sectors engage in microelectronic fabrication. (These results hold even when the giants are removed.)

The data also indicate that firms use the cluster's core skills in an increasing range of activities. In part, this can be revealed by looking at the sector and sub-sector data from the OED. For instance, Figure 1 is based on all firms with more than 100 employees:

Figure 1. The Ottawa cluster

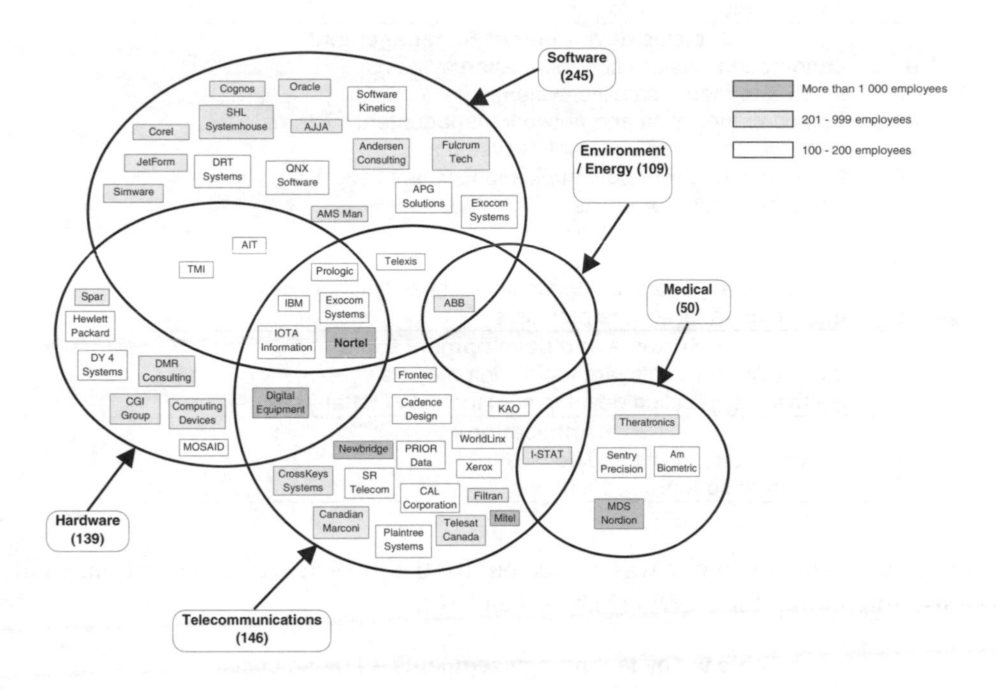

The size of the small circles reflects the size of the individual companies. Companies are grouped into sector circles according to their activities; companies falling into two or three sectors are in the appropriate intersections. One can clearly see the way in which firms are spreading from the core sectors into other kinds of activities by applying their core competencies in new ways. In fact, the exploitation of new opportunities is carried out by: *i)* extending the activity of existing companies, as described above; and *ii)* creating new companies to exploit an emerging niche, but nevertheless employing many people with core-sector skills.

This important point may also be made by looking at an entirely different set of studies recently commissioned by the National Research Council. One such study (Canadian Marketing Group, 1997) grouped the region's telecommunications firms into the following "key technology segments" (Table 8).

Each segment has several local firms and when compared with the firms in similar clusters, this data gives an excellent idea of the cluster's competencies.

Table 8. Key technology segments – Telecommunications and computing

	Technology building blocks
1	Fibre optics and optoelectronics
2	Integrated circuits, boards and components
3	Broadcasting equipment and services
4	Computer & peripherals manufacturing
5	Telecommunications equipment
	Systems development & management
6	Land-based wireless communication systems
7	Fixed and mobile satellite systems
8	Telecommunication and network management systems
9	Defence command and control systems
10	Enterprise/end user communication systems
11	Network security products
12	Robotics and intelligent systems
13	Visual information systems
14	Systems integrators and technology consulting services
15	Internet/intranet service providers
	Systems and development tools
16	Internet/intranet software tools and products
17	Software application development tools and databases
	Applications
18	Multimedia products and services
19	Corporate software applications

The following year a similar exercise was carried out for the life sciences and environmental industries (SOMOS Consulting Group Ltd., 1998) (Tables 9 and 10).

Table 9. Key technology segments – Life sciences

	Health care and medical technology
20	Telemedicine/Remote access health care
21	Health-care software applications
22	High-technology medical devices
	Biomedical and biotechnology research
23	Neurology
24	Cardiology
25	Oncology
26	**Biomedical engineering and contract research**
27	**Biomedical and biotechnology equipment and supplies**

Six of the eight key technology segments in the medical sector (telemedicine, health-care software, medical devices, neurology, cardiology, and equipment) have direct ties to the region's core sectors. For example, telemedicine combines medicine with telecommunications.

Table 10. Key technology segments – Environmental technology

28	Environmental software and information systems
29	**Chemical and physical process technology**
30	Environmental engineering and contract research
31	**Environmental systems, equipment and instrumentation**

Similarly, two of the four key environmental segments (environmental software and environmental engineering) have direct ties to the core sector. In addition to their technological links to the core sector, these firms exploit the core sector's management style, entrepreneurial practice and lender resources. For example, the President of the World Heart Corporation comes from a telecommunication company.

Finally, as one would expect from the region's concentration in telecommunications, software and IT, alliance capitalism is ingrained in local firms. Reflecting this, the OED questionnaire enquires about desired affiliates. Firms can volunteer up to five kinds of alliances. Firms' replies note partners such as R&D, distributors, strategic alliances, joint ventures, licence agreements, etc. The numbers reveal that almost half the firms are seeking to expand their affiliates:

Table 11. Ottawa study – Desired affiliates

Responses	Number of firms
none	441
1	193
2	161
3	94
4	34
5	20

Equally revealing, alliances are so important that companies seldom answer specific questions on current alliances.

The picture that emerges from a study of available data on Ottawa firms confirms the presence of a strong cluster culture. It has significant mass in terms of firms, labour force, research and sales. It has firms that can compete worldwide. Activities are expanding into other areas related to its core skills. It uses relationship capitalism techniques which enable small firms to participate effectively in world markets. Moreover, although the data do not yet show this, many of the more productive types of behaviour are self-reinforcing. To a certain extent, regional firms are spiralling up in a virtuous growth curve.

4.2. The local high-technology infrastructure

At the present time, the Ottawa cluster enjoys a deep, broad public knowledge infrastructure with many kinds of institutions. The following matrix (Table 12) is based on the Belgian approach to measuring the public infrastructure.[10]

There are well over 30 relevant institutes of many types:

- Broker agencies concentrate on building alliances. Ottawa's are extremely effective; partly because they are revenue dependent, and partly because firms increasingly need them to co-ordinate common cluster initiatives.
- Federal laboratories engage in research. Policy now encourages them to form business ties with joint research, contract research and incubator services. We should note that there are Ottawa laboratories in areas where the cluster is not strong.
- Many industry associations are located in Ottawa. CATA is particularly important to local companies.

Table 12. The Ottawa cluster – Public knowledge infrastructure

Agency	Type	Sectoral focus							Policy focus		
		none	core	Tel	IT	S-W	Health	Env	S&T	Dif	App
Canarie Inc.	Broker			x					x		
NRC – Regional Innovation Office for National Capital Region	Broker		x						x	x	x
Ottawa Centre for Research & Innovation	Broker		x						x	x	x
Ottawa Economic Development	Broker	x								x	x
Ottawa Life Sciences Council	Broker						x		x	x	x
Canada Centre for Remote Sensing	Fed Lab							x	x	x	
Canmet Energy Technology Centre	Fed Lab							x	x	x	
Communications Research Centre	Fed Lab			x					x	x	
Defence Research Establishment Ottawa	Fed Lab		x						x		
NRC – Institute for Biological Sciences	Fed Lab						x	x	x	x	
NRC – Institute for Information Technology	Fed Lab		x						x	x	
NRC – Institute for Microstructural Sciences	Fed Lab				x				x	x	
NRC Institute for Chemical Process and Environmental Technology	Fed Lab							x	x	x	
Canadian Advanced Technology Association	Ind Ass		x						x	x	
OPCOM	PrResI				x				x		
Telecommunications Research Institute of Ontario	PrResI			x					x		
RMOC – Economic Affairs	Broker	x									
Ottawa Life Sciences Technology Park	Res Park						x		x	x	x
Carleton University	University		x						x		
University of Ottawa	University		x						x		
Loeb Research Institute	Hosp Rec						x		x	x	
CHEO Research Institute	Hosp Res						x		x		
University of Ottawa Eye Institute	URI						x		x	x	
University of Ottawa Heart Institute	URI						x		x	x	
Institute of Biological Sciences, U. O.	URI						x		x	x	
Federal Development Bank	Ven Cap	x									x
Algonquin College	Voc Train		x								x

- Two provincial research agencies are located in Ottawa. They are not traditional institutes, but mostly work as R&D brokers.
- RMOC (Regional Municipality of Ottawa-Carleton) plays a key development role. They understand the importance of an attractive regional lifestyle. RMOC helped found OED and OCRI.
- Research parks and incubators help start-up companies.
- The two universities are increasingly responsive to cluster needs for training and research, as are the community colleges like Algonquin.
- Hospital research agencies are industry focused and depend on licences and other arrangements for a significant part of their income.
- University research institutes have similar industry ties.
- The Federal Development Bank's (FDB) local office helps fund start-up companies.

The second part of the table indicates which parts of the Ottawa Cluster the different institutions primarily serve. The final section indicates which of the three main policy objectives: knowledge creation, knowledge diffusion and appropriability, the institutes focus on. Both of these sections indicate broad coverage in the region.

Compared to other countries, many services provided by governments are also (or entirely) provided by firms in Ottawa. This includes some of the 44 testing and measurement firms, 14 vendor-certified IT training firms, and many other activities, especially in consulting. Without being dogmatic, one can argue that placing services in the private sector can often be efficient. There is less of a cultural barrier and service providers must be responsive.

Led by the local broker agencies, cluster firms have increasingly ambitious objectives. Examples include the local management of the international airport and the increasing rationalisation of local training. The airport is another example of federal policy, in this case the withdrawal from airport management, having a local impact. However, the local business community influenced the devolution in a way that better served local firms by opening up new routes, for instance, to California, and negotiating US customs examination in Ottawa, reducing travel time to US destinations by hours. Relations with the local universities and community colleges now include planning, course design and co-operative assignments where students work in a firm as part of their studies. Local companies are learning how to achieve common goals.

Lastly, the local cluster also enjoys a supporting periphery. The media is now tuned to cluster developments:

- The major newspaper has a weekly high-tech report and has recruited leading business reporters.
- Two weekly business papers report on the cluster.
- Radio and TV follow cluster developments regularly.

Thus, whenever there is a major economic event, like a new budget, the local media assesses its impact on the cluster. In fact, many of the cluster firms are followed with the kind of attention formerly reserved for sports teams.

The region continues to focus on quality-of-life issues. There are active associations of high-technology workers and a growing, distinct, set of high-technology bars and clubs. All sorts of normal community services like realtors, stores, lawyers, etc., have developed a cluster focus. Feedback loops are emerging. For example, Nortel is piloting its ISDN[11] network in Ottawa because the region has a dense, computer-literate clientele. The cluster draws on more than its firms and its knowledge infrastructure.

5. The Ottawa cluster: policy lessons

Based on the above, how do we answer the three questions with which this chapter began?

To start with, let us summarise the evolution of the Ottawa cluster. In doing so, we will break it out into three periods, review each period's critical success factors, and note any policies which made things happen. In looking at policies, it is important to look at their intent.

Table 13. The Ottawa cluster – Stages of growth

Success factor	Policy/intent
Birth (1947 to 1960) Initial firms are formed	
Expansion of high-tech procurement	Rationalise government
Much government research and procurement centred in Ottawa	Government centralisation then unquestioned
Infancy (1961 to 1985) A distinct sector (culture) evolves	
Northern Telecom centralises research in Ottawa	• US antitrust action • Canadian protection for domestic telephone market
Pattern of vigorous local start-ups takes hold	Permitted/supported by existing regulations and attitudes
Expansion from telecommunications into related core activities (software and microelectronics)	None. The logic of this development is commercial
Adolescence (1986 to 1998) A cluster is recognised	
Market for core sector experiences unprecedented growth	Governments start to dismember telecommunication monopolies (not the only factor)
Cluster (and technology in general) achieves political visibility and becomes a major policy focus	
Local broker agencies emerge	A local political achievement, not a major federal objective
Cluster increasingly learns "how to", first for firms, and then as a community	
Federal laboratories develop business services	Federal technology policy objective
Universities expand business ties	Federal and provincial technology policy objective

Most of the content of this table stems from the previous parts of this chapter, and need not be further explained. Its use, however, is to highlight that in the first two stages, while many policies impacted on the cluster, policy makers were either unaware or uncaring. Only after the cluster had become significant and, incidentally, the importance of the high-technology industries generally recognised, did the well-being of cluster firms become a distinct policy objective. Even then, given Canada's federal structure, federal policy did not focus on Ottawa firms, but the country's telecommunication, software and microelectronic industries.

1. *What were the sources of the cluster's growth before it attracted local policy attention?* Births were accidental in that firms grow based on the existing alignment of resources and opportunities. Although, in hindsight, certain policy initiatives can be identified as important, they were seldom intended to be so. At this stage certain properties of the national innovation system are overwhelmingly important. For example, Canada's R&D tax credits and acceptance of start-ups was important to Ottawa. Infancy, the time when copy-cat firms emerge, operates in a similar policy shadow. Northern Telecom's expansion in Canada was not due to Canadian policy, and its location in Ottawa was partly accidental. Again, in the absence of purposeful policies, sound framework policies

are important. For example, one reason for the recent and vigorous expansion of the Ottawa medical and environmental/energy sectors is the increasing interest in business ties by local universities and laboratories. These ties have been a federal and provincial policy priority for more than a decade.

Clearly, a combination of unintended circumstances and national framework conditions launched the Ottawa cluster. Moreover, due to federal constraints, even now, federal policy towards the Ottawa region remains relatively hands-off.

2. *Once started, how successful has local policy been in consolidating and accelerating the cluster?* Several factors indicate the success of local policies. First, the number, variety and strong ties between cluster firms and the knowledge infrastructure indicate growing and useful interaction. Second, cluster firms are increasingly effective in promoting common objectives with broad returns. Third, the fast growth of the medical and environment/energy clusters may in part be due to their ability to tap into local support. It took years for firms and the knowledge support infrastructure to learn how to work together.

3. *What are the general lessons from the Ottawa experience that might apply to other local clusters?* The cornerstones of policy must therefore be:

   - Strive for sound framework policies so that the greatest number and variety of births can emerge and clusters (local, regional or national) begin to coalesce.

   - Early support for cluster infrastructure, including broker-type agencies. In this way, the support agencies will grow with their firms and thus be ready to assume some of the more difficult, but productive activities, earlier. (I use cluster in a general sense here, not just that of a geographic cluster like Ottawa.)

This policy agenda has clear consequences for policy research.

First, we need to better understand the impact of national innovation systems on firms, sectors and regions. It is clear that national systems profoundly affect clusters of all kinds, but impacts are still too difficult to predict. We do not know the extent to which new sectors can be supported without harming existing strengths.

Cluster support agencies seem to be productive in two ways: assisting cluster firms and fostering cluster-wide initiatives. Appropriate national policy and the identification and promotion of infrastructure activities can encourage these agencies. Members of the knowledge infrastructure could perhaps benefit from their own broker institution.

NOTES

1. This is an essay in the traditional sense of the word. I expect to be studying the Ottawa cluster for some years, over which time I hope to learn more. Here, I try to sketch a complete picture, which means that there are many gaps and instances where opinion and speculation have to stand, temporarily I hope, for fact.

2. In this chapter, Ottawa and "the region" and the "Ottawa cluster" all refer to this aggregation of municipalities.

3. Refer to the Regional Municipality of Ottawa-Carleton (RMOC) Web site at http://www.rmoc.on.ca/Public_Affairs/maprmoc.html.

4. Much of this material is taken from OED, 1998.

5 The reasons why Bell chose Ottawa are not clear. Sweetman notes the attraction of the government's research infrastructure , especially the NRC. In addition, local legend argues that Ottawa was a safe compromise between Toronto and Montreal – an amusing idea since that is partly why Ottawa became the nation's capital in the first place.

6. Steed and Nichol (1985) in a discussion of the same work.

7. Kenneth Arrow made this point in a talk at Industry Canada in 1997.

8. For further information on OED and its data, go to their Web site at http://www.futureottawa.com/.

9. Major Product Group codes are copyrighted by Corporate Technology Information Services, Inc.

10. See Capron, H. and W. Meeusen (1997), discussion document prepared for the meeting of the Focus Group on Institutional Mapping of NIS, 27 October.

11. ISDN is a super-fast Internet connection with speeds more than 50 times current modem links.

REFERENCES

Canadian Marketing Group (CMG) (1997), "Ottawa-Carleton Region: Telecommunications and Computing Industry".

Coopers & Lybrand (1997), "New Media: Ottawa's Emerging Hot Spot".

Davis, Steven J. and Magnus Henrekson (1997), "Explaining National Differences in the Size and Industry Distribution of Employment", NBER Working Paper No. 6246.

Doyle, Denzil J. (1991), "From White Pine to Red Tape to Blue Chips: How Technology-based Industries can Provide Continued Prosperity for the Ottawa-Carleton Region".

McFetridge, Donald C. (1992), "The Canadian System of Industrial Innovation", in R.R. Nelson (ed.), *National Innovation Systems.*

Nader, George A. (1976), *Cities of Canada*, Vol. Two, *Profiles of Fifteen Metropolitan Centres*, MacMillan of Canada.

Pricewaterhouse Coopers (1998), *The Ottawa Techmap*, Pricewaterhouse Coopers, Ottawa.

SOMOS Consulting Group Ltd. (1998), "Ottawa's Life Sciences and Environmental Industry".

Steed, Guy (1987), "Policy and High-technology Complexes: Ottawa's 'Silicon Vally North'", in F.E. Hamilton (ed.), *Industrial Change in Advanced Economies*, Croom Helm.

Steed, Guy and Linda Nichol (1995), "Entrepreneurs, Incubators and Indigenous Regional Development: Ottawa's Experience", unpublished manuscript, October.

REFERENCES

Chapter 8

THE EMERGING INFORMATION AND COMMUNICATION CLUSTER IN THE NETHERLANDS

by

Pim den Hertog and Sven Maltha*
Dialogic, Utrecht

1. Introduction

Information and communication technology (ICT) is regarded as the key technology underlying innovation and change in a great number of industries. ICT is seen as the enabling technology of the knowledge-based economy. For that reason, diffusion and use of ICT by *using sectors* and end consumers is vital. However, ICT is not only an *enabler*, it also stimulates tremendous growth in those industries involved in the production of information and communication products. These firms vary from producers of hardware (computer companies, consumer electronics, including components), through telecommunication operators, cable operators, software developers and broadcasters to publishers, audio-visual companies, other content providers and all sorts of firms that play a mostly intermediary role – *e.g.* the new category of Internet service providers – in providing information and communication goods and services to industrial users and consumers.

However, this group of industries is not generally considered, or at least not recognised, as a cluster in its own right. This is true for the Netherlands, where in the mega-cluster approach, no cluster was identified for "information and communication" (I&C) (Roelandt *et al.*, 1999; Witteveen, 1997).[1] One explanation for this is that the goods and services provided by the industries in this group are generic in character. Not only are various industries involved in their production, but these products and services are used by more or less all industries; therefore no specific I&C cluster has yet been identified. Another explanation is that, so far, it would seem that (vertical and horizontal) networks – including links to knowledge-producing agents and customers – in the I&C industries are either lacking or are not sufficiently developed to justify the "cluster" label. However, at a time when politicians and policy makers are concerned about the competitiveness of the I&C industries and the lack of I&C capabilities, the phrase "I&C cluster" is frequently used without making sure that this cluster exists in economic terms. It is therefore justified to ask whether a functioning cluster does actually exist. In this context, the processes associated with industrial convergence (Section 2) lead us to believe that linkages within the I&C industries are increasing.

---

* The authors can be contacted at Dialogic, Wilhelminapark 20, 3581 ND, Utrecht, The Netherlands (e-mail: denhertog@dialogic.nl). This chapter is based on a more extensive study with the same title (den Hertog *et al.*, 1998). The authors are indebted to Hessel Verbeek (research student, Ministry of Economic Affairs) and Theo Roelandt (Ministry of Economic Affairs) who provided valuable input to the study.

The key question addressed in this chapter is whether a functioning Dutch I&C cluster exists? More specifically, we will focus on the following issues:

- What are the features of the Dutch I&C industries (composition, characteristics)?
- To what degree are the Dutch I&C industries linked and what is their innovation style? What policy instruments are used to improve/support innovation in the Dutch I&C cluster?

To answer these questions, we adopted the following points of departure:

- Firms do not innovate in isolation (Edquist, 1997). Firms, and especially innovative firms, are increasingly dependent on the knowledge and know-how obtained from other firms and institutions with which they co-operate. Increasingly, innovative interactions cross sectoral borders and result in new industrial alliances.
- Clusters are defined as networks of production of strongly interdependent firms (including specialised suppliers), knowledge-producing agents (universities, research institutes, engineering companies), bridging institutions (brokers, consultants) and customers related to each other in a value-adding production chain. Clusters play a role in generating and diffusing knowledge, and thus ultimately in innovation. Although clusters can be identified at various levels, the macro level – the level of the mega-cluster – is taken as our starting-point.
- In the field of I&C goods and services, clustering is exemplified by a process of technological, industrial, and eventually regulatory, convergence (Section 2). This is an ongoing process which serves to increase the availability of channels for the distribution of information, communication and entertainment services. Essentially, it covers the increasing overlap between IT, telecommunications and media.

This chapter attempts to answer the questions posed above using readily available statistics, studies and insights, as well as input-output and innovation survey data. Section 1 presents some of the characteristics of the major building blocks of this cluster, *i.e.* telecommunications, IT and media industries. This is followed by a brief description of the notion of industrial convergence and an introduction to the *grand value chain* of the "I&C cluster" (Section 2). Subsequently, some figures relating to this cluster are presented (Section 3). Section 4 uses I/O data to highlight the mutual linkages between the I&C industries. Section 5 examines innovation performance and the "innovation styles" of the I&C industries as well as the supposed I&C cluster as a whole using 1992 innovation survey data. In Section 6, the policy approach to the I&C industries is analysed, and the extent to which an I&C cluster policy exists examined. Finally, in Section 7, a number of conclusions are drawn.

2. The convergence process

Convergence plays a key role in the clustering of I&C firms and industries. Essentially, the notion of convergence covers the shift away from vertical industrial structures (a specific product supplied through a specific infrastructure using specific terminals), towards horizontal industrial structures (a specific product can be distributed over various infrastructures using multi-functional terminals). The development towards multi-functional I&C networks (technical convergence) fuels the process by which the separate value chains of telecommunications, IT and media industries (see Box 1 for a characterisation of these industries in the Netherlands) are gradually merging into a *grand I&C value chain* (industrial convergence).

Box 1. Key characteristics of the "building blocks" of the Dutch I&C cluster

Telecommunication industry

- Well-developed, but decreasing manufacturing capacity;
- Switch towards systems integration and R&D;
- Dominance of foreign multinational firms;
- SMEs focus mainly on imports and distribution;
- Well-developed telephony-based services (audiotext, call centres) and transaction-based applications (EDI, tele-banking);
- Relatively large number of Internet hosts and users.

IT industry

- High growth in terms of number of firms, turnover and employment;
- Predominance of selling hours (secondment, outsourcing) and unique events such as the Millennium and the introduction of the euro; lesser focus on product development and innovation;
- Small niches of relative strength (ERP, financial transactions, trade & distribution, Web design);
- IT firms increasingly perceived as knowledge diffusers.

Media industry

- Key role in gluing the I&C cluster together;
- Traditionally strong publishing & printing sector;
- Fast-growing advertising industry;
- Strong TV producer and technical facilitating broadcasting company;
- Fragmented AV industry;
- High penetration rate of cable TV and concentration/re-positioning tendency in cable TV industry.

In reality, this grand I&C value chain (Figure 1) is made up of numerous smaller value chains. The creation of I&C products can be seen as a process of creating *content* followed by subsequent steps such as packaging, service provision, infrastructure provision, transport, vending of the required terminals to consume I&C products and, finally, their consumption by the final consumer. Of course, this is a highly stylised representation,[2] but nevertheless the transition from Phase 1 to Phase 2 in Figure 1 illustrates the clustering process as more and more industries become involved in various parts of the I&C value chain.

Figure 1. Clustering of I&C industries

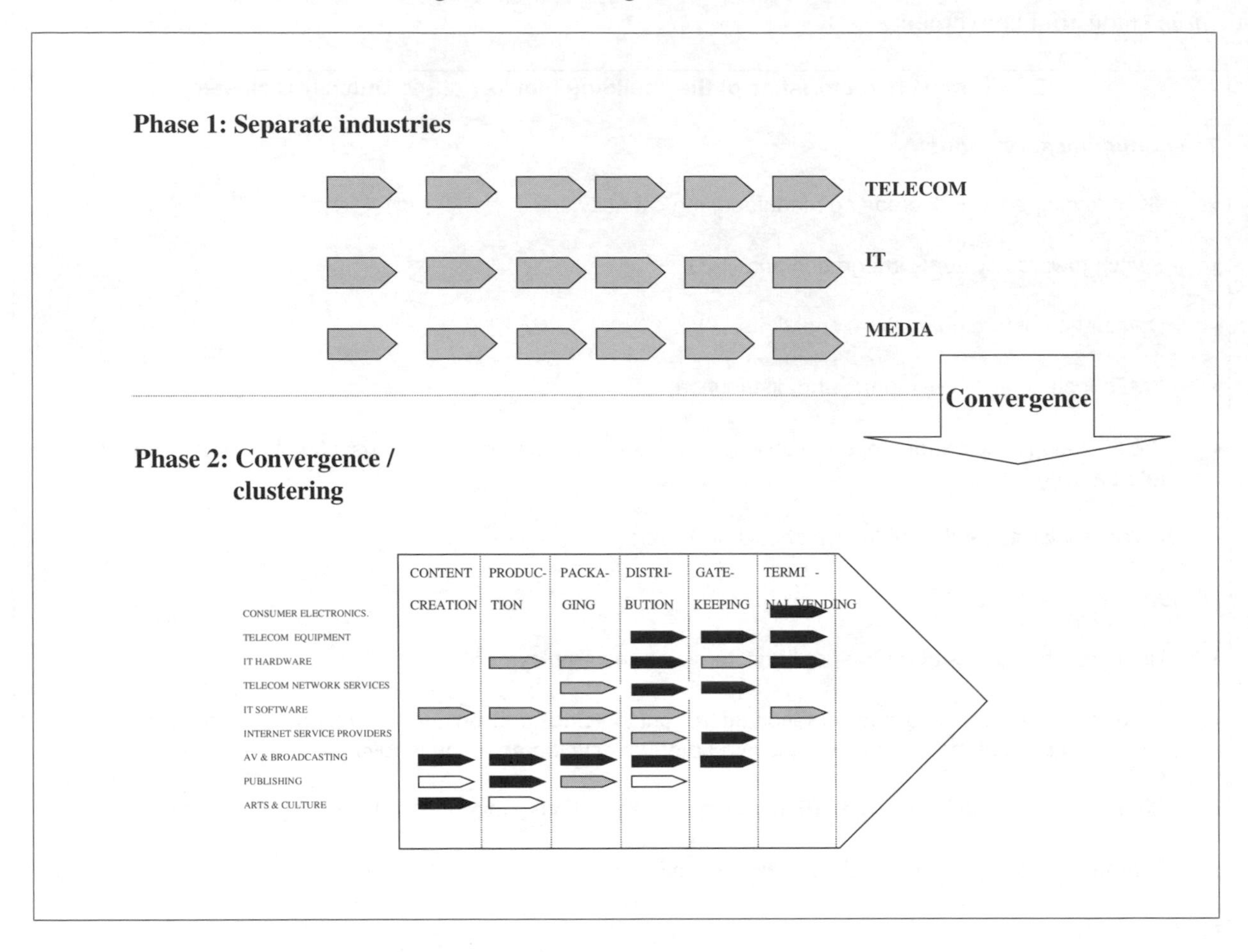

Note: The black arrows indicate relatively strong competencies in the Dutch context.

In Phase 2 of the convergence process, the I&C cluster is easier to perceive. A broad range of sub-sectors operate in each other's traditional domains. This results in greater competition within specific elements of the value chain, on the one hand, and the emergence of new – and sometimes unexpected – alliances or other forms of co-operation, on the other hand.

In Phase 3, the I&C cluster has matured, and specialisation and differentiation patterns appear (Figure 2). Market players bundle core competencies with respect to specific applications or services. This leads to the creation of new smaller value chains in which different actors from the grand value chain show stronger interlinkages. At the market level, this could eventually result in a process of divergence. However, it is clear that in the digital era there is no longer a clear borderline between the telecommunications, IT and media industries.

Figure 2. The specialisation and differentiation process in the grand I&C value chain: the example of digital TV

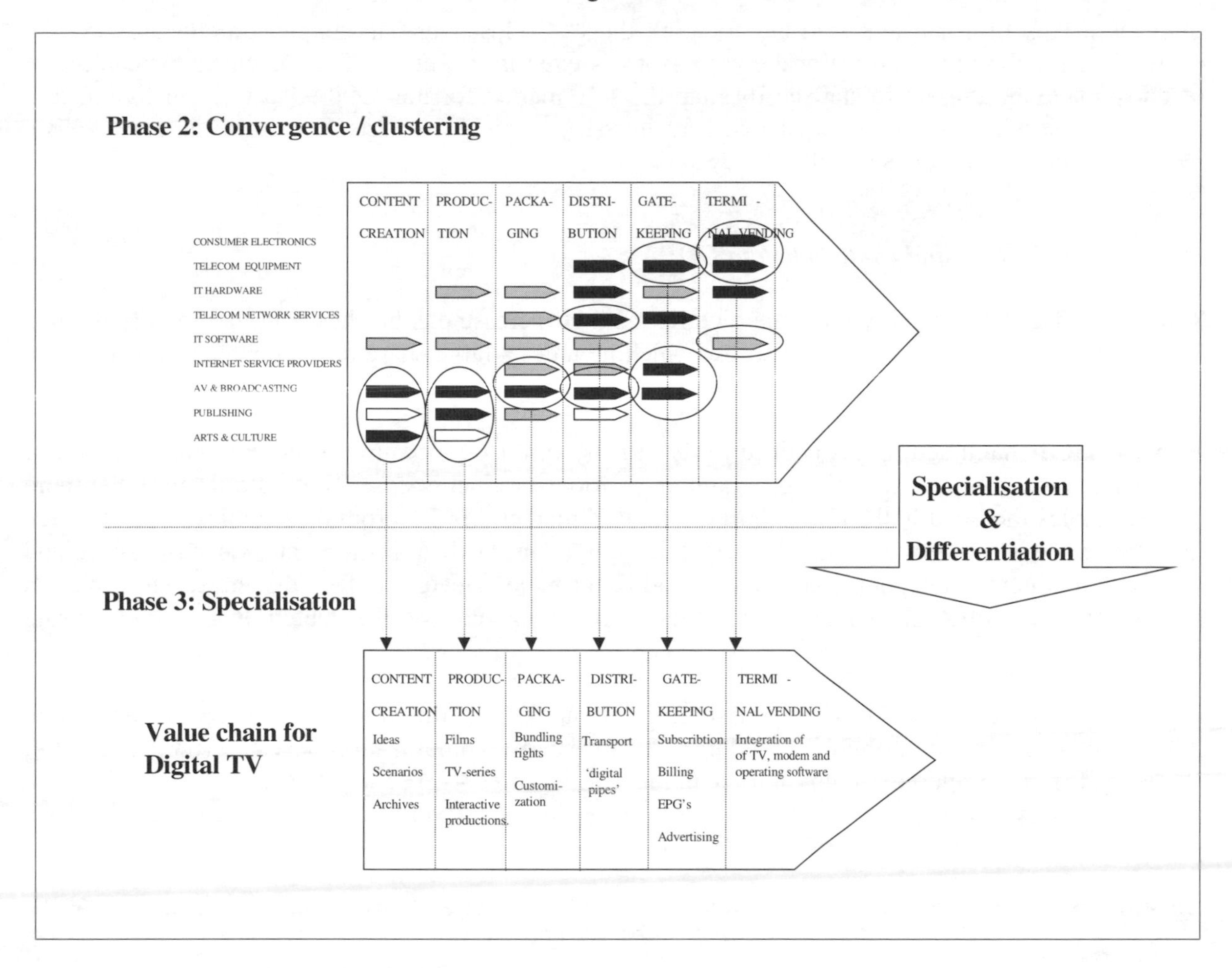

The process of industrial convergence and, subsequently, specialisation can be illustrated using the case of digital television (Figure 2). The introduction of digital television will require not only input from various players in the grand value chain, it will also lead to new roles for established players, thus furthering new competition patterns. This is even more the case as some roles can be performed by various players. Currently, both Philips and KPN Telecom are developing their own WebTV systems. KPN Telecom recently announced the introduction of Net.Box, a tool for connecting television and Internet through the telephone line. Cable operators and the terrestrial broadcasting organisation (Nozema) are preparing for digital transmission of both RTV and interactive services. The switch to digital broadcasting will have a strong impact on the other elements of the value chain, especially production, packaging and gate-keeping. Digital television will stimulate the market for pay-TV and on-demand-services. Content providers offering a "bouquet" of digital programmes and services (*e.g.* Canal +) will play a more dominant role in packaging, distribution and gate-keeping.

3. Composition and characteristics of the I&C cluster

There is a lack of consistent statistics for both the I&C cluster and its major sub-clusters; and the dearth of data allowing international comparisons is even more severe. The Dutch Statistical Office recently made an attempt to statistically map the ICT market for the 1990-95 period, including ICT goods, ICT services and (part of) the content industry, namely the "arts & culture industry" (CBS, 1997a).[3] This section draws mainly on CBS data.[4]

3.1. Production and contribution to GDP

Tables 1 and 2 show roughly the same picture: a slight increase can be shown in the broadly defined information and communication industries (which includes a large share of the content industry). The contribution of the service and content industries to this total is gradually increasing.

However, ICT manufacturing is defined fairly broadly (probably because of the dominant position of Philips), and can only be specified by examining various product goods.[5] These are listed in the more detailed tables included in the CBS report on the ICT market (1997). From these tables, it can be seen that, for example, control equipment, navigation equipment, medical instruments, various passive and active components, telephone equipment and information media (CDs, diskettes, etc.) play a prominent role, reflecting some of the traditional strongholds of the Dutch ICT manufacturing industry.

ICT services are composed of computer services (about 32% of the total production value and about 27% of this sector's contribution to GDP – 2.24%) and communication services (68% and 73%, respectively). It is important to note that in almost all industries – especially the service industries such as banking, business services and trade – software is (partly) developed in house. Over the past six years the contribution of ICT services to GDP has increased from 2.63% to 3.01%.

Table 1. Total production value of ICT manufacturing, ICT services and A&C (content) industries, 1990-95
1990 prices, NLG million

	1990	1991	1992	1993	1994*	1995*
ICT manufacturing**	21 729	21 959	21 800	20 974	21 804	23 272
ICT services	19 092	20 247	20 878	21 191	22 108	23 499
Total ICT	40 821	42 206	42 678	42 165	43 912	46 771
A&C industry	30 289	30 905	30 913	30 793	32 234	33 150
Total ICT + A&C industry	71 110	73 111	73 591	72 958	75 146	79 921
Total Dutch economy	949 245	976 340	990 749	991 944	1 019 234	1 045 786
Total ICT + A&C as a percentage of the total Dutch economy	7.49	7.48	7.42	7.36	7.37	7.64

* Estimates.
** Electrotechnical and electronics, classified according to the "old" industry coding scheme (SBI '74).
Source: Based on CBS, 1997b, pp. 60-61.

The arts & culture (A&C) industry is included because it represents an important share of the (organisers) of content that can be distributed through I&C infrastructures. This industry includes publishing and printing, advertisement agencies, broadcasting activities, but also libraries and museums. The reason for including A&C in the I&C industries is that, increasingly, A&C activities are interlinked with those of the telecommunication and IT industries. The A&C industry is clearly dominated by the publishing and printing industry (approximately 60% of production value), which is well developed in the Netherlands. Two other segments – advertising and broadcasting – show above-average growth rates in terms of production value and contribution to GDP. The contribution of the A&C industry as a whole to GDP amounts to approximately 2.2%.

Table 2. Contribution of ICT manufacturing, ICT services and A&C (content) industries to GDP
1990 prices, NLG million

	1990	1991	1992	1993	1994*	1995*
ICT manufacturing**	10 039	9 927	9 707	9 890	10 565	11 089
ICT services	13 632	14 758	15 812	15 717	16 273	17 304
Total ICT	23 671	24 685	25 519	25 607	26 838	28 393
A&C industry	11 227	11 389	11 502	11 559	10 649	10 993
Total ICT + A&C industry	34 898	36 074	37 021	37 166	37 487	39 386
Total Dutch economy	516 550	528 280	538 980	543 090	561 500	573 520
Total ICT + A&C as a percentage of the total Dutch economy	6.75	6.82	6.86	6.84	6.67	6.86

* Estimates.
** Electrotechnical and electronics, classified according to the "old" industry coding scheme (SBI '74).
Source: Based on CBS, 1997b, pp. 60-61.

The total production value of the ICT industries (including content) as a percentage of total production value amounts to approximately 7.5%. Some fluctuations can be noted, but these may have been caused by price movements (especially deflationary developments) rather than by changes in volume. In a similar vein, it can be observed that the contribution to GDP of the ICT industries taken as a whole is fairly stable, at around 6.75%.

3.2 *Exports*

Trade figures (Table 3) are incomplete as import figures for services are lacking.[6] For ICT goods, there is a trade deficit – although an examination of the more detailed tables shows that for some product groups there are considerable trade surpluses, *e.g.* electronic components, equipment for telephony, and compact discs. This, again, reflects strongholds in the Dutch ICT manufacturing industries. For some product groups, trade figures are considerably higher than those for production value (by a factor of up to five or six for some goods such as copiers and computers), emphasizing the *transito* function of the Netherlands for these goods. In addition to the presence and strong performance of Amsterdam's airport and Rotterdam's harbour as gateways to Europe, this function is reflected in the traditionally strong presence of trading companies and the establishment of a number of European distribution and service centres over the last decade. Exports of ICT services increased at an annual average of 5% (in current prices). The A&C industries are clearly more focused towards national markets. In addition to the trade surplus for goods produced by the publishing and printing industry, a remarkable surplus for photo-chemical and photo-technical products can be noted.

Table 3. ICT goods and services and A&C (content) industry goods: imports, exports and trade balance, 1990-95

Current prices, NLG million

	1990	1991	1992	1993	1994*	1995*
ICT goods**						
Imports	24 952	26 594	26 255	27 728	30 030	32 850
Exports	20 046	21 463	22 501	24 435	26 438	29 733
Balance	-4 906	-5 131	-3 754	-3 293	-3 592	-3 117
ICT services						
Imports						
Exports	1 833	2 034	2 103	2 214	2 351	2 297
Balance						
Goods A&C industry						
Imports (only goods)	3 149	3 323	3 423	3 159	Unknown	Unknown
Exports	4 246	4 691	4 910	4 652	Unknown	Unknown
Balance (only goods)	868	1 130	1 237	1 239	Unknown	Unknown
Total ICT + A&C goods						
Imports						
Exports	26 125	28 188	29 514	31 301		
Balance						
Total Dutch economy						
Imports	25 5828	267 386	269 174	257 502	275 907	297 785
Exports	279 746	293 086	294 879	293 180	313 978	338 259
Balance	23 918	25 700	25 705	35 678	38 071	40 474

* Estimates.

** Including exports of services of advertising agencies (NLG 229 million in 1990, NLG 263 million in 1995).

Source: Based on CBS, 1997b, pp. 20, 30-31.

3.3. *Employment*

Employment figures show a mixed picture. Employment in the information and communication industries taken as a whole decreased slightly over the period 1990-95. ICT manufacturing is decreasing quite substantially. This is not completely compensated by the increase in employment in the ICT services sector and the slight increase in the A&C industry. The higher production volumes, combined with sharply reduced employment levels in the ICT manufacturing industry, point to automation in the production of ICT goods (and components). In addition, it should be noted that the production of telecommunication hardware – and especially the production of telecommunication network equipment which is traditionally well established in the Netherlands – is now essentially related to software production (*e.g.* intelligent network functions). Nevertheless, the magnitude of the ICT manufacturing industry is still considerable. Employment in ICT services increased by 9.2% over the six-year period. Although it cannot be seen from the fairly general figures included here, this increase is the result of a 50.3% increase in employment in computer services, and a 7.5% decrease in employment in communication firms. This decrease mainly reflects the effects of the gradual transformation of the state-owned telecommunication operator KPN Telecom into a market-driven (privatised) telecommunication provider. At the same time, the character of jobs at KPN Telecom have changed. For example, newly hired staff tend to be highly educated with a background in sales and marketing, IT, finance and economics, while job losses are mainly concentrated in activities such as network construction and maintenance (Mansell *et al.*, 1995, p. 76). Within the A&C industry, a slight increase in employment can be noted for the period 1990-95. The majority of employees work in the printing and especially publishing industry, although the share of the latter is decreasing. This is more than compensated by the 49% increase in employment in the advertising industry – most likely the result of the multiplication of commercial radio and TV stations – and the broadcasting industry.

Table 4. Employment in ICT manufacturing, ICT services and A&C (content) industries, 1990-95
Thousands of man years

	1990	1991	1992	1993	1994*	1995*
ICT manufacturing**	105.9	97.2	89.9	85.8	79.7	78.8
Annual growth rate		-5.6%	-7.4%	-4.4%	-7.1%	-1.1%
ICT services	112.4	118.2	121.8	120.0	119.0	122.7
Annual growth rate		5.2%	3.0%	-1.5%	-0.8%	+3.1%
A&C industry	126.1	127.9	130.1	128.6	128.9	133.8
Annual growth rate		1.4%	1.7%	-1.2%	0.2%	3.8%
Total Dutch economy	5 203	5 273	5 328	5 322	5 304	5 380
Annual growth rate		1.35%	1.04%	-0.11%	-0.34%	1.43%
Total ICT + A&C industry	344.4	343.3	341.8	334.4	327.6	335.3
Total ICT + A&C as a percentage of the total Dutch economy	6.62	6.51	6.42	6.28	6.18	6.23

* Estimates.
** Electrotechnical and electronics, classified according to the "old" industry coding scheme (SBI '74).
Source: Based on CBS, 1997b, pp. 10, 13, 29, 34.

In the following sections, we will first look at linkages among I&C industries using I/O data, and then at innovation styles using innovation survey data.[7]

4. Linkages in the I&C cluster

Clustering, or inter-firm networking, results in above-average interlinking and exchange of (intermediary) inputs between firms in the cluster. An attempt was made to map these linkages in the "I&C cluster" using I/O analysis. First, the most relevant I&C industries were selected. Taking the categories used in the CBS study as a starting-point (presented in aggregated form in the preceding section), these were translated into the I/O table categories used in the 1992 214 x 214 I/O-table.[8] The resulting 18 industries were grouped, and the degree to which intermediate goods and services of these industries were supplied by I&C industries or non-I&C industries assessed. The I/O analysis provided an insight into the degree to which the 18 I&C industries are dependent on foreign sources for their intermediate inputs as well as the degree to which their intermediate and final goods and services are exported. The main results of the analysis are reproduced below.

4.1. Four I&C sub-clusters

Some of the clear trends which emerge from the analysis are shown in Figure 3 which depicts the main links within the I&C industry. It should be noted that, although the individual industries differ considerably in size (shown in brackets in the boxes), some clear linkages and sub-clusters can be traced. For example, the content industries are not only quite impressive in absolute terms, they also seem to be closely linked to each other. Clearly, the advertising industry links the more traditional publishing and printing sub-cluster to the entertainment industry. Likewise, the I&C hardware industry is an established sub-cluster, although its links to, especially, the printing and publishing sub-cluster are not particularly well developed. The content industries have relatively stronger intra-industry linkages (the "diagonal" in the I/O table) compared with the I&C hardware and I&C service industries. I&C services show a more confusing picture. Communication services are, for obvious reasons, fairly closely linked to the other sub-clusters, while this is less the case for the computer services industry. However, it should be borne in mind that the figure depicts the situation for 1992. More recent data would probably reveal a more integrated computer services sector as convergence of, especially, IT and telecommunications has really taken off over the last few years.

Figure 3. I/O linkages among various I&C industries

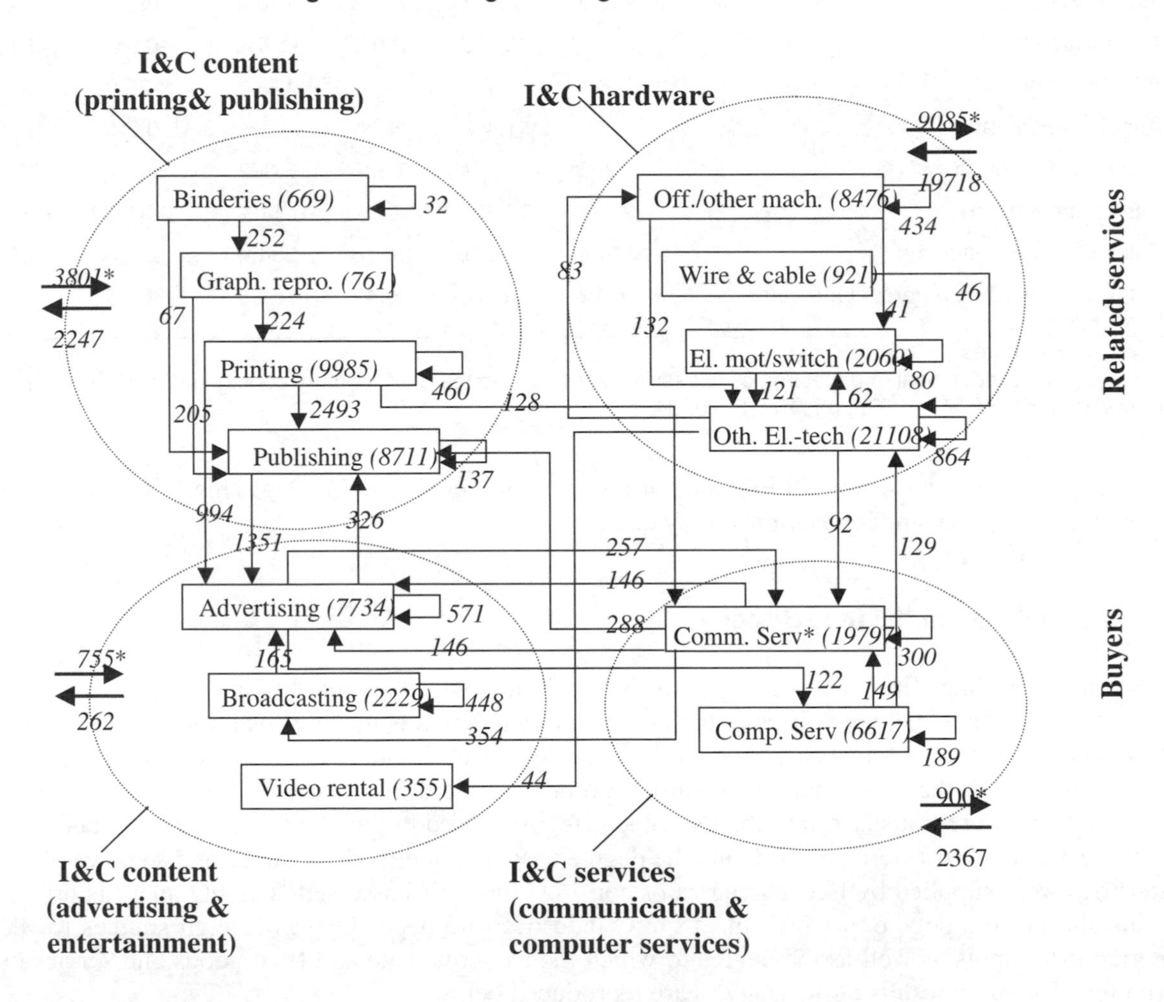

Table 5 shows, in another way, how dependent the individual I&C industries are on the I&C cluster in terms of intermediate inputs received from ("I&C use") as well as intermediate inputs supplied to the I&C cluster as a whole. In general, the rather high scores underline the dependence on or interlinkages with the I&C cluster as a whole. At the same time, important differences can be observed. It appears that some I&C content industries (publishing, advertising, broadcasting) are fairly dependent on intermediary inputs supplied by the I&C cluster as a whole (up to 75% in the case of publishing); this suggests that content industries are linked in more ways (and not only as content providers) to the I&C cluster. Printing and printing-related industries, as well as other content industries such as video rental, libraries and again broadcasting, supply a relatively large share to the I&C cluster as a whole. I&C hardware industries, I&C service industries and advertising clearly show lower percentages, indicating that their intermediate products and services are used more widely within the economy. It is quite possible that an I&C industry depends on other I&C industries for its inputs, while for its intermediate outputs, it does not. Typical supplying ICT hardware industries such as wire & cable industries or manufacturers of measurement and control equipment, appear to be least integrated in the I&C cluster.

Table 5. Relative importance of use and supply of I&C intermediate goods and services per I&C industry

I/O code	I&C INDUSTRY	I&C USE (total deliveries of intermediate goods & services used by the I&C industries as a % of all intermediate goods and services used, per I&C industry)	I&C SUPPLY (total intermediate goods & services supplied to the I&C industries as a % of all intermediate goods and services supplied, per I&C industry)
2401	Printing (daily newspapers)	47.44	49.28
2402	Graphical reproduction industry	20.23	84.24
2403	Publishing	74.26	43.98
2404	Binderies	32.35	67.59
3208	Office machinery & other machinery	33.42	38.14
3301	Electrical, wire & cable industry	10.00	21.46
3302	Electrical motors & switches installation	42.21	26.37
3303	Other electrotechnical industry	36.43	28.30
3602	Measurement & control equipment	24.49	15.63
3607	Photo & film laboratories	20.75	21.52
4601	Communication services, railways*	23.18	14.40
5003	Computer services	27.37	14.05
5005	Advertising	68.16	20.80
5602	Libraries, museums, etc.	26.26	90.24
5603	Broadcasting	62.87	57.70
5604	Film-related organisations	22.78	37.07
5605	Cinema	46.85	0
5606	Video rental	46.40	100

* This category is (unfortunately) very broad. About 20% of the activities in this category can be attributed to KPN Telecom (formerly PTT Telecom); postal services are excluded.

Source: Figures based on 1992 I/O table.

Analysis of the links between individual I&C industries and non-I&C parts of the economy also shows a mixed pattern.[9] For example, some I&C industries are relatively more embedded than others. Publishing, printing, advertising, and to a lesser extent communications and broadcasting, seem to show relatively more (or stronger) links with I&C industries than do some of the other I&C industries. Analysing the inputs or industries from which deliveries are received, we see that apart from key inputs such as paper for printing, car maintenance for computer services or artists for advertising, some general categories and apparently expensive inputs such as accommodation (or the renting of it), temporary labour and services like accountancy, banking and cleaning, are among the more regular industries (most important to the user). Apart from other I&C industries and key industries such as business consultancy and banking for computer services, major users of products from the I&C industries are wholesale and retail trade (outlet channels) and public services.

4.2. Import dependency and export orientation

Import dependency and export orientation of I&C industries can also be assessed using I/O data. From Table 6 and Figure 3, it can be observed that the printing, printing-related industries and the industries grouped in the "I&C" hardware manufacturing category display a fairly high *import dependency* rate; on average, they depend on foreign sources for 50% of their intermediary inputs.

Table 6. Import dependency and export orientation of individual I&C industries

I/O code	I&C INDUSTRY	IMPORT DEPENDENCY (Imports of intermediary goods and services as % of all intermediary goods and services supplied per I&C industry)	EXPORT ORIENTATION (Exports of intermediary and final goods and services as % of total production per I&C industry)
2401	Printing (daily newspapers)	56.42	11.88
2402	Graphical reproduction industry	48.05	12.61
2403	Publishing	10.90	10.03
2404	Binderies	42.95	13.60
3208	Office machinery & other machinery	51.16	55.31
3301	Electrical, wire & cable industry	54.74	39.52
3302	Electrical motors & switches installation	45.93	44.03
3303	Other electrotechnical industry	56.05	62.25
3602	Measurement & control equipment	43.68	70.73
3607	Photo & film laboratories	42.70	4.91
4601	Communication services, railways*	13.54	4.84
5003	Computer services	9.44	21.28
5005	Advertising	10.43	3.23
5602	Libraries, museums, etc.	8.53	0
5603	Broadcasting	2.88	0.22
5604	Film-related organisations	26.85	3.14
5605	Cinema	10.48	0
5606	Video rental	24.70	0
	All (I&C and non-I&C) industries	20.76	26.73

* This category is (unfortunately) very broad. About 20% of the activities in this category can be attributed to KPN Telecom (formerly PTT Telecom); postal services are excluded.

Source: Figures based on 1992 I/O table.

The majority of the I&C content sub-cluster and the I&C communication & computer services sub-cluster receive their intermediary inputs mainly from national sources. In terms of export orientation, which cannot directly be compared with import dependency as final goods and services are included, the various I&C industries show even more blatant differences. Only the I&C manufacturing industries manage to export an impressive share of their production. The export orientation of the other three I&C sub-clusters can hardly be considered well developed, even in those sectors in which some well-known multinational firms operate, such as publishing, communication services and computer services. Of course, some industries, such as some of the entertainment industries, by definition, target final consumers in national markets. Nevertheless, it is rather remarkable that, for example, computer services focus to a large extent on national markets and – at least in 1992 – were unable to export more of their products. It would be most interesting to see if the situation has changed since. Import and export figures for the four sub-clusters identified are combined in Figure 3. In all sub-clusters, imports exceed exports (even if final products are included). Even if we take into account the relatively large size of the I&C manufacturing sub-cluster, the export performance of the other three I&C sub-clusters can hardly be considered well developed.

5. Innovation style of the I&C cluster

In addition to the I/O analysis presented above, we also examined the 1992 innovation survey data.[10] We assessed the "innovation style" and innovation performance of the I&C industries, both individually and as a cluster. In fact, the analysis performed can be seen as a variation on that for the ten "official" mega-clusters (see Witteveen, 1997; Roelandt *et al.*, 1999). As mentioned above, until now the I&C cluster has not been treated as a separate cluster. We therefore "artificially" grouped the various I&C industries and assessed the degree to which they differed from non-I&C industries. Aspects covered in the analysis and which, taken together, provide a picture of the innovation characteristics (*i.e.* innovation style) include:

- level of innovation activity;
- formalisation of R&D function;
- co-operation in R&D networks (internal/external, public/private, regional/international);
- participation in "networks of innovation";
- use of the knowledge infrastructure.

Some of the more interesting results are presented in Figures 4, 5 and 6, and will be briefly discussed below.

5.1. Innovativeness of the I&C cluster

Figure 4 illustrates that there are some notable differences between I&C and non-I&C firms in terms of level of innovation activity and the degree to which R&D is formalised. The number of firms that see themselves as not being innovative is considerably lower for I&C firms. To put it more directly, I&C firms tend to be more innovative than non-I&C firms. At the same time, this innovative behaviour does not translate into more firms with more R&D departments. On the contrary, the number of innovative firms without R&D departments is considerably higher for the I&C firms than for the non-I&C firms (28.1% *vs.* 17.2% of all firms). Innovation does not automatically follow from (formalised or not) R&D.[11] This is true not only for service industries such as publishing & printing, advertising and broadcasting, but equally for most of the I&C manufacturing industries. Computer services score on average in terms of innovation and also perform R&D, but to an important degree these activities are not performed in a R&D department.

Figure 4. Formalisation of innovative behaviour of I&C *vs.* non-I&C industries, 1992

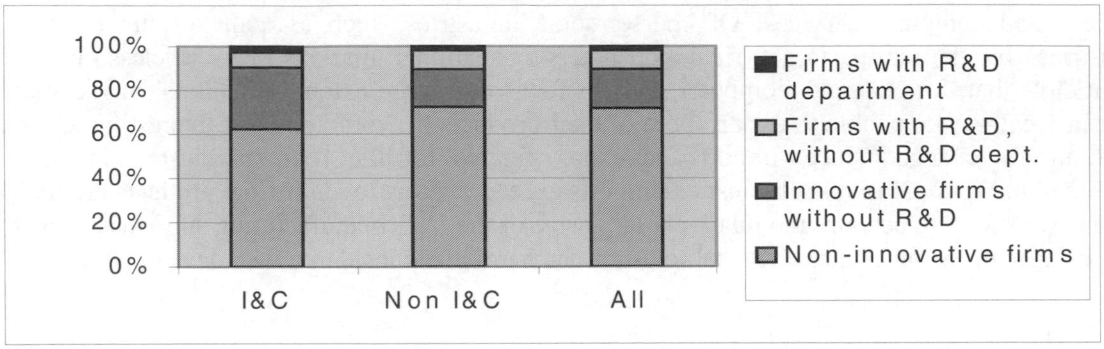

100%
80%
60%
40%
20%
0%
I&C
Non I&C
All
Firms with R&D department
Firms with R&D, without R&D dept.
Innovative firms without R&D
Non-innovative firms

Source: SEO/Ministry of Economic Affairs.

5.2. *Networks of innovation in the I&C cluster*

Firms seldom innovate in isolation and innovation is as much about networking – defined as establishing semi-permanent relationships – as it is about R&D spending. The majority of innovative firms rely on external sources of knowledge in addition to their internal knowledge sources, and therefore to a certain extent participate in networks of innovation (OECD, 1999). Using the DeBresson typology, six degrees or categories of innovation networks can be distinguished (OECD, 1999). Firms are categorised as belonging to one of these categories depending on the importance attributed to various sources of information for innovation. At one end of the continuum, innovation networks are described as weak if external sources of information for innovation are reported as unimportant or absent. At the other extreme, innovation networks are characterised as complete if various private and public sources of information are used in combination for innovation purposes.

Figure 5 shows the degree to which I&C and non-I&C firms participate in networks of innovation. I&C firms apparently participate more often in innovation networks than do non-I&C firms. Where I&C firms participate, they tend to participate complete types of "networks of innovation". Compared to non-I&C firms, I&C firms use "suppliers and clients" as a source of information for innovation twice as much as do non-I&C firms (15.2% *vs.* 7.6%). They also score considerably better for the two most complete innovation networks. Especially for the most complete network of innovation category in which many types of information are used, including public and private external research (including private consultancies), I&C firms score considerably better than do non-I&C firms (7.2% *vs.* 2.6%). In conclusion, the more "complete" networks of innovation are relatively well represented.

This finding could be interpreted to support the suggestion that lower levels of formalised R&D are compensated for by a higher involvement and participation in more complete innovation networks. Given the changes affecting the I&C industries, the fact that I&C innovations often cut across fields of knowledge that used to be separate and the pivotal role played by clients in implementing I&C products and services, probably explain why "networks of innovation" in the I&C industries as a whole tend to be more developed.

Figure 5. Networks of innovation: I&C *vs.* non-I&C industries, 1992

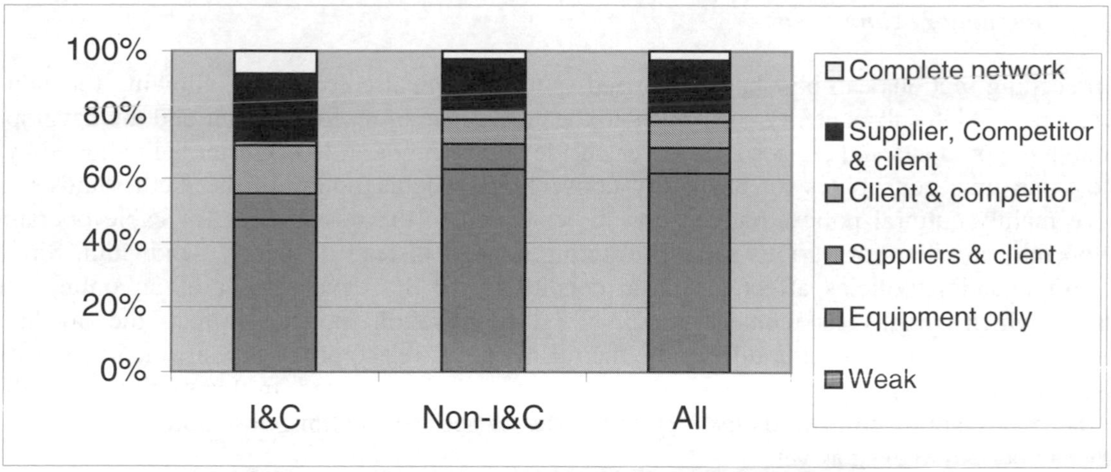

100%
80%
60%
40%
20%
0%
I&C
Non-I&C
All
Complete network
Supplier, Competitor & client
Client & competitor
Suppliers & client
Equipment only
Weak

Source: SEO/Ministry of Economic Affairs.

5.3. Use of the knowledge infrastructure

Figure 6 shows the degree to which industries are using the public and private knowledge infrastructure[12] as a source of information for innovation. Although networks of innovation tend to be more complete in I&C industries, the actual use of the knowledge infrastructure is lower compared with non-I&C industries. There may be various reasons for this. One could simply argue that I&C industries are less capable of linking up with the knowledge infrastructure. However, it might equally be the case that, at that time, the knowledge infrastructure was less well equipped in terms of available expertise to act as a source of innovation to the I&C industries. A closer look at the data for individual industries – not reproduced here – reveals that I&C services and content industries showed below average scores. This might indicate that the interaction between these softer I&C industries and the relevant I&C knowledge infrastructure was underdeveloped. In this perspective, the recent establishment of, for example, the Telematics (Top) Institute and, more generally, the policies aimed at furthering the Electronic Superhighway, seem to address a need. These topics will be addressed in the following section.

Figure 6. Use of knowledge infrastructure (public, private or consultancy): I&C *vs.* non-I&C industries

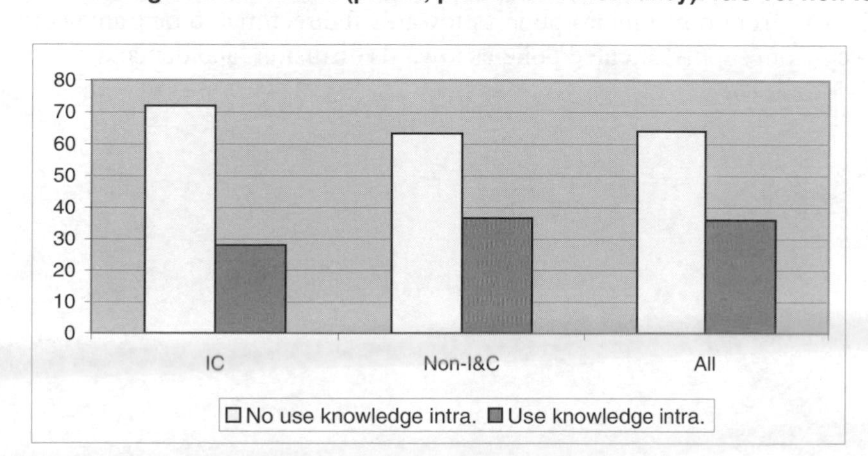

6. I&C cluster policy making

6.1. Regulatory convergence

The preceding sections can best be summarised as technological convergence allowing for industrial convergence which subsequently contributes to the emergence of an I&C cluster and the development of innovative products and services. However, the development of an I&C cluster is facilitated by, and at the same time contributes to, regulatory convergence. Media policy in the Netherlands used to involve mainly cultural policy, but can now be extended to the possibilities for cable operators to develop into providers of various services offering content instead of merely bandwidth. Similarly, telecommunication policies affect the basic conditions for the development of an online service industry which mainly is content-based. IT standardisation policies affect the media and telecommunication industry. Put differently, the realms of IT, telecommunications, media, innovation, and science & education policies are becoming increasingly intertwined. In this section, we will deal with the policy approach towards the I&C industries to see if something resembling an I&C cluster policy can be said to exist as yet.

As can be seen from Figure 7, a whole array of individual policy measures and schemes could be labelled "I&C policies". Although the emphasis applied by individual countries or the mixture of policy instruments tends to vary considerably (den Hertog and Fahrenkrog, 1993), most countries dispose of similar instruments. Although supply-oriented instruments are still in place and new ones continue to see the light of day (creation of centres of excellence, support for basic research, decisions to invest massively in computers in schools), these measures are complemented by instruments that aim to diffuse ICT knowledge and encourage user communities to articulate their needs and wishes in developing ICT applications for their respective businesses. Technology transfer, feasibility, demonstration and pilots are the buzz words, especially linked to SME policies. Budgets for ICT applications in such areas as education, traffic and transport, health, and agriculture, as well as for the functioning of government itself, are higher than for direct support to ICT technology *per se*. At the same time, old-fashioned industrial support to individual national champions is declining due to cuts in budgets for direct specific support or to the gradual introduction of competition and market dynamics into sectors that were dominated by monopolists in the past. In addition to creating favourable framework conditions, activities such as ensuring competition, adapting intellectual property rights to the age of the Electronic Highway, and performing foresight studies are seen as increasingly important (see below). In summary, ICT policy is no longer mainly sectoral technology policy, but is increasingly becoming innovation policy aimed at creating the conditions for the provision of high-quality, reasonably priced ICT services, competition among various service providers and allowing massive application in the using industries, including government itself. There is a clear trend away from direct/micro policies towards indirect/macro or framework policies, and a similar trend away from supply-oriented policies towards diffusion- and demand-oriented policies.

Figure 7. Basic scheme to characterise ICT policies

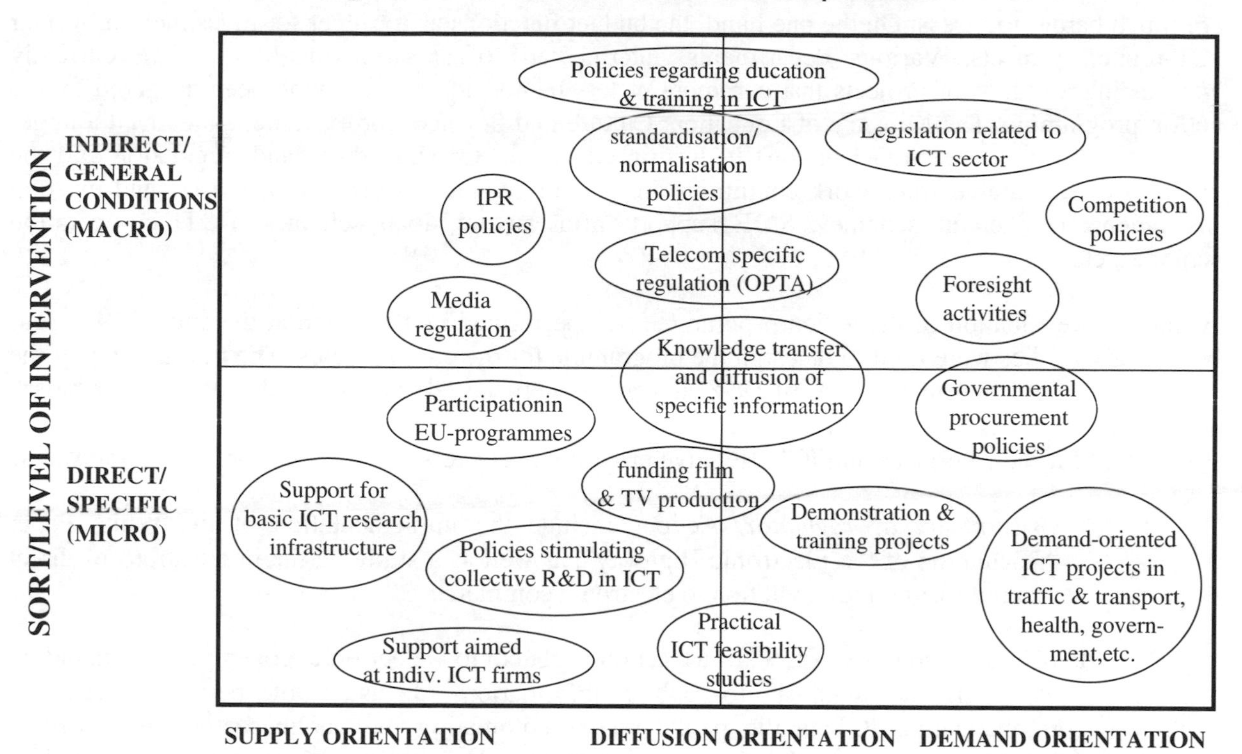

Source: Based on den Hertog and Fahrenkrog, 1993.

6.2. *Policies for the Electronic Superhighway as implicit I&C cluster policy making?*

In the Netherlands, no formal ICT cluster policies exist as ICT is not (yet) recognised as a cluster. A number of individual support schemes exist (Figure 7), and the regulatory regime for telecommunications still differs somewhat from the more open and competitive world of IT, while the policy implications of traditional media such as broadcasting and publishing entering the multimedia age are not yet clear. Nevertheless, a trend towards greater policy convergence can be noted. Probably the best example is the set of policies initiated under the umbrella of the *Action Programme Electronic Superhighway,* first introduced at the end of 1994. This can be seen as implicit ICT cluster policy making.[13] A brief inventory of the most relevant policy measures and initiatives outlined in the White Paper on cluster policy in the Netherlands (Ministry of Economic Affairs, 1997) is presented below.

The Action Programme Electronic Superhighway was introduced at the end of 1994 by the Ministries of Economic Affairs, Traffic and Water Management, the Interior, Education, Culture and Science.[14] The Ministry of Justice joined in 1996, followed by the Ministry of Social Affairs and Employment in 1997, and other ministries are involved in individual projects. The programme aims to make the Netherlands the "information gateway to Europe". Most remarkable about the programme is that it is interactive; most of the interested parties participate actively in giving shape to the various action plans. In addition to policies geared towards (liberalisation of) infrastructure, policies aim at creating favourable framework conditions, supporting the introduction of new electronic services and (increasingly) at the organisational and societal implications of introducing ICT on a large scale.

Although funding is limited to approximately NLG 90 million annually, the exact budgets involved are much harder to assess. On the one hand, the budget functions as a trigger for investment in ICT or ICT-related projects. Various departments, intermediary organisations and firms are currently announcing all kinds of projects that can more or less be considered as having been triggered by the action programme. The Ministry of Education, Culture and Science, for example, announced a large-scale investment (over NLG 1 billion) in ICT in education. On the other hand, regulation and the creation of favourable framework conditions are part of regular government activities and existing instruments (tax credit schemes, SME support infrastructure, loan schemes, R&D co-operation schemes, etc.).

A mid-term evaluation of the action programme was performed in 1997, and at the end of 1997 four benchmark studies were used to redesign the programme for the coming years. The new accents of the programme were presented in April 1998 and are largely based on the results of the benchmark studies. In the years to come, the programme will focus on development of services, development of the content for these services and ICT use. Five programme clusters are foreseen for the coming years:

a) *Government as regulator.* Actions include the implementation of the White Paper "Regulation of the Electronic Highway", as well as a study on the implications of direct and indirect taxing in relation to electronic commerce.

b) *ICT and government.* Various actions related to "electronic government", including electronic accessibility to public information, diffusion of best practices for implementing ICT in the public sector, automating the public sectors' back office. Further activities, such as creating trusted third-party services provided by the public sector and foresight studies on ICT in health and social services, are planned.

c) *Knowledge and accessibility.* This activity is mainly aimed at increasing the availability of relevant ICT knowledge through training and education, implementing the National Research Agenda, Informatics, an assessment of the accessibility and transparency of the ICT knowledge infrastructure for industry, etc.

d) *Innovation of ICT in the market secto*r. Although some existing elements, such as KREDO and promotion and demonstration projects, have been prolonged, new elements include the implementation of the action plan on electronic commerce,[15] and probably most remarkably, the introduction of the concept of "twinning" (Box 2).

e) *Telecommunication infrastructure.* Activities include formulating a position on the convergence issue, experiments in telecommunication services (*e.g.* digital video broadcasting), formulating a vision on broadband research networks (Internet-2), and the role of the government in creating international broadband networks.

Box 2. The "twinning" initiative

This initiative of the Ministry of Economic Affairs was launched at the beginning of 1998 to support start-up firms active in the field of information and communication technologies. The scheme aims not only at new-firm creation, but also to facilitate expansion of a considerable number of these new start-ups and to ensure that they are competitive on the European and world ICT market. There are four basic elements to the initiative:

- *Twinning fund for start-up companies.* Some NLG 21 million (NLG 10 million provided by government, and some NLG 11 million provided by venture capital firms) has been made available to ICT start-ups.

- *Twinning growth fund.* This fund aims to help young ICT firms expand. The Ministry of Economic Affairs plans to contribute NLG 25 million to the fund, and other capital providers have expressed interest in participating: the European Investment Fund (EIF) and the National Investment Bank (NIB), among others.

- *Creation of twinning centres.* A number of twinning centres are planned, in each of which approximately 45 firms will find not only facilities, but also an innovative climate. The first two centres will be located in Amsterdam and Eindhoven, and others are planned.

- *A network of mentors.* Young ICT firms can benefit from coaching by experienced ICT entrepreneurs and managers that have a good track record in the ICT industry and can help these young firms enter foreign markets. The network is headed by Mr. Roel Pieper, member of the board of directors of Philips International.

Although the scheme is not yet fully operational, the first signs are promising. The Ministry of Economic Affairs is expected to invest about NLG 70 million. Private venture capital firms are supportive of the initiative, and most importantly about 200 ICT start-ups have shown interest in participating (30 have already provided business plans). In addition, several municipalities are competing to have a twinning centre located in their area.

The scheme has been set up as a commercial activity, to be operated by Twinning Holding, a firm in which government will hold 100% of the shares. The new Minister has already suggested that Twinning Holding will probably be privatised in the near future.

Compared to the first phase of the programme, a number of new accents and initiatives have been introduced, and some new approaches will be explored. A budget of NLG 90 million is available for 1998. The larger share of the budget is available for lines *d)* (NLG 43 million) and *c)* above (NLG 35 million). For lines *b)* and *e)*, the joint budget amounts to NLG 12 million. However, the budgets for the five programme clusters may be increased in the future.

The multitude of policies related to the Action Programme were not explicitly presented as cluster policies. However, if the measures, programmes and projects are classified according to the three roles highlighted in Dutch policy making, it can be observed that all three roles are covered. In our view, therefore, these policies can be characterised as implicit cluster policy making. In addition, the Electronic Superhighway policies comply with what we believe to be essential elements of cluster policy making, namely:

- ♦ The cluster policy approach is an intrinsically open-minded, interactive policy strategy that takes shape along the way.

- ♦ The cluster approach does not necessarily require a complete set of new instruments and huge budgets. It can also imply using (or adapting) existing tools to attain a specific clearly defined goal or to provide incentives through regulation and the provision of appropriate framework conditions.

- ♦ Cluster policy making is, to a large extent, aimed at adapting the existing institutional framework in which innovation is embedded. In areas where there are traditional strengths, there tends to be greater readiness to make the necessary changes, thus facilitating further specialisation.

Table 7. Policy measures related to the Action Programme Electronic Superhighway using the Dutch cluster policy categorisation

Creating favourable framework conditions	Broker policies	Government as a demanding customer
• Liberalisation of telecommunication infrastructure (new Telecom Law) • Liberalisation media market (new Media law) • Establishment of OPTA/Cie for the media • Regulations on computer crime, harmful content, IPR, encryption, trusted third parties) • Research programme on IT and law • Implications of electronic commerce for direct/indirect taxation • "Trusted third party" arrangements • Foresight studies on ICT in health and social services • ICT education and training Telematics Centre of excellence	• Credit facilities & R&D support (KREDO, including regular schemes such as WBSO, BTIP, etc.) • Awareness campaign (SpOED) • Information service (INFORME), knowledge transfer and demonstration, *e.g.* Media Plaza • Extensive consultation with firms, *e.g.* on development of electronic services • Software action plan 1996-2000 (SWAP 2000) • Accessibility/transparency of the ICT knowledge infrastructure • General policies such as IC network, BTS • Platform Electronic Commerce • Twinning network/twinning centres/twinning funds	• Electronic accessibility of the public sector (*e.g.* Project Communication Government–Citizens) • Demonstration projects in the public sector (OL 2000, ON21) • Telecommunication experiments (*e.g.* digital video broadcasting) • Project on ICT in schools (network, equipment, services, training, etc.) Traffic and Watermanagement Net • Health-Net

Note: These policy measures are not exhaustive. For a detailed discussion of the three roles for cluster policy making, see Roelandt *et al.* (1997) and the White Paper on clusters (Ministry of Economic Affairs, 1997).

7. Conclusions

Having presented the main findings of our study, the following more general conclusions regarding the Dutch I&C cluster can be drawn.[16]

♦ Information and communication technology is not only an enabling technology applied in almost all industries, it also encourages change. ICT has stimulated the tremendous growth in the number of Dutch industries that are actually involved in the production of I&C goods and services. In conclusion, strong intra-cluster relations in the production of goods and services are combined with strong inter-cluster relations after the production of the goods and services.

♦ There is every indication that a Dutch I&C cluster is emerging. Through the process of convergence and the linking of content industries to more traditional IT and telecommunication industries, I&C industries are increasingly becoming part of a fully fledged I&C cluster.

- The scale of the I&C cluster differs for the various I&C sub-clusters. For intermediary inputs, the import dependency of all sub-clusters is rather high, indicating that industrial networks extend beyond national borders. However, the export orientation of the I&C service sectors in particular is not very well developed; this supports the general assumption that manufacturing and service industries differ in terms of their international orientation.

- The observation that firms do not innovate in isolation clearly applies to firms belonging to the I&C cluster. Not only do these firms participate more often in networks of innovation, they also show a preference for more intensive forms of co-operation. This fact provides an additional stimulus to investigate the interface between I&C firms and the knowledge infrastructure as this interface seems to be less well developed.

- The implicit I&C cluster policies presented in this report highlight the shift away from direct intervention towards more indirect incentives in the area of information and communication policies. These more indirect forms do not necessarily require substantial budgetary outlays.

In summary, despite globalisation and macroeconomic convergence, there is still room for national and regional specialisation. The current specialisation patterns of the Dutch I&C cluster (for example, its leading position in some telecommunication sub-markets, parts of the content and I&C service industry related to trade and transactions) clearly reflect the importance of historical specialisation patterns,[17] and the need to nurture these strengths, even in increasingly global markets such as the market for I&C goods and services.

A comparison with more recent data (input-output data, new innovation survey) on the I&C cluster as they become available, will allow us to update the current picture of the Dutch I&C cluster and to provide a more dynamic picture of the ongoing convergence process. This could provide an indication of how successfully firms, sub-clusters of co-operating firms and policy makers are adapting to the new networked economy.

Technical Annex

CLASSIFICATION OF THE INFORMATION & COMMUNICATION (I&C) INDUSTRIES[1]

CBS 1997 study	
ICT manufacturing activities	
22300	Reproduction of van recorded media
30000	Manufacture of office machinery and computers
31300	Manufacture of insulated wires and cables
32100	Manufacture of electrical components
32200	Manufacture of telecommunication and broadcasting (RTV) transmission equipment
32300	Manufacture of audio and video equipment
33200	Manufacture of electrical measuring and controlling equipment
ICT service activities	
64000	Post and telecommunication services
72000	Computer services and IT consultancy
Art & cultural activities	
22119	Publishers
22219	Printing
71400	Video retail
74400	Advertising
74800	Other business services[2]
92100	Film and video related activities
92200	Radio and broadcasting services
92400	Press and news agencies, journalists
92500	Museums and libraries

1. In order to analyse the innovation survey data, these categories were translated into the standard industrial classification '93 and '74 used by CBS. The categories were then linked to the (different) I/O categories used in Section 4 of this chapter. These conversion tables are included in the more extensive report with the same title on which this paper is based. It can be obtained from the authors upon request.
2. An attempt was made to separate out photo development centres, taking as a starting-point photography and development of films and photos.

NOTES

1. Although the key technology is ICT, this chapter focuses on the industries that shape ICT. As the focus is on the cluster of industries, including those industries providing content, rather than on the technology, the term information and communication (I&C) industries or I&C cluster has been used.

2. In reality, some chains become obsolete (mainly physical outlets of products and services as these are replaced by electronic services) and new chains come on stream (for example, new categories such as Internet service providers, telecommunication service companies reselling capacity, or resellers of mobile phones).

3. Similar reports were produced in Canada and Finland. The content industries in these two countries were defined more broadly. Figures should be used and interpreted with caution as comparisons with other sources such as EITO and OECD reveal important – and partly unexplained – differences.

4. For more detailed tables on the Dutch ICT market, see CBS (1997b).

5. According to the OECD, in 1995 11.6% of value added in Dutch manufacturing can be attributed to ICT-producing industries (using different classifications, ICT includes office and computing machinery; radio, TV and communication equipment and scientific and professional instruments); this means that the Netherlands ranks among the countries with the highest specialisation in ICT industries.

6. Software products are barely visible in the Dutch trade statistics. Imports of packaged software cannot be assessed.

7. This partly implies entering *terra incognita*. We stayed as close as possible to the original 18 economic "activities" defined by the CBS as ICT industries (see the Technical Annex for details). Another limitation was that the most recent innovation data dated from 1992; the data are thus fairly old for industries that are developing at such a fast pace. However, using the 1997 innovation data that will be published at the end of this year and as more recent I/O data become available, we hope to be able to update the findings presented in Sections 4 and 5 of this chapter.

8. For more detail, see the Technical Annex. Although a 1993 I/O table is available, 1992 data was used as the innovation survey data used in Section 5 are from 1992 as well.

9. See den Hertog and Maltha (1998) for more details.

10. Basically, the same selection of industries was used as in the I/O analysis. See the Technical Annex for details. The firms included in the current analysis in the I&C cluster were originally placed in four mega-clusters: media, metal-electro, transport and commercial services.

11. A finding that is mainly reported in the more recent research on innovation in services. See, for example, Gallouj and Weinstein (1997) and Bilderbeek *et al.* (1998).

12. The consultancy industries are believed to play a role as diffusers of new knowledge to their clients comparable to the role traditionally associated with the public knowledge infrastructure. This notion of a "second (private) knowledge infrastructure" in addition to the formal "first (public) knowledge infrastructure" and the various mixes of the two is further developed in den Hertog and Bilderbeek (forthcoming) and Bilderbeek *et al.* (1998).

13. This summary is largely based on various White Papers such as the *Action Programme for the Electronic Superhighway – From Metaphor to Action* (December 1994) and *Boven NAP – herijking*

van het Nationaal Actoeprogramma Elektronische Snelwegen (April 1998), and other documents related to individual activities, progress reports and benchmark studies. Most noteworthy, from the perspective of cluster policy making, is the report, *Netherlands' ICT Twinning Centers and Investment funds. Building the Mind-set and the Skill Base for the Information Society* (February 1998) by Booz, Allen & Hamilton in co-operation with the Ministry of Economic Affairs. See Box 2.

14. The first phase of Action Programme 6 (originally 7), defined a number of programme lines. For an overview, see de Hertog and Maltha (1998).

15. Which, in turn, is an amalgamation of actions such as raising awareness, establishing an electronic commerce platform, developing tools to help firms implement electronic commerce and looking into the possibility of electronic tendering by the government.

16. An international benchmarking of I&C clusters could considerably enhance insights into the systemic imperfections of the Dutch I&C clusters.

17. Supporting the notion that clusters can be interpreted as reduced-scale national innovation systems.

REFERENCES

Bilderbeek, Rob, Pim den Hertog, Goran Marklund and Ian Miles (1998), "Services in Innovation: Knowledge-intensive Business Services (KIBS) as Co-producers of Innovation", SI4S Synthesis Paper No. 3, STEP, Oslo.

Booz, Allen & Hamilton in co-operation with the Ministry of Economic Affairs (1998), *Netherlands' ICT Twinning Centres and Investment Funds. Building the Mind-set and the Skill Base for the Information Society*, Amsterdam/The Hague.

Booz, Allen & Hamilton (1998), *Benchmarkstudie elektronische diensten: op weg naar de informatie-maatschappij. Internationaal vergelijkend onderzoek t.b.v. de herijking van het actieprogramma Elektronische Snelwegen*, Onderzoek in opdracht van het Ministerie van Economische Zaken, Amsterdam/The Hague.

CBS (1997a), *Kennis en economie 1997*, Voorburg/Heerlen.

CBS (1997b), *ICT-markt in Nederland (1990-1995)*, Voorburg.

Edquist, Ch. (ed.) (1997), *Systems of Innovation. Technologies, Institutions and Organizations*, London.

Gallouj, F. and O. Weinstein (1997), "Innovation in Services", *Research Policy*, 26, pp. 537-556.

Hertog, P. den and R. Bilderbeek (forthcoming), "The New Knowledge Infrastructure: The Role of Knowledge-intensive Business Services in National Innovation Systems", report prepared in the framework of the SI4S project, to be published in I. Miles (ed.), *Services, Innovation and the Knowledge-based Economy*.

Hertog, P. den and G. Fahrenkrog (1993), "Adoption: Can Policy Help. Experiences and Trends in Five EC Member States", Discussion paper, TNO Centre for Technology and Policy Studies, Apeldoorn (STB/93/25).

Hertog, P. den and Maltha, S.R. (1998), *Convergentie, concurrentie en divergentie. Op weg naar gezonde infrastructuurconcurrentie voor multimediadiensten*, VECAI, The Hague.

Mansell, R., A. Davies and W. Hulsink (1995), "The New Telecommunications in the Netherlands. Strategic Developments in Technologies and Markets", Rathenau Institute, The Hague.

Ministry of Economic Affairs (1994), *Action Programme for the Electronic Superhighway – From Metaphor to Action*, The Hague.

Ministry of Economic Affairs (1997), *Opportunities through Synergy. Government and the Emergence of Innovative Clusters in the Market*, White Paper, The Hague.

Ministry of Economic Affairs, Ministry of Traffic and Water Management, Ministry of Education, Culture and Sciences, Ministry of the Interior, Ministry of Justice, Ministry of Social Affairs and Employment (1998), *Boven NAP. Herijking van het Nationaal Actieprogramma Elktronische Snelwegen*, The Hague.

OECD (1999), *Managing National Innovation Systems*, Paris.

Roelandt, Theo, Pim den Hertog, Jarig van Sinderen and Norbert van den Hove (1999), "Cluster Analysis and Cluster Policy in the Netherlands", this volume.

Witteveen, W. (1997), *Clusters in Nederland, een verkenning van het bestaan van clusters van toeleveranciers en afnemers en de implicaties voor overheidsbeleid*, Ministry of Economic Affairs, The Hague.

Chapter 9

INNOVATION PROCESSES AND KNOWLEDGE FLOWS IN THE INFORMATION AND COMMUNICATIONS TECHNOLOGIES (ICT) CLUSTER IN SPAIN

by

Cristina Chaminade*
Universidad Autónoma, Madrid

1. Introduction

Recent theoretical work on technical change and innovation has stressed the role played by external sources of knowledge and interactive learning. Since the emergence of the interactive model described by Kline and Rosenberg (1986), the complex process of knowledge acquisition and transfer has become central to the analysis of how firms innovate. In their production processes, firms need to interact intensively with a wide spectrum of institutions. The majority of these interactions take place within the same sector or around a group of highly integrated technologies.

The starting-point of this chapter is the theoretical and, in some cases, empirical evidence that the competitiveness of any system of innovation relies on the capacity to transfer and exploit the available knowledge in the system (David and Foray, 1995). Innovation stems from the re-combination of existing knowledge. The better the access to external sources of knowledge, the more competitive the firm (and the cluster) will be. For policy-making purposes, this implies that enhancing the knowledge flows between the actors in an innovation system will improve the competitiveness of the system and thus of the whole economy.

Compared to the traditional sectoral approach, the cluster approach focuses on the networks between firms, knowledge infrastructure, bridging institutions, suppliers and customers, usually within the same technological area (see Chapter 1 of this volume). In this sense, it has some elements in common with the technological system (Carlsson, 1995), and with the sectoral system of innovation approach developed in modern innovation theory. The literature on modern innovation theory suggests that the competitiveness of a cluster can be enhanced by improving its knowledge acquisition and knowledge transfer mechanisms. For policy purposes, this finding is extremely important as it implies that improving knowledge flows can enhance the innovative level of any given system of innovation. Thus, in a cluster where innovation is considered key for competitiveness, this goal can be achieved through policies focusing on stimulating knowledge flows within the system.

---

* This chapter describes the results of a research project directed by Prof. M. Paloma Sanchez of the Universidad Autónoma, Madrid.

This chapter presents the case of the electronics and telecommunication cluster in Spain. This cluster plays a strong and double role in the Spanish economy: it is a key sector in its own right and, in addition, the technologies of this sector, which is at the heart of the knowledge-based economy, have a significant impact on other sectors of the economy. For these reasons, this cluster and more specifically, the new information and communication technologies (ICT), are considered as key areas in most industrial and technological public support programmes.

2. Composition of the ICT cluster

2.1. Identification of the cluster: results of I/O analysis

The first stage in the analysis is to identify the cluster. To do so, we used both input/output tables and qualitative analysis. I/O tables look at trade linkages among different industries. The qualitative analysis complements the quantitative analysis, providing information on the flows between the industries in each category. The main limitation of I/O analysis is the level of aggregation of the data: 56 sectors. The methodology used here is that developed by Monfort and Dutailly (1983), and adapted by Roelandt *et al.* (1999).[1]

The analysis focuses on the categories most closely related to the ICT cluster. "Office machinery and computers" and "Telecommunications" were judged to be the two main categories in which the industries within the sector were suitable for analysis. Figures 1 and 2 present the main results.[2]

The main supplier of the "Office machinery and computers" category is electrical manufacturing, with the main users being health services (market- and non-market-oriented) and the public services. For "Telecommunications", the main supplier is the broad group of services to enterprises, while the main users are financial and banking services, and hotels and restaurants. The first conclusion from the analysis is that the public sector plays an important role in the cluster structure. The I/O analysis did not consider intra-sectoral trade flows. This implies that, due to the high level of aggregation of the tables, the analysis does not provide information on the role of intra-sectoral linkages, but looks only at inter-sectoral trade flows. We estimated the amount of the intra-sectoral flows. For the "Office machinery and computers" sector, the ratio of what can be labelled self-demand accounts for 45% of the total amount. Although electrical manufacturing represents 48% of the remaining trade flows of the sector, the main suppliers are firms within the sector. For the communications category, the ratio is 25%.

2.2. Composition of the Spanish ICT cluster

Following the "product" approach to cluster definition, we considered as part of the cluster all industries involved in the production of goods and services based on the new information technologies, including electronics and telecommunications. Following this definition, most national and international studies agree that the sector is composed of the following industries:

- consumer electronics;
- electronic components;
- professional electronics;
- telematics (telecommunications and computing);
- service providers (OPST).

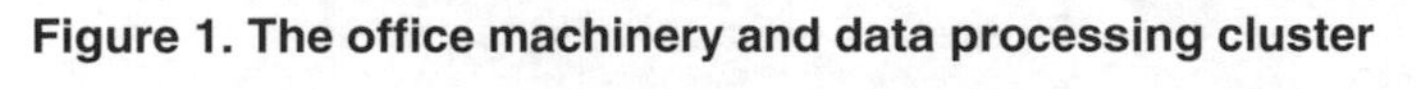
Figure 1. The office machinery and data processing cluster

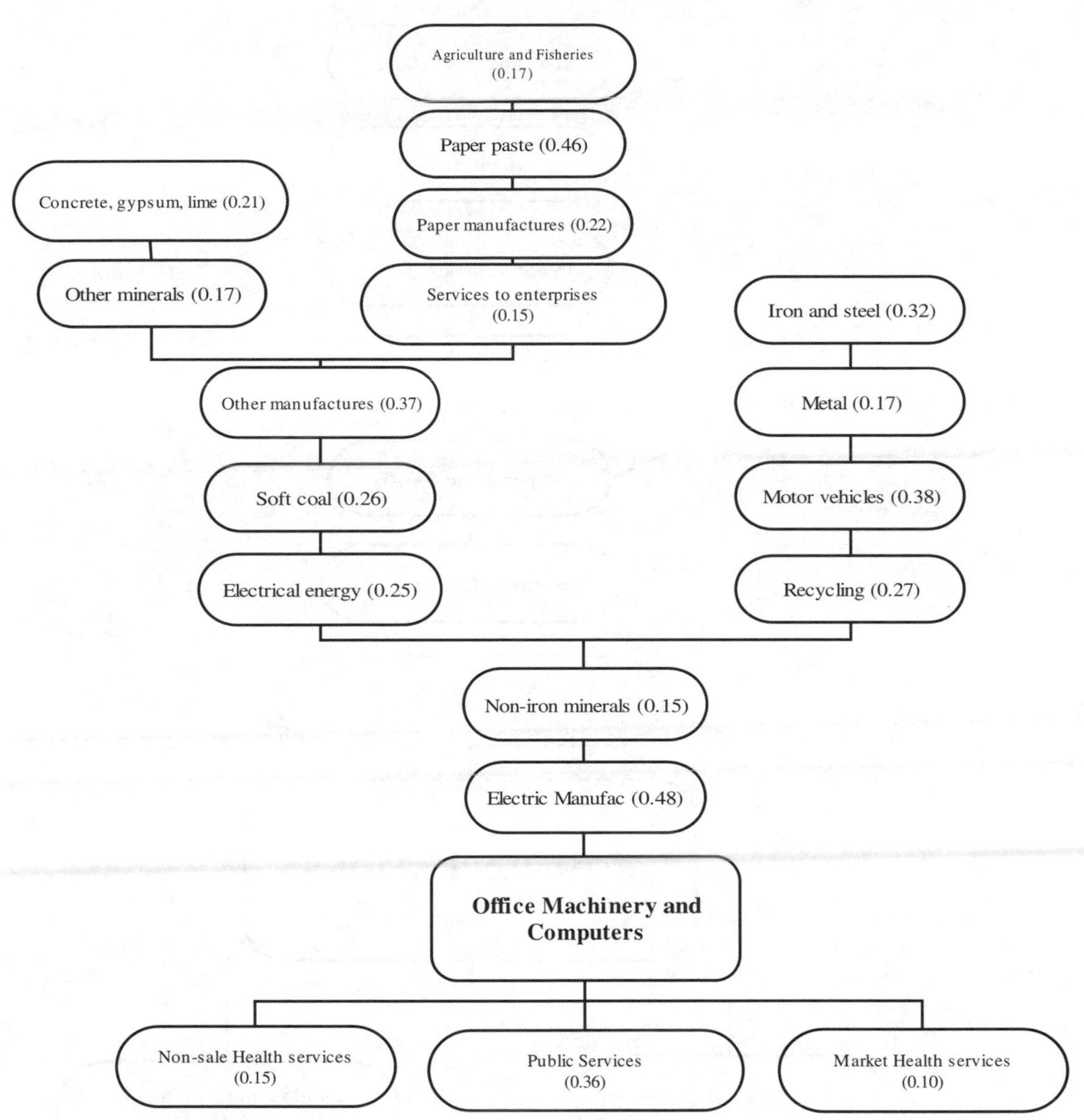

Figure 2. Communications sector

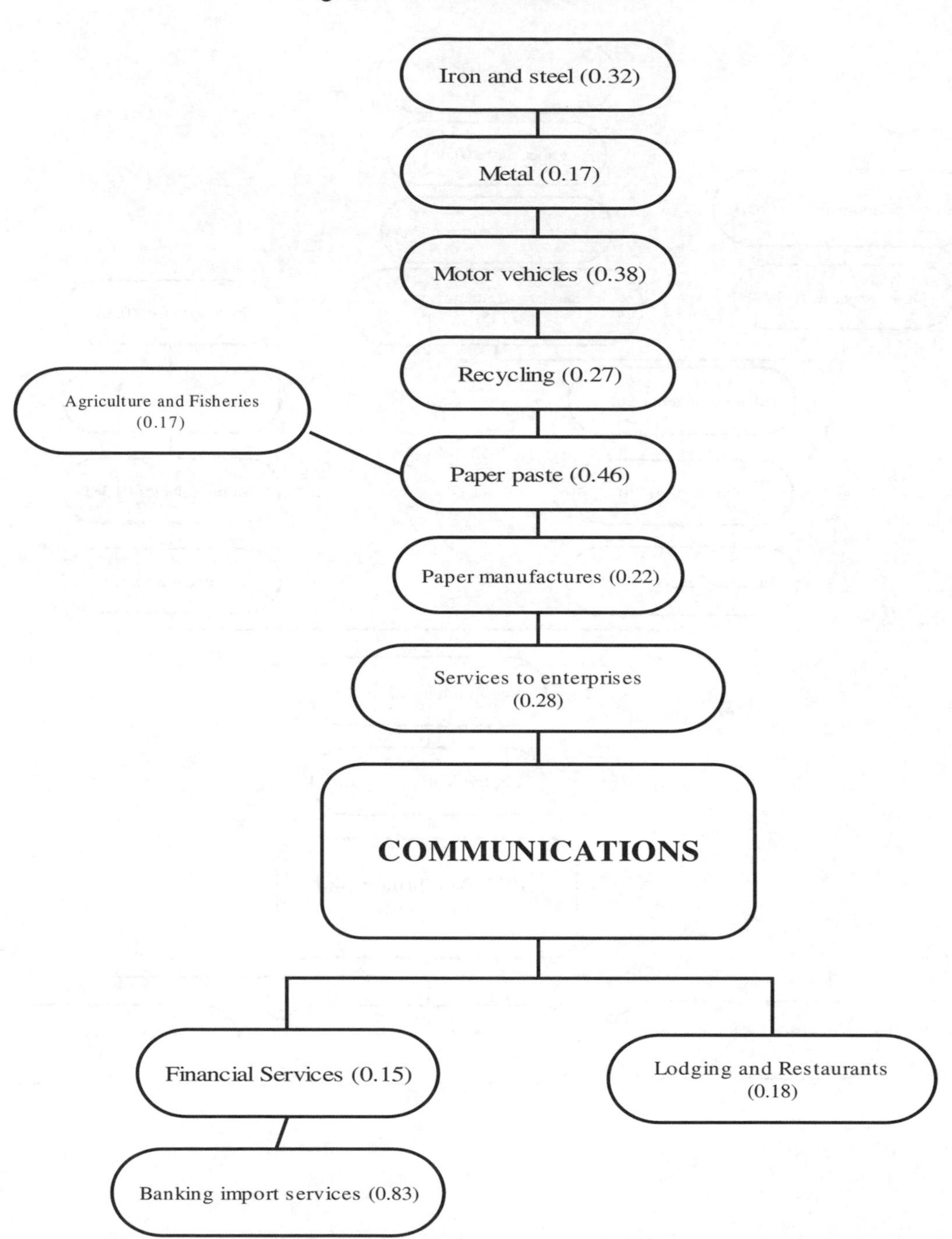

To complete the picture, we conducted a small survey[3] among different experts on clusters in order to obtain further information on the main producers, suppliers and users of each of the industries in the cluster. Figure 3 provides more detailed information on the main users and suppliers of the cluster and on the internal linkages between the different industries.

Figure 3. Composition of the Spanish ICT cluster

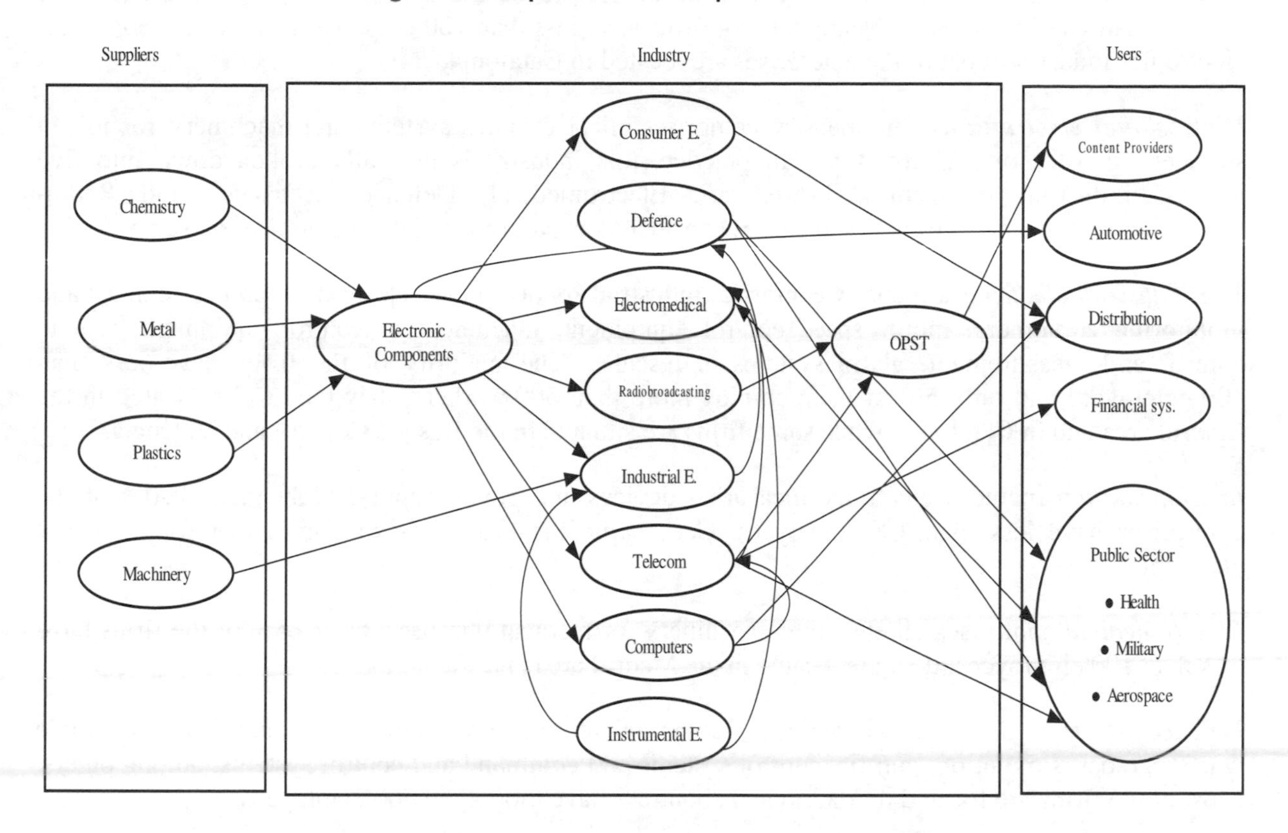

The analysis highlights the complexity of internal networks. As the interviews showed, these networks are based not only on flows of high-tech products and services but also on the intensity of the knowledge flows among the different actors.[4] The linkages are established on the basis of highly interconnected technologies. The last decade has been characterised by the progressive convergence of these technologies. Outside the cluster, as suppliers and users, we identify the chemicals, metal manufacturing, industrial equipment and the automotive cluster (which is one of the main users of electronic components), distribution, financial services and the public sector, which bears out the results obtained from the I/O analysis.

3. Main features of the Spanish ICT cluster

As pointed out above, the cluster encompasses six main industries: Consumer electronics, Electronic components, Professional electronics, Telecommunications, Information processing and Service providers. Each of these industries is briefly described below:

Consumer electronics: The composition of this industry is currently being redefined to include personal computers and telephones. Hitherto, it was made up of electronic goods for domestic use only, *i.e.* mainly TV, video and hi-fi. Over half of the manufacturers included in this group employ less than 50 workers; this percentage goes to the 70% if enterprises with less than 100 employees are included. Manufacturers are highly concentrated in Catalonia. In Spain, this cluster is characterised by the weighty presence of multinationals such as Sony, Samsung, Grundig, Panasonic, Sanyo, Philips, Kenwood, Nokia, Pioneer, Sharp or Thomson.

Electronic components: Includes all electronic devices to be incorporated in the final products of other industries of the cluster. Nearly 80% of firms have less than 100 employees. As for the consumer electronics industry, most of the enterprises are located in Catalonia.

Professional electronics: This industry comprises all electronics systems and machinery for use in services, industry or infrastructures. In practice, this industry is normally broken down into five groups: Industrial electronics, Instruments, Electromedical, Defence electronics and Radio-broadcasting (this latter group is sometimes considered as part of the Telecommunication sector).

The *industrial electronics* industry comprises industrial robots, power electronics, process control and monitoring equipment, motor speed control equipment, machine tool control equipment, remote control and measurement, alarm systems and scales. The majority of the firms have less than 100 employees and only 5% of them employ more than 500 workers; they tend to be located in the Madrid area and in Catalonia, while some firms are situated in the Basque Country and Valencia.

Instrumentation includes electronic measuring devices and power supply. More than a 90% of the enterprises have less than 100 employees. Geographic localisation is similar to that for industrial electronics.

Electromedical comprises all electronic machinery for medical purposes. Over 85% of the firms have less than 100 employees; they are located in the Madrid area and Catalonia.

Defence electronics comprises avionics, communications, navigation and equipment, electronic warfare, radar, simulation, gun fire control systems and command and control systems, among others. Most of the firms are located in Madrid and about 2% have more than 500 employees.

Finally, *Radio-broadcasting and TV* basically consists of radio-broadcasting transmitters, TV transmitters, TV and FM transposers, closed circuit TV, audio/video professional equipment materials, Hertzian links, etc. As mentioned above, according to the criteria used, these industries can either be included in Professional electronics or in Telecommunications. The geographical distribution is similar to that for Industrial electronics and 80% of the firms have less than 100 employees.

Telecommunications: Includes all equipment for diffusing and treating information, images and sound. A distinction can be made between telecommunication equipment and services. The majority of the firms are located in Madrid and Catalonia. More than 40% of these firms have more than 100 employees.

Information processing: Includes equipment, software and services related to the edition, storage and retrieval of information. Traditionally, this industry is sub-divided into two groups: hardware and software. Firms in this industry are mainly located in Madrid and Barcelona; the majority are small and medium-sized enterprises. Few have more than 500 employees (with the exception of a few multinationals such as IBM, AT&T [now Lucent Technologies], Ericsson, Fujitsu, Hewlett Packard, Siemens or Sony).

Service providers: Includes cable, mobile and satellite communication services, data transmission, networks and value-added services. Firms are mainly located in Madrid and the average size of firm is higher than in any other group.

The economic indicators of the cluster as a whole were analysed and compared to the Spanish economy, based on an ANIEL annual report (1998). It was found that the ICT cluster has significant weight in the economy in terms of both production and demand. Also, the average growth rate of the cluster is significantly above that for the Spanish economy as a whole.

4. Innovation patterns and knowledge flows in the ICT cluster: results of the analysis

This section is based on the results of a survey of the ICT cluster. The survey was conducted in two phases. During the first phase, information was collected through a questionnaire which was sent to 123 enterprises. The response rate was 46.4% (57 valid answers). The sample was highly representative. Production of the 57 enterprises (ESP 1.211.894 million) accounted for 92% of the total production of the cluster (ESP 1.318.466 million) in 1996.

During the second phase, five in-depth interviews were held with the managers of Ericsson, Amper, ELIOP, Telefonica and Alcatel. The conclusions of these interviews were used in the interpretation of the questionnaire results.

4.1. Description of the sample industries

Figure 4 shows the average size of the firms in terms of number of employees. Clearly, 65.5% of the firms have fewer than 250 employees, and only 14.5% employ more than one thousand. This is a clear reflection of the Spanish industrial structure, which is dominated by small and medium-sized firms.

Figure 4. Average size of the firms in the sample
Number of employees

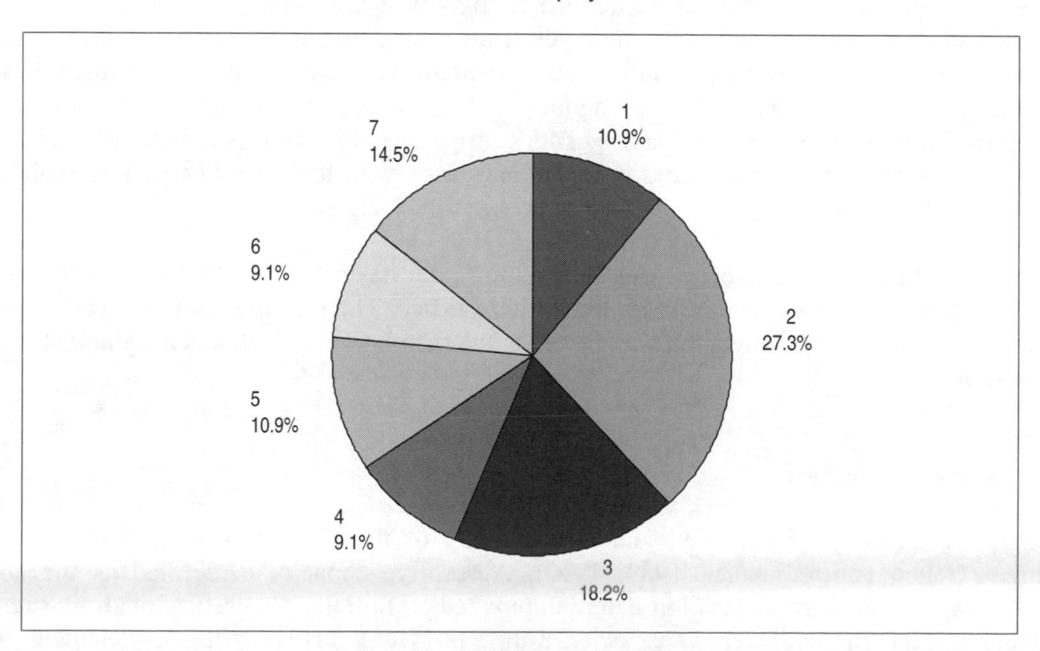

Note: **1.** Less than 19 employees; **2.** 20-49 employees; **3.** 50-99 employees; **4.** 100-249 employees; **5.** 250-499 employees; **6.** 500-999 employees; **7.** More than 1 000 employees.

However, if turnover is considered, the picture changes significantly. Figure 5 shows that more than 63% of the firms in the survey have a turnover of over ESP 1 000 million (roughly USD 7.14 million).

Figure 5. Distribution of surveyed firms, by turnover

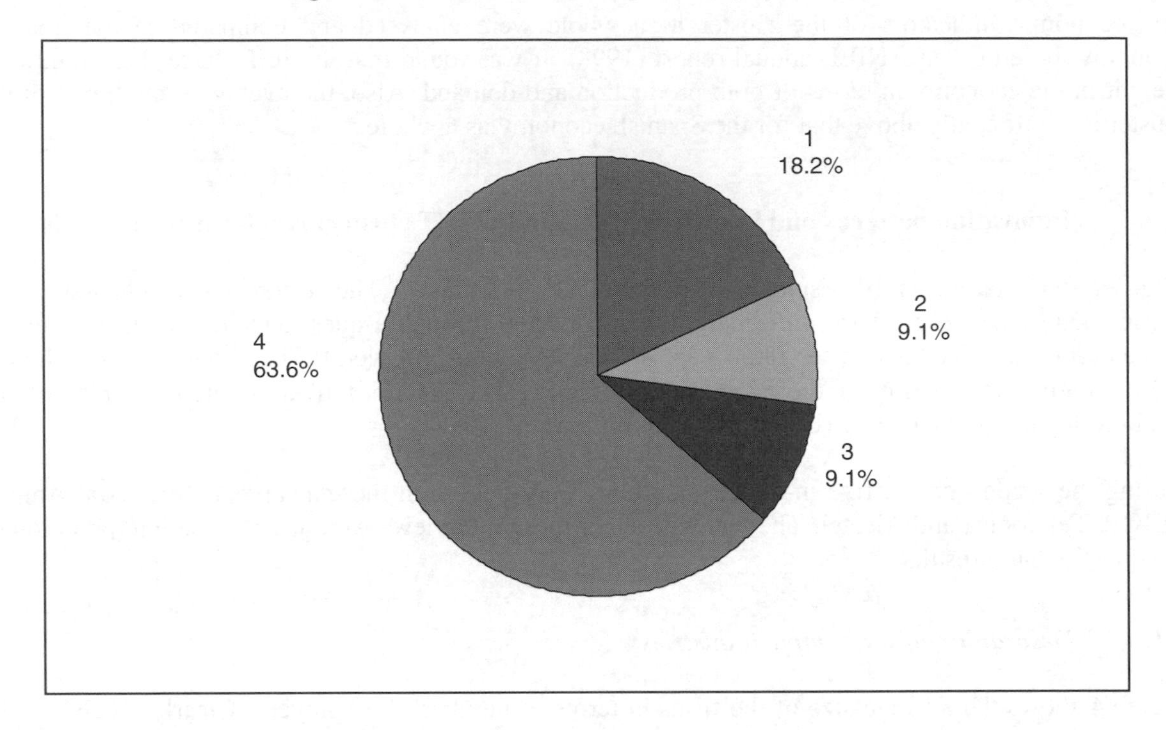

Note: **1.** Less than ESP 250 million (USD 1.8 million); **2.** Between ESP 251-400 million (USD 1.8 to 2.8 million); **3.** Between ESP 401-1 000 million (USD 2.8 to 7.1 million); **4.** More than ESP 1 000 million (over USD 7.1 million).

Geographically, the majority of the firms are located in Madrid and Catalonia, and to a lesser extent in the Basque Country. The majority of the surveyed firms were domestic (63.3%), but the proportion of multinationals compared to other industries in the economy is relatively high: more than 35% of the firms in the cluster are multinationals. Among these firms, the majority are subsidiaries: of the 35% of multinational firms, only 9.1% are Spanish, 14.5% originate other European countries and the rest (12.7%) are from abroad. Domestic and multinational firms with their head offices in Spain account for more than 72% of firms.

A large number of the multinational firms located in Spain have their R&D departments elsewhere. Also, their competitive strategy is defined in their head office. The managers interviewed stressed that these two factors necessarily affect the way in which the firm interacts with its environment, as will be shown Section 4.3.

4.2. Innovation patterns

The absorptive capacity of the firm, and consequently of the cluster, is built upon the stock of knowledge (Cohen and Levinthal, 1989; 1990). Absorptive capacity is defined as the ability to efficiently examine, acquire and exploit external knowledge. Initially, theoretical analysis stressed the role played by R&D in the absorptive capacity of the firm. But R&D is not the sole element: training, innovation, type of organisation or even the efficiency of communication channels all affect the absorptive capacity of the firm. In this section these elements will be examined in the light of the questionnaires and interviews.

If we define the innovative firm as one that has introduced new products, processes and services into the market during 1997, the majority of the firms in the cluster can be considered as innovative. Only a small proportion (5%) of the firms are not innovative. This introduces an important bias into the analysis: since the share of innovative firms in the economy as a whole is only 10.71% (INE, 1998), we can conclude that this cluster is highly innovative.

As is shown in Figure 6, during 1997 most of the firms introduced into the market new products that were both new to the market and new to the firm as well as processes that were mainly new to the firm. Only a small proportion of firms introduced processes that were new to the market or innovations in services.

Figure 6. Distribution of firms by type of innovation

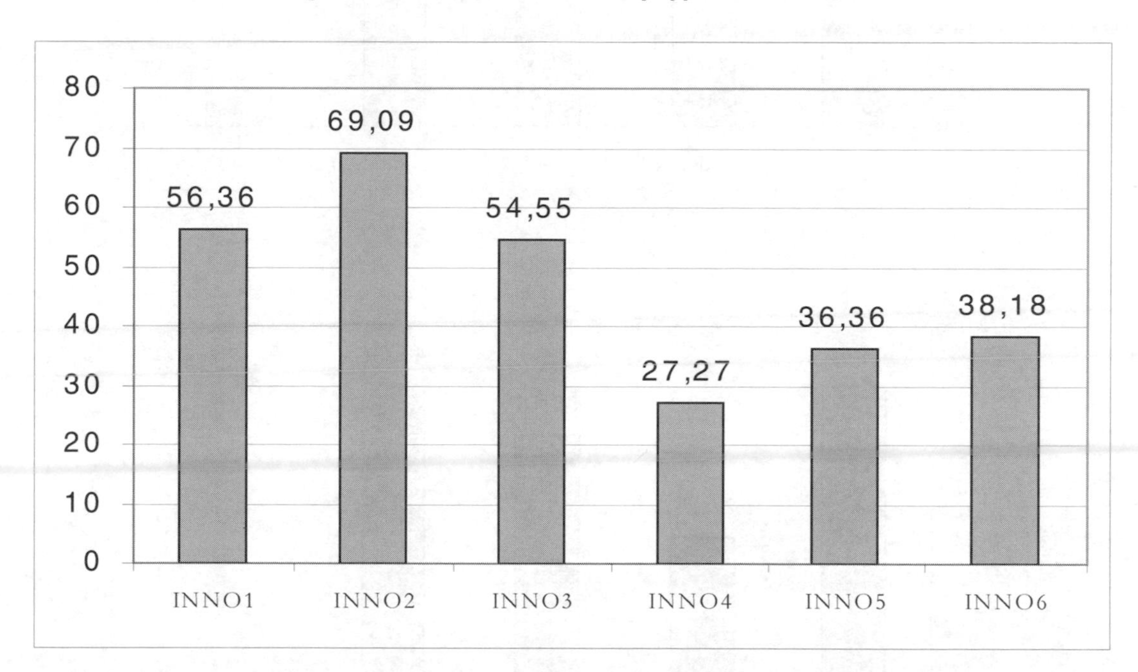

Note: INNO1: New products, new to the firm; INNO2: New products, new to the market; INNO3: New processes, new to the firm; INNO4: New processes, new to the market; INNO5: New services, new to the firm; INNO6: New services, new to the market.

The innovative level of these firms is not news. Several studies, such as that conducted by Pavitt *et al.* (1996) based on the Community Innovation Survey, have shown that the share of sales of new products compared to the industrial average of the Radio, TV and communications sector in Spain was the highest of all the European countries considered (Benelux, Denmark, France, Germany, Ireland, Italy and the Netherlands).

In introducing innovations into the market, most of the surveyed firms went through organisational changes. In our sample, nearly 71% of the firms introduced organisational changes over the 1996-97 period. Among these, the most significant changes related to the structure of the firm and to human resources.

Figure 7. Type of organisational innovation

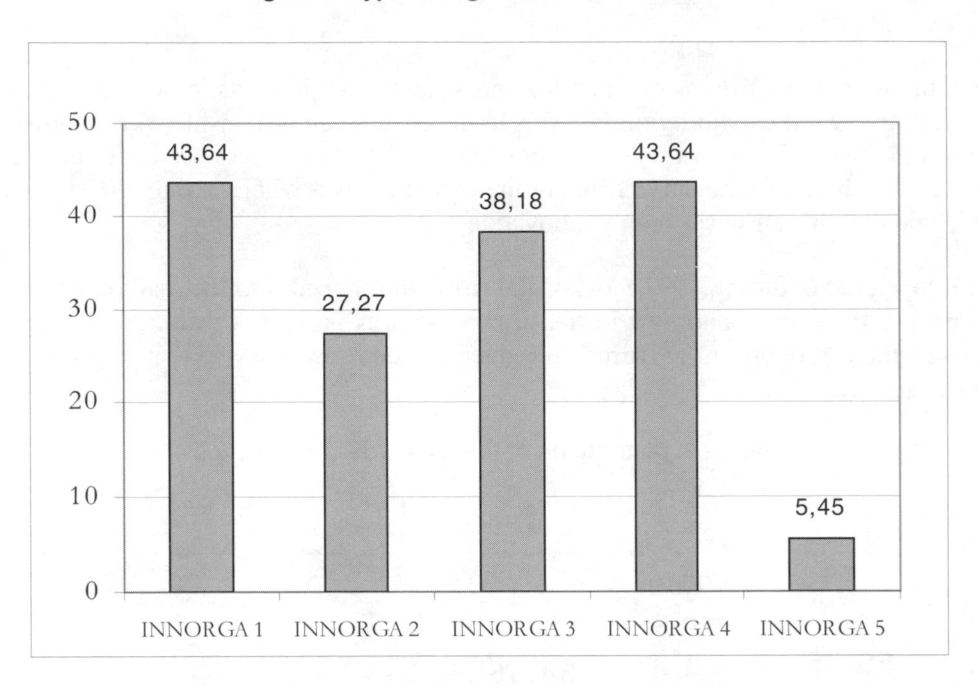

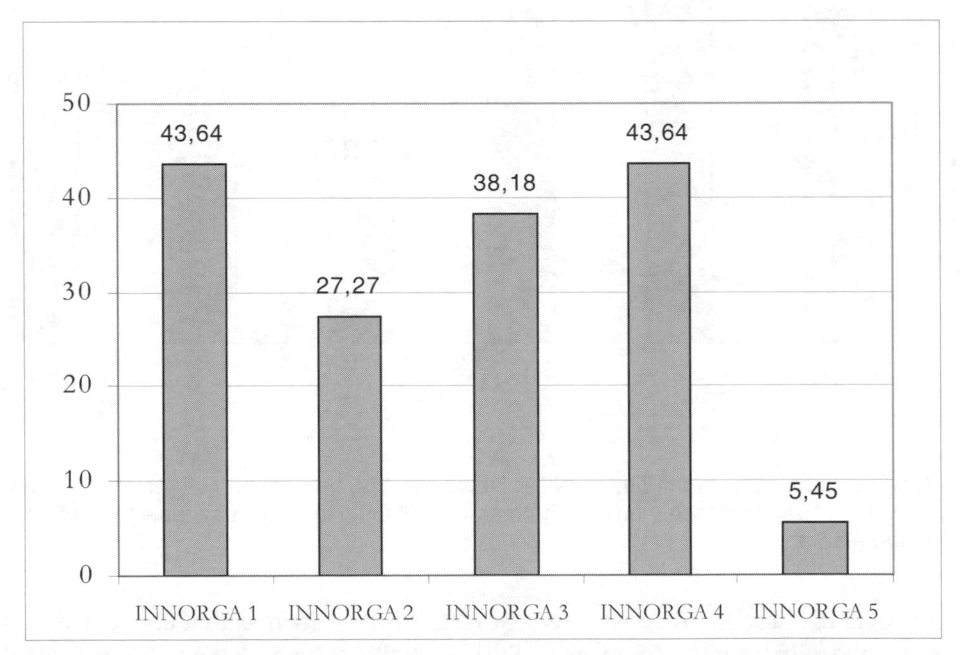

Note: INNORGA1: Human resources; INNORGA2: Strategy; INNORGA3: Information systems; INNORGA4: Structure; INNORGA5: Other (process management and administrative services).

The economic crises of the beginning of the 1990s, together with the recent liberalisation of the telecommunications sector, have provided an incentive for firms to innovate internally. Firms have had to adapt rapidly to changing conditions in order to maintain and sometimes even increase their market share.

Table 1. Evolution of R&D activities in the ICT cluster, 1991-95

		1995	1994	1993	1992	1991
Total	**R&D expenditures**					
	in relation to its GDP	8%	7%	8%	8%	7%
	Total R&D expenditures	63647	47000	45000	50000	50000
	Salaries	n.d.	61%	60%	57%	57%
	Equipment	n.d.	24%	25%	29%	29%
	Other	n.d.	15%	15%	14%	14%
	R&D total employment					
	in relation to total staff	10%	n.d.	9%	8%	8%
	Total employment	3769	n.d.	3800	3924	3965
Consumers	**R&D expenditures**					
	in relation to its GDP	3%	3%	4%	4%	4%
	Total R&D expenditures	3800	3600	4200	4952	5217
	Salaries	60%	60%	62%	n.d.	n.d.
	Equipment	15%	11%	12%	n.d.	n.d.
	Other	25%	29%	26%	n.d.	n.d.
	R&D total employment					
	in relation to total staff	5	n.d.	4,5	5	5
	Total employment	220	n.d.	194	310	330
Components	**R&D expenditures**					
	in relation to its GDP	5.0%	4.2%	4.9%	6.3%	6.1%
	Total R&D expenditures	6000	4370	4200	5100	5350
	Salaries	65.00%	63.10%	61.30%	57.20%	56.00%
	Equipment	28.00%	23.80%	28.20%	31.10%	35.60%
	Other	7.00%	9.10%	11.70%	11.70%	13.40%
	R&D total employment					
	in relation to total staff	7.50%	7.10%	6.90%	7.30%	6.90%
	Total employment	451	418	410	460	480
Professional	**R&D expenditures**					
	in relation to its GDP	11%	10%	9%	9%	10%
	Total R&D expenditures	9700	8537	8070	8235	9842
	Salaries	57%	6%	59%	59%	50%
	Equipment	15%	14%	11%	13%	25%
	Other	28%	80%	30%	28%	25%
	R&D total employment					
	in relation to total staff	15%	15%	13%	14%	12%
	Total employment		825	859	1204	999
Telematics	**R&D expenditures**					
	in relation to its GDP	10%	12%	9%	8%	7%
	Total R&D expenditures	44169	47400	36450	25895	28537
	Salaries	n.d.	n.d.	n.d.	50.00%	57.20%
	Equipment	n.d.	n.d.	n.d.	12.50%	14.30%
	Other	n.d.	n.d.	n.d.	27.50%	28.70%
	R&D total employment					
	in relation to total staff	10.50%	9.50%	9.50%	9.40%	7.00%
	Total employment		2351	2415	2595	2154

Absolute values in million pesetas
* Ratios refer to the total amount of the sub-sector.

Source: ANIEL.

These changes have had a positive effect on the absorptive capacity of firms since they increase the ability to exploit internally the knowledge and technology acquired externally. The firm needs to promote internal knowledge flows, personal contacts, human resource mobility and, in summary, the creation of diversity and the emergence of new ideas based on the recombination of existing knowledge.

In terms of R&D investment, the ICT cluster spends more than all the other sectors in the economy. In 1997, the cluster invested ESP 124 000 million (USD 885.7 million) (ANIEL, 1998) in R&D activities, significantly above the amounts spent by the other sectors and higher than the average for the economy. From this point of view, the cluster is the most innovative in the economy. Table 1 resumes trends in R&D activities between 1991 and 1995.

In summary, in terms of the stock of knowledge, the cluster is highly innovative. The share of innovative firms is nearly 95%, while the average for the Spanish economy is a mere 10.7%. In most of the firms surveyed, the introduction of new processes, products and services has run parallel with changes in internal organisation, mainly related to structure and human resources. Finally, in terms of R&D, the cluster is clearly in a lead position.

4.3. Knowledge flows

This section will focus on the external relationships of the firm. Firms were asked not only about the significance of different institutions and the mechanisms of knowledge acquisition but also about the frequency of interactions with each of them. The institutions and mechanisms considered in the analysis are shown in Table 2.

Table 2. Main institutions and mechanisms considered in the analysis

EXTERNAL RELATIONSHIPS OF THE FIRM		DIFFICULTIES COOPERATING WITH UNIVERSITYES AND TECHNICAL INSTITUTES	
A	Tech. Assistance Univ.		
B	Training in Univ.		
C	H. Res. Temp. exchange	A	Timing
D	H. Res. from Univ.	I	Financial constraints
E	Public Research Institutes	L	Lack of agreement in the design of the project
F	Coop. R&D Univ.	D	Lack of specific demand
G	Academic Public.	H	Is not interesting for the firm
H	Other firms dif. ind	B	Is not interesting for the Univ
I	Consultancy	G	No tradition in cooperation
J	Other units	J	Lack of knowledge of the Univ
K	Informal contacts Univ.	K	Difficulties in finding a partner
L	Tech. Inst.	C	Lost of know-how
M	Conferences	F	Appropiate nfrastructures
N	Other firms same ind.	E	High coordination costs
O	Publications		
P	Suppliers		
Q	Users		
R	Trade fairs		
	IMPORTANCE		FREQUENCY
1	NO IMPORTANT	1	NEVER
2	OF LITTLE INTEREST	2	SELDOM
3	NORMAL	3	FREQUENTLY
4	SIGNIFICANT	4	ALWAYS
5	VERY SIGNIFICANT		

Suppliers and users are considered to be the main source of knowledge feeding into the innovative process, as shown in Figure 8. Firms frequently collaborate with suppliers and users. Trade fairs seem to be one of the main mechanisms facilitating interaction and, in general, personal contacts and publications are the principal means of acquiring the knowledge required for innovation.

A related question referred to the main obstacles for innovation. As will be seen in Section 6, a high proportion of firms pointed to the lack of qualified demanding customers. That is, even if users are considered key actors in the knowledge acquisition process, firms appear to be dissatisfied with the information they obtain from them.

Figure 8. External relationships in the Spanish ICT cluster

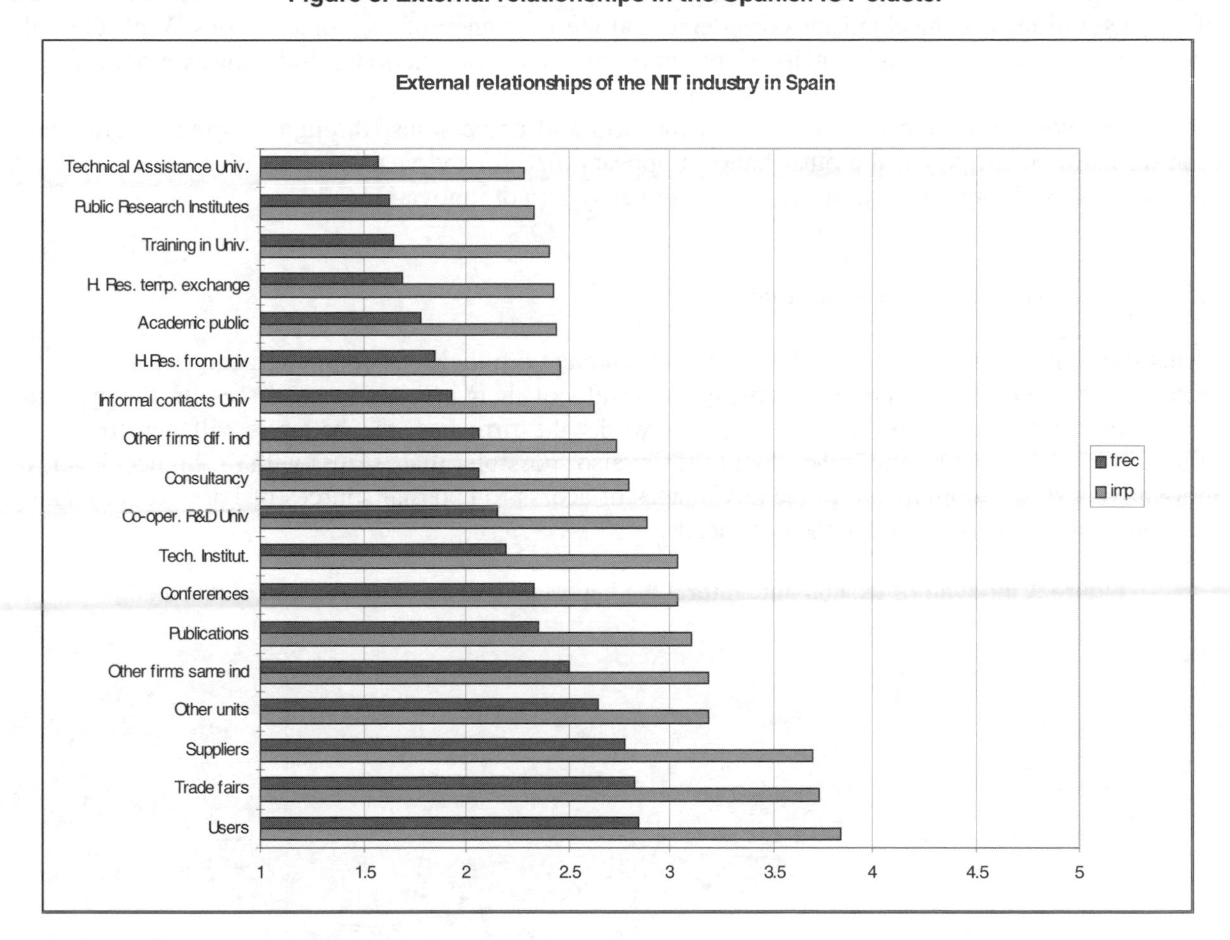

On the other hand, the results reveal weaknesses in firms' relationships with universities, technological centres and bridging institutions. In this sense, it could be said that the economic system has evolved apart from the academic system. This is, from our point of view, one of the main constraints for the development of the sectoral system of innovation.

One main explanation for this mismatch is that the various national R&D plans have largely promoted basic research in universities and research centres without sufficient encouragement to collaborate with firms in their R&D projects. In this sense, it could be claimed that academic research has not evolved closely enough to firms' needs. Although this situation is changing as policy programmes increasingly demand the participation of the future users of the invention (*i.e.* a shift to more market-oriented policy), the process will be slow as it involves two very different cultures (this point was stressed by the firms which responded to the survey).

4.4. *Multinational vs. local firms*

An interesting exercise is to compare the behaviour of national *vs.* multinational firms. Multinational firms are an important source of knowledge in any system of innovation and this is especially relevant for the ITC cluster in which nearly 35% of the firms are multinationals.

The results of our research illustrate the importance given by national and multinational firms to each of the institutions and mechanisms considered and the frequency of their interactions. With the only exception of suppliers and other units of the same enterprise, multinational find their external linkages with the local system of innovation less important than do national firms. In contrast, they co-operate more often with consultants, other units of the firm and universities (through training programmes) than do national firms. On the other hand, it appears that, for the vast majority of items considered, local firms have more interactions with the national system of innovation.

4.5. *Innovators vs. non-innovators*

Firms rarely innovate in isolation. They need to interact with their environment, acquiring knowledge from external sources and commercialising the results of their innovative processes. In the systemic approach to innovation, the more external networks the firm has, the better it will perform in its innovative activity. On the other hand, it is also possible that firms with a higher level of innovativeness are more aware of the advantages of access to external sources of knowledge, making it easier to select the knowledge the firm needs.

Figure 9. Innovators *vs.* non-innovators: the importance of external sources of knowledge

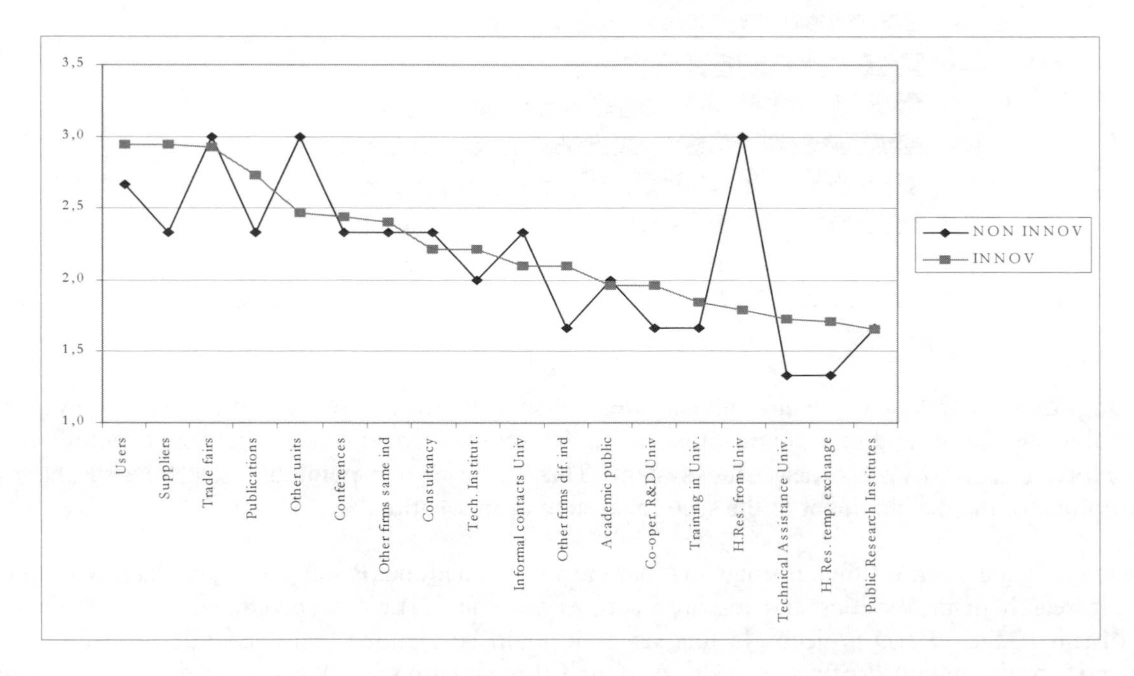

Figure 10. Innovators *vs.* non-innovators: frequency of interactions with external sources of knowledge

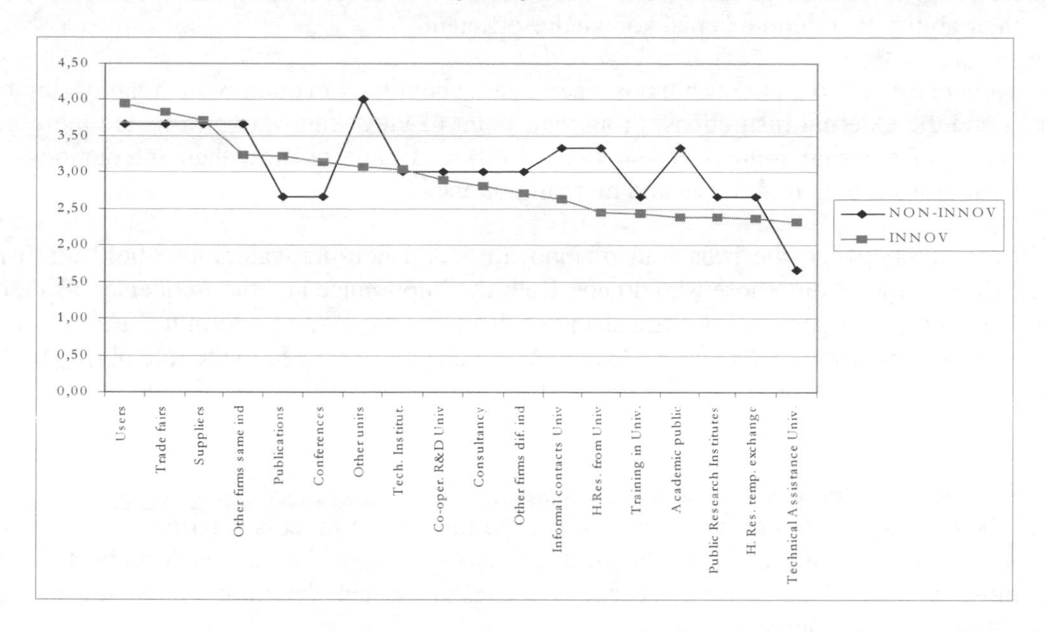

If we take innovative firms as those that had introduced new products, processes or services into the market during 1997, the results are not conclusive. Figures 9 and 10 show a very weak relationship between whether or not a firm is innovative and the way in which the firm interacts with its environment.

However, if, on the other hand, the definition of an innovative firm is changed to encompass firms that had invested in innovative activities during 1997 (*i.e.* the criteria is now the input rather than the output), the results are clear (Figure 11). In this case, the importance of each of the external sources is higher for the innovative firm, and the frequency of interactions is also higher.

Figure 11. Innovators *vs.* non-innovators according to inputs: importance of external sources of knowledge

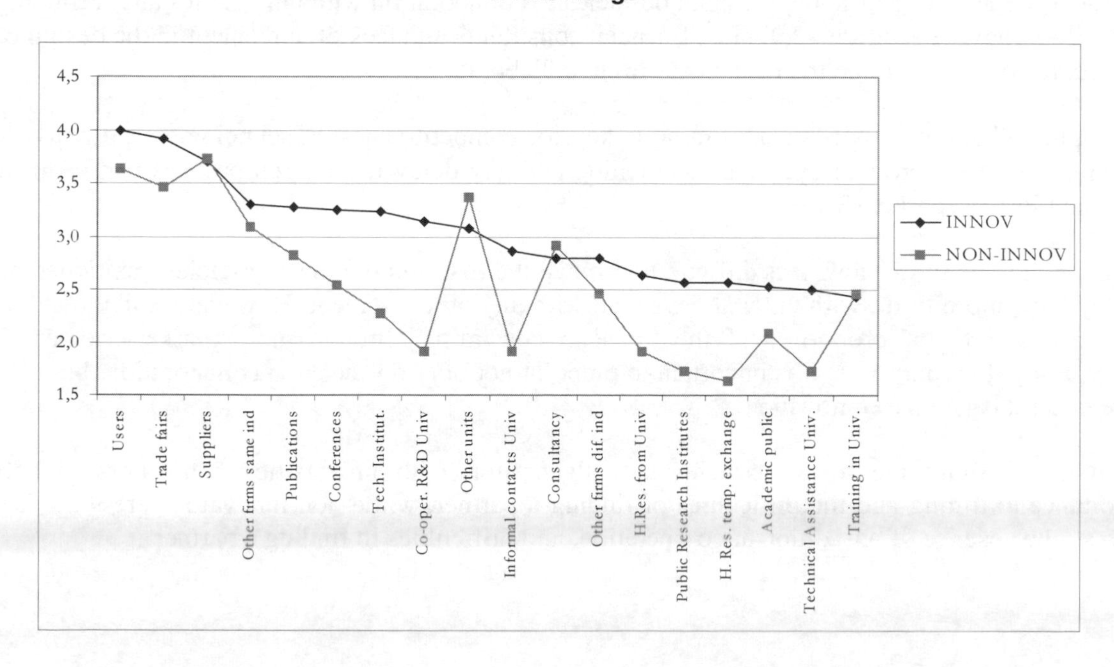

These results suggest that firms that invest in an innovation strategy improve their absorptive capacity and thus their ability to exploit external knowledge efficiently.

The managers interviewed were unanimous when asked about the direction of the relationship between innovators and the external institutions. From their point of view, innovative firms are more aware of the importance of external sources of knowledge and, as a consequence, their interactions with the external institutions are more frequent and more important.

These differences between the behaviour of innovators and non-innovators also hold for firms that undertake R&D projects and those who do not. Both the importance and the frequency of interactions are higher among R&D-performing firms than non-R&D firms. R&D-performing firms interact with more frequency than do non-R&D firms. Again, this result can be related to the role played by R&D in the absorptive capacity of the firm. The greater the firm's investment in R&D activities, the more it learns, thus increasing its ability to identify useful knowledge, recombine it in a useful manner and exploit it.

From the above analysis, it is clear that inputs into the innovative process determine the way in which firms interact with their environment. The most logical explanation for this result seems to be the link between those inputs and the absorptive capacity of the firm, which determines to what extent the firm can profit from external sources of knowledge.

From the managers' point of view, easy access to external sources of knowledge does not make the firm innovative. Rather, it is strategy – the presence of a dynamic entrepreneur – that makes the firm innovate. The firm has to build up capacity to access these external sources, the ability to identify that knowledge and to capture it and, finally, to exploit it. This is only possible if the firm has invested in its absorptive capacity.

5. Obstacles to collaboration with universities and research centres

Regarding the apparent mismatch between the manufacturing and academic systems mentioned in Section 4, we asked firms about the main obstacles to collaboration with universities and technological centres. As shown in Figure 12, timing, financial constraints and lack of agreement in the design of the project are considered to be the main obstacles to collaboration.

In a rapidly changing environment, timing is key for competitiveness. Technologies and markets are evolving very fast, product cycles are shortening and any delay in research projects can result in an important loss of market share.

From a policy point of view, it is difficult to remove the first and the third obstacles mentioned above as they have more to do with cultural gaps than with any other element. However, policy makers can remove a significant proportion of the financial constraints through more market-oriented R&D policy. For policy purposes, it is important to pinpoint not only obstacles to collaboration, but also the more general barriers to innovation.

When asked about the main obstacles to collaboration with universities, innovators give more importance to timing and the difficulties of finding a partner, while non-innovators stress the lack of interest, the absence of a tradition of co-operation, and difficulties in finding a partner.

Figure 12. Main obstacles to collaboration with universities and research centres

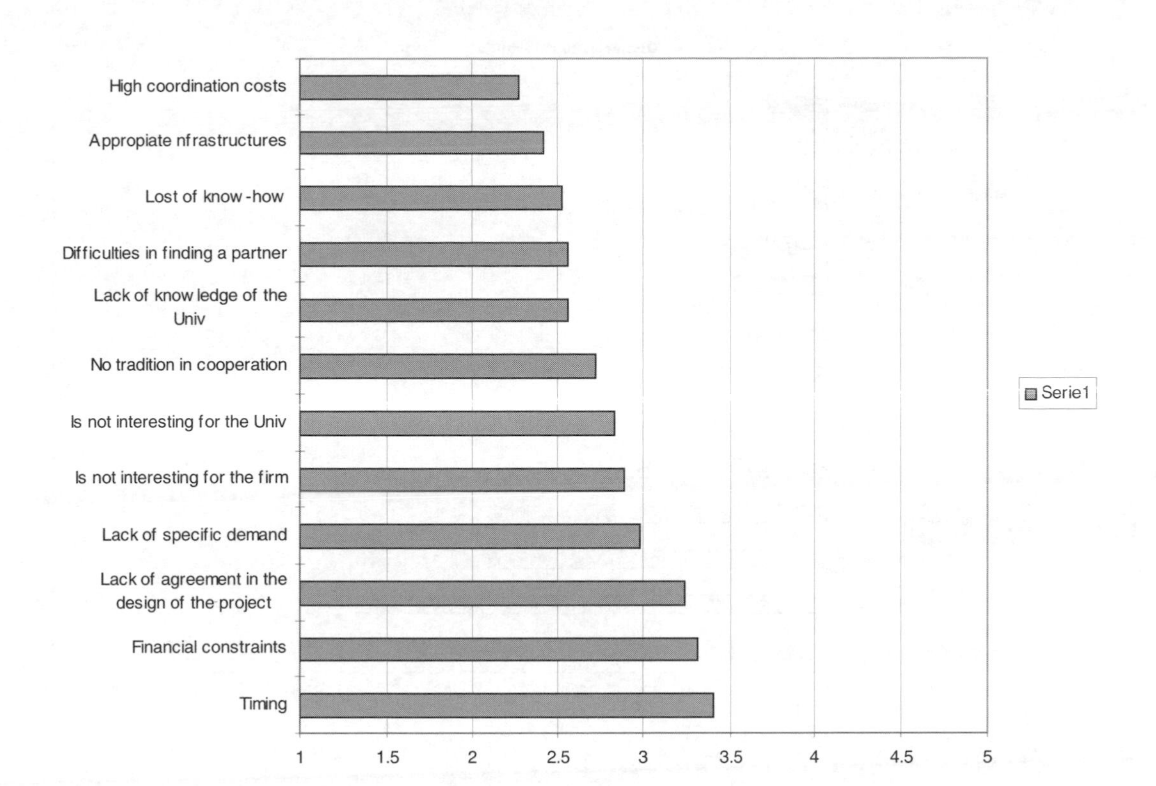

5.1. R&D vs. non-R&D-performing firms

In relation to R&D, the outcomes of our survey suggest that R&D-performing firms consider all kind of obstacles to be more important than do non-R&D firms. The main explanation for this could be that because R&D firms collaborate more frequently with universities and research centres, they are more aware of the difficulties of collaborating with other institutions in R&D projects. The main obstacles are the absence of appropriate infrastructures, lack of specific demand and, again, timing. This again is the result of different cultures, in the sense that basic research is not usually subject to terms or objectives, while the applied research and development normally conducted by firms is.

Some of the difficulties described above could clearly be overcome by an efficient network of bridging institutions providing two-way information: on the results of the innovative process in universities and research centres; and on firms' needs to researchers in academic institutions. This will be discussed in more detail in the policy section.

6. Incentives to innovation

The above section discussed the obstacles to collaboration with universities and research centres. To obtain information on the innovative behaviour of the cluster so as to be able to draw policy conclusions, we asked the firms what they thought they needed to become more innovative. The replies are shown in Figure 13.

Figure 13. Incentives to innovation

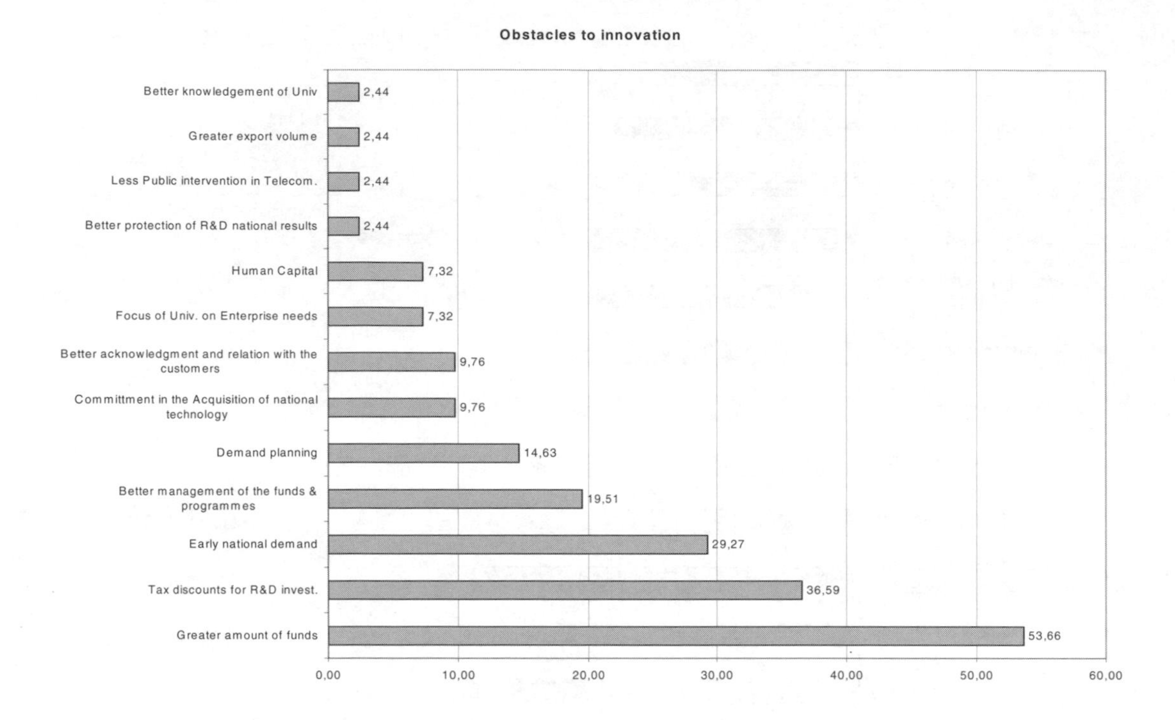

The main elements relate to the supporting financial system and to demand. In the case of financing, firms complain about the lack of co-ordination among the various public R&D programmes, overwhelming bureaucracy, the short-term vision of the programmes, and the lag between the time the firm applies for project participation and the date it is accepted or rejected. Firms believe that public support to innovation in the cluster could be improved through fiscal measures.

This result shows that entrepreneurs have a misconception of Spanish R&D tax incentives since a 1996 OECD report states: "Spain, Canada and Australia have extensively applied tax incentives in their innovation policies for many years. These three countries have maintained their top position over time in terms of the generosity of the R&D tax provisions".

The second group of factors has to do with demand conditions. The managers feel that national demand is very weak. This is an important constraint to enterprise development and thus to the resources devoted to innovation. Despite globalisation, national demand is fundamental for competitiveness.

7. Main policy conclusions

The cluster of the ICT industries in Spain is a very complex network of producers, users and suppliers, but with very weak links with what could be considered the suppliers of knowledge. Suppliers and users are the main source of the knowledge used by firms in their innovative processes.

In relation to the users of the technology, Spanish demand is characterised by its low technological content. In other words, consumers are not very concerned about the technological features of the goods and services. This is a major obstacle to innovation and, thus, to technology transfer.

One of the main users of this cluster is the public sector. The weight of the administration as user varies among industries. For Professional electronics, the public sector is one of the main consumers. On the other hand, Consumer electronics is more oriented towards the private sector.

For the first group of industries, public procurement could be usefully used to promote innovation. A recent Spanish study (Molero and Marín, 1998) showed that the public procurement policy could be used more effectively as a tool for innovation policy. This point should be reconsidered in those industries highly dependent on public sector demand.

The weakness of the relationships between the industrial and academic systems is, without doubt, a serious obstacle to the development of the system of innovation. If, for whatever reason, firms do not access the public knowledge base, then this situation should be addressed by cluster policy, as it constitutes an important barrier for the innovative system. It is especially important in a cluster where innovation is the key factor for competitiveness.

Two factors seem to explain this mismatch: on the one hand, the basic character of university research; on the other, the weakness of the bridging institutions. In the case of university research, only a few decades ago, the level of R&D investment of Spanish institutions was very weak. To overcome this situation, the Spanish Government devoted a large amount of funds to R&D activities in universities and research centres. It was believed that for Spain to reach the level of the more developed countries, it was necessary to invest in basic research. This criterion has led to a notable increase in Spanish scientific output. However, the link with the industrial system has not been sufficiently encouraged. For the cluster, this means that a clearer policy to promote the development of co-operative research projects is needed. The third national R&D plan introduced some changes in this direction. From the managers' perspective, this orientation should continue to be strengthened in the coming years.

On the other hand, weaknesses in the Spanish bridging institutions can also help to explain the gap between the academic and industrial systems. One of the most complete studies on bridging institutions in Spain is that directed by Ignacio Fernandez de Lucio (1996). This study presents information on the number, type, objectives, activities, and instruments of interaction most frequently used. It distinguishes between four different environments, comprising four basic types of institution:

- *Scientific environment,* the main activity of which is the creation of new knowledge for the system. The main institutions are the universities and public research institutes. The bridging institutions are of three kinds: OTRIS (Transfer Office for Research Results), Fundación Universidad-empresa (Enterprise-University Foundation), as well as specialised scientific bridging institutions.

- *Technological environment,* made up of technological institutes, engineering services and entrepreneurial R&D units. Its bridging institutions include: technical centres for training and consultancy, technical services, consultants, and technological institutes.

- *Producer environment,* producing value-added goods and services. Its main institutions include technological parks and a range of entrepreneurial bridging institutions (such as sectoral associations).

- *Financial environment,* generating the financial resources needed for the innovative activities carried out by the various actors of the system. The main bridging institutions are of two kinds: venture capital institutions and the public administration.

One of the most interesting conclusions of the study directed by Fernandez de Lucio is that bridging institutions are responsible for only 15% of total technology diffusion in Spain. Technology flows between the scientific, technological and the producer environments are very low, mainly due to the rivalry that exists between the bridging institutions of the scientific (SBI) and technological (TBI) environments. This has led to a situation in which TBI, instead of being nearer to firms' needs, tries to render the same services as SBI. Thus, there is no specialisation, and there is an overwhelming lack of bridging institutions serving enterprise needs.

This is also clear for the ICT cluster. The surveyed firms all admit that the present network of bridging institutions does not cater to firms' needs. In 1986, the government implemented and financed the OTRI network. A decade later, it is clear that a thorough revision of the role of the bridging institutions needs to be carried out.

Finally, the analysis of the external relationships of firms seems to support the idea that inputs to innovation (*e.g.* R&D investment) determine the way in which the firm interacts with its environment. The more the firm invests in innovation strategies, the more it interacts with external sources of knowledge.

This implies that stimulation of the absorptive capacity of the firm should be a cornerstone of any policy aiming at promoting innovation in the cluster. This is particularly important for SMEs whose absorptive capacity is clearly limited by the scarcity of qualified human resources, R&D investment, training, etc. In this sense, training policies are a necessary complement to S&T cluster policy.

NOTES

1. Broadly, the analysis identifies the main forward and backward linkages of the cluster. The starting point is the intermediate matrix of the input/output table. From the total acquisitions (sales) of the sector, we subtracted the amount due to the same sector to obtain the total inter-cluster trade flows. Rows show the users of the cluster, while columns refer to suppliers of the cluster under consideration. We then estimate the ratio between each user (supplier) and the total inter-cluster trade flows already calculated. In the Spanish case, we selected users and suppliers with a ratio higher than 0.15.

2. The corresponding ratio is shown in brackets. For example, in Figure 1, 36% of the sales of the sector goes to the public services, while the sector acquires 48% of supplier goods from the electrical manufacturing sector (excluding intra-sectoral linkages).

3. At the Spanish National Association of Electronics and Telecommunications (ANIEL), we interviewed managers of departments of Consumer Electronics, Electronic Components, Professional Electronics, Telematics and Service providers.

4. At the end of the study, five in-depth interviews were conducted with staff of Alcatel, Ericsson, ELIOP, Telefónica and Amper.

REFERENCES

ANIEL (1998), "Memoria del sector electrónico español", ANIEL, Madrid.

Arrow, K. (1962), "Economic Welfare and the Allocation of Resources for Invention", in R. Nelson (ed.), *The Rate and Direction of Inventive Activity*, Princeton University Press, Princeton, pp. 609-629.

Bijker, W. *et al.* (eds.) (1992), *The Social Construction of Technological Systems*, MIT Press, Cambridge, Massachusetts.

Buesa, M. (1994), "La política tecnológica en España: una evaluación en la perspectiva del sistema productivo", *Información Comercial Española*, No. 726, pp. 161-178.

Carlsson, B. (1995), *Technological Systems and Economic Performance: The Case of Factor Automation*, Kluwer Academic Publishers, Dordrecht.

Cohen, W. and D. Levinthal (1989), "Innovation and Learning: The Two Faces of R&D", *Economic Journal*, No. 99, pp. 569-596.

Cohen, W. and D. Levinthal (1990), "Absorptive Capacity: A New Perspective on Learning and Innovation", in Burgelman *et al.* (1998), *Strategic Management of Technology and Innovation*, (2nd edition), Irwin, Boston, pp. 541-558.

CICYT (1995), *III Plan Nacional de I+D (1996-1999)*, CICYT, Madrid, p. 47.

Dahmén, E. (1970), *Entrepreneurial Activity and the Development of Swedish Clusters 1919-1939*, Irwin, Illinois.

David, P. and D. Foray (1994), "A Conceptual Framework for Comparing National Profiles in Systems of Learning and Innovation", report prepared for the OECD, Paris.

David, P. and D. Foray (1995), "Interactions in Knowledge Systems: Foundations, Policy Implications and Empirical Methods" in *STI Review*, No. 16, OECD, Paris, pp. 14-68.

DeBresson, C. (1989), "Breeding Innovation Clusters: A Source of Dynamic Development", *World Development*, Vol. 17, No. 1, pp. 1-16.

Dosi, G. (1982), "Technological Paradigms and Technological Trajectories", *Research Policy*, Vol. 11, No. 3, pp. 147-162.

Fernandez de Lucio, I. (1996), "Inventario de estructuras de interfaz en España", Universidad de Valencia, Valencia.

Freeman, C. (1994), "The Economics of Technical Change", *Cambridge Journal of Economics*, Vol. 18, No. 5, pp. 463-514.

Gille, B. (1978), *Histoire des techniques*, Gallimard, Paris.

Hertog, P. den *et al.* (1995), "Assessing the Distribution Power of National Innovation Systems. Pilot Study: The Netherlands", Apeldoorn (Holland), mimeo.

Hirschman, A. (1958), *The Strategy of Economic Development*, Yale University, New Haven.

Hughes, T.P. (1983), *Networks of Power. Electrification in Western Society 1880-1930*, Johns Hopkins, Baltimore.

Instituto nacional de estadística (INE) (1995), *Estadística sobre actividades en investigación científica y desarrollo tecnológico (I+D) de 1993*, INE, Madrid, p. 162.

Instituto nacional de estadística (INE) (1996), *Boletín mensual de estadística*, INE, Madrid.

Instituto nacional de estadistica (INE) (1998), *Encuesta sobre innovación tecnológica en las empresas 1996*, INE, Madrid.

Kline, S. and N. Rosenberg (1986), "An Overview of Innovation", in Landau and Rosenberg (eds.), *The Positive Sum Strategy*, National Academy of Sciences, Washington, DC.

Lázaro Lafuente, M. (1994), "Tecnologías de la información y de las comunicaciones: presente y futuro", *Economía Industrial*, March-April, pp. 35-42.

Lundvall, B.-Å. (1992), *National Systems of Innovation: Towards a Theory of Innovation and Interactive Learning*, Pinter, London, p. 342.

Marceau, J. (1994), "Clusters, Chains and Complexes", in M. Dogson and R. Rothwell, *The Handbook of Industrial Innovation*, Edward Elgar, Cheltenham, pp. 3-12.

Molero, J. and M. Buesa (1995), "Configuración productiva y capacidad de innovación en la industria española", *Información Comercial Española*, No. 743, pp. 59-84.

Molero, J. and F. Marin (1998), "Las compras públicas y la innovación en España", *Cuadernos COTEC*, No. 7, April, pp.1-16.

Monfort, M.J. and J.C. Dutailly (1983), "Les filières de production", *INSEE Archives et Documents*.

Nelson, Richard (1959), "The Simple Economics of Basic Research", in N. Rosenberg (ed.) (1971), *The Economics of Technological Change*, Penguin Books, Harmondsworth, p. 478.

Nelson, R. (ed.) (1993), *National Innovation Systems. A Comparative Analysis*, Oxford University Press, New York, p. 541.

Nelson, R and S. Winter (1974) "Neo-classical *vs.* Evolutionary Theories of Economic Growth", *Economic Journal*, Vol. 84, pp. 886-905.

Nelson, R. and S. Winter (1982), *An Evolutionary Theory of Economic Change*, Harvard University Press, Cambridge, Massachusetts.

OECD (1994), "Technology Flows in National Systems of Innovation", report prepared for the OECD, Paris.

OECD (1996), "Embodied Technology Diffusion: An Empirical Analysis for Ten OECD Countries", *STI Working Paper 1996/1*, OECD, Paris.

OECD (1996), "R&D Tax Treatment in OECD Member Countries: A Comparison", report prepared for the OECD, Paris.

Pavitt, K. *et al.* (1996) "Innovation Outputs in European Industry: Analysis from CIS", paper presented at the International Conference on Innovation Measurement and Policies, Luxembourg.

Porter, M. (1990), *The Competitive Advantage of Nations*, Macmillan, London.

Roelandt, T. *et al.* (1999), "Cluster Analysis and Cluster Policy in the Netherlands", this volume.

Roelandt, T and P. den Hertog (1998), "Draft Synthesis Report on Phase 1 OECD Focus Group on Cluster Analysis and Cluster-based Policy", mimeo.

Rosenberg, N. (1982), *Inside the Black Box*, Cambridge University Press, Cambridge.

Sanchez, M.P. *et al.* (ed.) (1990), "Indicadores de ciencia y tecnología. Indicadores de actividades complementarias de I+D y los indicadores de las nuevas tecnologías. La visión de un conjunto de expertos españoles", *Documento de trabajo No. 10*, Instituto de Sociología de las Nuevas Tecnologías, Madrid, p. 47.

Smith, K. (1990), "Measuring Investment and the Diffusion of Technology. New Innovation Indicators: Conceptual Basis and Practical Problems", report prepared for the OECD, Paris.

Smith, K. (1994), "Interactions in Knowledge Systems: Foundations, Policy Implications and Empirical Methods", STEP Group Report, Oslo, p. 35.

Soete, L. and A. Arundel (eds.) (1993), *An Integrated Approach to European Innovation and Technology Diffusion Policy*, MERIT, Limburg.

Chapter 10

INDUSTRY CLUSTERS: A METHODOLOGY AND FRAMEWORK FOR REGIONAL DEVELOPMENT POLICY IN THE UNITED STATES

by

E.M. Bergman
Vienna University of Economics and Business

E.J. Feser
University of North Carolina

1. Introduction

Industry clusters have become an increasingly popular focus in local and regional development policy circles of OECD countries, particularly within the United States. Lacking a national-level NIS strategy, empirical studies of industry clusters are far more frequently conducted at the sub-national level, often patterned after Porter's (1990) model of competitive advantage (Nelson, 1993; Ohmae, 1995). For the development official (or the independent consultant to state and local development agencies), at least some knowledge of how to identify industry clusters present in a local economy is now essential. City planners, policy analysts, regional economists and others involved in regional development activities in the United States often view their collection of analytical methodologies and techniques as a "toolkit" for economic development planning and policy. Industry cluster analysis has lately assumed a central place in that toolkit.

But this new and shiny "policy hammer" increases the risk that everything begins to look like a nail. It therefore comes as no surprise that enthusiastic applications of industry cluster analysis to regional economies has at times generated ambiguous results or yielded unfocused policy implications. Key features of the industry cluster concept (*i.e.* the performance of a given industry must be viewed in concert with the performance of related sectors and institutions, that proximity between businesses enhances exchanges of knowledge and technologies, and that co-located businesses in a shared value chain can become a source of national competitive advantage) have been more fully exploited in some economic analyses than in the design of policies based on such analyses.

For example, one can use a variety of methodologies to detect linked industries and institutions and the strength of their presence in national or regional economies (Hill and Brennan, 1998; Rosenfeld, 1997; Sternberg, 1991; Held, 1996; Doeringer and Terkla, 1995; DeBresson, 1996; DeBresson and Hu, 1999; Roelandt *et al.*, 1999; Enright, 1996; Jacobs and de Jong, 1992; Peneder, 1995). One can also examine, as Porter and others have demonstrated, that countries or regions with well-developed clusters tend to out-perform places with few identifiable clusters, although this particular analytical approach verges on tautology since cluster identification hinges mainly on their easily visible success. Even so, identifying and implementing policies to promote cluster development where none exist or to expand existing clusters has been quite difficult. Some of this difficulty arises from the application of inappropriate analytical methods to inform and support the formulation of specific policies.

This chapter outlines an approach to regional policy making that relies upon broad industry cluster ideas and analyses them with robust methods and data. The approach adopts a slightly different perspective on the link between cluster analysis and economic development policy than is common in the literature. Rather than beginning with a general objective of identifying industry clusters, and assuming that policy implications follow logically, we begin with a specific (and rather traditional) policy objective. We then use cluster analysis to determine better ways of pursuing that objective. In our approach, cluster analysis is mainly supportive of development strategies that strengthen the value-chain connections between firms, strategies that are typically already in place in some form in most regions. The goal of the cluster analysis itself is to permit policy officials to acquire unique and quite detailed insight into the basic features of the regional economy by emphasizing industry linkages and interdependence among firms.

We illustrate our approach with reference to a major policy initiative undertaken by the lead technology agency in one southern US state, North Carolina. In 1996, the agency contracted with us to produce a study of the state's manufacturing industry clusters in order to identify the principal channels through which production technologies tend to spread and diffuse. The study produced two reports: an overview of national industry clusters (which are by definition North Carolina's or any state's *potential* clusters), including a state-wide analysis of the regional distribution of those clusters, and a guidebook designed to assist local development agencies apply the findings in their own policy making process. The agency commissioned a second study in 1997 that produced more detailed cluster analyses for each of the state's seven economic development planning regions. Those reports are now being used by the lead regional development agencies for a variety of development planning purposes.[1]

The following sections outline the logic of the original study as well as some of the findings. Given the policy question at hand, we used a macro-based approach to identify the principal manufacturing trading partners in the US economy. This provided us with a set of national industry clusters, based on manufacturing firms' value chains. We then used detailed micro-level data of enterprises in North Carolina to examine their relative representation in North Carolina's national-level value chains. Although North Carolina industries are unlikely to track the US economy in its particulars, national value chains serve as useful benchmarks for highlighting North Carolina's unique cluster concentrations and regional distributions. The micro-level data set permits close study of the evolution and growth of different value chain composition and geography.

2. Industry clusters and regional policy

In most cases, the use of industry cluster concepts in regional policy making begins with a general study of existing clusters. One starts with a particular definition of the concept "industry cluster". The definition might be based on Porter's four-part model of national competitive advantage (the "diamond") or a more narrow view of industrial interdependence based on technological linkages. Whatever the specific definition, it effectively determines the methods and techniques of cluster analysis. A cluster definition that emphasizes informal, "untraded" ties between businesses probably dictates a qualitative study of visible cluster members (Rosenfeld, 1997; Gollub *et al.*, 1997), while a definition focused on buyer-supplier relationships implies a detailed quantitative analysis of I/O matrices that trace documented trade flows among many and occasionally unexpected industries (Roelandt *et al.*, 1999; Enright, 1996; Feser and Bergman, forthcoming). Policy implications drawn from the study findings and used to design development strategies and implementation measures intended to establish or expand specific clusters of greatest economic promise. Figure 1 is a simplified diagram of the logic of the approach.

Figure 1. Common approach to cluster analysis and policy

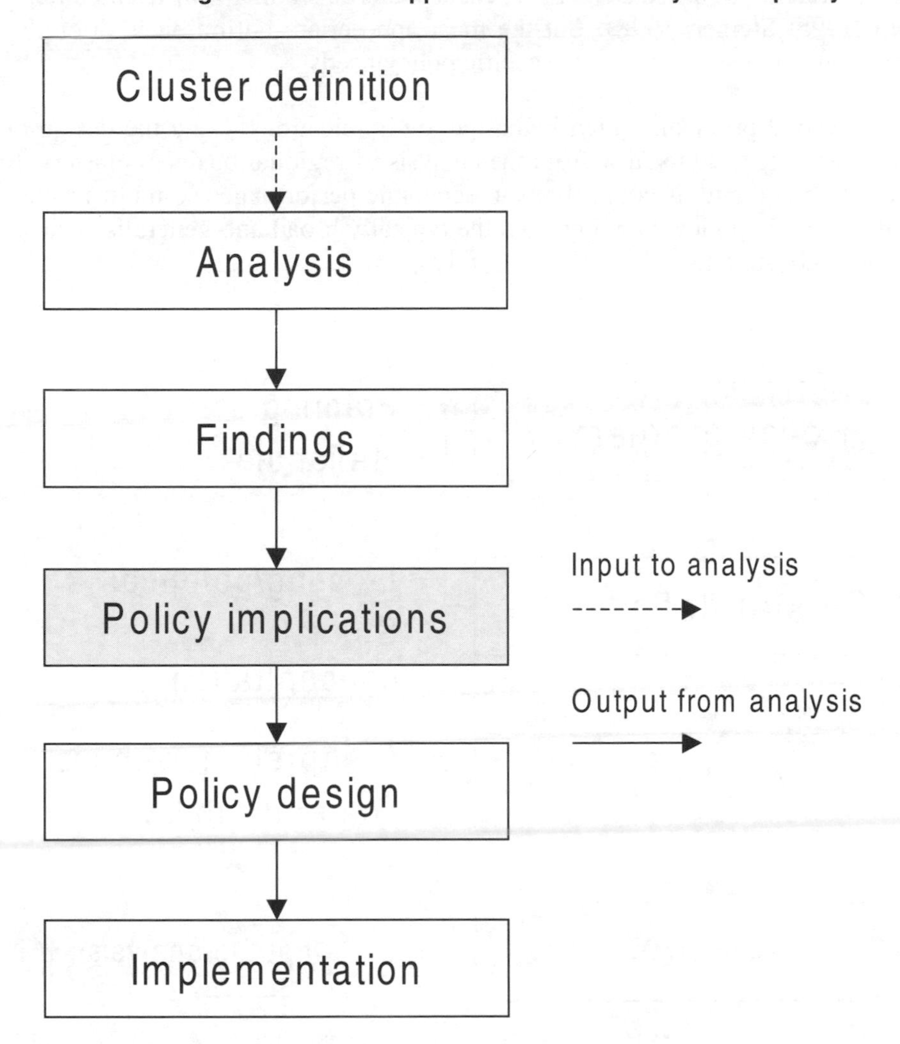

There are two principal problems with the sequence in Figure 1. First, contrary to conventional wisdom, starting with the cluster definition itself provides little guidance for narrowing the scope of inquiry in a meaningful way. Making clear at the outset what one means by "industry cluster" has obvious advantages, but an *a priori* definition leaves the analyst with an extremely broad set of parameters in which to work. This results from the very generality of the theories and ideas behind the industry cluster concept.

For example, should the focus be placed on existing industry clusters or clusters that are emerging or are likely to emerge? Are clusters necessarily localised within a country? Porter suggested that they often (but not always) are. In an increasingly global economy, many enterprises' key trading and non-trading partners may be located in wholly different countries and regions. If some of the most important interdependencies between firms are non-local, how does the analyst examine this feature of clusters in way that is meaningful for *regional* policy making? Are informal ties between businesses most critical or are trading relationships? The answer likely depends on whether one is talking about innovation or efficient production. The ambiguities surrounding the cluster concept (and other related concepts such as industrial districts), proper definitions, and their relationships to regional economic performance are the subject of a large literature (Asheim and Isaksen, 1997; Feser, 1998a, 1998b; Harrison, 1992; Heidenreich, 1996; Isaksen, 1997; Jacobs and de Man, 1996; Kaufman *et al.*, 1994; Park and Markusen, 1995; Steiner, 1998). But the most appropriate definition is often driven by a normative mandate to pursue one or more very specific policy needs.

This brings us to the second problem, which is that policy implications – and the design of specific policy interventions – are expected to flow from the analysis of regional business clusters. But again, the cluster concept is so broad and its connections to economic performance so multi-faceted that it is often difficult to infer specific policy meaning from the typically broad and generalised cluster studies churned out by a host of consultants.

Figure 2. Alternative approach to cluster analysis and policy

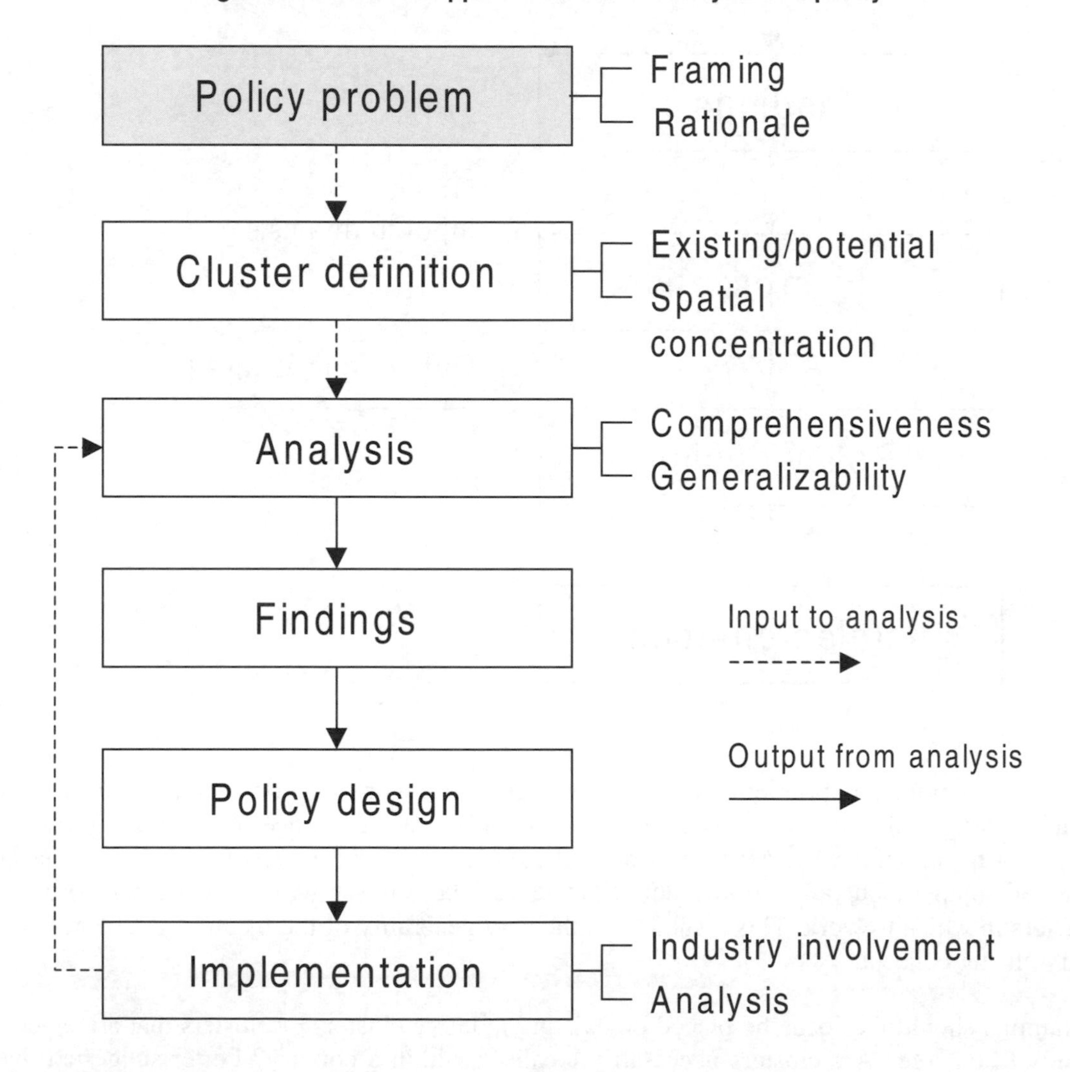

Figure 2 proposes an alternative means of integrating cluster concepts into the regional policy making process. Concepts of industrial interdependence and competitive advantage are used to inform and address specialised policy needs that are identified in advance of the cluster analysis. The framing of the policy problem effectively provides a set of restrictions on the cluster definition, which, in turn implies an appropriate methodology given existing data constraints and resources. Findings related to the policy issue are more easily extracted from the more focused analysis. That helps provide better guidance for the design of specific policy initiatives.

An additional feature of the sequence in Figure 2 places the concern for policies and their implementation before the choice of analytical methodology. For example, there is increasing recognition in US economic development circles that business leaders must be part of the policy process from the beginning if they are to take an active role in the policy initiative (as they often must in cluster-type policy initiatives such as network formation). Thus, some cluster studies emphasize heavy industry involvement and guidance (perhaps through focus groups) over more traditional quantitative and analytical approaches.

3. Industrial modernisation policies drive cluster analysis

Figure 2 essentially summarises our approach to the technology agency's directive in 1996. The agency was concerned with a specific policy problem: how to diffuse advanced production technologies efficiently among businesses in a manufacturing economy that is dominated by a least-cost competitive ethos. North Carolina's rapid growth from the 1960s through the 1980s was fuelled initially by the re-location of branch plant facilities from high-wage, unionised locations in the industrial mid-western and north-eastern parts of the country. Although the state has gradually established a solid high-technology base (principally centred upon Research Triangle Park) and banking presence (in the Charlotte region), its economy remains disproportionately specialised in traditional sectors that remain under unrelenting pressure from low-cost, overseas producers (*e.g.* textiles and apparel). In this environment, encouraging producers to invest in, adopt and utilise best practice production technologies can be an exceptional challenge.

In an earlier study of technology adoption practices among producers in the state's nominal automotive supply chain, we found that smaller and often more rural producers tended to be less aggressive in adopting new manufacturing techniques (Bergman *et al.*, 1995). Reasons cited included lack of information about advanced technologies and inadequate access to sources of capital that do not dilute control over the firm. More passive or traditionally oriented firms are satisfied with the existing market, and are not interested in pursuing an aggressive growth strategy through investment in technologies that will open new markets, even though such complacency will be fatal in certain traded industries.

On the other hand, we also found that producers presently in the vehicle supply chain tend to adopt and use technologies at a significantly higher rate. Evidence suggested that this was partly because final market vehicle assemblers are essentially "forcing" adoption of new methods by their suppliers as well as serving as a source of information about best practice techniques. Also important are increasingly strict international certification requirements (*e.g.* ISO 9000). In short, we had good evidence of a powerful diffusion effect that spread competitive production technologies *through the supplier chain*, a well-known view that continues to receive considerable support from the growing research on buyer-supplier relations.

One implication for our analysis is the critical importance of interindustry trade channels for the diffusion of new technologies. The technology agency agreed and sought our help to identify linked producers in the state to better target technology adoption programmes. The agency also hoped to identify mismatches between production and input supply, *i.e.* local industries that could be served by local supplier sectors that may exist (but do not supply local producers, perhaps due to quality control problems) or could be established through co-ordinated regional development strategies. In this way, the agency hoped to build and expand value chains by better targeting development resources.

In this context, both the policy needs and rationale for studying industry clusters based on value chains were clear. Our focus became one of identifying key buyer-supplier chains in the state, those that currently exist and those that may be emerging. Focusing strictly on existing activity would prevent us from observing gaps in particular clusters that might prevent efficient technology diffusion. Alternatively, attention to emerging or "potential" clusters would provide a means of identifying key growth points in the economy, particularly those sectors which are likely to grow because of unique locational advantages offered in certain North Carolina regions. To examine emerging clusters, we essentially needed a benchmark against which to compare the North Carolina manufacturing economy. As we will also demonstrate, this benchmark or "template" can also be used in other advanced OECD countries with similar interindustry trading patterns.

Other policy needs included attention to the geographic distribution of linked sectors and the need to examine the manufacturing economy comprehensively. The agency also planned to use the results primarily to improve state- and regional-level economic development analysis and planning; industry involvement was not critical and therefore we could adopt a primarily analytical approach. Indeed, one of our objectives was to build a better analytical capacity for state and local development planning, one that incorporated industry cluster ideas.

4. Clustering logic: industrial and regional *trade* cluster approach

Comparative production and trading advantages increase any region's (or nation's) ability to focus on what can be produced competitively for export, and to purchase at lower cost from others what cannot be made competitively. Regions are then freed to specialise in a narrower but still trade-linked range of goods produced in ever larger volumes for sale in national-to-global markets. Increasing returns to scale permit growth of large-scale enterprises in specialised regions to produce goods more efficiently and cheaply, and to ship production easily throughout open markets. Increasingly, industries with strong trade ties co-locate in the same or nearby regions to take advantage of common production factors, creating afresh the familiar pre-conditions for industrial clusters and districts.

Although most firms compete fiercely as independent entrepreneurs, few prosper nor can long survive in absolute isolation. Competitive firms *rely* on their suppliers to make timely deliveries of high-quality intermediate inputs and services. They also *depend* on the continued incorporation of key technological innovations and product improvements by suppliers of inputs, production machinery, and other forms of capital equipment. Finally, even firms not trading directly with one another rely upon informal exchange of information, knowledge, and skilled labour that routinely flows between technologically-related industries, a practice many now acknowledge as a source of important unpriced externalities that improve individual and collective competitiveness. *Groups* of producers in multiple sectors thus share *collective fortunes*.

An important distinction is made between ***industrial*** and ***regional trade clusters***. Industrial trade clusters are derived from average US *macro* I/O purchasing patterns observed among groups of producers (in different sectors) that engage directly or indirectly in trade. These macro-based industrial clusters may then be applied as "templates" to benchmark available *micro* data of specific sub-national areas, *e.g.* North Carolina manufacturing firms, and the resulting constellation of firms and industries constitutes a regional trade cluster. Broad concepts of agglomeration economies and industrial co-location familiar to regional development scholars are particularly intriguing when applied to the broader questions of theory and policy stimulated by recent policy interests in industrial clusters, particularly the opportunity to apply well-established methods to useful macro and micro data sources.

Studies of clustering behaviour among firms can be said to proceed mainly from either micro or macro perspectives. Micro-based studies rely upon direct observations of firms having strong, often frequent, economic connections that co-locate within fairly close proximity. Reasons for such co-location may be found in uniquely supplied local resources or historical accidents of advantage that stimulate firm success because of inherent productive efficiencies. These richly detailed studies are compellingly familiar, even to casual observers, although they are so uniquely framed that their lessons and generalisations are often difficult to distil or to apply in other regions. However, such studies typically limit their view to clusters observed at fairly small geographical scales (the "industrial district"), thereby overlooking key linkages that "visible clusters" may have with other firms in a broader regional context. They also remain entirely blind to other clusters of co-located firms, where "proximity" covers a much wider spatial reach than is immediately apparent to locally oriented observers. This further implies that significant clustering phenomena may go unrecognised by "street-light sampling", *i.e.* restricting one's view to points where vision is easiest.

Macro-based studies of the type used in this study start instead by examining networks of known connections and interactions that channel various flows between all industrial firms in an economy. These might consist of communication links, transportation contacts, technology diffusion and exchange, or trade flows, as primary examples. One then applies suitable analytical techniques to flow data that represent repeated transactions among specific subsets of all firms and industries that constitute a value-chain cluster. The most common flows are inter-sectoral factor purchases and product sales, as captured by industrial I/O tables. While macro I/O studies capture highly probable intra-cluster trade dependencies, these clusters remain meta-level abstractions without much visible grounding, at least until detailed micro-data are employed to portray regional specifics.

Our analytical approach blends elements of both levels of inquiry: national I/O-based industrial cluster templates, into which micro-level firm data are embedded for specific regions of North Carolina. Micro-level data supplies information about every manufacturing firm, including its identification number, total employment, total wages, branch *vs.* single firm status, and four-digit Standard Industrial Classification (SIC). The longitudinal data set applies to postal code (about 1 100 in total) or specific address co-ordinates for two observation years (1989 and 1994). Wage and SIC detail permit further descriptive inferences about cluster members, *e.g.* estimates of output, value added and technology intensiveness; the two-year observations permit longitudinal understanding of underlying processes of cluster formation and change.

We next summarise briefly and cite fully our I/O cluster estimation technique that yields 23 national industrial clusters, each constituted by a set of three/four-digit SICs. These are the "cluster templates" that permit further, detailed regional trade cluster identification and analysis. Figure 3 summarises our methodological approach to the cluster analysis and its application to North Carolina and its economic development planning regions. Each step in the process utilises different analytical techniques, from factor analysis of the US I/O table to GIS-based analysis of each cluster's presence in North Carolina. The result is a detailed set of procedures and data for applying cluster concepts to regional development questions in the state.

Figure 3. Object of the original cluster study

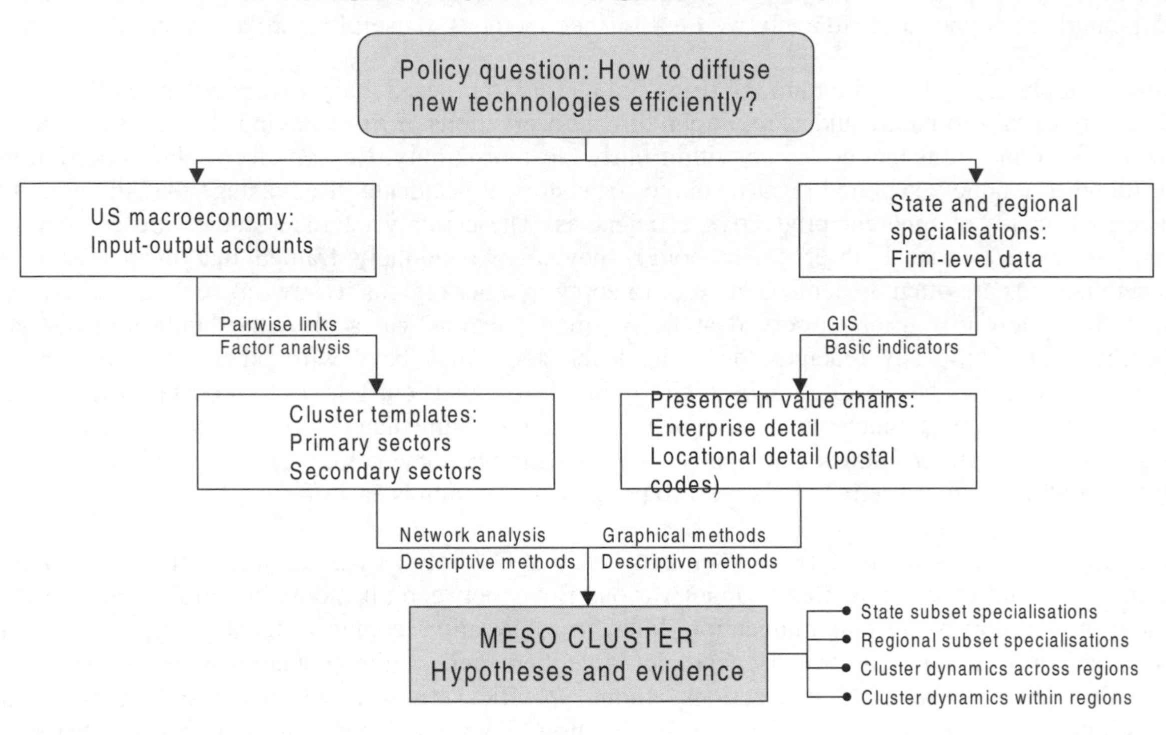

5. US industrial trade clusters: the macro analysis[2]

Earlier studies employed a range of methodologies, including graph theory, triangulation and factor analysis to sort industries into groups based on I/O linkages. For this study, measures of direct and indirect linkages computed from interindustry trade information for each sector were treated as variables in a principal components factor analysis. Data are from the detailed 1987 benchmark US I/O accounts, released by the Bureau of Economic Analysis in late 1994. The derived components were rotated to a varimax solution to facilitate interpretation, where the decision regarding the number of rotated components was made based on the relative proportion of variance explained by each component, the size of the associated eigenvalues, and scree plots. Multiple analyses were conducted using alternative assumptions regarding the number of rotated factors. The results were then compared for consistency and interpretability before a final set of clusters was selected.

The generated set of loadings effectively provide a measure of the relative strength of the linkage between a given industry and a derived factor, where the highest loading industries on a given factor are treated as core members of an industrial cluster. In interpreting the factor analysis results to identify specific industrial clusters, we tried to reconcile several competing objectives. Our primary objective was to derive a set of clusters based on the most significant linkages as revealed in the I/O data matrix. Accordingly, the concern was to identify the industries with the tightest linkages to each cluster (*i.e.* the highest loading industries for each factor), regardless of whether those industries are also tightly linked to another cluster. A secondary objective was to identify, to the degree possible, a set of mutually exclusive clusters in the sense that each sector would be assigned to only one cluster. Such a result would facilitate cross-cluster comparisons of size and growth rates. A tertiary objective was to investigate the linkages both between clusters as well as between industries within each cluster. Such linkages are sometimes revealed by an examination of sectors that are only moderately or weakly related to each cluster, thus competing with the first objective.

The final set of clusters represents a compromise. Each cluster contains a set of "primary" and "secondary" industries. *Primary industries* for a given cluster are those sectors that achieved their highest loading on that factor *and* whose highest loading was .60 or higher. For example, SIC 277 (greeting cards) achieved its highest loading on the sixth factor (what we interpret below as the printing and publishing cluster), and, since the loading (.90) is greater than or equal to .60, 277 is classified as a primary industry for that cluster. *Secondary industries* for a given cluster are those sectors that achieved loadings on the cluster equivalent to or greater than .35 but generally less than .60. For example, SIC 3652 (pre-recorded records and tapes) achieved a loading of .54 on the sixth factor and is thus classified as a secondary industry for printing and publishing. For some clusters, the set of secondary industries also includes industries with loadings exceeding .60 but that achieved their highest loading on a different cluster. While SIC 2677 (envelopes) achieved a loading of .68 on the sixth factor, it achieved a still higher loading on factor 1 (interpreted as the metalworking cluster). Therefore, it is classified as a primary industry in metalworking and a secondary industry in printing and publishing cluster. As a general rule, primary industries are those that are most tightly linked to a given cluster while secondary industries are those that are less tightly or moderately linked. Considering only primary industries yields a set of mutually exclusive industrial clusters that can be used for cross-comparison purposes. A fuller account of the procedures and findings is available in Feser and Bergman (forthcoming) and Bergman *et al.* (1996).

The final set of 23 clusters includes clusters in three broad industrial categories: heavy manufacturing (*e.g.* metalworking, vehicle manufacturing, chemicals), light manufacturing (wood products, printing and publishing, textiles), and food goods and tobacco (packaged foods, canning and bottling, feed products, tobacco). Table 1 lists the full set along with the number of component three- and four-digit Standard Industrial Classification (SIC) sectors. The number of different two-digit SIC codes (aggregations of the three- and four-digit codes) are also listed to indicate the industrial breadth of many of the clusters. The clusters essentially represent a reordering of the US manufacturing classification scheme into groups of related industries rather than groups of sectors that produce similar products (as represented by the SIC classification scheme). In our application to the North Carolina case, we use the US clusters as templates to study the pattern of related manufacturing industries within the state and its sub-regions.

6. Regional trade clusters: micro application to North Carolina regions

This section integrates the industry cluster templates with micro-level evidence of specific firms in each sector and regional cluster. It is here that we combine macro templates and micro evidence, much like any template is used. This is accomplished with micro data for approximately 10 000 North Carolina manufacturing firms. Table 1 lists each cluster's employment in North Carolina by totals and percentages.

The last two columns of the table also compare the relative concentrations of each cluster in North Carolina with the United States to see how closely North Carolina's industrial structure fits the US cluster templates. This comparison reveals the state's well-known disproportionate concentrations in wood products (mainly furniture and household components), knitted goods, fabricated textile products, and tobacco. But the figures also show surprisingly large shares of vehicle manufacturing, electronics and computers, chemicals and rubber, and metalworking.

Table 1. Benchmark manufacturing clusters: North Carolina employment in each cluster, 1994

Cluster	Primary and secondary			Primary industries only				
	Detailed sectors	2-digit SIC sectors	Employ-ment	Detailed	2-digit SIC	Employ-ment	% of manufacturing employment NC	US
Metalworking	116	10	132 755	93	9	91 451	10.6	20.7
Vehicle manufacturing	58	16	168 744	35	13	129 607	15.0	15.2
Chemicals & rubber	48	14	106 831	20	6	32 658	3.8	3.2
Electronics & computers	38	8	97 287	25	6	66 972	7.7	11.9
Packaged foods	44	5	61 372	21	1	12 381	1.4	3.6
Printing & publishing	32	8	72 591	21	5	47 730	5.5	11.1
Wood products	23	6	85 520	16	2	77 607	9.0	4.6
Knitted goods	23	5	279 728	13	3	187 341	21.6	6.6
Fabricated textile products	22	9	211 858	12	7	82 288	9.5	2.8
Non-ferrous metals	14	4	11 825	8	3	1 327	0.2	0.7
Canned & bottled goods	12	2	8 463	6	2	8 043	0.9	1.5
Leather goods	9	1	2 680	6	1	1 555	0.2	0.4
Aerospace	10	6	8 929	5	2	2 551	0.3	4.2
Feed products	10	2	22 378	5	1	2 232	0.3	0.4
Platemaking & typesetting	14	7	15 512	4	3	1 576	0.2	0.5
Aluminum	9	4	7 788	4	3	2 901	0.3	0.9
Brake & wheel products	9	4	9 353	4	3	1 932	0.2	0.6
Concrete, cement, & brick	8	2	7 489	3	1	3 735	0.4	0.7
Earthenware products	8	1	1 955	5	1	974	0.1	0.2
Tobacco products	4	1	19 015	4	1	19 015	2.2	0.2
Dairy products	6	1	2 920	3	1	1 238	0.1	0.6
Petroleum	5	2	409	3	2	53	0.0	0.5
Meat products	5	2	29 309	2	1	7 788	0.9	1.2
Non-loading	n.a.	n.a.	n.a.	n.a.	n.a.	82 498	9.5	7.9

n.a. = not applicable.
Source: North Carolina ES-202 files. See Bergman, Feser and Sweeney, 1996.

We take full advantage of the detailed, longitudinal firm-level data for two years to analyse the cluster features of North Carolina's regions (Figure 4). The following sub-sections reveal essential cluster characteristics in terms of technological intensiveness, plus shares of small and independent establishments. Cluster composition by region and the spatial tightness of cluster firms are evaluated. Finally, the dynamics of firm loss and gains in clusters, plus the growing or declining presence of clusters in regions, provide a glimpse of the trajectories of regional clusters.

Figure 4. North Carolina development planning regions

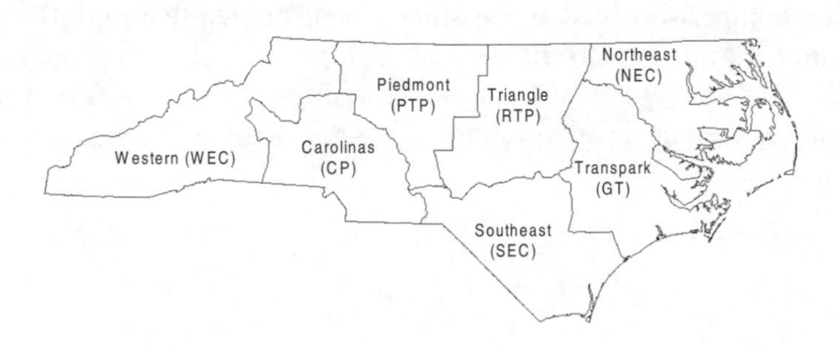

Regional cluster characteristics

Three of the most useful and readily available industry sectoral characteristics are technology classification, establishment size, and independent firm status. Technology initiatives may focus on firms producing high-technology goods or using high-technology production equipment that serve as diffusers in their value chains. Where this is the case, clusters predominantly made up of high-technology producers (*e.g.* aerospace) may represent the most important policy assets. Technology modernisation programmes might seek to leverage inter-firm synergies that encourage technology upgrading among firms (both high- and low-tech) in the same buyer-supplier chain. Since high-technology sectors in a given buyer-supplier chain are often the key developers, demanders, and/or disseminators of advanced technologies and methods, it is then useful to identify potential clusters in the state where such sectors are over- or under-represented (relative to US benchmarks).

Technology policies are also frequently targeted to smaller plants, based on evidence that such enterprises face greater obstacles to modernisation (*e.g.* resource constraints, expertise, etc.). For quite similar reasons, a further distinction may be made between branch plants (of multi-establishment units) and single establishment firms, where the latter are assumed to have comparatively fewer resources at their disposal. These three criteria are used in this section to further describe and characterise the set of potential North Carolina clusters for the purposes of selecting policy targets.

Table 2. Percentage of high-technology output and employment by cluster
Primary/primary and secondary industries, 1993/94

Cluster	Primary industries only				Primary and secondary			
	Output		Employment		Output		Employment	
	US	NC	US	NC	US	NC	US	NC
Metalworking	36	56	36	54	43	62	42	59
Vehicle manufacturing	63	36	36	27	52	32	32	25
Chemicals & rubber	94	94	84	83	69	63	45	50
Electronics & computers	90	85	90	85	77	72	73	68
Packaged foods	0	0	0	0	8	10	6	8
Printing & publishing	1	1	1	1	9	11	5	7
Wood products	0	0	0	0	1	1	1	1
Knitted goods	0	0	0	0	8	8	3	4
Fabricated textile products	23	3	21	3	10	1	7	1
Non-ferrous metals	6	21	11	33	3	3	5	4
Canned & bottled goods	0	0	0	0	0	0	0	0
Leather goods	0	0	0	0	0	0	0	0
Aerospace	97	56	98	62	97	85	97	89
Feed products	0	0	0	0	47	68	64	77
Platemaking & typesetting	35	19	27	16	20	15	18	15
Aluminum	80	87	75	75	43	43	34	28
Brake products	0	0	0	0	19	27	15	20
Concrete, cement, & brick	0	0	0	0	2	0	3	0
Earthenware products	0	0	0	0	0	0	0	0
Tobacco products	0	0	0	0	0	0	0	0
Dairy products	0	0	0	0	0	0	0	0
Petroleum	99	18	89	16	97	3	81	2
Meat products	0	0	0	0	0	0	0	0
Non-loading (primary) sectors	38	42	31	30	--	--	--	--
Total	**46**	**29**	**35**	**23**	--	--	--	--

Note: Figures are percentage cluster activity classified as high, moderately high, or somewhat high technology (definitions from NC ACTS and the North Carolina Employment Security Commission.
Source: NC and US data are from 1994 and 1993, respectively (NCESC and USLBS).

High-technology sectors. Table 2 reports the shares of output and employment in sectors classified as high technology by cluster for both the United States and North Carolina. Output in several US industrial trade clusters is predominantly in sectors that are characterised as high-tech at some level. When primary industries alone are included in cluster definitions (second column, Table 2), the share of output in sectors classified as high technology meets or exceeds 80% in petroleum, aerospace, chemicals and rubber, electronics and computers, and aluminium clusters. Several other clusters are responsible for low to moderate shares of high tech output: vehicle manufacturing (63%), platemaking and typesetting (35%), metalworking (36%), and fabricated textile products (23%). Fourteen out of 23 clusters, including the five food products clusters, knitted goods, non-ferrous metals, wood products, printing and publishing, tobacco, cement and brick, brake products, and earthenware products involve very little or no high-technology output.

A comparison of the relative distribution high-tech output in North Carolina *vs.* the United States suggests some under- and over-representation of high-technology sectors in the state's industrial trade clusters. The ratio of high-to-standard technology production in the North Carolina chemicals, electronics and computers, and aluminium clusters is on a par with, or exceeds (in the case of aluminium), the ratio for the United States as a whole. Although confirmation is not possible without a detailed look at the component sectors in each potential cluster, the aggregate profile indicates that at least some of the critical high-technology links are present in the state's extended buyer-supplier chains. Although North Carolina's traditional manufacturing base operates at generally lower levels of technology, individual clusters or product chains consist of very high-technology sectors. The percentage of high-technology production in North Carolina's metalworking cluster, for example, well exceeds its US benchmark. As is shown below, the majority of state-wide activity in this cluster is in the higher-tech, higher-wage industrial machinery sectors, rather than basic metals production and fabrication.

Conversely, the share of high-tech production in the comparatively very small North Carolina aerospace and petroleum clusters is well below US averages; the few establishments in these clusters are producing largely standard rather than high-technology components in these buyer-supplier chains. More importantly, among some larger potential clusters with moderate shares of high-tech activity at the national level, the North Carolina vehicle manufacturing, fabricated textiles, and platemaking and typesetting clusters produce significantly lower shares of high-tech output relative to the United States To the degree that buyer-supplier relations do influence technology-adoption behaviour, the fact that some high-technology links in these chains are under-represented in the state could limit any local inter-firm influences encouraging technology upgrading among cluster members.

Establishment size and structure. Size and branch plant status have consistently proven key indicators of the level and rate of advanced process technology adoption among manufacturing plants in scientific studies. Numerous survey-based studies have found that large branch plants in nearly every major manufacturing industry adopt new technologies faster and to a greater degree than their smaller counterparts (Bergman *et al.*, 1995). Smaller producers have fewer of the necessary resources, both financial and human, to effectively integrate complicated new technologies into their production regimes. Alternatively, the owners of some smaller businesses may be reluctant to invest in technology upgrading if such investment requires some dilution of their equity in and control of the firm. Identifying those sectors with a predominance of smaller manufacturers, particularly those at the smallest end of the size scale, is thus one preliminary means of narrowing down areas of potential demand and need for competitiveness initiatives. Size, in effect, serves as a very rough proxy for level of modernisation, and indirectly, of a need for some form of technology assistance.

Table 3 lists the shares of both small and single (*vs.* branch plant) establishments in each cluster. Among the largest North Carolina clusters, the wood products, printing and publishing, and metalworking clusters are each made up predominantly of very small firms and establishments. In each case, close to 80% of businesses employ fewer than 50 workers. With the average share of branch plants at just 12%, these clusters are also largely composed of single-establishment enterprises. The clusters with the lowest shares of small plants are knitted goods (41%), packaged foods (52%), fabricated textile products (55%), chemicals and rubber (55%), and vehicle manufacturing (58%). With the exception of vehicle manufacturing, close to one-third of the establishments in each of these clusters are branch locations of multi-location firms.

Table 3. Plant size and status by cluster, North Carolina, 1994

Cluster	Primary industries only			Primary & secondary		
		Percentages			*Percentages*	
	Total	Single	<50 employees	Total	Single	<50 employees
Metalworking	2 183	88.0	76.1	2 613	87.8	74.6
Vehicle manufacturing	1 356	78.1	57.8	2 283	80.1	62.6
Chemicals & rubber	332	71.1	54.8	1 017	75.7	53.9
Electronics & computers	509	81.3	62.3	905	82.8	60.1
Packaged foods	120	71.7	51.7	433	73.0	54.0
Printing & publishing	1 668	87.8	79.3	2 229	87.6	78.5
Wood products	2 025	87.6	80.2	2 144	87.4	79.5
Knitted goods	1 485	62.4	40.9	1 998	63.2	42.4
Fabricated textile products	606	73.4	54.5	1 751	66.2	46.9
Non-ferrous metals	42	88.1	83.3	94	76.6	59.6
Canned & bottled goods	70	67.1	42.9	85	71.8	49.4
Leather goods	22	63.6	45.5	50	84.0	64.0
Aerospace	27	74.1	55.6	71	70.4	50.7
Feed products	84	72.6	69.0	159	75.5	62.3
Platemaking & typesetting	94	93.6	85.1	342	88.6	80.4
Aluminum	46	84.8	63.0	87	75.9	48.3
Brake products	46	93.5	80.4	605	97.7	92.2
Concrete, cement, & brick	173	33.5	32.4	322	57.1	54.3
Earthenware products	53	90.6	81.1	68	86.8	77.9
Tobacco products	26	46.2	3.8	26	46.2	3.8
Dairy products	14	21.4	14.3	33	45.5	36.4
Petroleum	10	70.0	70.0	19	73.7	73.7
Meat products	101	89.1	75.2	149	74.5	59.1
Non-loading (primary) sectors	533	69.8	88.7	--	--	--
Totals/cluster average	**11 625**	**79.7**	**67.5**	**--**	**79.5**	**64.7**

Source: NC Employment Security Commission and authors' calculations. See Bergman, Feser and Sweeney, 1996.

Regional and spatial distributions

Regional shares of state-wide cluster employment reveal some very general properties about gross spatial distribution, and the strength of this approach is revealed by the greater detail afforded about cluster concentrations within the state. Table 4 reports first the share of state-wide estimated output by cluster across the seven development planning jurisdictions. These data document substantial differences between the eastern and western halves of the state in terms of relative cluster concentrations.

Table 4. Regional share of total North Carolina cluster output
Includes primary and secondary sectors, 1994

Cluster	Western (WEC)	Carolinas (CP)	Piedmont (PTP)	Triangle (RTRP)	South-eastern (SEC)	Transpark (GT)	North-eastern (NEC)
Metalworking	10.6	34.9	23.3	12.5	7.1	9.7	1.9
Vehicle manufacturing	11.6	37.2	22.9	8.3	9.9	8.9	1.2
Chemicals & rubber	12.3	18.9	16.5	24.4	10.7	10.5	6.7
Electronics & computers	8.9	23.7	13.2	46.0	3.5	4.0	0.6
Packaged foods	16.7	20.4	11.1	13.9	11.9	11.4	14.5
Printing & publishing	17.9	28.5	17.4	9.8	8.5	8.2	9.7
Wood products	19.6	20.3	24.4	11.3	8.5	9.8	6.0
Knitted goods	11.5	32.0	29.4	8.7	8.9	6.7	2.7
Fabricated textile products	11.9	34.9	27.5	8.8	8.1	5.7	3.1
Non-ferrous metals	2.4	56.1	13.6	15.1	1.9	10.8	0.0
Canned & bottled goods	7.4	23.2	40.1	10.9	11.2	2.7	4.5
Leather goods	11.5	47.7	26.9	9.6	0.8	3.3	0.1
Aerospace	1.8	34.4	21.6	26.0	13.7	1.9	0.5
Feed products	6.5	6.1	15.6	52.0	4.2	15.7	0.0
Platemaking & typesetting	39.6	14.8	6.7	3.9	10.3	1.0	23.8
Aluminum	3.6	39.8	19.7	19.5	10.6	0.2	6.5
Brake products	15.4	48.9	15.6	5.1	3.5	9.9	1.7
Concrete, cement, & brick	8.4	35.9	18.5	16.0	16.7	3.0	1.5
Earthenware products	17.2	34.3	2.4	29.5	4.2	12.3	0.1
Tobacco products	0.0	13.0	80.0	4.5	0.1	2.3	0.0
Dairy products	7.5	10.5	21.9	5.5	0.5	54.1	0.1
Petroleum	1.4	69.8	8.4	5.0	12.8	2.6	0.0
Meat products	9.8	14.3	5.7	14.0	27.2	20.2	8.8

Source: North Carolina Employment Security Commission and authors' estimates. See Bergman, Feser and Sweeney, 1996.

Several major clusters are heavily concentrated in North Carolina's most urbanised regions, primarily the Carolinas and Piedmont regions. Over 60% of estimated output manufactured by North Carolina enterprises in fabricated textiles, knitted goods and vehicle manufacturing clusters is produced in these two urbanised regions, as is 58% of state-wide metalworking output. The third largest share of production in these clusters (except metalworking) originates in the Western region. Electronics and computers are even more heavily concentrated in the central part of the state, with nearly one-half of state-wide output of this cluster produced in the Research Triangle. Another 25% is produced in the Carolinas region. Barely detectable levels of manufacturing activity in this high-technology cluster are found in eastern North Carolina (the Southeastern, Transpark, and Northeastern regions).

Direct regional comparisons of relative specialisation in the particular sectors that comprise each region's clusters are illustrated in Figures 5 and 6. Selected industrial trade cluster templates along with each constituent sector's employment level are illustrated for the Research Triangle and Southeast regions. A template is configured for all possible sectors in the cluster, starting with its core sectors positioned at 12 o'clock, all others arrayed clockwise in declining order of importance, as measured by their correlation with the overall cluster. Employment data is then embedded for any sector present in a region along its appropriate radian. A cursory glance at the templates shows that either of these regions has only a subset of sectors for each cluster, and they are typically quite different sectors. The selective presence of specific sectors reveals how specialised regional economies within the same state can become, even after controlling for the same cluster.

Figure 5. Regional intra-cluster specialisations, electronics and computers

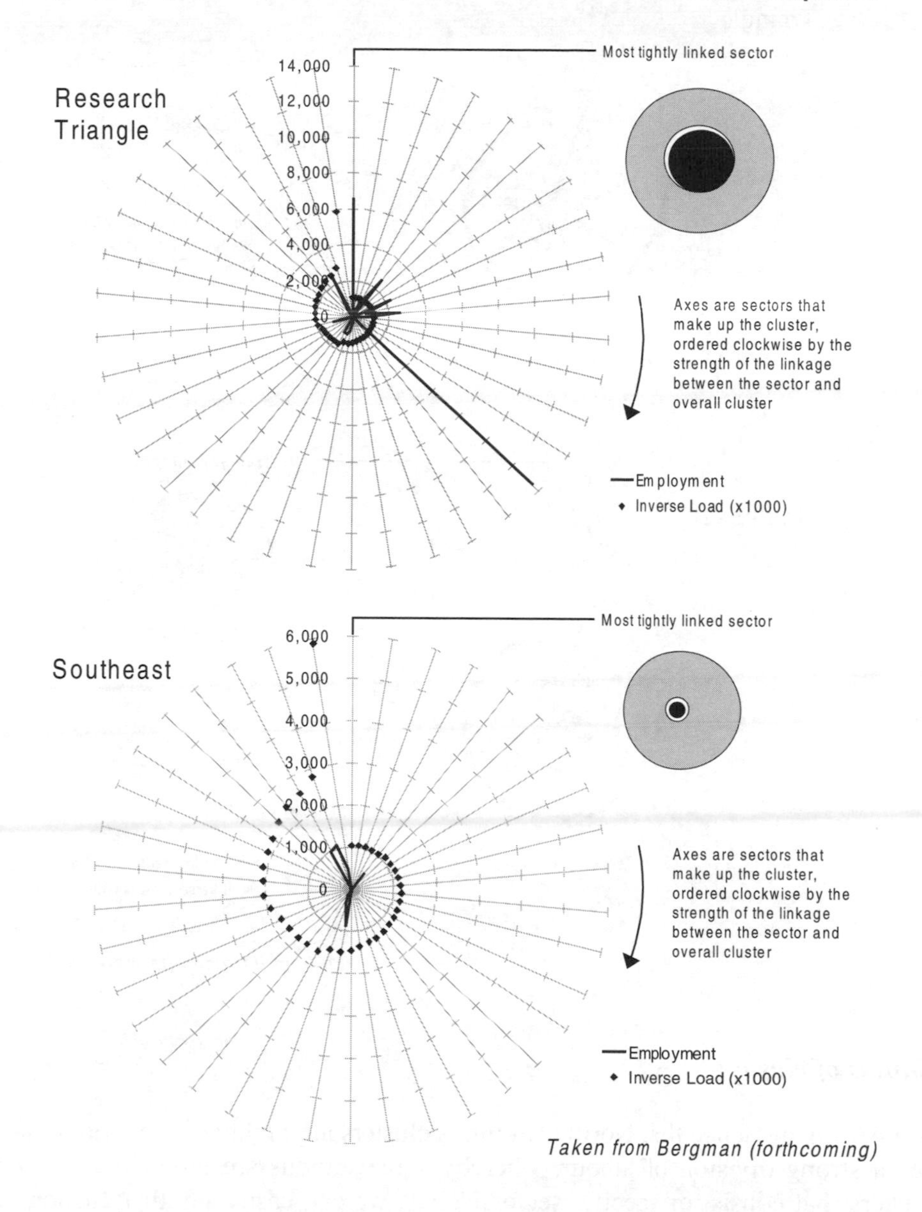

Taken from Bergman (forthcoming)

The "bulls-eye target" located next to each template shows the relative importance of each cluster in the regions. The full circle represents 100% of each region's total manufacturing (the Research Triangle's larger full circle indicates proportionately more total manufacturing employment in all its clusters). The white inner circle represents any cluster's share of total regional cluster employment (note the Triangle's relatively much larger electronics cluster), while the black inner circle represents the most tightly trading sectors *within that cluster* and the white halo represents the non-core sectors that trade with more than one cluster. The Research Triangle region's relatively small textile cluster consists of higher proportions of core sectors than is true for the Southeastern region's relatively larger textile cluster, which shows a much larger white halo of cross-trading sectors.

Figure 6. Regional intra-cluster specialisations, metalworking

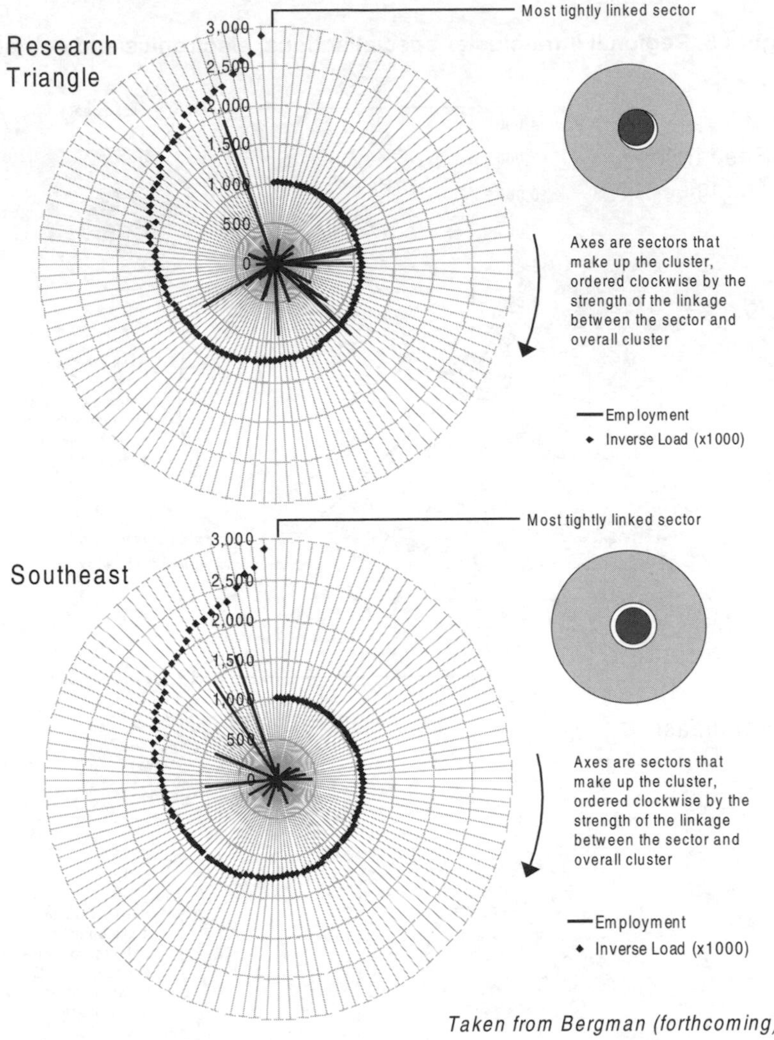

Research Triangle

Most tightly linked sector

3,000
2,500
2,000
1,500
1,000
500
0

Axes are sectors that make up the cluster, ordered clockwise by the strength of the linkage between the sector and overall cluster

Employment
Inverse Load (x1000)

Southeast

Most tightly linked sector

3,000
2,500
2,000
1,500
1,000
500
0

Axes are sectors that make up the cluster, ordered clockwise by the strength of the linkage between the sector and overall cluster

Employment
Inverse Load (x1000)

Taken from Bergman (forthcoming)

Spatial tightness of clusters

We have compelling evidence that North Carolina's clusters are highly concentrated in certain regions. This implies a strong division of labour, whereby some regions are more likely to attract and host growing clusters that consist of specific sectors. Thus, we can easily say that clusters have tightened and concentrated within regions. But can we infer also that cluster firms have tightened their spatial connections within regions, perhaps by locating more closely together to capture cluster spillovers due to relative proximity?

This question requires far more detailed analyses of cluster-member firm locations *vis-à-vis* other cluster firms and non-cluster firms. Can cluster-member firms be convincingly shown to "bunch" together more (or less) tightly than they bunch together with average firms? And how close together (or spread apart) are such bunches?

The results from our analyses provide the basis for a "spatial-economic test" of this question. The test involves use of a case-control design to test whether certain types of manufacturing firms (*i.e.* cluster firms) are more spatially concentrated than we might expect given the general geographic pattern of all firms in the state. In the present case, all plants associated with a given industry cluster (say motor vehicles) are used as cases and a matched sample of all other manufacturing firms is drawn as controls. The difference in concentration between the two, as measured with standard statistical geography techniques, provides evidence of spatial concentration or dispersion at different spatial scales for the firms in the economic cluster. The full methodology and results are outlined in Feser and Sweeney (1998).

Findings for three regional clusters with distinct degrees and types of spatial tightness are particularly interesting: vehicle manufacturing, printing and publishing, and wood products. In the case of vehicle manufacturing, firms are more tightly concentrated than at all spatial, although clustering is most significant at scales of two to six kilometres. It seems that the just-in-time pressures known to characterise this industry cluster may force greater than average spatial tightness over a wide range of distances. The slow convergence toward average spatial concentration may also reflect the many different sectors that comprise this cluster, ranging from highly urban, skill-intensive sectors to fairly rural, standardised production sites, which together typically spread along connecting interstate highways and major transportation corridors.

Similar in pattern, yet still unique, is the printing and publishing cluster. It also begins higher than average spatial concentration, but peaks earlier at about 12 km, and has nearly converged to average concentration at 50 km. This is clearly a highly urban industry, where shorter radial distances are the rule, often with face-to-face contacts necessitated by frequent design or delivery requirements.

Wholly distinct is the wood products cluster. This distinctive pattern shows that cluster members are *far more dispersed relative to each other* than they are to other firms. From 7 km onwards, firms become increasingly dispersed (relative to the average). This is one practical consequence of wood products being a natural resource-based industry cluster, where proximity to high-weight, moderate-value inputs automatically disperses its firms in remote places of resource availability. Worth noting is the second pattern of barely positive concentration over extremely short distances, which still remains far less concentrated relative to other firms. This may reflect wood product locations near non-wood product firms in smaller crossroads and market towns.

From these findings, it is evident that firms in some clusters are indeed far more closely co-located with each other than with other non-cluster firms. This implies that cluster externalities and advantages exceed those available to all other firms that enjoy available urbanisation externalities. Greater spatial tightness also implies stronger face-to-face possibilities, and the diffusion of technology, knowledge and general learning that is possible through such spatially permitted contacts. This finding confirms industrial folklore about the role of localised suppliers and machinery vendors in many industries, particularly the needle trades industries, but it also gives support for certain spatially centred and provided services to firms in such clusters. And we can detect points of relative concentration at distances that support other known industrial location tendencies, such as corridor-located motor vehicle supplier chains, urban-oriented printing clusters, or highly dispersed locations of clusters dependent upon natural resource distributions. Finally, some logical value-chain clusters can be far more dispersed than average firms. This implies comparatively high degrees of spatial looseness and independence, not tightness or contact intensity. Wood producers are visibly dispersed, but so too are such textile groups as the knitted goods cluster and fabricated textile cluster. Access to the highly intense inputs (natural resources or low-skill workers) that permit these highly dispersed structures may prove resistant to spatially based cluster remedies launched on their behalf.

The co-location tendencies should give pause to those advocating cluster-enhancing policies based on routine assumptions that all cluster firms now (or are willing to) locate within very short distances in small town, "industrial district" or urban settings. At the very least, we must go well beyond convenient descriptions drawn from streetlight samples of inter-firm clustering before industrial policy prescriptions are dispensed with confidence.

Firm dynamics within clusters

Regional cluster composition tends toward the types of concentration shown above, and is in continuous flux; some clusters and constituent sectors expand due to local advantages or related factors while others decline for opposite reasons. High growth rates are restricted mainly to the smallest clusters, although some large clusters such as motor vehicles, metalworking, and electronics and computers also gained employment. An overall glimpse of which clusters are growing or declining in importance gives a useful impression, but taken alone it offers very little sense of whether such change is due to the altered prospects of existing firms or to the shifting membership of new firms replacing the failures.

Table 5. Employment growth, plants existing in 1989 and new plants
North Carolina, primary industries only, 1989-94

Cluster	Employment change		Employment growth (%)		
	Plants existing in 1989	New plants	Plants existing in 1989	New plants	Net
Metalworking	-10 991	19 217	-13.2	23.1	9.9
Vehicle manufacturing	-18 734	24 345	-15.1	19.6	4.5
Chemicals & rubber	-1 801	7 348	-6.6	27.1	20.5
Electronics & computers	-6 438	9 373	-10.1	14.6	4.6
Packaged foods	-2 220	2 762	-18.8	23.3	4.6
Printing & publishing	-3 052	7 951	-7.1	18.6	11.4
Wood products	-11 837	14 031	-15.7	18.6	2.9
Knitted goods	-66 893	45 789	-32.1	22.0	-10.1
Fabricated textile products	-25 121	15 020	-27.2	16.3	-10.9
Non-ferrous metals	-1 653	485	-66.3	19.4	-46.8
Canned & bottled goods	-2 921	1 070	-29.5	10.8	-18.7
Leather goods	-1 368	226	-50.7	8.4	-42.3
Aerospace	-326	2 002	-37.3	228.9	191.7
Feed products	-297	476	-14.5	23.2	8.7
Platemaking & typesetting	-834	165	-37.2	7.4	-29.8
Aluminum	-2 257	868	-52.6	20.2	-32.4
Brake products	-871	842	-44.4	42.9	-1.5
Concrete, cement, & brick	-320	678	-9.5	20.1	10.6
Earthenware products	-436	390	-42.7	38.3	-4.5
Tobacco products	-6 134	659	-25.0	2.7	-22.4
Dairy products	-957	114	-46.0	5.5	-40.5
Petroleum	-17	28	-40.6	65.6	25.0
Meat products	-93	2 546	-1.7	47.7	46.0
Non-loading (primary) sectors	-10 522	19 471	-14.3	26.5	12.2
Totals	**-176 094**	**175 856**	**-20.3**	**20.3**	**0.0**

Source: Estimates of employment change due to existing and new plants are derived from matching NCESC ES-202 files. See Bergman, Feser and Sweeney, 1996.

Table 5 offers additional clues about the forces of overall cluster change that can be traced to underlying performance differentials in constituent firms. The most impressive point is *that no cluster expanded employment from its existing 1989 group of base firms; every cluster lost employment from such firms*. These losses ranged from as little as 1.7% in meat products to over 50% in the aluminium or leather goods clusters. Although the employment base of established firms declined in every cluster, the overall net change in total employment for all clusters (last row and column cell of Table 5) fell by far less than 1%.

This means that considerable growth came from *newly formed cluster firms;* some of it was very explosive new-firm growth. As examples, aerospace cluster employment grew from its 1989 base by 229% (its growth percentage rate was six times the loss percentage rate), while meat products grew 47.7% (its growth was 27 times its loss). About half of the industrial clusters grew more from new firms than lost employment from their original base. Other clusters lost more from their original employment base than was regained from new-firm growth: non-ferrous metals lost 66% of its 1989 base employment and gained back only 19% from new firms (its growth was 0.40% of losses); dairy products lost 46% of base and regained only 6% of new-firm employment (growth here was barely 0.12% of losses).

Regional cluster share dynamics

Have the losses and gains at the firm level triggered regional reconfigurations of clusters as well? To answer this question, we will compare share values in the same two periods for selected industrial clusters in every North Carolina region. We present evidence for four illustrative clusters (metalworking, motor vehicles, electronics, and food); each region is represented on the figure along a geographically oriented axis marked by increasing share percentages that radiate outwards. Shaded connections at each axis represent a region's share of total state-wide cluster employment, while connecting lines mark the region's share of five-year state-wide cluster growth. To read the figures, one first evaluates the shape of shaded shares to determine in which North Carolina region(s) each cluster is initially most heavily concentrated. If North Carolina's heavily shaded (concentrated) regions are becoming more specialised, then they should have equal or greater shares of five-year growth, as indicated by the connecting lines. If a region is losing its state-wide share, then the five-year growth share will be less than its initial total share.

For example, Figure 7 indicates increasing shares of metalworking and vehicle manufacturing clusters in the three western-most regions, all of which are very close to the major north-south highway systems, down which the automobile industry has steadily spread from Michigan and Ohio to Tennessee and Kentucky. Figure 8 reveals a centre-to-eastern state orientation for the chemical cluster (mainly Research Triangle region pharmaceuticals), and a pronounced strengthening of food product clusters in the Northeastern and Transpark regions. These clusters are simultaneously strengthening in some regions, while weakening in others during this five-year period, exactly as one might expect from an increasing returns model view of regional specialisation (Krugman, 1991; Martin and Sunley, 1996). Quite clearly, firms are seeking places with the best comparative advantages for themselves, and indirectly as well, in terms of advantages due to proximity of close trading partners. It is as though industrial districts and regions are shaping and reforming so rapidly as to be nearly visible to the naked eye.

Figure 7. Cluster dynamics

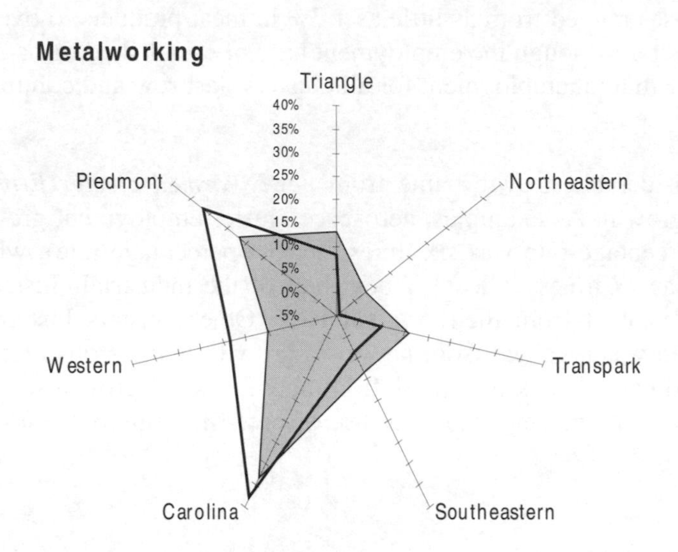

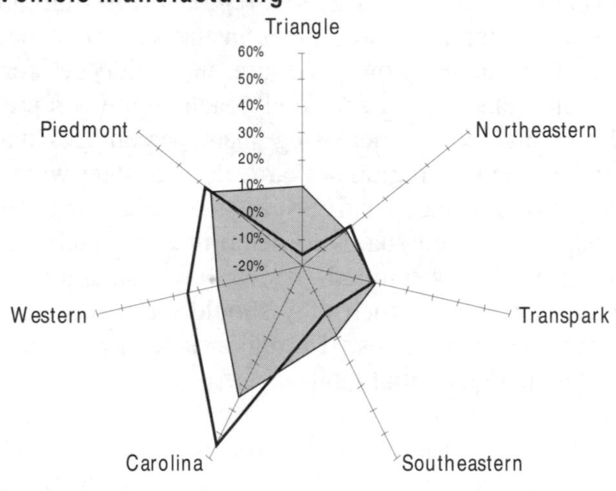

Note: shaded area = percentage share of state-wide cluster output, 1989; outline = percentage share of state-wide cluster growth, 1989-94.
Source: NC Employment Security Commission and authors'calculations.

Figure 8. Cluster dynamics

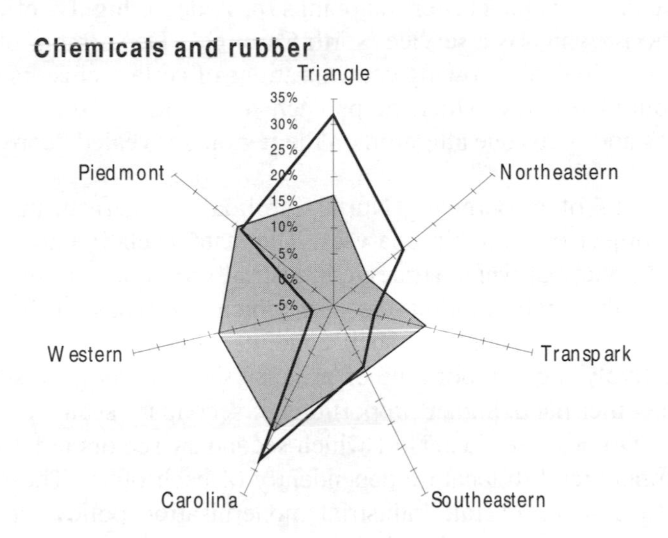

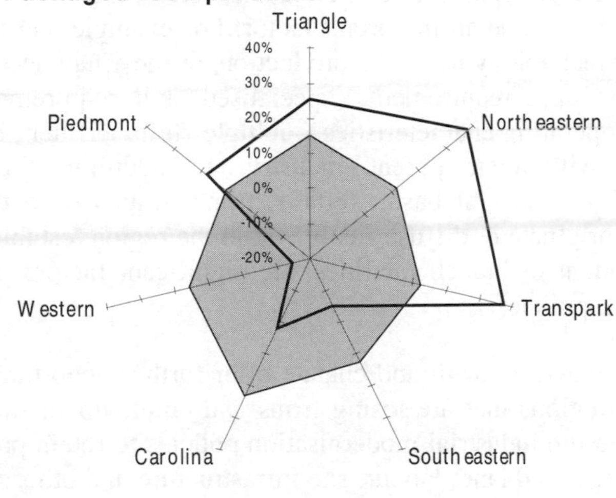

Note: shaded area = percentage share of state-wide cluster output, 1989; outline = percentage share of state-wide cluster growth, 1989-94.
Source: NC Employment Security Commission and authors'calculations.

6. Policy framework applications in North Carolina

This section advocates a policy-driven approach to the analysis of industry clusters and policy applications, and much of the evidence and analysis discussed earlier resulted directly from the original policy inquiry concerning the diffusion of modern production technologies. However, as each new analytical finding was presented to the sponsor, the original purpose became expanded to a wider range of collateral policy issues of interest to the client. For example, value-chain clusters proved very helpful to regional development agencies that must focus policy attention and scarce resources on key industrial subgroups. Policy targeting is much more efficient when the full industrial base can be segmented into coherent objects of programmes and initiatives of maximum value to its members, particularly if they also tend to co-locate in space. At the same time, relocating firms want new sites close to their suppliers, so value-cluster mappings provide a highly effective way to compare alternative plant locations, which is a service North Carolina's Department of Commerce is now able to offer. Dynamic changes indicate growing concentrations of certain clusters in favourable regions, or their loss in less-favoured regions, which helps each region assess the local factors that are most responsible for changes and focus due attention on the region's revealed "core competence".

The principal policy focus of modernising North Carolina's manufacturing base received the most attention and was the object of the main analyses. The state's relative concentration in each cluster provided the necessary review of which strategic industrial groupings are internally linked along their value chains, which is the main conduit through which all types of information flow. Modern production technologies will flow through such conduits to other cluster firms in the region *if* the cluster consists of relatively high proportions of primary sectors *and if* cluster members co-locate in tight proximity. On the other hand, higher proportions of secondary sector firms permit modernisation flows to spread across two or more clusters of which secondary sectors are a member, or two or more regions if cluster members tend to locate independently of each other. These are extremely valuable principles on which to base an overall industrial modernisation policy, and specific regional data permits their application to a variety of situations.

North Carolina's regions clearly concentrate in certain sectors of a given cluster, which means that the *sector characteristics* also become an important factor. For example, certain sectors have inherently higher shares of R&D or technology-intensive production, or they may need common inputs (*e.g.* high electrical or natural gas input requirements, specialised skill requirements, unique infrastructure demands), or particular operating characteristics (multiple shifts, effluent or discharge), all of which are closely correlated with their parent industry. In addition, the predominant *enterprise characteristics* of a region's sectoral base exerts a strong influence on the likelihood that modern production technologies originate or diffuse easily within the region: establishment sizes, and whether they are mainly independent or branch facilities, are significant factors in the design of regionally appropriate modernisation policies.

Finally, the dynamics of cluster growth and change offer further opportunities to fine-tune industrial modernisation policies. Regions that are losing firms and employment shares of some clusters may tempt policy makers to design industrial modernisation policies to retain production, but this could be counterproductive if the released land, labour, site infrastructure and other resources are rapidly being deployed by another cluster's dynamic firms. Specific firm dynamics also reveal the extraordinary importance of new-firm start-ups and arrivals to maintain a healthy cluster. Here again, policy makers might be tempted to design policies intended to rescue ailing firms, when the key to sustained success lies instead in diffusing competitive production technologies to newly entering and well-functioning firms.

NOTES

1. Copies of the reports are available from the North Carolina Alliance for Competitive Technologies, 1110 Navajo Drive, Suite 510, Raleigh, North Carolina, 27609.

2. This section is drawn from Feser and Bergman (forthcoming).

REFERENCES

Asheim, B.T. and A. Isaksen (1997), "Location, Agglomeration, and Innovation: Towards Regional Innovation Systems in Norway?", *European Planning Studies,* Vol. 5, No. 3, pp. 299-330.

Bergman, E.M (forthcoming), "Industrial Trade Clusters: Seeing Regions Whole", *Proceedings*, Theories of Regional Development Workshop, Uddevalla, Sweden.

Bergman, E.M., E.J. Feser and J. Scharer (1995), *Modern Production Practices and Needs: North Carolina's Transportation Equipment Manufacturers*, UNC Institute for Economic Development, Chapel Hill, North Carolina.

Bergman, E.M., E.J. Feser and S.H. Sweeney (1996), *Targeting North Carolina Manufacturing: Understanding the State's Economy Through Industrial Cluster Analysis*, UNC Institute for Economic Development, Chapel Hill, North Carolina.

DeBresson, C. (1996), *Economic Interdependence and Innovative Activity*, Edward Elgar, Cheltenham.

DeBresson, C. and X. Hu (1999), "Identifying Clusters of Innovative Activity: A New Approach and a Toolbox", this volume.

Doeringer P.B. and D.G. Terkla (1995) "Business Strategy and Cross-industry Clusters", *Economic Development Quarterly*, Vol. 9, No. 3, pp. 225-237.

Enright, M.J. (1996), "Regional Clusters and Economic Development: A Research Agenda", in U.H. Staber *et al.* (eds.), *Business Networks: Prospects for Regional Development*, Walter de Gruyter, Berlin.

Feser, E.J. (1998a), "Enterprises, External Economies, and Economic Development", *Journal of Planning Literature*, Vol. 12, No. 3, pp. 283-302.

Feser, E.J. (1998b), "Old and New Theories of Industry Clusters", in M. Steiner (ed.), *Clusters and Regional Specialisation*, pp. 18-40, Pion Limited, London.

Feser, E.J. and E.M. Bergman (forthcoming), "National Industry Clusters: Frameworks for State and Regional Development Policy", *Regional Studies.*

Feser, E.J. and S.H. Sweeney (1998), "A Test for Spatio-economic Clustering", draft manuscript, Department of City and Regional Planning, University of North Carolina at Chapel Hill, North Carolina.

Gollub, J. *et al.* (1997), "Cluster-based Economic Development: A Key to Regional Competitiveness – Case Studies", National Technical Information Service PB98-117088, Springfield, Virginia.

Harrison, B. (1992), "Industrial Districts: Old Wine in New Bottles?", *Regional Studies*, Vol. 26, No. 5, pp. 469-483.

Heidenreich, M. (1996), "Beyond Flexible Specialization: The Rearrangement of Regional Production Orders in Emilia-Romagna and Baden-Württemberg", *European Planning Studies*, Vol. 4, No. 4, pp. 401-419.

Held, J.R. (1996), "Clusters as an Economic Development Tool: Beyond the Pitfalls", *Economic Development Quarterly*, Vol. 10, No. 3, pp. 249-261.

Hill, E.W. and J. Brennan (1998), "A Methodology for Identifying the Drivers of Industrial Clusters: The Foundation of Regional Competitive Advantage", forthcoming, *Economic Development Quarterly*.

Isaksen, A. (1997), "Regional Clusters and Competitiveness: The Norwegian Case", *European Planning Studies*, Vol. 5, No. 1, pp. 65-76.

Jacobs, D. and M.W. de Jong (1992), "Industrial Clusters and the Competitiveness of the Netherlands", *De Economist*, Vol. 140, No. 2, pp. 233-252.

Jacobs, D. and A.-P. de Man (1996), "Clusters, Industrial Policy and Firm Strategy: A Menu Approach", *Technology Analysis and Strategic Management*, Vol. 8, No. 4, pp. 425-437.

Kaufman, A., R. Gittell, M. Merenda, W. Naumes and C. Wood (1994), "Porter's Model for Geographic Competitive Advantage: The Case of New Hampshire", *Economic Development Quarterly*, Vol. 8, No. 1, pp. 43-66.

Krugman, P. (1991), *Geography and Trade*, MIT Press, Cambridge, Massachusetts.

Martin, R. and P. Sunley (1996), "Paul Krugman's Geographical Economics and its Implications for Regional Development Theory: A Critical Assessment", *Economic Geography*, Vol. 72, No. 3, pp. 259-292.

Nelson, R.R. (1993), *National Innovation Systems: A Comparative Study*, Oxford University Press, New York.

Ohmae, K. (1995), *The End of the Nation State: The Rise of Regional Economies*, Free Press, New York.

Park, S.O. and A. Markusen (1995), "Generalizing New Industrial Districts: A Theoretical Agenda and an Application from a Non-Western Economy", *Environment and Planning*, Vol. 27, No. 1, pp. 81-104.

Peneder, M. (1995), "Cluster Techniques as a Method to Analyze Industrial Competitiveness", *International Advances in Economic Research*, Vol. 1, No. 3, pp. 295-303.

Porter, M.E. (1990), *The Competitive Advantage of Nations*, Free Press, New York.

Roelandt, T., P. den Hertog, J. van Sinderen and N. van den Hove (1999), "Cluster Analysis and Cluster-based Policy in the Netherlands", this volume.

Rosenfeld, S.A. (1997), "Bringing Business Clusters into the Mainstream of Economic Development", *European Planning Studies*, Vol. 5, No. 1, pp. 3-23.

Steiner, M. (1998), *Clusters and Regional Specialisation*, Pion Ltd., London.

Sternberg, E. (1991), "The Sectoral Cluster in Economic Development Policy: Lessons from Rochester and Buffalo, New York", *Economic Development Quarterly*, Vol. 5, No. 4, pp. 342-56.

Chapter 11

CLUSTER AND NETWORK DEVELOPMENT PROJECTS IN DEVELOPING COUNTRIES: LESSONS LEARNED THROUGH THE UNIDO EXPERIENCE

by

Giovanna Ceglie, Michele Clara and Marco Dini*
Private Sector Development Branch, UNIDO, Vienna

1. Introduction

The guiding principle of UNIDO's approach towards SMEs is that *small-scale manufacturing enterprises can play a key role* in triggering and sustaining economic growth and equitable development in developing countries. However, this potential role is often not fulfilled because of a particular set of problems relating to the size of SMEs. Individually, SMEs are often unable to capture market opportunities which require large production quantities, homogenous standards and regular supply. By the same account, they experience difficulties in achieving economies of scale in the purchase of inputs (equipment, raw materials, finance, consulting services, etc.). Small size also constitutes a significant hindrance to the internalisation of functions such as training, market intelligence, logistics and technology innovation – all of which are at the very core of firm dynamism. Furthermore, small scale can also prevent the achievement of specialised and effective internal division of labour which, according to classical economic theory, fosters cumulative improvements in productive capabilities and innovation. Finally, because of the continuous and fierce struggle to preserve their scarce profit margins, small-scale entrepreneurs in developing countries are often locked into their routines and are unable to innovate their products and processes and look beyond the boundaries of their firms to capture new market opportunities.

Through networking, individual SMEs can address the problems related to their size and improve their competitive position. On account of the common problems they all share, small enterprises are in the best position to help each other. Through horizontal co-operation (*i.e.* with other SMEs occupying the same position in the value chain), enterprises can collectively achieve scale economies beyond the reach of individual small firms and can obtain bulk-purchase inputs, achieve optimal scale in the use of machinery and pool together their production capacities to satisfy large-scale orders (Pyke, 1992). Through vertical integration (with other SMEs as well as with large-scale enterprises along the value chain), enterprises can specialise on their core business and give way to an external division of labour (Marshall, 1920).

---

* The opinions, figures and estimates set forth in this chapter are the responsibility of the authors and do not necessarily reflect the views or carry the endorsement of the United Nations Industrial Development Organisation (UNIDO).

Inter-firm co-operation also gives rise to a collective learning space, an "invisible college" (Best, 1998), where ideas are exchanged and developed and knowledge shared in a collective attempt to improve product quality and occupy more profitable market segments. Lastly, networking among enterprises, providers of entrepreneurial services (*e.g.* training institutions or technology centres) and local policy makers can help to shape a shared local development vision and give strength to collective actions to enhance entrepreneurial strategies.

This chapter is an attempt to reflect upon UNIDO's experience in promoting networking as a strategy to develop small-scale enterprises. Section 2 provides the rationale for the approach, while Section 3 illustrates real cases of networking development projects drawn from some of the countries where the approach is currently being implemented. Section 4 emphasizes the key components of a methodology that have emerged as a result of a five-year experience in project implementation. The concluding section reflects upon some of the key lessons learned and highlights what emerge as the most significant issues that could be usefully considered in further applications.

Prior to a closer examination of the main elements of UNIDO's experience, a working definition for the concept of "networks", "clusters" and "networking" needs to be introduced. In this chapter, the term *network* refers to a group of firms that co-operate on a joint development project – complementing each other and specialising in order to overcome common problems, achieve collective efficiency and conquer markets beyond their individual reach. The term *cluster* is used to indicate a sectoral and geographical concentration of enterprises which, first, gives rise to external economies (such as the emergence of specialised suppliers of raw materials and components or the growth of a pool of sector-specific skills) and, second, favours the rise of specialised services in technical, administrative and financial matters. Such specialised services create a conducive ground for the development of a network of public and private local institutions which support local economic development by promoting collective learning and innovation through implicit and explicit co-ordination.[1] Lastly, the verb *networking* refers to the overall action of establishing the relationships characterising both networks and clusters. In this chapter, therefore, networking development projects indicate those projects aimed at promoting the development of clusters and networks.

2. Origins of the cluster/network-based approach

Evidence of well-performing SME clusters has been extensively reported in the literature (Goodman *et al.*, 1989; Pyke *et al.*, 1990; Sengenberger *et al.*, 1990; UNCTAD, 1994, among others). In many performing clusters, such as the Italian industrial districts, inter-firm networking primarily emerged spontaneously as the result of the peculiar historical and social environment surrounding the SMEs (Brusco, 1982; Piore and Sabel, 1984; Beccattini, 1990; Best, 1990). Spontaneous networking has also been observed in some developing countries (Schmitz, 1990; Navdi, 1995), but appears to be relatively uncommon. Even less common is the spontaneous emergence of other features of successful clusters, such as institutions promoting collective learning and innovation.

In spite of the potential benefits for the enterprises, therefore, evidence shows that inter-firm co-operation and the other features of successful clusters do not always emerge spontaneously. Three factors are among the main ones significantly hindering this process: *i)* the significance of the transaction costs that need to be borne to identify suitable network partners and to forge relationships; *ii)* the imperfect market functioning for the provision of crucial inputs for networking development such as information and innovation; and *iii)* the high risk of "free riding" that is especially faced in contexts where the legal framework to back up joint endeavours is relatively underdeveloped.

The available literature vividly bears out that the intervention of an "external agent" acting as a catalyst to facilitate the emergence of clusters and networks can greatly reduce the significance of the above factors. Among cases of developing countries, Navdi (1995) provides interesting examples of successful interventions aimed at fostering co-operative relations within SME clusters drawn from the experience of Brazil, Mexico and India. Along the same lines, Humphrey and Schmitz (1995) describe the main features of the Chilean PROFO (*Proyectos de Fomento*) programme consisting of a carefully designed set of public incentives which has stimulated the establishment of approximately 450 SME networks with significant results in terms of increased SME profitability and sales (Dini, 1998).

Taking stock of these experiences and of the general reflection on the clustering and networking phenomena, UNIDO has promoted a new technical assistance programme for SMEs. This is characterised by an emphasis on the promotion of efficient systems of relations between enterprises and between enterprises and institutions which allow enterprises to overcome their isolation and reach new collective competitive advantages beyond the reach of individual small firms. The programme[2] also emphasizes the development of local institutions to act as facilitators of the networking process, or "system integrators". These should support the emergence of a joint entrepreneurial vision involving the whole business system – composed of firms, their suppliers, buyers and support institutions – and be able to enact that vision through common development projects. Indeed, it is this emphasis on the whole business system – and not on the individual enterprise – that constitutes the main difference between networking programmes and other traditional technical assistance programmes.

While highlighting this difference, it should also be pointed out that networking projects are not in competition with other technical and financial incentives but, on the contrary, usually enhance their use by the enterprises. The collective projects generated by the networks, in fact, demand technical and financial inputs for their execution. This demand is not directly satisfied by the networking project (as it is not part of its core functions) but is channelled, by the "system integrators", to other available providers of technical/financial services. In this way the relationships between the enterprises and the local service providers can be optimised while the usage rate of the services is increased. If gaps are detected in the support system, the networking project can take steps to overcome them by initiating institution building (such as in the case of Jamaica illustrated in the next section).

3. From theory to practice

In order to illustrate UNIDO's experience and its crucial aspects, the following cases have been selected which exhibit some of the more significant features of the projects implemented over the last five years.

The case of *Honduras* represents a project which evolved from the creation of SME networks into the establishment of a specialised institution (CERTEC) acting as a networking promotion agency. The case of *Nicaragua* illustrates three main points: first, how, as a result of its successes with network brokering, a project has achieved significant influence on policy making at the national level; second, the importance of local capacity building; and third, how the principles of economies of scale and scope inform the delivery of support measures. The case of *Mexico* highlights a project promoting vertical integration arguing for the direct involvement of large-scale manufacturers into suppliers' upgrading efforts. And, finally, the case of *Jamaica* presents an example of entry at the top institutional level (whereby the effort to bring about cluster-based development lies at the operational core of the national SME support agency) and a case of creation of specialised service centres (for garments, fashion, furniture, etc.).

3.1. Honduras: evolution of a networking project

In 1993, the Honduran Government requested UNIDO to design and implement a technical assistance project for the development of the SME sector. Due to the unsupportive institutional environment which characterised the SME sector at the onset of the project, the project focused directly on the enterprise level, relying on a group of eight national consultants with engineering and management skills under the guidance of the local UNIDO director.

Initially, the national consultants[3] concentrated on identifying groups of enterprises with similar characteristics and growth constraints and helped them to establish common development projects. The beneficiary enterprises were micro and small enterprises with an average staff of between two and 15 people. The enterprises were selected based on either personal knowledge of the consultant or through the assistance of local producers' associations (*Asociación Nacional de Mediana y pequena Industria de Honduras* – AMPIH, and *Asociación Nacional de Industriales* – ANDI) or through other local institutions such as INFOP (*Instituto Nacional de Formación Profesional*), the local training institute. Following a visit to the enterprises selected, a series of weekly joint discussions were organised by the consultant to support the group of entrepreneurs in analysing their problems, identifying common solutions and outlining a common work plan.

The work plan envisaged a division of tasks among network members and, often, group saving schemes to establish a common fund to finance common activities. The consultants also assisted in the implementation of the work plans, calling upon other local SME support institutions to provide specialised inputs. Among the most active participating institutions were PASI (*Programa de Apoyo al Sector Informal*) for the provision of credit and INFOP (*Instituto Nacional de Formacion Profesional*) for the provision of training. The close contact with other institutions also helped in channelling the entrepreneurs' demand for different and improved services, leading to a lasting upgrade of the locally available business services.

Over its five years of operation, with an investment of approximately USD 680 000 (contributed by the Government of the Netherlands), the project has established 33 networks with common development projects involving some 300 enterprises. Common projects focused on, for instance, joint purchasing of raw material, joint establishment of shops to retail finished products, launching of new production lines, product or process specialisation, sharing of large orders (including public procurement), and creation of new enterprises which complement existing production facilities. It should be noted that the assistance given by the project to the networks has consisted entirely of technical assistance; no funding of working capital or investment, whatsoever, was granted. On the financial side, the project acted as an intermediary between the networks and the financial institutions to help the enterprises meet the requirements for obtaining loans. One example of a network developed by this project is described in Box 1.

In most cases, the collective projects have launched new businesses for the networks, increased the revenues of the participating enterprises and generated new employment. A recent in-depth evaluation of six networks, selected at random among the 33 mentioned above, showed a positive trend for all basic performance indicators. For instance, comparing the data at the beginning of the project with current data, sales increased between 35% and 200%, employment increased between 11% and 50% and investment in fixed assets increased between 10% and 100%.

Box 1. The Emasim metalworking network

Emasim is a group of eleven enterprises in the metalworking sector in Tegucigalpa. Their average staff is four workers. At the beginning of the project, the entrepreneurs were invited to participate in a training course at INFOP to improve their technical capabilities. It was through this course that the entrepreneurs started to get to know one another better and, with the help of a project consultant, began analysing their problems while searching for common solutions.

The consistent supply and cost of raw materials was identified as the most urgent problem. In response, a common raw material supply centre was created, a common loan obtained from PASI and an internal revolving fund established to be used by members of the network as working capital. Based on the progress achieved through this initiative, the network members expanded their co-operation to the production level, by exchanging tools, identifying and sharing large orders (for instance in metal construction and maintenance works at supermarkets and banks) and examining ways to complement their production processes. In order to diversify production and target new market segments, the network decided to collectively invest in new, larger equipment and to establish a separate independent enterprise to manage the new equipment and provide services to the network members. Among the quantitative results registered in this network, it can be noted that, to date, collective sales have increased by 200% (in comparison to total individual sales prior to the establishment of the network), employment increased by 15% and fixed assets by 98%.

As the project implementation advanced, two interlinked themes emerged, namely: *i)* how to increase the impact of the project by creating additional networks – thus benefiting more entrepreneurs – and by accelerating their development process; and *ii)* how, over time, to guarantee the sustainability of the networking promotion effort.

In an attempt to address the first theme, a three-point strategy was adopted. First, the project consultants started *training other "network brokers"* in order to diffuse network creation capabilities and multiply results. New network brokers were selected from locally active institutions (especially entrepreneurial associations) and from other technical assistance projects. More and more local institutions are currently demanding the network-broker training service. Second, the project consultants drafted a *network development methodology* to facilitate the transfer of knowledge to new brokers in order to accelerate their learning process. At the same time, working instruments were devised to assist and facilitate their work. One such instrument is described in Box 2. Third, *the scope of networking was broadened* to include: *i)* the development of vertical networks involving relations between small enterprises and larger ones; and *ii)* the development of clusters where the emphasis shifted from the pure entrepreneurial strategy of the horizontal networks to a strategic vision of local development involving local institutions and local governments.

Box 2. The Network Evaluation Tool (NET)

As a result of the network development experience acquired in the Honduras project, the Network Evaluation Tool (NET) was developed. This tool is structured on the basis of a matrix which intersects network development indicators with network development stages in order to measure the level of network development. The development indicators used are: *group cohesion, group organisation, capacity of problem analysis, capacity of strategic planning, production and organisational changes, changes in economic variables,* and *relations with the external economic environment.* The development stages, as described in Section 4, are: promotion *and motivation, strategic planning, pilot projects, strategic projects* and *self-management.* At each intersection of development indicators and development stages, the results the network should achieve are described. Achievement, partial achievement or "non-achievement" of the results are translated into scores that, at the end of the application, indicate the level of network progression. This score is then graphically represented permitting the assessment of a network's evolution over time and comparisons with other networks for benchmarking purposes. The tool is also a useful instrument to constantly assess and redesign the network development methodology. It provides feedback to network brokers on their own work so that they, in turn, can adjust the services rendered to enterprises accordingly.

In an attempt to address the second theme of long-term sustainability, a process of project "privatisation" was implemented. A foundation was established whose employees are the team of national consultants and whose members are local private and public institutions. The CERTEC (*Centro de Recursos y Tecnologìa*) foundation started working within the UNIDO project in 1997. After one year of operation, during which USD 60 000 in revenues was generated, representing more than 50% of total annual costs, the institution became independent from UNIDO and is now managing its own budget and strategy.[4]

Finally, two elements of the Honduras experience are worth highlighting. First, it should be stressed that the functions of the national team evolved from *direct assistance* to the enterprises to *higher-level functions* of training other intermediaries (network brokers), improving the intervention methodology and devising new integration modalities. This resulted in a substantive multiplier effect. For example, between 1997 and 1998, CERTEC trained 71 brokers who have since organised 59 networks with the participation of 1 200 enterprises.

Second, the type of funding required by projects such as the one in Honduras needs to be taken into account. The financing of a networking project (or of an institution like CERTEC) has to draw from a combination *of public and private funds*. In the case of CERTEC, a tripartite funding is envisaged for the next years of operation, namely: *i)* service fees, *i.e.* funds generated by the sale of services to the enterprises (networking services) and to institutions (associations, local and central government for services such as training of network brokers); *ii)* membership fees; and *iii)* public funds which, in the case of CERTEC, will be contributed by an international donor for the next three years.

As will be seen from Section 5, the investment of public funds in institutions such as CERTEC is justified by the fact that CERTEC aims at implementing development measures for the SME sector, a sector which is predominantly populated by enterprises that are not, as clients, in a position to fully fund the operating costs of the organisation. Pushing CERTEC to survive under purely commercial conditions would tempt the foundation to look for wealthier clients, thus giving second priority to the demands of the small enterprises for whom the institution was created.

3.2. Nicaragua: broadening the scope of networking

The Nicaragua project started in 1995 with PAMIC (*Programa Nacional de Apoyo a la Micro Empresa*, which later became INPYME, *Instituto Nacional para la Pequena y Mediana Empresa*) as the local counterpart. During the first phase of the project, the strategy has been similar to that described in the Honduras case. Some 20 networks (horizontal) were created (one example is described in Box 3) by a team of seven national consultants assisted by short-term international consultants.

The main difference *vis-à-vis* the Honduras case is that the Nicaragua project has had, since the beginning, a public counterpart. This has had three consequences. First, the project has benefited from *easier entry into local policy dialogue and formulation*. As a result, the project has had leverage in proposing the networking strategy as a key SME development strategy. Networking promotion has now become one of the main axes of the government's support to the private sector. Second, the project has also played a prominent role in *inter-institutional co-ordination* and has had greater access to local people and resources (channelled through the counterpart). Third, the Nicaraguan project, from the onset, displayed a much *clearer prospect of sustainability*. The long-term prospect is that the project, with its team, will be taken over by the counterpart and will become one of its strategic branches. For the time being, however, the project has retained its autonomy in operational terms and is managed independently (although in close consultation with the counterpart). This will guarantee that the project team acquires the necessary skills and experience to ensure the required maturity in its dialogue with the counterpart and with the public sector in general.

Box 3. The EcoHamaca Handicraft Hammock Sector

EcoHamaca is a network of eleven enterprises operating in the handicraft hammock production sector. While the network members compete with one another in the local market, they now collaborate in an attempt to break into foreign markets. Prior to UNIDO's assistance, none of the local producers had direct exporting experience. Through the project, the producers were assisted in standardising their production in order to collectively reach quantities suitable for export and, at the same time, improve the quality and design of the products and the pricing systems. The group selected an ecologically friendly strategy, focusing on changing the wood used for the poles (from cedar wood, which is close to extinction, to other more abundant exotic species) and the dying substances from chemical to natural ones. This strategy proved successful since it permitted the group to penetrate important markets such as the EU and the United States. To date, the producers have exported on eight different occasions to destinations such Finland, Peru, Sweden and the United States, and over 3 000 hammocks are exported on average every month. In order to consolidate results and further common work, the group has acquired legal status and has hired a manager whose tasks presently include the identification of more formal training schemes for the workers, the research of other technical and financial assistance inputs from a variety of local SME support institutions and strengthening of their marketing strategy. EcoHamaca is now present on Internet.

Further and complementary to the above, three points are worth emphasizing with regard to this project (which has now entered its second three-year phase, with a budget of USD 1.3 million financed by the Government of Austria).

First, the project is actively promoting *inter-institutional networking* at two levels. Through the establishment of an inter-institutional committee, the project shares activities with local providers of financial and technical services (including other multilateral and bilateral donors). This committee is presently co-operating on a variety of issues including improving loan access for SMEs, co-ordinating the design and application of evaluation criteria for SME assistance intervention, transferring network development methodology to other service providers and executing specific joint projects in localities or sectors of common interest (for instance wood and furniture in Masatepe, leather and shoes in Leon, etc.). On a more comprehensive scale, the project has been invited to assume an important role in the National Committee for Competitiveness and Sustainable Development – formed by high-level policy makers and main representatives from the private sector as well as the main economics university of Nicaragua – and is contributing to the design of an overall SME development policy.

Second, an important component of this project is *local capacity building*. The national consultants working on the project are local professionals with no international experience and no direct knowledge of cluster or network practices or policies. The project is therefore investing in training in order to upgrade and specialise their skills; this should result in improved services to the enterprises. Training is provided by international consultants and UNIDO staff via thematic seminars and on-the-job training. As further explained in Section 5, other forms of training are also being implemented such as joint learning programmes abroad on best practices in cluster/network development.

Third, as was also the case in Honduras, the project is now *diversifying its activities* to include, in addition to the promotion of horizontal networks, training of new network brokers, promotion of industrial integration along production chains (SME/large enterprise subcontracting with emphasis on supplier upgrading) and promotion of industrial districts (in Nicaragua the term "industrial district" is used to mean clusters as defined in the introduction of this chapter). This evolution has come about quite naturally while realising that economies of scale and scope can extend beyond the boundaries of horizontal networks. As exemplified in the following box on Masaya, the task is to find "the right equilibrium" in scale and scope of the joint action and to aim the common development projects to achieve maximum economic efficiency and return. In Masaya, this has translated into a progressive evolution of the scale of the common project from networks to sector to cluster to national level, as schematically described in Box 4.

Box 4. Masaya Handicrafts

Masaya is a town south-east of Managua with a strong handicraft tradition. One of the main local products is hammocks. Initially, the project assisted *networks* of hammock producers to upgrade their products for export (see Box 3). While implementing the network's projects, it became evident that the main factor influencing hammock prices was the cost of the cotton yarn used as raw material. After studying the relationship of the cost of cotton yarn to quantities purchased, it became clear that the best prices could be obtained for quantities greater than those required by the single networks, *i.e.* co-ordinating the purchase of yarn at the level of the whole hammock *sector* of Masaya. The brokers, therefore, focused on creating a local purchase centre offering raw material to a large number of Masaya hammock producers.

Another important factor which was identified for improving the performance of the hammock sector was design. In response to this, the project is working towards improving design according to market trends and creating new products (hanging chairs, deck chairs, cribs, etc.). The design of the new products, in order to be successful, is being done in co-operation with the wood and furniture sector. At the same time, since the most interesting market for this line of products is the export market, the need to build up an export promotion strategy has arisen. The resulting launch of the common brand, "made in Masaya", is being developed to promote local identity accompanied by activities to increase the quality of local products. This brand will be extended to all handicraft products from Masaya and therefore to the entire *cluster.*

Finally, other initiatives will have a *national* dimension. For instance, actions to facilitate export transactions which are part of the export strategy for Masaya, will obviously extend their effect to all Nicaraguan enterprises.

The conclusion emerging from the experience in Masaya is that the concepts which guide the implementation of network/cluster-based projects is demand orientation and creative solution design. Brokers should look at the entire business system, tap all available resources and design the intervention in order to take maximum advantage of economies of scale and scope.

3.3. *Mexico: promoting vertical integration*

In the second half of 1997, the Mexican Confederation of Industrial Chambers (CONCAMIN), the *Fundación para la Transferencia Tecnológica a la Pequeñas y Medianas Empresas* (FUNTEC), the United Nation Development Programme (UNDP) and UNIDO, gave birth to the Programme of Industrial Integration (PII). Through a flexible and decentralised set of initiatives, the PII aims at stimulating and supporting local projects to promote networks of SMEs as well as subcontracting networks between small- and large-scale enterprises.

The two projects initiated in the states of Chihuahua and Jalisco over the first six months of the programme focus on this second feature[5] – aiming at increasing the competitiveness of local SMEs by stimulating a deeper and broader integration with the multinationals established locally.[6] In the case of Chihuahua, the entrepreneurial counterpart has been the association of *maquiladoras* firms. In the case of Jalisco, the participating association is the *Camara de la Industria Electrónica.* In both Chihuahua and Jalisco, the entrepreneurial counterpart covers one-third of the operating costs; another third is contributed by the state government, while the remaining third is funded by the PII.

Both projects are in the process of establishing two technical centres (Centres for Supplier Development) with the following aims: *i)* helping enterprises to identify subcontracting opportunities; *ii)* co-operating with the technical personnel of leader firms in the definition of support programmes targeted at upgrading the capabilities of the identified subcontractors; and *iii)* identifying and channelling technical support, training and loans (when required) from locally available institutions to the subcontractors to assist them in meeting the needs of the main contractors.

While both centres are still in their initial phase, some lessons can, nevertheless, be derived from the experience gained during their design and initiation:

- Despite the well-known scepticism that many foreign multinationals, and especially *maquiladoras* firms, have towards local producers, the fact that leader firms are playing an important role in both centres in terms of direct (financial) support and sensitisation of other partners, proves that in Mexico large-firm openness towards establishing linkages with small firms is improving.

- The benefits of a Centre for Supplier Development for the client enterprises are two-fold. First, such a centre can co-ordinate the demand for the goods and services of the main contractors. It then becomes possible to achieve significant economies of scale that not only lower the prices of production inputs but that can also justify new investments by the subcontractors to meet the demands of a pool of leader firms. Second, such a centre can co-ordinate supply and help establish horizontal networks among subcontractors. This type of action is crucial where there is a significant gap between the leader firms and their subcontractors – especially in terms of production capabilities, technology and management. The creation of a network, with the task of organising and improving the offer of a group of subcontractors, can provide fulfil an intermediate position that is frequently missing in the supply chain.

- In spite of the above-mentioned advantages, the idea of a centre implies a collective action by the main contractors which can often be extremely complex. The establishment of a consensus on the design of the centre and the co-ordination of the technical inputs for its management are all initiatives characterised by significant transaction costs which are often high enough to freeze or radically slow down the development of any collective project. It is precisely the reduction of such transaction costs that justify the existence of a PII whose main added value is, therefore, to speed up the decision-making process at the enterprise level, minimise the time wasted in negotiations, promote the emergence of a consensus and co-ordinate the contributions at various levels.

- The experience of the Mexican project indicates that a support measure focusing on subcontractors has the best prospect for maximising its impact when the leader firms participate not only in its funding, but also commit their own technical personnel to the selection of the potential subcontractors and design of the support initiatives. This type of participation ensures not only that the initiatives are genuinely demand-led but also the transfer of the knowledge base accumulated by the main-contractors to the subcontractors.

- Lastly, it needs to be noted that, in spite of numerous similarities, the two centres initiated in Jalisco and in Chihuahua are profoundly different from the traditional subcontracting exchange schemes that operate in many countries with the aim of linking the demand and the supply of subcontracting services. The centres in Jalisco and in Chihuahua do not operate on the notion that the main obstacle to the creation of such links is an information failure (which is at the basis of traditional types of subcontracting exchanges). While instruments that tackle the information gap are used (such as the creation of databanks on demand and supply), the centres mainly concentrate upon technical support initiatives to address the basic problems of capacity failure and difficulty in establishing relationships based on trust.

3.4. Jamaica: an example of institutional networking

The Jamaica project is another example of entry at the institutional level. The project, which was initiated in 1994 (the second phase started in 1997 for a duration of three years; the total budget for the two phases of approximately USD 1.5 million was contributed by the United Nations Development Programme – UNDP) was requested by the Jamaican Government to assist the public development agency, JAMPRO, in implementing a support strategy for the local SME sector.

The Productivity Centre, located within JAMPRO, is the focal point for project implementation. Unlike the other projects described, the activities of the Jamaican project are directly implemented through the staff of the Productivity Centre. An international chief advisor, funded by the project, has been requested by JAMPRO to assist the local team. Two main features of this project are noteworthy: institutional capacity building and network promotion. Institutional capacity building consisted of: *i)* strengthening the capabilities of the Productivity Centre to act as a networking promotion agency; and *ii)* creating specialised centres, co-ordinated by JAMPRO, to provide "real services" to SMEs. As a result of the project, the Productivity Centre now performs the following functions: identifying SME needs and designing the public institutional answer to meet these needs; networking and co-ordinating actions with other local institutions active in SME-related fields (such as HEART, the national training agency, community colleges, University of West Indies, vocational schools, specialised service centres, etc.) favouring streamlining and specialisation of services; acting as an information hub on issues related to SMEs; acting as network broker. Specialised centres have been created/upgraded (mainly within existing institutions) by the project in fields such as garments and fashion, furniture, food processing, handicrafts, and in the metalworking sector. The centres provide technical services to entrepreneurs (see Box 5 as an example) and act as "second-level" networking institutions linking the entrepreneurs with other service providers for services they do not offer.

Box 5. Network support system for the fashion industry

An institutional support network has been established, involving educational, training and technical institutions, to help Jamaican SMEs operating in the fashion sector. At the heart of the network is the JAMPRO Design Centre which, through its fashion division, offers the following services: information on fashion trends, advice to manufacturers on design improvements using CAD systems, linkages between manufacturers and local and foreign designers, and information on suppliers of inputs for the fashion industry. Other important actors in the networks are the two Apparel Technical Centres -- one in Kingston and one in Montego Bay – which provide training and technical assistance to producers in areas such as computerised pattern-making and grading, product development and flexible manufacturing systems. These centres have both the functions of diffusing best manufacturing practices and stimulating SMEs to network for joint purchases of raw material, joint marketing, etc.

What should be emphasized about this project is that the entry at the top institutional level has guaranteed highest local ownership of the initiative and, in turn, good prospects for sustainability. Moreover, JAMPRO, in its role as system co-ordinator, is ideally positioned to articulate a coherent structure of services for SMEs. The project now faces the challenge of helping the support institutions to study and implement a coherent fee structure to recover at least part of the service costs from the client enterprises.

4. Methodology

On the basis of the experiences described above, and UNIDO's current overall involvement in cluster/network-related projects (including projects in eleven countries), it is possible to draw some conclusions on the methodological steps and principles which characterise UNIDO's networking initiatives. Four phases, which represent distinct intervention levels, need to be distinguished: *i)* promotion of networks; *ii)* restructuring at the firm level; *iii)* improvement of the institutional environment; and *iv)* improvement of the dialogue between the public and private sectors. These phases do not require strict adherence. On the contrary, as the case studies amply demonstrate, their sequencing and relative importance must be fine-tuned in accordance with the surrounding environment.

4.1. The promotion of networks[7]

The experience of UNIDO proves that it is possible to initiate and develop effective relationships among independent entrepreneurs based on collaboration and production integration even where the entrepreneurs had no previous knowledge of each other. The central element for the development of a network is the creation of a sufficient level of trust through a process of mutual learning which can be suitably stimulated and guided by an external agent (the network broker) trained to perform such a function.

In somewhat simplified terms, it could be argued that the mutual learning process has the following two features. First, it is an *empirical process* based on trial and error within which theoretical and conceptual elements necessarily play a limited role. In order to create a relationship based on trust, entrepreneurs need to be exposed to an interactive process starting with "role assignment" leading to "criticism based on the analysis of the results" and finally "reassignment of responsibilities" within which they can assess, empirically, the capability and commitment of their partners. Second, the process is an *incremental* one because it is assumed that, lacking any previous experience with trust, the group needs to act gradually; it will therefore start by undertaking initiatives with a low level of risk for the participating enterprises and only subsequently shift to more complex projects as mutual trust is build up.

In practical terms, through UNIDO's experience in the field, five different phases have been identified in the establishment of an effective and viable network of enterprises: *i)* promotion and motivation; *ii)* strategic planning; *iii)* pilot projects; *iv)* strategic projects; and *v)* self-management (Rabellotti 1998).

The *promotion and motivation phase* consists of a set of promotional initiatives which need to be launched to contribute to: *i)* the identification of a critical mass of SMEs sharing similar growth constraints; *ii)* their sensitisation to the benefits of networking; and *iii)* the emergence of groups and group leaders. In this first stage, the network brokers normally organise large, open meetings to introduce the principles of networking and to point out their possible applications. As a result of this promotional initiative, entrepreneurs group around issues (problems and/or opportunities) they have in common. There appears to be no optimal selection criterion for enterprises to be part of the same network. The entrepreneurial characteristics that appear to be most conducive to collective actions, and which need to be stimulated by the network brokers, are a willingness to learn and an openness to discuss and develop relationships with other people. Similarly, there appear to be no general rules concerning size or location of the groups. Nevertheless, it should be noted that geographical concentration and limiting the number of members reduce co-ordination costs. The viability of a collective project depends, in other words, on the trade-off between the critical mass of enterprises necessary to sustain the joint action and, inevitably, its co-ordination costs.

Once groups have emerged, it becomes possible to move to the *strategic planning phase* which involves the following elements: *i)* analysis of common problems and opportunities; *ii)* establishment of a common work plan; and *iii)* group organisational structure. For the identification of common problems and opportunities, it is necessary that the network brokers carry out an in-depth analysis of the growth constraints of the enterprises and of their causes and that they do not rely exclusively upon the perception of the entrepreneurs themselves. Often, the entrepreneur is biased towards short-term needs, for instance shortage of working capital, without realising the causes of those needs which could be, in the case of working capital, inappropriate cash-flow management. A crucial component in delineating a group work plan is to reach a consensus concerning a definition of the evaluation criteria of the collective action to be applied in the short, medium and long term. Such criteria need to be both quantitative (as in the Honduras case) and qualitative, and must be easily understood, computed and, needless to say, must be in line with the objectives that the group has selected. An important function of the network brokers is to inject the group with a mind-set founded on continuous improvement based on periodical evaluation of the results obtained and setting up of new objectives. From this point of view, the monitoring system needs to be perceived by entrepreneurs as a useful tool in evaluating the performance of their partners (and of the network brokers) and to keep track of the evolution of the project, while also evaluating the return on investment and time. From the viewpoint of the network brokers, the criteria are the key instruments to evaluate performance of the network and decide on whether or not to continue assistance. Finally, it is during this phase that the group selects its legal status and the rules which govern its internal organisation, such as the key features of its representative bodies (function, duration, etc.), the fines to be levied upon "free riders" and the affiliation fees. These rules can be characterised by different degrees of formalisation but, above all, they need to be both thoroughly transparent and readily understood and implemented.

The strategic planning phase opens the door for the implementation of a *pilot project phase* through which co-operation should start bearing concrete results for the participating enterprises. In general, the projects undertaken during this phase are of a commercial and/or promotional kind: joint participation in fairs, joint purchase of raw material, design of a collective catalogue, etc. The idea is to generate visible results (although of a short-term nature) in order to engender optimism and trust and consolidate the network's willingness to further co-operation.

Where successful, pilot projects are expected to give way to *strategic projects* – focusing on specialisation and complementation at the production level. Strategic projects commonly involve one or more of the following components: *i)* an increase in the degree of specialisation by process and by product of the network members; *ii)* the provision of common facilities through the creation of new enterprises (as in the case of EMASIM described above); or *iii)* the launch of new product lines and common brands (as in the case of EcoHamaca also described above).

The final stage of the network-building exercise, the *self-management phase*, coincides with the group of enterprises earning greater autonomy from the network brokers and the capability of independently carrying out further joint activities. Self-management is not always an easy step and it has been observed that networks often tend to rely on broker's assistance for a longer time than initially envisaged. To avoid dependency, two rules apply. The first is that the work plan established by the network members and the broker must have a specified time frame. In this way, the network members know from the beginning that they can count on the broker only for a limited period of time and must use this time wisely. The second rule is that the fees which are normally charged to the network for the assistance given by the broker, and which are quite low at the beginning, must be progressively increased to encourage network autonomy and, from the broker's point of view, allow investment on new target beneficiaries.

The last element, which is worth stressing in such a process of enterprise network establishment, is the role and profile of the network leaders. In the initial phases of group establishment, the network brokers are the real leaders. As groups mature, the function of the network brokers must shift towards softer co-ordination and a progressive transfer of responsibilities from the network brokers to the entrepreneurs must be ensured. Often, in order to counterbalance the reduction in the assistance provided by the broker, networks contract a manager to assist in the implementation and upgrading of the work plan (see the EcoHamaca case as an example).

4.2. Restructuring at the firm level

In addition to engendering a collective competitive advantage, network creation often also brings about a transformation within the individual member enterprises aimed at adapting their production and organisational capabilities to the requirements of the common objectives. If, for instance, the network embarks upon process specialisation whereby the network members subcontract with each other, the individual enterprises will be pushed to improve their internal organisation to respect the quality standards, production schedules and pricing levels decided by the group. Group pressure will stimulate individual enterprises to fully commit themselves to implementing the necessary improvements and will sanction members for failing to achieve the common objectives.

While networks can generate positive changes in the individual enterprises, the opposite also holds true: enterprise restructuring can greatly contribute to improving network prospects. Therefore, the objective of individual improvements should be kept in mind by the network broker who should help to orient the efforts of the enterprises and liase with the various institutions operating in the surrounding environment.

4.3. Improving the institutional environment

Two types of institutions are involved in the UNIDO network programmes: those which are direct actors in project implementation (which have a primary and proactive role); and those which play an indirect role (*i.e.* which support the implementation of actions designed by the first type of institution). The "cluster/network brokers" and the "networking unit" belong to the first type of institutions. The cluster/network brokers play a pivotal role at the level of direct assistance to enterprises. They are the agents (institutions and consultants) who facilitate the generation of the networks. The networking unit plays the strategic role within the networking projects, it: *i)* bears the responsibility for designing and promoting the networking strategy in a given country; *ii)* identifies the sectors/regions to be addressed depending on their potential; *iii)* carries out extensive awareness building among small-scale enterprises and local institutions; *iv)* trains network brokers; *v)* manages the available funds, searching and implementing a sustainability strategy; *vi)* monitors the development and impact of the networking initiative; and *vii)* provides feedback to the various actors involved.

At the beginning of a networking project, the functions of the networking unit and of the network/cluster broker are usually assumed by the same institution/team of professionals. As the scale of activities of the project increases and the need for specialisation grows, the two functions are progressively split and assumed by different actors, as described in the case of Honduras.

The external institutions, on the other hand, essentially support the realisation of the networks' work plans, thus requiring a wide range of technical and financial services. It is the task of the networking unit to ensure that networks can obtain the assistance they require from their surrounding environment. In a relatively weak institutional environment, this task implies upgrading the capacity of specialised service centres or, in some cases, bringing about their establishment, as was the case in Jamaica.

4.4. *Improving the dialogue between the private and public sectors*

Finally, a fundamental component of a networking/cluster project concerns the establishment of co-operative relations between the public and private sectors. The aim of such relations is to promote the emergence of a co-ordinated industrial policy and identify, develop and implement coherent actions to support the entrepreneurial effort.

In each of the UNIDO projects described, the creation of a public/private Project Advisory Committee, or participation in existing co-ordinating bodies such as the National Committee for Competitiveness in Nicaragua, have contributed to sensitising policy makers to the benefits of clusters and networks, thus favouring the internalisation of the key principles of networking development within the strategy of public SME support agencies. At the same time, this co-ordination allows the projects to convey to policy makers issues of concern to the private sector (such as reforestation policies in Nicaragua to guarantee a regular supply of raw material to the local furniture sector; banning illegal imports of leather goods in Honduras; improving credit access to the SME sector in Nicaragua). On each of these issues, the projects contributed to elaborating proposals for consideration by the public authorities.

5. Lessons learned

The experiences gained during five years of UNIDO involvement in network/cluster-related projects permit certain conclusions to be drawn. The nine "lessons" presented below do not purport to be a "summa" of prescripts to apply in networking projects, but rather a selection of observations that may prove useful in designing future projects:

1. An important principle in the design and implementation of networking projects is *demand orientation*. In UNIDO's experience, project strategies must be flexible and vary from network to network and from cluster to cluster depending on the nature of the constraint/objective of the target population. One important requirement is that the intervention must be designed after a thorough analysis of target beneficiaries' needs as well as of the surrounding economic environment from which resources can be tapped to satisfy those needs.

   Consideration should be given to the type of demand orientation used in these projects. While projects are initiated on the basis of a beneficiary's demand, beneficiaries should be helped to formulate these demands based on an analysis of their growth constraints and of the underlying causes. In this sense, demand orientation is not passive but proactive, with the brokers playing an important role in the strategic planning process of the enterprises. Especially in developing countries, where small enterprises have a weak capacity to develop a strategic response to market challenges, this approach has proven the most suitable in UNIDO's experience.

2. Three principles guide UNIDO's work with respect to networks, namely, they need to be: *i)* business-oriented; *ii)* production-grounded; and *iii)* targeted at SMEs. *Business orientation* refers essentially to two components: first, networking must aim to visibly improve the economic situation and prospects of participating SMEs; second, it must grant the group a new competitive advantage which the individual enterprises could not obtain alone. While the first point might seem obvious, it has repeatedly been shown that networking can be interpreted as pure exchange of information or as an end in itself rather than as a means to achieve concrete economic advantages. In the Honduras case, for instance, it took great efforts to change network meetings from social events to

business talks. A further step is to translate business talks into action and ensure that the actions are profitable and lead to positive structural, as opposed to temporary, changes in the enterprises. The second point emphasizes the fact that, while other technical assistance schemes promote the network concept as purely applied to groups of enterprises participating in the same activity, in UNIDO's approach a network should also have a further scope. Although common activities are useful, as in the case of joint training which reduces the fixed cost of training, in UNIDO's experience networks should also aim at generating a new competitive advantage translating into the generation of new business without which the networks do not fulfil their whole potential (as in the case of EcoHamaca where the participating SMEs were able to enter the export market thanks to the joint action).

The second principle, *focus on production,* points to the importance of process and product innovation and structural improvement as opposed to, for instance, an increase in sales resulting from an occasional participation in a trade fair. While activities like information exchange and joint participation in fairs are important parts of a network work plan, they are not the end objective of UNIDO's approach which is rather to improve the business prospects of the SMEs producing long-term changes in their production capability and organisation. It might surely be argued that a new market opportunity stemming, for example, from the joint participation in a fair, might spur the creation of networks and the development of co-operative relationships among members. In UNIDO's experience, however, such a transition rarely occurs automatically. In some projects, the networks have been exposed to market opportunities (especially for export) which they could not fulfil due to a lack of organisational capabilities and productive capacity. Supporting a network, therefore, should involve not only the search for new market opportunities, but also provide the assistance required to restructure the network's production organisation to respond to new markets in a timely manner, with the right quantities and quality.

Finally, the *focus on SMEs* refers to the fact that, even though networks may involve other partners (such as large-scale firms, retail chains, etc.), the primary beneficiaries need to be the SMEs. For instance, in the case of Mexico, while multinational industries are among the main actors involved in the project, the focus is clearly on supplier development and local development.

3. *Networking is a multidimensional concept and does not only apply to enterprises.* Institutional networking, networking between the private and public sectors, country networking (as in the case of the Joint Learning Programme outlined in point *5)* below) are equally important concepts in UNIDO activities. The idea is to specialise and co-operate to the maximum extent, so that each actor in the economic system can dedicate itself to its core functions and perform them to the best of its abilities. In practice, this principle translates into the natural evolution of the networking units which, as described in the cases of Honduras and Nicaragua, specialise into strategic functions, decentralising the implementation functions to other network brokers after an initial period during which all such functions are centralised. By the same token, this principle implies a suitable division of labour among network brokers and other service providers, whereby the network brokers do not pretend to solve all problems of the enterprises but help the enterprise to identify other service providers which may be of assistance – as is the case in Jamaica where JAMPRO is working towards diffusing the specialised function to other institutions.

4. The key resources in networking initiatives are the *people* involved (policy makers, brokers, other service providers). With this in mind, it is important to discern four factors that can increase the likelihood of project success: people's ownership, empowerment, skills and incentives. At all levels, project actors must:

   – *Own the project* and feel that it is their interest to carry it out successfully. To this end, it is important to adequately invest in raising awareness, at all levels, to involve local actors in project design, and encourage their continual feedback for improving project implementation.

   – *Be empowered to act.* In other words, all the actors involved must have the leverage, credibility and resources to play their role. If, for instance, counterpart institutions do not have credibility *vis-à-vis* the beneficiaries, project activities will not have the desired impact.

   – *Have the right skills to act.* In addition to an appropriate academic and professional background, the skills of network brokers must encompass such invaluable "extra-curricula" skills such as the capacity to build teams, thorough knowledge of local social rules and an openness to establish contacts. Network brokers must possess a rare combination of technical background, business mentality and "social sensitivity" to produce market-feasible projects for collective benefit.

   – *Have the right motivation and incentives.* The issue at stake here is that, in addition to the leverage and skills to act, network brokers must also have the right motivation to look for clients and help them to improve their businesses. UNIDO's experience indicates that appropriate incentive schemes can enhance brokers' motivation and channel their efforts into projects that hold the possibility of higher impact and longer-term gains for the networks. However, what types of incentives work best in achieving the desired results, and what types of results should be encouraged? On the latter issue, while incentives anchored to the financial gains of the assisted networks may seem to be a sound idea, there lurks the danger that this could bias the choice of projects/firms, leading network brokers to select relatively "easy" targets (*i.e.* larger enterprises) or promote relatively "short term" activities with quick returns rather than longer-term but more structural changes. In UNIDO's experience, the incentives must be anchored not only to the financial performance of the networks but also to more comprehensive criteria involving qualitative assessment. The qualitative assessment is based on the achievement of the objectives indicated in the work plans agreed upon by the network broker, the network and the overall project co-ordinator (depending on the case, this can be the UNIDO project manager or the director of the project counterpart).

   Regarding the nature of the incentive, in UNIDO's experience the most effective incentive for network brokers has been training – such as the study tours discussed below in point *6)*. Study tours, and the possibility to learn about successful experiences in other countries and regions, have proven to be a very positive stimulus to improving performance, especially among young professionals. A less tangible, but equally effective motivator, is the existence of a framework that allows network brokers to work together and exchange ideas, thus fostering a sense of teamwork. The positive atmosphere created when such teamwork is encouraged and the sense of "not going it alone" not only applies to enterprises in a network but also is key in supporting, encouraging and motivating the brokers.

5. The importance of investing in people has been emphasized in point *4)* above. However, one of the critical ways to support these key actors through the provision of the *necessary training and exposure to best practices* warrants further expansion. The importance of continuous training, as well as the need to diffuse information related to best practices to orient networking agents' decisions, is crucial. In UNIDO's experience, the kind of training that has emerged as most valuable and effective in transferring knowledge on the "nuts and bolts" of networking, is to rely heavily on concrete cases of successful networks and clusters and let networking agents hear directly from other agents who have implemented successful networking projects. To this end, UNIDO has elaborated on the idea of the "Joint Learning Programmes", aimed at providing first-hand exposure of cluster and network agents from developing countries to successful cluster/network experiences. To date, this programme has been run in the Emilia Romagna region of Italy, focusing on the experience of Italian industrial districts. This programme will be expanded to the overall European experience through invitations to other countries to participate. A second programme is planned in Chile, based on Latin American network/cluster promotion experiences. In addition to specific training, a series of working tools, such as "NET" (Box 2), are being systematised in order to facilitate the work of the network brokers and accelerate the transfer of knowledge to new networking agents. Other instruments are being developed such as a practical manual for network brokers and a set of monitoring and evaluation indicators for networking projects. All these instruments are constantly evolving and are meant to stimulate creative thinking rather than impose rigid boundaries.

6. A combination of *private and public investment* appears to be the best way to finance network development initiatives. The main elements militating against an exclusive reliance on the market is that networking projects aim at balancing market failures (as described in the introduction of this chapter); therefore, the market cannot be expected to entirely cover their costs. Such a realisation should not, however, lead one to believe that networks need to rely entirely on public funding. The elements which diminish the appeal of exclusive reliance upon public funding are first, the limits it is likely to impose on the accountability of project managers to market feedback and therefore beneficiaries' satisfaction; second, the fact that beneficiaries' co-financing ensures selectivity of beneficiaries on an objective basis (or, from another angle, less discretionality by the service provider in targeting one beneficiary or another). Finally, the balance between private and public funding does not need to remain the same over time: as the initiative progresses and its impact is more visible, funding normally changes in favour of a higher market share.

7. Evaluation criteria for *networking projects need to be carefully designed,* as illustrated by the Honduras example. While quantitative evaluation indicators are always useful, there are three aspects to consider: *i)* the scarcity of reliable and comparable data on the performance of small firms; *ii)* the understandable unwillingness of entrepreneurs to release confidential data about their businesses; and *iii)* purely quantitative measures often fail to take into consideration results such as institution building, as well as indirect results such as those resulting from the work of second- (or third-, etc.) generation brokers. On the other hand, in spite of the difficulties related to the quantitative measurement of project impacts, the collection of objective data is essential not only for evaluating the return on the investment made by the donor but also to disclose the possibility of charging private sector beneficiaries who, understandably, want to know with a certain degree of objectivity what benefits they can expect from buying certain services. In UNIDO projects, a combination of qualitative (related to the specific

objectives of cluster/network work plans) and quantitative criteria (of the type mentioned in the country cases) is used to evaluate networking projects.

8. The introduction of the elements of market cost recovery should be pursued as early as possible in order to avoid the risk that the beneficiaries become accustomed to full subsidies and that the *enterprises become dependent on the project activities*. Progressively increasing the share of the cost that enterprises have to cover is one a way to reduce such a risk. It is the task of the network brokers to lead the networks towards a process of self-management (described in Section 5) and to develop an autonomous capacity to identify new collective strategies, implement the joint projects and liase with SME support institutions. By the same token, in UNIDO's experience, the long-term impact of projects may be endangered unless networking institutions/cluster-brokers can free themselves from dependence on the continuous assistance provided by UNIDO and develop an autonomous "strategic thinking" capability to continually improve and upgrade their services in line with the dynamics of the entrepreneurs.

9. Lastly, there is no *single, predefined path to be followed* in the implementation of cluster/network promotion initiatives that can be effortlessly replicated across countries, regions and industrial sectors. Cluster/network support initiatives need to be flexible and in tune with the characteristics of the environment where SMEs operate. While the elements that comprise the intervention are always those described in Section 4 (network, firm, institution and policy), the "dosage" and "sequence" need not be the same for all projects and all countries. In UNIDO's experience, a bottom-up approach, centred on fostering an entrepreneurial vision and supporting local actors' initiative to realise that vision, appears to be the best approach.

NOTES

1. This definition takes into consideration Humphrey and Schmitz, 1995.

2. The term *programme* here indicates a technical assistance framework implemented through country *projects.*

3. In the projects described in this chapter, the consultants promoting networks/clusters are also called "network/cluster brokers". The terms "consultant" and "broker" will therefore be used interchangeably.

4. Revenues have been generated by selling services to the networks and to client institutions especially for training of network brokers.

5. According to the document programme, 12 projects will be initiated over the three years of the Programme.

6. Multinationals in Mexico have very little interaction with local subcontractors. In the case of Chihuahua, for example, the integration level (which is slightly higher than the national average and has grown over the last few years) barely reached 3% in 1997.

7. This section will primarily focus on horizontal networks of enterprises.

REFERENCES

Becattini, G. (1990), "The Re-emergence of Small Enterprises in Italy", in Sengenberger *et al.* (1990), *The Re-emergence of Small Enterprises: Industrial Restructuring in Industrialised Countries*, ILO, Geneva.

Best, Michael H. (1990), *The New Competition: Institutions of Industrial Restructuring*, First Harvard Press, Great Britain.

Best, Michael H. (1998), "Cluster Dynamics in Theory and Practice with Application to Penang", *UNIDO Report.*

Brusco, S. (1982), "The Emilian Model: Productive Decentralisation and Social Integration", *Cambridge Journal of Economics*, Vol. 6, No. 1, pp. 167-184.

Dini, M. (1998), "Proyectos de Fomento – Chilean Experience Promoting the Implementation of SME Networks", paper presented at the UNIDO Joint Learning Workshop, Bologna, 28 September-3 October.

Goodman, E., J. Bamford and P. Saynor (1989), *Small Firms and Industrial Districts in Italy*, Routledge, London.

Humphrey, J. and H. Schmitz (1995), "Principles for Promoting Clusters and Networks of SMEs", *UNIDO Discussion Papers*, No. 1, Vienna.

Marshall, A. (1920), *Industry and Trade*, Macmillan, London.

Navdi, K. (1995), "Industrial Clusters and Networks: Case Studies of SME Growth and Innovation", *UNIDO Discussion Paper*, Vienna.

Piore, M. and C. Sabel (1984), *The Second Industrial Divide: Possibilities for Prosperity*, Basic Books, New York.

Pyke, F., G. Becattini and W. Sengenberger (1990), *Industrial Districts and Inter-firm Co-operation in Italy*, International Institute for Labour Studies, Geneva.

Pyke, F. (1992), *Industrial Development through Small-firm Co-operation,* ILO, Geneva.

Rabellotti, R. (1998), "Helping Small Firms to Network", *Small Enterprise Development*, Vol. 9, No. 1, pp. 25-34.

Schmitz, H. (1990), "Small Firms and Flexible Specialisation in Developing Countries", *Labour and Society*.

Sengenberger, W., G.W. Loveman and M.J. Piore (1990), *The Re-emergence of Small Enterprises: Industrial Restructuring in Industrialised Countries*, ILO, Geneva.

UNCTAD (1994), *Technological Dynamism in Industrial Districts: An Alternative Approach to Industrialization in Developing Countries,* United Nations, New York.

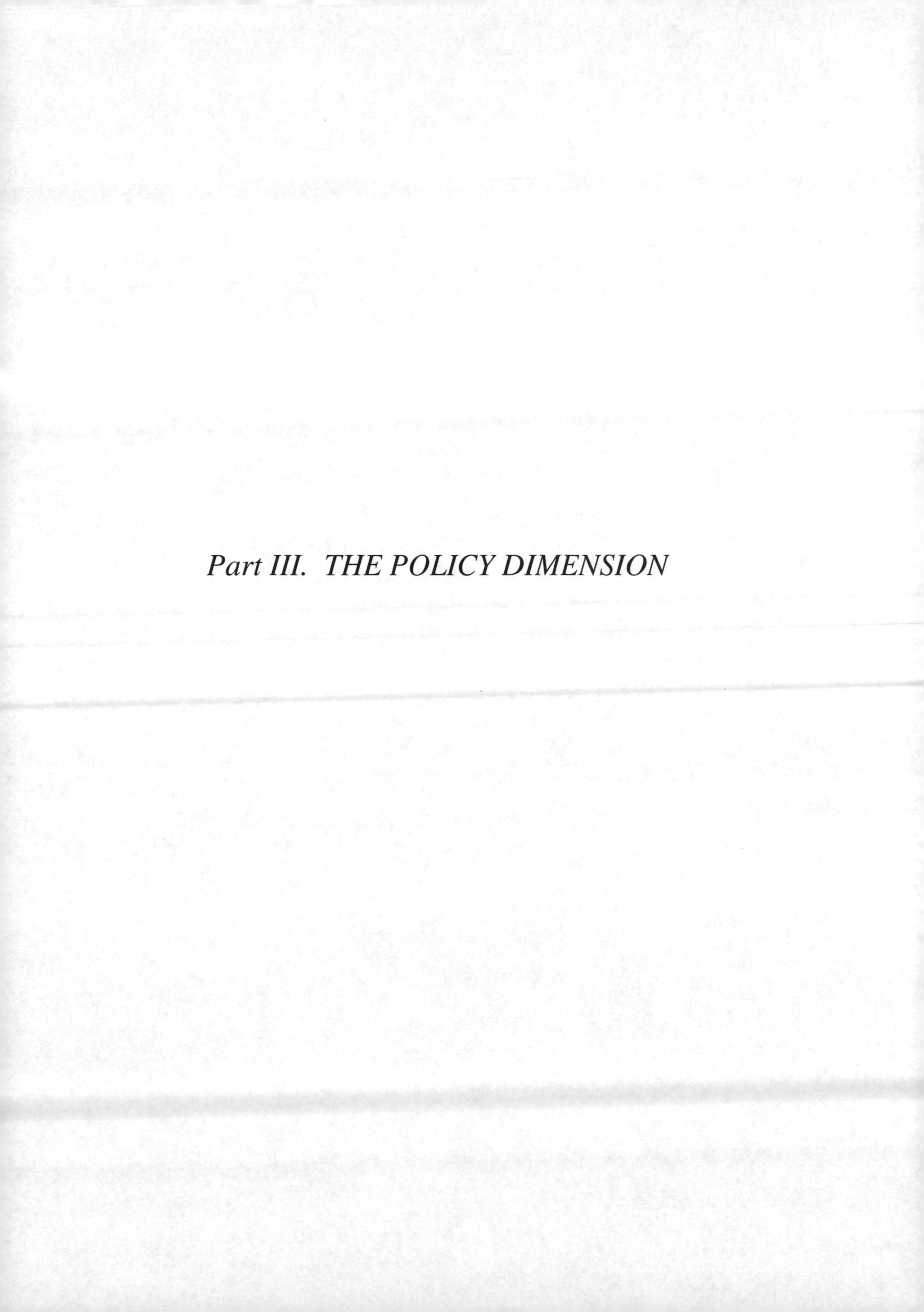

Part III. THE POLICY DIMENSION

Chapter 12

STUDIES OF CLUSTERS AS A BASIS FOR INDUSTRIAL AND TECHNOLOLOGY POLICY IN THE DANISH ECONOMY

by

Ina Drejer, Frank Skov Kristensen and Keld Laursen
The IKE Group, Aalborg University

1. Introduction

The purpose of this chapter is to provide an overview of the cluster studies carried out in Denmark over the last two decades, both as analytical "tools" in their own right, and as the basis for industrial and technology policy measures.

Cluster studies have, in recent years, become the cornerstone of Danish business and industry policy making, although attempts to identify clusters of production date back to the early 1980s. The latest development in Danish cluster studies has been a movement towards identifying innovative clusters at different levels. The definition, or rather definition*s*, of clusters applied in this chapter vary according to focus of the study and the level of aggregation, and the chapter does not propose a unique definition of a cluster as a unit of analysis.

In the early studies, a cluster – an industrial complex – was defined according to supply-and-demand linkages in the production structure. This led to at least four "meso"-based studies, and three related "micro"-based studies that focused on important linkages in sectors in which Denmark is strongly specialised. Production linkages and, more importantly, comparable policy framework conditions were used to delimit these clusters which continue to be used in Danish business and industry policy making today, *i.e.* in the form of "resource areas".

The main focus of this chapter centres on these resource areas: their methodological and theoretical foundation; the statistical and methodological problems involved in identifying the clusters; and their use in policy formulation and implementation.

2. The first cluster studies

This section will present cluster studies performed at different levels of analysis. Figure 1 presents an overview of different levels of analysis for cluster studies and their corresponding analytical focus. The Danish case studies have primarily been carried out at the meso/industry and the firm level. As an introduction to the presentation of the four main types of cluster studies in Denmark, Table 1 provides an overview of the studies, their level of analysis, the areas covered, as well as the main focus of each study. Section 2.1 will describe in further detail meso-level studies of linkages between industries within industrial complexes, while Section 2.2 identifies clusters at the intersection between the meso and micro level of national strengths, based on strategic inter-firm linkages. The Porter studies and resource areas will be covered in Section 3.

Figure 1. Cluster approaches at different levels of aggregation

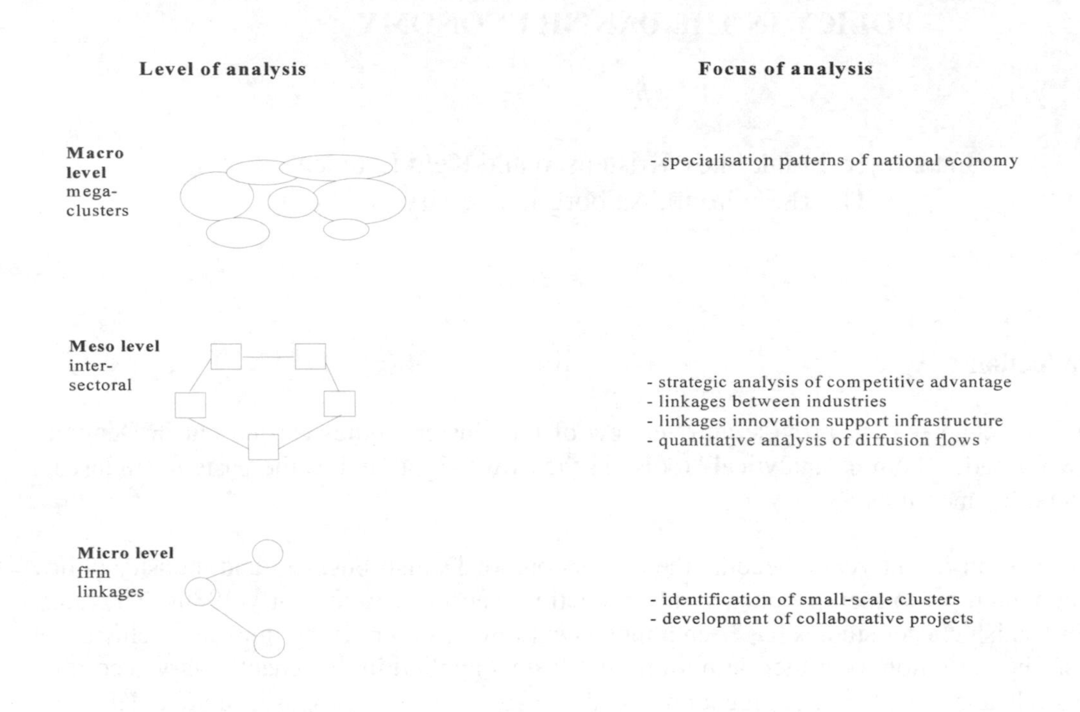

Source: Roelandt and den Hertog, 1997.

2.1. *Industrial complexes*

A series of studies of so-called industrial complexes were carried out in Denmark in the early 1980s. These studies were part of a project financed by the Danish Technology Council and focused on the development and diffusion of new technology throughout the Danish economy, and more specifically on the ways in which the use of microelectronics influenced central economic variables such as the balance of payments and employment.

Four industrial complexes were studied: the agro-industrial complex; the textile complex; the environmental complex; and the office machinery complex.

Table 1. **An overview of cluster studies in Denmark**

Name	Level of analysis	Areas covered	Period	Focus
Industrial complexes	Meso/industry	Primarily agriculture and related industry, but efforts were also made to study the textile complex, the office machinery complex and the environmental complex	Early 1980s	Vertical and horizontal linkages, and their importance for the development, diffusion and use of new technology
Micro-based cluster studies	Firm	Separate studies of: ♦ Electro-medical instruments; ♦ Furniture; and ♦ Pharmaceuticals	1992-96	Inducement mechanisms to innovation
Porter studies	Firm → Industry	♦ The agro-food cluster; ♦ The shipping cluster; ♦ The technical cluster; ♦ The pharmaceutical/ biotechnology and medical cluster; ♦ The mink cluster	1988-90	Up- and downstream value-chain relations, institutional settings, firms' surroundings
Resource areas	Meso/industry	In principal, the whole economy, was originally divided into eight areas: ♦ Services; ♦ Agro/food; ♦ Construction; ♦ Environment/energy; ♦ Transport/communication; ♦ Medico/health; ♦ Consumer goods; ♦ Tourism/leisure ♦ A residual: "General suppliers"	1993 →	Mutual interdependence or common relations related to production requirements, common factor conditions

The concept of industrial complexes, and the related concept of "production verticals" is based on the notion that the linkages between, on the one hand, firms developing new technology expressed in components, machines and production systems and, on the other hand, firms using this technology, are at the core of the economic system. These linkages are crucial for the development, diffusion and use of new technology. The theoretical foundation for industrial complexes is found in Dahmén's (1988) development blocks – complexes of industrial interrelations – which can be analysed in terms of the dynamics of these interrelations as they evolve over time.[1] The concept of development blocks belongs to the field of "Schumpeterian dynamics" and is central to industrial economics. *Transformation* is a central factor, focusing on changes in broad aggregates as expressions of underlying changes through time, both within and between micro-entities. Examples of transformation include the introduction of new methods of production and marketing; the appearance of new markets and marketable products and services; the exploitation of new sources of raw materials and energy; the scrapping of "old" methods of producing and marketing products and services; the disappearance of "old" products and services (Dahmén, 1988, p. 4).

The introduction of microelectronics is related to almost all the above-mentioned transformation processes, especially the processes concerning production methods, including the organisation of production and the development of new products and services. This implies that, although an industrial complex cannot in a narrow sense be perceived as an innovative cluster, there is an obvious linkage to innovation and new technology through the focus on technological development, in general, and microelectronic development, in particular.

The four complexes

The four complexes analysed, although by no means representative of the whole economic system in a statistical sense, each represent different types of relations between producers and users of new technology.

Different methodologies, all related to vertical linkages between users and producers, were applied in identifying the complexes.

The *agro-industrial complex*, the subject of the most detailed study, was mainly identified through the use of input-output (I/O) tables of the national economy. Sectors either receiving a relatively large fraction of their input from, or delivering a relatively large share of their output to, the core sector of the complex (primary agricultural production) are considered to be part of the complex. Sectors which are only indirectly connected to the core areas of the complexes are also included (identified by the use of Leontief inverse I/O matrices). In order to capture the flows of capital goods, however, other means than I/O tables have to be included. The agro-industrial complex is the largest integrated complex in the economy, with a production value almost as large as that of all the other manufacturing sectors taken together in the observed period. Therefore, production related to agriculture played and continues to play a major role in the Danish economy, in terms of both consumer and investment products (Lundvall *et al.*, 1984). The agro-industrial complex also illustrates the importance of the home market for international trade specialisation. One component of the complex is the "dairy vertical". An important part of the vertical consists of the linkage between users of dairy equipment (large Danish dairies) and manufacturers of machinery for use in this sector. According to the ISIC classification, dairy exports belong to the food, drink and tobacco sector, whereas dairy equipment is included in non-electrical machinery – two sectors in which Denmark is heavily specialised. An important part of the knowledge base of these sectors is created in the interaction between the two, resulting in a co-evolution between the two sectors which tends to produce international competitiveness in both fields.

A major aim of the study of the *agro-industrial complex* was to analyse whether it was possible to identify either the presence or possibility of technological dynamics, crucial for liberating the complex from its vulnerable specialisation in standardised products in a stagnating market. The study concluded that qualified and demanding users had played an important role in the development of new technology. However, signs of an increasing inequality in competencies between producers and users of new technology – which could have negative effects on production, exports and employment – could be identified. Concerning possible new technological dynamics, biotechnology was identified as having by far the greatest potential with regards to renewal in the agro-industrial complex, not simply regarding the supply of innovation from outside the complex, but also in terms of using biotechnology for practical purposes inside the complex in both primary agriculture and processing industries. However, new technology unless accompanied by appropriate organisational change cannot provide a solution; this led to a recommendation for a strengthened "sector" or "complex policy" aimed at improving the vertical relations between sectors as opposed to specific "microelectronic policies" or "technology policies" (Lundvall *et al.*, 1984, pp. 126-148).

The second complex studied was the *textile complex,* defined as the textile-producing industry and its main suppliers and users. Even though this complex builds on an I/O approach, lack of detail in the aggregation of the I/O tables made it impossible to use the tables in the definition of this complex. As opposed to the agro-industrial complex, machinery is almost negligible in the textile complex, which makes the textile complex almost identical to the textile and clothing industry (Thøgersen, 1986). Pasinetti's (1981) production-based model of technological dynamics, linking economic growth to structural change, together with the product-life-cycle theory, are the main theoretical starting-points. The main conclusion is that, although the complex as a whole is fairly mature, resulting in a breakdown of the linkages in the complex, certain new or emerging product fields could be identified (*e.g.* carpets), indicating that the interactions could still give rise to dynamic effects. Technological development in this complex has shown the characteristics of a mature set of industries with a clear tendency towards standardisation. New products and processes, created through the interaction between the agents at different levels in the complex, are necessary to fight the decline in domestic production and employment in the textile complex.[2]

Lack of statistical data was a major obstacle in the analysis of the *environmental complex*, which was to a large extent based on interviews, in particular with experts on wastewater treatment. The environmental complex was defined as made up of users and producers of environmental technology as well as intermediaries, *i.e.* suppliers of information and advice about new technological opportunities (Gregersen, 1984). The *environmental complex* differs from the two complexes described above in that the public sector is a main user as well as an important regulator. In this regard, the study of the complex provided new insights into the role of the public sector in the building up and maintenance of national competencies through its actions as a competent user. However, as a complex or cluster study, it is atypical.

The analysis of *office machinery* is only marginally related to the complex approach. The basis of the analysis was a questionnaire-based survey on electronic data processing and office automatisation in Danish municipalities. The basic difference between the analysis of the three complexes described above and the analysis of office machinery is, that in the latter case, the focus is on a specific type of user as opposed to a focus on the interdependence between producers and users. The producers are indirectly present in the analysis, however, since the office technology used by the municipalities to a large extent consists of systems developed by the public company Kommunedata ("Municipality Data"). The fact that both the users and the producers belong to the public sector offers a new perspective on the interdependence between the two types of actors (Brændgaard *et al.,* 1984). The main conclusion is that the introduction of office machinery in the municipalities was not driven by a desire to cut jobs, whereas the possibilities for ensuring that the introduction of new technology is employment-neutral in the future are more uncertain.

The analysis of the four complexes, through their variety in focus and method, shows some general characteristics of the technological changes experienced in the different complexes, and some characteristics of the influence of these changes on Danish employment and competitiveness.

Methodological considerations

The above-mentioned differences in approach and focus of the industrial complex studies can be considered both as a methodological problem and strength. Although user-producer linkages are the main determinants of an industrial complex, no general way of identifying complexes was developed.

In terms of comparability and consistency of definition, this creates problems. On the other hand, the variety of approaches does supply a complementarity in the facets of the different analyses which can be of great value when dealing with a relatively new research area. The study of the agro-industrial complex is the most consistent in relation to a cluster approach, and it is also in this study the one in which the method is most clearly defined and developed. Some considerable problems arise in using I/O data. The level of aggregation is to a large extent given in advance – which is why the I/O method proved to be inadequate in the study of the textile complex – a level which might not be appropriate for identifying the most important user-producer linkages. The lack of dynamics in the analysis is also apparent. In addition, the studies of industrial complexes are not representative of the Danish economy, rather each represents an area which *in isolation* is an interesting object for analysis. However, the four complexes cannot be perceived as *the most* important clusters in the Danish economy; their value lies in the thoroughness of the individual studies rather than in the combination of the four studies.

2.2. *Micro-based cluster studies*[3]

Related studies of important linkages facilitating the distribution of knowledge in the innovation system, were carried out at the micro level, in three different sectors in which Denmark has a strong specialisation (Dalum, 1996). This section illustrates the differentiated nature of the knowledge base in manufacturing and, in this context, the differentiated importance of specific knowledge across sectors. The section will briefly describe the creation and distribution of knowledge in a specialised-supplier sector (electro-medical instruments), a supplier-dominated sector (furniture), and a science-based sector (pharmaceuticals).

Electro-medical instruments

The ISIC sector *instruments*, is one in which Denmark is specialised in terms of R&D expenditure and where Denmark has a relatively high R&D intensity compared to a selected group of OECD countries.[4] Lotz (1993) demonstrates the historical importance of the interaction between medical instruments and an advanced domestic hospital sector. One example of the importance of the interaction is the most successful Danish firm in this area, namely Radiometer, where internal R&D conducted since 1935 has provided a basis for close interaction with Danish hospitals, especially *Rigshospitalet* (the State University Hospital) in Copenhagen. One of this company's major innovations (apparatus for measuring the level of pCO_2 in the blood) was actually invented by a head of department of clinical chemistry at *Rigshospitalet* in the early 1950s, but was transformed into an innovation at Radiometer. Today, the interaction with hospitals (especially in the Copenhagen area) continues, in order to maintain competitiveness by means of distributing user-knowledge from hospitals to specialised suppliers in the instrument sector.

Furniture

Wood, cork and furniture is another sector in which Denmark is specialised in terms of value added, but with a lower level of R&D intensity, compared to the OECD-9. In other words, the sector seems to be competitive, although it has a comparatively low R&D intensity. This apparent paradox is explored by Maskell (1996). Wooden furniture production consists of two distinctive and technologically distinct processes – the process of manufacturing the furniture (wood cutting, drilling, grinding and shaping), and the process of painting it (the coating process, including smoothing, painting or lacquering, priming, drying/evaporating of fumes, polishing, etc.). In the manufacturing process, the exchange (distribution) of knowledge does not take place through the development of capital equipment since 90% of the machinery is imported, mainly from Germany and Italy. The same goes for the machinery used in the painting process (mainly imported from Italy). However, while the industry works with more or less given process technologies, part of the manufacturing process can be "moulded" or adapted to give a leading edge. This part includes lacquer and paints which are adapted in conjunction with domestic manufacturers. Another important contribution to the knowledge base comes through an *agglomeration effect*, reflecting the fact that local and specialised educational institutions play an important role, together with (local) mobility in the labour market.

Pharmaceuticals

The pharmaceuticals sector is one of the few science-based sectors strongly present in Denmark. In terms of growth in Danish share of OECD-9 R&D in 1980-91, the pharmaceuticals sector accounted for nearly 50% of total Danish growth, and from a more static point of view the sector accounted for nearly 24% of total Danish R&D in 1991. By far the largest Danish producer in this sector is the world's largest manufacturer of insulin for diabetics Novo Nordisk (Laursen, 1996). From a historical point of view, it is remarkable that major breakthroughs in terms of new and radically better insulin products have been achieved inside the firm's R&D department, often in collaboration with foreign scientists (mainly American). Thus, largely firm-specific knowledge has been accumulated over nearly three-quarters of a century, with technological linkages (dynamic synergy effects) between different products an essential feature. Nevertheless, the importance of the presence of a strong national science base needs to be highlighted. A particularly strong Danish science base can be identified in this context, by taking the number of published papers per capita in life sciences; the number of papers published in Denmark was about 20% higher than the US figure, and about 70% higher than the EU-10 average over the period 1981-86. However, generally speaking, Denmark is highly ranked in all the major science fields in addition to the life sciences (mathematics, physical sciences, engineering and chemistry), both in terms of number of papers per citizen, and measured as mean citation per paper.

Although basic research is increasingly becoming globally accessible, since it has a strong public good element, this is not the full agenda. Recent research by Hicks *et al.* (1994) shows that publications produced by Japanese companies (basic research) tend to over-cite the national science system by approximately 30%; this suggests that the economic benefits are geographically and linguistically localised, since they are embodied in persons and institutions, and thus mainly transmitted through personal contacts. Similar findings have been made by Narin and Olivastro (1992), showing that national patents cite national science and *vice versa*. A strong position in basic research is therefore economically important at the national level, because it provides research training, state-of-the-art development and use of research techniques and instrumentation, and access to high-quality international networks (Gibbons and Johnston, 1974; Pavitt, 1993). In addition, basic research provides an important country-specific resource to science-based firms, providing recent results from national as well as international state-of-the-art research as an input to commercial research. These benefits accrue, not only through the research conducted by the scientists of a given country, but (mainly, at least in a small-country case) through the increased ability to assimilate results of basic research conducted by other countries, an ability which, in turn, partly depends on the home country's ability to perform high-quality basic research. In the Novo Nordisk case, major breakthroughs nearly always took place at foreign universities. In this context, the *research skills* developed at Danish universities have been of utmost importance in assimilating and commercialising inventions made abroad. In the context of *state-of-the-art development and use of research techniques and instrumentation*, comprehensive mathematical molecular models should be mentioned. Another potential impact of basic research was found abundantly throughout the entire history of Novo, namely *ready access to high-quality international scientific networks*, a story which began with the Nobel Prize winner and originator of insulin production in Denmark, August Krogh in 1923, and now encompasses contacts with "centres of excellence" in biotechnology, situated in California.

Thus, a continued commitment to basic research is of central importance to the competitiveness of this sector. So far, little research has been conducted applying (at a detailed level) bibliometric methods to assess the continued viability of the science base in the Danish system.

The knowledge bases

This section has shown that sources of technology differ across different development blocks. Table 2 sets out some of these differences, and serves to illustrate that the important knowledge base of sectors in a system may well reside in the interaction with other parts of the system.

Smith (1995) distinguishes between three different areas of production-relevant knowledge related to different levels of specificity. The first level is the *general scientific knowledge base*, which consists of very differentiated fields of knowledge with widely varying relevance for industrial production. The fields with the closest connections to major industrial sectors are to be found within areas such as molecular biology, physics, genetics and inorganic chemistry. The second level covers knowledge bases at the level of the *industry or product field*. At this level, industries often share particular scientific and technological parameters, and know-how relating to technical functions, performance characteristics, the use of materials, etc., is shared. The third level concerns the knowledge bases of *individual firms*. At this level, we are dealing with only one or a few technologies which are well integrated into the firms and form the basis of their competitive position. Due to the high level of specificity of technology at this level, there are clear limits to the firms' competencies. Thus, they must be able to access and use knowledge from outside the core area of the firm.

A characteristic feature of the three levels is that the level of specificity, as well as the intangible elements of the knowledge base, decreases as the level of aggregation increases.

In the studies presented in Section 2.1 and this Section 2.2, we are primarily dealing with industry- and product-field-related knowledge bases but, as exemplified in the study of pharmaceuticals above, also the national science base (level 1) and the combination of firm-specific and external knowledge (level 3) are included, *i.e.* all levels of knowledge are relevant at each level of analysis, but in different "proportions".

Table 2. The most important knowledge bases for three particularly strong Danish sectors

Sector	Important external knowledge bases (interaction with)	Level of cumulativeness in R&D	Importance of scientific knowledge	Importance of technological service systems
Furniture	Domestic producers of lacquer and paint	Low	Low	Some
Medical instruments	Domestic hospitals	High	Some	--
Pharmaceuticals	National and international science bases	High	High	--

As shown in this and the preceding section, interaction or interdependence plays an important role in defining a cluster, both within and between industrial sectors. The next section will show that this notion has survived in more recent policy-oriented industrial cluster studies. The industrial complexes can be perceived as the "forerunners" of the "resource areas" introduced by the Danish Ministry of

Business and Industry in 1993. The fact that the agro-industrial complex was the most consistent is underlined by the survival, or rather revival ,of the complex as a "food products resource area". The other complexes studied are all part of larger resource areas.[5]

3. The Danish Porter studies and resource areas[6]

The industrial complexes and "micro" studies of Danish clusters mentioned in Section 2, have all been either a direct input or a reference point in developing the resource areas. However, the most direct influence has been that of the Danish Porter studies. Denmark was one of ten countries that participated in Porter's analysis of clusters of competitive advantages. The studies were later used as the empirical foundation for the theory presented in Porter (1990), *The Competitive Advantage of Nations*. In the context of Porter's book, it is interesting that that part of Porter's analytical framework concerning "demand conditions" (see below, for a description) was inspired – among other studies – by a study (Porter, 1990, p. 86) in the context of international trade, of the previously discussed Danish agro-industrial complex (Andersen *et al.*, 1981). The Danish Porter studies have underpinned the resource area notion in two ways: first, the ideas behind and methodology used in the resource areas are highly related to the Porter studies; and second, and perhaps more importantly, some of the key researchers and consultants who worked on the Porter studies, later worked with or advised on the resource areas.

3.1. The Porter studies

Porter, who gained widespread recognition for his work on competitive advantages and firm strategy, was engaged by the US authorities to analyse growth and competitiveness. In relation to these issues, Porter found that because the wrong questions were being asked, the wrong answers were being returned (Porter, 1990, p. 3). Until the mid-1980s, industrial and trade policy in the United States had mainly considered macro and factor conditions in a Heckscher-Ohlin tradition. Porter argued that it was not countries or industries in different countries that competed, but firms. Therefore the focus of analysis should be on firms. Further, trade specialisation could not be explained satisfactorily by factor conditions, neither could the fact that there seemed to be clusters of successful firms in particular regions. Porter initiated a study in ten countries (including Denmark)[7] where ten or more industries in each country were analysed. The qualitative analysis focused on the upstream and downstream value-chain relations of the firms, the institutional setting, the firm's surroundings and the firm itself. The theoretical approach was to employ the well-known "Porter diamond" (Porter, 1990). Furthermore, the analysis should have a historical angle since "path dependency" is often important in explaining why a region or country hosts a particular set of firms working in the same sector.

A salient feature of Porter's analysis is that it focuses on a broad set of factors influencing the competitive advantage of nations, rather than focusing on one particular explanation. However, the approach has been the target of criticism, emphasizing that the connections between the level of the industry and the level of the nation is unclear, when Porter draws his conclusions. Thus, Porter himself tries to answer some of the "wrong questions" he initially warns of (Dalum, 1992).

The Porter methodology

Innovation plays a crucial role in Porter's understanding of a firm's competitive advantages; both product and process innovations are central in creating new markets or gaining and sustaining market shares. A central observation is that there would seem to be clusters of firms in a country or a region that are "doing well" in the same business. Chemistry in Germany, pharmaceuticals in Switzerland, semi-conductors in the United States and later Japan, and mobile phones in Scandinavia, are just some of the many examples of this type of clustering.

The home base, where firms allocate the bulk of their resources to R&D, is seen as central to firms' competitiveness, and it is argued that it is crucial for economic growth that a nation provide an attractive home base. This implies that it is not enough to analyse the firm in order to explain the firm's competitiveness. Rather, the question is why and how regions have developed as centres of excellence in a particular industry. If an understanding of the processes behind these developments can be achieved, this can give a hint as to these centres can be created. The creation of attractive home bases or centres of excellence is the stated goal of industry and business policies.

Porter uses his diamond (which includes the firm, the sector, related sectors, the home market, factor advantages and the state) to describe the central interactions and relations in an economic system: The strategy of the *firm*, together with the firm's management and its organisation and routines, are of course important, since in the end it is the firm that must gather and use the knowledge and the factors of production, in an effective way. Regarding the *sector,* formal and informal co-operation, as well as rivalry between firms in the sector, is important. Also important is the strength of the organisations and institutions which support the interests of the sector as a whole. The presence or absence of *related sectors* that are internationally competitive, and either supply or adopt technology in a way that stimulates a cumulative and interactive process, is also influential. This is related to product and process innovations where lead users or producers of knowledge are seen as key. The relative size of the *home market* and the quality of demand also plays a crucial role. *Factor advantages* are both traditional factor endowments and the infrastructure, human resources and technological ability of the country. The *state* plays a role through regulations related to the sector or to business in general, in addition to investments in, for example, infrastructure. The state also plays a central role as an advanced user.

These broad elements are the focus of the analysis and it is the interaction between different elements that determines competitive ability. As mentioned above, history matters, and many of the clusters we see today have been developed over long time horizons. This implies, first, that cumulative processes have a strong influence and, second, that advantages take a long time to build up, even if the answers to how to build these advantages are known *a priori*.

Porter used the diamond in a qualitative and historical analysis of each sector in each country included in the study. From these studies, Porter hoped to find some patterns to show how different interactions and developments in different sectors sometimes give rise to competitive advantages for firms in these particular sectors or clusters.

3.2. The Danish studies

The Danish Porter studies were carried out by a consortium of researchers from various universities and business schools together with several private consultants. The work on clusters in Denmark was carried out from 1987 to 1991, and was reported by Pade (1991). This report was preceded by a central publication by Møller and Pade (1988) on industrial success and competitive factors in nine Danish sectors.

Fifteen sectors were analysed in the Danish Porter project: the dairy sector; slaughterhouses; mink producers; the consumer fish industry; agricultural machinery industry; the biotech industry; the pharmaceutical industry; electro-medico equipment; telecommunications; engineering; the environmental industry; furniture; shipping (sea); cleaning services; and the mobile phone industry.

On the basis of these sector studies, five clusters with a high competitive ability were identified: the agro-food cluster; the shipping cluster; the technical cluster; the pharmaceutical/biotech and medico cluster; and the mink cluster. For methodology and results, see Pade (1991) or Porter (1990).

In general, the rather extensive analysis pointed towards different strengths and weaknesses in the clusters and in Danish industry in general. The relatively small home market was seen as a problem, together with the general small firm size in Denmark. Market size hampered competition between large rivals, and small firms had problems in reaching export markets. On the other hand, the small firms had the advantage of flexibility and dynamics. The relatively high quality of the workforce was viewed as a strength, together with a relatively well-developed infrastructure. Furthermore, in some instances the state had been able to procure innovations.

The results of the studies stimulated intense debate on the competitiveness and dynamics of Danish firms, and also on the appropriate policies to support development. The more fundamental question of what industrial policies actually are, became a hot and widely discussed topic in Denmark.

4. The Danish resource areas

The Danish Business Development Council (*Erhvervsudviklingsrådet* – EUR), which is responsible for advising the Danish government on business policies, took up the idea of clusters as a new perspective on business policies, and initiated the analysis of resource areas in Denmark. Since the Porter studies, there had been a number of ongoing analyses of clusters, areas or blocks in several Danish ministries and agencies. In particular, the Danish Agency for Development of Trade and Industry (*Erhvervsfremme Styrelsen*) had carried out several studies. Also, the Danish Ministry of Finance carried out studies of Danish industry employing a cluster terminology. In 1992, the Ministry of Finance presented an analysis of "blocks", in which they showed that the four blocks analysed were responsible for 60% of Danish exports, and in general played a strategic role in the Danish economy. The four blocks analysed were the agro/food, construction, naval, and health blocks. Later, the resource areas carried on these four blocks in four out of the total of eight areas. In a sense, therefore, the initiation of the resource areas was the outcome of a cumulative process in which the results from the Danish Porter studies, and the Porter industry-policy way of thinking, played a central role.

4.1. A theoretical and analytical approach to the resource area studies

In one sense, the methodological and theoretical foundation of the resource area is the same as that used in the Porter studies. The Porter diamond was the theoretical cornerstone in the analysis of the relations and interplay in the resource areas, and the methodology used to analyse the connections and flows was qualitative with a historical perspective.

The Danish Business Development Council (EUR), through consultations with participants in the Porter studies, identified eight resource areas and a residual area labelled "general suppliers".

However, the methodology does sometimes differ from that used in the Porter studies and in cluster studies in general. First, it was decided that the whole of Danish industry should be included in the eight areas or the residual, although the public sector was not incorporated. In the Porter studies, some 40% of all Danish firms were represented in the clusters. In other words, the resource areas have a wider scope than the Porter clusters and clusters in general.[8] Second, where the Porter clusters were all defined by their coherency in interaction, some of the resource areas were also seen as clusters since they shared the same demand side. In these areas, the demand structure was the common denominator. Thus, it is explicitly stated that the deviation from earlier industry divisions lies in the focus on the end product. Third, when cluster analysis is employed, certain factors or variables are chosen in order to identify the possible clusters. This was not the case with the analysis of the resource areas, which was relatively fixed from the outset.

The eight original resource areas were: services; agro/food; construction; environment/energy; transport/communication; medico/health; consumer goods; and tourism/leisure.

The residual "general suppliers" could not be placed in any single area. Analyses were carried out in all of the eight areas by different groups of researchers and consultants. These analyses were all published and served as the foundation for further work on resource areas.

In addition to these qualitative analyses, a statistical group was established to produce statistics for the eight areas covering, for example, size, growth, employment, imports, exports, etc. In order to obtain these statistics, several special data runs were carried out on the Danish ISIC code at the six-digit level in order to aggregate the resource areas, which cut across the standard statistical aggregation. The work was carried out in a rather detailed manner, *e.g.* several firms were moved from one sector to another, and in many cases a firm's input and output were divided into different areas. It is important to stress that it is the firm that is the unit of analysis, and not the flow of goods between firms or sectors.

The method employed will be described below. There are ongoing developments in the methodology and in our understanding of the areas. Several analyses of these areas have been carried out: some are international comparisons to which we will refer later in this chapter; but foremost there is an ongoing and intense dialogue with representatives of firms, organisations and public institutions and ministries. In addition to leading to a number of policy recommendations, the work has also been used to develop and redefine the resource areas. In the following sections we address this new definition of the resource areas.

4.2. The resource areas as defined today

The six resource areas, as defined today, are listed below with a description of the changes from the original eight-area definition. Further, we touch upon the relationship with other cluster studies. Before listing the resource areas, the definition of a resource area, which was constructed in the process of changing from eight to six resource areas, is presented:

- ♦ A resource area consists of a broad range of products or services, which is relatively stable over time and has a considerable weight or size in the economy.
- ♦ A resource area is made up of sectors that are mutually interdependent or are in a common relation due to the requirements to produce the final product or service in co-operation.
- ♦ The firms in a resource area have the same needs in terms of factor conditions.
- ♦ There is one or more position of strength measured by trade performance in a resource area.

This is the official definition of a resource area, and employing this definition, the six areas as well as the general supplier residual are described as follows:

- *Food.* This resource area covers agriculture, fishery, dairy, slaughterhouses, production of agricultural machines, fishing boat yards, dairy factories as well as supplies to these, such as cooling machinery, thermostats, etc. This area, as defined today ,is very similar to the agro-industrial complex described in Section 2.1. In relation to the Porter studies, it contains four of the sectors which were found to be internationally competitive.

- *Consumer goods and leisure.* This resource area covers, for example, production of clothing, production of electrical equipment, production of furniture, hotels and bedding, culture, retailing, etc. Here we find the furniture sector which was also found in the Porter studies and in the micro-based studies of particularly strong Danish sectors.

- *Construction/housing*, covering construction, construction engineering, construction materials, retail of construction materials, entrepreneurial companies, crafts related to construction, cleaning, housing administration. Here we find the cleaning sector analysed in the Porter studies.

- *Communication*, including printing, printing machinery, media and communication equipment, media, mail and telecommunications, communication services, retail of communication equipment, etc. This area also includes the mobile phone industry, identified both in the Porter analysis and by Dalum (1995).

- *Transport and supplying industries*, covering shipyards, production of other transport machinery, energy equipment producers and suppliers, automobile services, road/sea/air/railway transportation, heating/electrical/gas supply, retail and trade with fuel and trash, environmental equipment producers and suppliers, etc. Here we find the environmental sector identified in the Porter studies, and the environmental complex found in the industrial complex study.

- *Medico/health,* covering pharmaceuticals, medico techniques and aid as well as pharmacies. Here we find the pharmaceuticals sector and the medico-technical sector identified in the Porter analysis.

- *General supplier businesses,* which is an aggregation of sectors producing and supplying goods that either are so general in nature that they cannot be assigned to a particular area, or produce highly specialised equipment to several sectors in different resource areas. These general suppliers include the metals industry, other production industry operational services, consultancy and retail.

There were several reasons for the redefinition, only some of which are presented in the following. The new areas incorporate the public sector, which was not the case in the past. In some of the areas, publicly produced goods are vital parts of the area and these are accounted for. As can be seen from the above list of the new areas, services is no longer a separate area. This is because the service sector was very heterogeneous, and in most cases provided services that are closely related to all other areas.

Another visible change is that communications is now an area of its own whereas before it was part of telecommunications/transport. This modification was the outcome of many studies and much intense dialogue with key persons. First, it was found that a combination of communication and transportation does not function coherently. Second, it was evident that communications was growing rapidly in volume as well as in importance.

The generation of statistics to enable an analysis of the resource areas over time and which could be used to analyse the resource areas in an international context, was an important issue in the Ministry of Business and Industry. The previous statistics were generated on the basis of ISIC codes, which are no longer a common international standard. The basis for aggregation is now the NACE codes (the level from which the areas are aggregated is the four-digit NACE code). In the original work on statistics, a great deal of resources were used in moving some firms from one sector to another, and in dividing the output and input from some firms and sectors into different resource areas. With the new aggregation this is no longer necessary as a four-digit-level NACE sector is placed in a single resource area.

Of course, employing this method causes some problems in relation to the reliability of the statistics, to which we shall return below. With the new statistics, it is now possible to follow the development of the resource areas. Figure 2 shows the relative sizes of the current resource areas according to number of employees and value added.

In addition, it is now possible to compare resource areas in Denmark with areas defined in the same way in other countries. A full comparison of all six areas plus the residual would make little sense but, for example, a comparison of the food areas in Denmark and the Netherlands could be used as a benchmarking exercise since both countries are specialised in food and have possible similarities.

Figure 2. Relative sizes of Danish resource areas, 1995

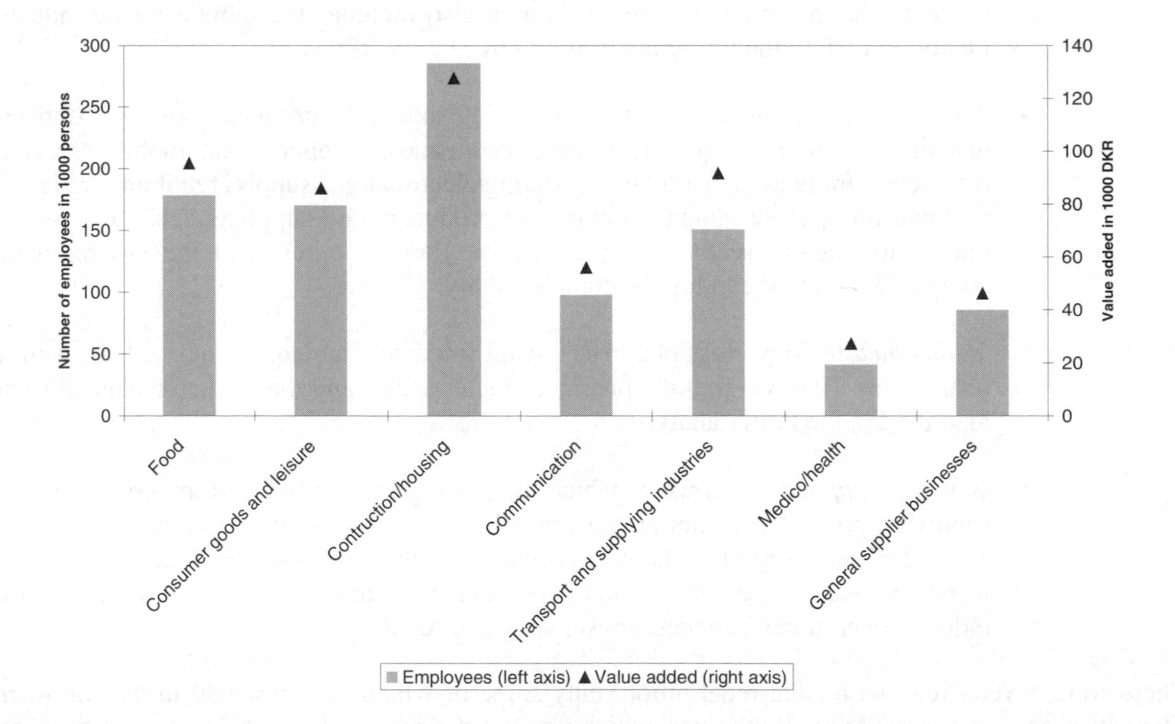

4.3. *The working method in the resource areas*

The focus and scope of this chapter is on the applied methodology, with the aim of further developing the methodology and ideas underlying the resource areas. This, of course, is not the main aim of the day-to-day work of the Ministry of Business and Industry. The resource areas, and the applied working method and analysis, are primarily targeted at formulating policies and identifying problems, future threats, and firm strengths, which in turn will lead to new policy initiatives. This is important to bear in mind, especially when discussing the methodology in relation to the focus and scope of the present chapter. Still, it is the knowledge created in the day-to-day work that is used to develop the resource areas. In the following sections, we briefly describe the work methods used in the resource area framework.[9]

A reference group has been set up for each of the resource areas. The Ministry of Business and Industry has invited representatives from firms, organisations and related ministries to participate in these reference groups. The reference groups are stable over time and are in a sense the focal point of the work. On the basis of dialogue and the various analyses, the reference groups highlight critical policy conditions. In some case, these policy conditions require further analysis and discussion before policy initiatives can be recommended. This further work is carried out in work groups constituted around these specific tasks. The ideas and initiatives from the working groups are then fed into the political process. In principle, the initiatives are fed into the political process by the representatives, ministries or organisations that are best placed to support the initiatives. The working groups are then wound up as they work on an *ad hoc*, task-specific basis.

Vital to the work on the resource areas is the fact that it is made public, so that anyone with an interest can contribute to or follow the work. The work focuses on the problems or the wishes of the firms in the resource areas, and firm employees and organisational representatives participate directly in the reference groups as well as in the working groups.

A number of the critical conditions and policy initiatives are related to roles traditionally undertaken by ministries other than the Ministry of Business and Industry, *e.g.* the Ministry of Research, the Ministry of Education or the Ministry of Employment. This calls for greater co-ordination among the different ministries.

In 1997, a status rapport was made on the results of the work on resource areas (Industry, 1997). Since the resource areas were introduced, 35 working groups have been initiated, of which 22 have completed their work, and 13 are still working. In these groups, 513 people from firms, organisations, institutions and the ministries were involved. The working group dialogues have led to 152 suggestions or initiatives, of which 66 have been implemented, either completely or partly. The initiatives cover a broad range; some could be carried through without any change in laws, others called for a change in laws or administration, yet others had an effect on the fiscal budgets of ministries.

There are ongoing attempts to develop the theoretical foundation and the understanding of the resource areas, since the dialogue and the various analyses of problems and critical policy conditions are also used to develop the resource areas. As mentioned above, some comparative analyses of policies, institutions and interactions have been carried out in other countries.[10] Often the countries are chosen because of their strengths in specific areas, or because particular policy initiatives are thought to be of specific importance. The policy initiatives in the countries analysed are used as input in decisions on how best to set up policies aimed at a specific area. In general, the studies are studies of best practice or benchmarking studies.

An example of such a comparative analysis is that carried out in the food resource area, concerning the policies and conditions related to biotechnology (A/S, 1997). The development, knowledge and use of biotechnology was identified as important for the future development in the food resource area (as was foreseen in the industrial complex study, Section 2.1 above). The comparative analysis of policies and conditions was carried in Germany, the Netherlands, the United Kingdom and the United States. The analysis focused explicitly on the conditions set by regulations in order to benchmark these conditions. The qualitative analysis was based on interviews with key persons in the respective countries. A brief summary of the outcome of the comparisons led to the following recommendations:

- It is imperative to achieve greater public understanding of biotechnology issues. This needs to be supported by multi-faceted public information and debate.

- The regulations related to the use and testing of biotechnology are critical for the development of the field, and for the protection of consumers. A specific recommendation was that common regulations be established in the EU countries.

- Research is the vital input to the field, and this should be strengthened in Denmark, both through the allocation of additional more resources and through better co-ordination of research. In addition, it is proposed that research in strong, existing, research environments be further supported.

4.4. A discussion of methodology

In any discussion of the strengths and weaknesses of the methodology used in the analysis of resource areas, it is important to note that many different methodologies are employed. The central method used in the initial analysis were the qualitative and historical studies of developments and interactions in the respective resource areas. Another often-used methodology is to measure positions of strength. Positions of strength are based on trade specialisation; the general idea underlying the method is to use a Balassa index (Balassa, 1965) to determine in which sectors trade specialisation can be found. To summarise, dialogue with key persons is crucial, together with comparative analyses. In addition, the statistical exercise is used as a yardstick.

In the following sections, the methodology discussion will: *i)* relate to the initial eight-area breakdown and the subsequent breakdown into six areas; *ii)* discuss the strengths and weaknesses of the qualitative approach; *iii)* discuss the strengths and weaknesses of the statistical exercise; and *iv)* discuss whether or not other methods could usefully contribute to the current methodologies.

As mentioned above, the resource areas, in their present form, cover the entire Danish economy (which is divided into six areas plus a residual). Initially, there were eight areas and a residual. Coherency, interdependence and the need for common policy conditions are stressed as factors in the definition of a resource area. In other studies of clusters in Denmark and elsewhere, the scope of a resource area is considerably wider. During the process, both in relation to the initial division, but also in connection with the redefinition, the resource areas were criticised for including more clusters in a single resource area, leading to a situation where coherency can be questioned in some areas. Second, because of the broad scope, it is questionable whether the interdependencies between firms and sectors in a resource area are of equal strength and importance.

Turning to the qualitative and historical studies of the relationships in a resource area, the strengths are that most of the relationships are intangible in the sense that they are difficult to measure statistically. Many important relations will not show in firm-level statistics or I/O tables. Especially when a cluster, as in the Porter and resource area tradition, is understood as a highly complex interaction of several factors, causal relations cannot be measured; only the *ex post* outcome of the relations can be measured. Therefore, qualitative interviews are perhaps the only way to obtain information on possible causal relations. In that context especially the comparative studies are very useful since the importance of factors is more easily determined in the comparisons. The weakness is, of course, that in a complex world, factors other than those stressed could also be important; qualitative analysis is often criticised for its subjectivity due to the fact that the issues cannot be measured.

The statistics generated, and the methods used to generate them, have advantages and disadvantages. The benefits are that it is now possible to make time series comparisons and international comparisons across areas. The weakness lies in the fact that the method employed cannot capture all the dynamics in the areas, and especially across the areas. Further, there is a possibility that when four-digit NACE sectors are aggregated into the six (plus one) areas, some "failures" occur. Some firms and many sectors deliver to different areas, but are only measured in one. Also, highly product-diversified firms are only considered in one area. The decisions as to which sectors belong to which resource areas have never been empirically tested, and therefore the coherency of their placement can be called into question. However, by using firm-level statistics rather than, for example, I/O tables, the analysis can be made at a more disaggregated level. This is important since the resource areas and clusters in a Porter sense cut across traditional sector aggregations.

In the context of resource areas, it would appear that I/O analysis has never been carried out to supplement the other methods employed. Denmark has relatively disaggregated I/O tables (130 sectors since the shift to SNA '93). The disadvantages are, as mentioned above, that I/O tables measure relations as flows of goods and services. In other words, only the tangible relations are included, since any relation or interaction that is not an economic transaction is not measured. With the use of I/O analysis it is possible to determine whether the transactional relation which is argued to exist in the different resource areas, is actually present. One could argue that if the 130 sectors were aggregated into the six resource areas, they should have a relatively high internal flow of goods if the value chain perspective is employed.

An analysis of inputs and outputs in the 130 sectors could be used to empirically test the placing of an sector in a specific resource area. I/O analysis could then be applied to divide input/output from one sector into the respective areas, by measuring the path of its outputs. Since it is the flow of goods or transactions that is measured, one could make a dynamic analysis which could also be used to identify emerging clusters. Further, when I/O tables are merged with data on, for example, education, public and private R&D or patent statistics, it is possible to measure the importance of embodied knowledge flows (Drejer, 1998).

5. Some conclusions

This chapter has described how a framework (resource areas) for making technology and industrial policy, founded on theoretically-based cluster studies, has been implemented and modified in Denmark during the 1990s.

The chapter has further shown that the cluster concept has been, and continues to be, related to very different types of studies and properties of the observed objects. If a single factor linking all studies and types of clusters were to be extracted, it would have to be the knowledge base, which is explicitly or implicitly present in all of the above-mentioned clusters. If a definition of a cluster is to be proposed in this concluding section, it has to relate to this common knowledge base: *a cluster is a group of firms, an industry, or a group of industries, which exists in relation to a strong knowledge base*. The knowledge base does not necessarily imply high technology, as with the case of agro/food and beverages, although a distinct knowledge base is still present. In the agro/food case, the knowledge base is developed through the existence of a home market with close linkages between users of equipment and manufacturers of machinery for use in the cluster. An important part of the knowledge base is created in the interaction between users and suppliers, resulting in a co-evolution between the respective industries which tends to produce international competitiveness in both fields. In the resource areas, the knowledge bases are expressed as positions of strength, and because of the generally very broad definitions, some resource areas have more than one position of strength. This is, for example, the case in the consumer goods and leisure area, which includes production of electrical equipment and traditional, more design-based industries such as furniture, as well as some services.

Different methods of identifying clusters have been discussed in this chapter. One way is to apply measures of trade specialisation by identifying areas of comparative advantage in international trade. Such a procedure was, for example, followed in the Porter-related studies. Other variables, such as value added, production and R&D, have also been used. Another approach has been to identify vertical relations in the economy. Such relations can be identified at the micro level, mainly through case studies, while I/O analysis is a widely used methodology at the industry level.

Policy has played an important role in this chapter, since the majority of the cluster studies were carried out with specific policy aims as a major driving force. However, it is evident that theoretically based studies and practical policy actions do not always sit easily together. While theoretically based studies aim to provide clarity and coherence in analysis, policy making is concerned with "muddling through" the complex reality. An example of this trade-off can be found in the history of the Danish resource areas, discussed in this chapter. On the one hand, it can be said that the areas are to some extent based on theoretically based cluster studies. On the other hand, each resource area acts as a framework for dialogue between firms and the public authorities. Hence, it would not be wise in a policy context to exclude some firms from certain sectors because such sectors were not identified as a cluster or as a part of a cluster. Thus, this trade-off should be acknowledged, so that a balance between pragmatic policy making (with more than a single aim) on the one hand, while not losing the theoretical foundation, on the other, can be maintained.

NOTES

1. Dahmén originally introduced the development block concept in his dissertation (1950, 1970).

2. As history has shown, the domestic production, and thus employment, was not maintained, and if an analysis of the textile complex were to be carried out today, it would not be possible without including production located abroad in low-wage countries. The consequences of the moving out of production for the vertical linkages in the complex are not discussed in the present chapter.

3. This section draws on Laursen and Christensen (1996).

4. This group includes Canada, Denmark, Finland, France, Germany, Italy, Japan, the United Kingdom and the United States, and is referred to as the OECD-9.

5. Textiles are mainly part of the "Consumer goods/leisure resource area"; environment is mainly relevant to the "Transport/utilities resource area"; while office machinery is placed in the "Communications resource area".

6. The authors would like to thank Birgit Kjølbye and Mette Kaae Hansen from The Danish Ministry of Business and Industry, and Britta Vegeberg from The Danish Agency for Development of Trade and Industry, for useful discussions and comments in relation to this section.

7. The other countries were Germany, Italy, Japan, Korea, Singapore, Sweden, Switzerland, the United Kingdom and the United States.

8. The effort to include the whole of industry in the eight areas resulted in a problem with overlaps between most areas.

9. A recent change in work practices applied within the resource areas is that the reference groups and the working groups have been terminated. Instead, the problems identified by these groups are now dealt with on an *ad hoc* basis. This change reflects the fact that the most pressing problems were solved and the generation of new focuses has been, if not totally stopped, then at least slowed down. In a sense, the change towards a more unstructured *ad hoc* practice was decided to prevent the creation of new rigid structures for understanding industry policy. Further, some of the more recent projects within the resource area project, such as the biotech project, cut across the boundaries of the resource areas. This may reflect the fact that the definition of the boundaries between resource areas can change over time due to the emergence of new business opportunities and newly established linkages between firms and institutions.

10. International studies have been carried out outside the auspices of the Danish Ministry of Business and Industry. An example of this is NUTEK (1995) which compares experiences of cluster studies and related industry policies in Denmark, Finland, the Netherlands and Norway. The experiences are used in a discussion of whether and how a Swedish cluster analysis should be carried out. The Danish model is emphasized because of it includes the whole of industry, rather than focussing on industries with particular strengths with regard to trade specialisation, as well as because of the large degree of policy relevance.

REFERENCES

Andersen, E.S., B. Dalum, and G. Villumsen (1981), "The Importance of the Home Market for the Technological Development and the Export Specialization of Manufacturing Industry", IKE Seminar, Aalborg University Press, Aalborg.

Balassa, B. (1965), "Trade Liberalization and 'Revealed' Comparative Advantage", *The Manchester School of Economics and Social Studies*, Vol. 32, No. 2, pp. 99-123.

Brændgaard, A. *et al.* (1984), "Besparelser eller beskæftigelse. En undersøgelse af danske kommuners anvendelse af EDB og ETB" ("Reductions or Employment. An Analysis of Danish Municipalities' Use of Electronic Data and Text Processing"), Aalborg University, Aalborg.

Dahmén, E. (1950), *Svensk industriell företagarverksamhet*, 2 vols, Lunds Universitet, Lund.

Dahmén, E. (1970), *Entrepreneurial Activity and the Development of Swedish Industry 1919-1939, American Economic Association Translation Series*, Irwin, Homewood, Ill.

Dahmén, E. (1988), "'Development Blocks' in Industrial Economics", *Scandinavian Economic History Review*, Vol. 36, pp. 3-14.

Dalum, B. (1992), "Export Specialisation, Structural Competitiveness and National Systems of Production", in B.-Å. Lundvall (ed.), *National Systems of Innovation. Towards a Theory of Innovation and Interactive Learning*, Pinter Publishers, London.

Dalum, B. (1995), "Local and Global Linkages: The Radio-communications Cluster in Northern Denmark", *Journal of Industry Studies*, Vol. 2, No. 2, pp. 89-109.

Dalum, B. (1996), "Growth and International Specialisation: Are National Patterns Still Important", paper prepared for a forthcoming book edited by Jane Marceau, Research School of Social Sciences, Australian National University, Canberrra, Australia, IKE Group, Aalborg University, Aalborg.

Drejer, I. (1998), "Technological Interdependence in the Danish Economy – A Comparison of Methods for Identifying Knowledge Flows", paper presented at the Twelfth International Conference on Input-Output Techniques, New York, 18-22 May, DRUID, Aalborg.

Gibbons, M., and R. Johnston (1974), "The Role of Science in Technological Innovation", *Research Policy*, Vol. 3, pp. 220-242.

Gregersen, B. (1984), "Det miljøindustrielle kompleks. Teknologispredning og beskæftigelse" ("The Environmental Complex. Technology Diffusion and Employment"), Aalborg University, Aalborg.

Hicks, D. *et al.* (1994), "Japanese Corporations, Scientific Research and Globalization", *Research Policy*, Vol. 23, No. 4, pp. 375-384.

Laursen, K. (1996), "Horizontal Diversification in Danish National System of Innovation: The Case of Pharmaceuticals", *Research Policy*, Vol. 25, No. 7, pp. 1121-1137.

Laursen, K., and J.L. Christensen (1996), "The Creation, Distribution and Use of Knowledge – A Pilot Study of the Danish Innovation System", Danish Agency for Trade and Industry, Ministry of Business and Industry, Aalborg and Copenhagen.

Lotz, P. (1993), "Demand as a Driving Force in Medical Innovation", *International Journal of Technology Assessment in Health Care*, Vol. 9, No. 2, pp. 174-188.

Lundvall, B.-Å., N.M. Olesen, and I. Aaen (1984), "Det landbrugsindustrielle kompleks" ("The Agro-industrial Complex"), Aalborg University, Aalborg.

Maskell (1996), "Vertical Integration and Competitiveness. The Danish Furniture Industry", mimeo, Danish Research Unit for Industrial Dynamics, Copenhagen.

Ministry of Business and Industry (1997), "Erhvervsredegørelse 1997" ("Status of Business and Industry 1997"), Ministry of Business and Industry, Copenhagen.

Møller, K., and H. Pade (1988), *Industriel Succes (Industrial Success)*, Samfundslitteratur, Gylling.

Narin, F., and D. Olivastro (1992), "Status Report: Linkage between Technology and Science", *Research Policy*, Vol. 21, pp. 237-249.

NUTEK (1995), "Klusteranalys och näringspolitik. Erfarenheter från Danmark, Finland, Holland och Norge" ("Cluster Analysis and Industry Policy. Experiences from Denmark, Finland, the Netherlands and Norway"), NUTEK, Stockholm.

Pade, H. (1991), *Vækst og dynamik i dansk erhvervsliv (Growth and Dynamics in the Danish Business Sectors)*, Schultz, Copenhagen.

Pasinetti, L.L. (1981), *Structural Change and Economic Growth*, Cambridge University Press, Cambridge.

Pavitt, K. (1993), "What do Firms Learn from Basic Research?", in D. Foray and C. Freeman (eds.), *Technology and the Wealth of Nations. The Dynamics of Constructed Advantage*, Pinter Publishers, London.

PLS CONSULT A/S (1997), "Dialog med Fødevarer. Muligheder og betingelser for brug af bio- og genteknologi i udvalgte lande" ("Dialogue with the Food Area. Possibilities and Conditions for the Use of Biotechnology and Genetic Engineering in Selected Countries"), *Background Analysis*, The Danish Ministry of Business and Industry, Copenhagen.

Porter, M.E. (1990), *The Comparative Advantage of Nations*, Free Press, New York.

Roelandt, T.J.A., and P. den Hertog (1997), "Mapping Innovative Clusters – Research Proposal and Discussion Note".

Smith, K. (1995), "Interactions in Knowledge Systems: Foundations, Policy Implications and Empirical Methods", *STI Review*, No. 16, Special Issue on Innovation and Standards, pp. 69-102, OECD, Paris.

Thøgersen, J. (1986), *Omstilling i tekstil- og bekædningsindustrien (Reconversion in the Textile and Clothing Industry)*, Aalborg Universitetsforlag, Aalborg.

Chapter 13

CLUSTER ANALYSIS AND CLUSTER POLICY IN THE NETHERLANDS

by

Theo Roelandt[*], Pim den Hertog[**], Jarig van Sinderen[***] and Norbert van den Hove[****]

1. The Dutch value chain approach

1.1. Introduction

Over the last decade, the cluster approach has attracted the attention of researchers and policy makers in various countries. Since the mid-1980s, the cluster concept has received a great deal of attention and different cluster concepts and approaches have been developed.[1] Cluster case studies have been conducted in some countries (Denmark, Finland, Italy, the Netherlands, Sweden and the United States) using Porter's diamond and network approaches as a framework for analysing the competitiveness of the local production structure (Porter, 1990).

The cluster concept focuses on the linkages and interdependencies among actors in a production chain in producing goods and services and in innovating. Clusters can be characterised as economic networks of strongly interdependent firms linked in a value-adding production chain. In some cases, clusters encompass strategic alliances with agents in the knowledge infrastructure, such as research institutes, universities, engineering companies and firms of consultants.

Following Schumpeter's description, innovation can be characterised as new combinations of existing knowledge and competencies, originating from different actors in the value chain. Modern innovation theory suggests that firms almost never innovate in isolation (Edquist, 1997). Successful innovation calls for close interaction and knowledge exchange with customers, competitors and specialised suppliers of machinery, services and inputs.[2] This challenges the traditional line of thinking in economic research, which primarily analyses horizontal relations among competing firms with similar activities, focusing on price competition, entry barriers and the individual firm. Today, competition is not about individual firms but rather about networks of dissimilar firms in the same value chain (clusters). A firm's competitiveness depends very much on its ability to participate in strategic production networks. This implies that inter-firm co-operation and competition go hand in hand.

---

* Dutch Ministry of Economic Affairs.

** Dialogic.

*** Dutch Ministry of Economic Affairs and Research Centre for Economic Policy, Erasmus University Rotterdam.

**** Ministry of Economic Affairs, Directorate for Technology Policy.

This paper reflects the personal views of the authors and not necessarily those of the government of the Netherlands.

1.2. The methodology

A number of empirical cluster analyses of Dutch industrial structure have been conducted since the mid-1980s. Although various methods exist for identifying clusters, the two main approaches used in the Netherlands are *i)* monographic case studies (mainly based on Porter's diamond); and *ii)* input-output (I/O) analysis, which aims to identify interlinkages and knowledge flows between industry groups.[3] In a sense, the two methods are complementary. Case study material reveals detailed information about actors, behaviour and strategies within the cluster network. I/O analysis allows the network structure to be illustrated and quantified by systematically linking main users and suppliers (based, in our case, on trade linkages). If the two approaches are combined, a picture emerges of the economic structure of a country in terms of networks of industries (at various levels of aggregation) and specialisation patterns.

Below we elaborate on these two approaches before briefly characterising the clusters identified at the national level (Section 1.3).

Monographic case studies[4]

The cluster concept, based on complementarity and interdependencies in production relations, has become quite pervasive in the Netherlands over the last decade, and cluster analyses have gained in popularity. In the Netherlands – as in various other countries – the process was fuelled by the publication of Porter's (1990) *The Competitive Advantage of Nations*. Regional economics (the industrial district literature), evolutionary economics and the developing of the national innovation systems paradigm provided a further base. In particular, Porter's work further triggered an interest in clusters in both academic and policy circles, providing an appealing and heuristic tool to deal with the issue of competitiveness.

In 1990, a study was published on *The Economic Strength of the Netherlands*.[5] In this study – based on a pre-publication manuscript of the book – Porter's model was applied to the situation in the Netherlands. This work yielded several new insights into the strength of the Dutch economy in a number of relatively small industry groups where technological upgrading and the quality of the network of supplying and supporting industries play an important role (cut flowers, yacht building, cocoa, bodywork construction). It also revealed the depth and breadth of network formation in Dutch agriculture and industry. This first Porter study for the Netherlands – commissioned by the Ministry of Education and Science and the Ministry of Economic Affairs – initiated a number of research projects in which traditional sectoral studies were gradually replaced by cluster studies focusing on competitive strengths. Some 60 industry groups were covered by monographic cluster case studies over the years. The Dutch government has announced plans to update and review these cluster studies in order to monitor the competitive advantage of the main Dutch clusters and to evaluate the impact of recent cluster policy initiatives (see Section 3 for details).

Over the years, the character of these cluster studies has undergone considerable change. Initially, the studies could be described as analytical devices to gain better insight into the competitive strengths of individual clusters. Gradually, the analyses have come to serve (at the meso level) as a starting point for strategic advice on:

- how to make individual clusters more competitive;
- identifying important knowledge issues;

- designing strategies for upgrading; and

- how to turn negative competitive dynamics into strategic co-operation and differentiation-based competition (see Box 1 for the example of the construction cluster).

Subsequently, the cluster studies increasingly served as an input to macro and micro-level issues. At the macro level, for instance, the work was used in discussions on industrial and innovation policy making and on how to improve the (mis)match between the public research institutions and higher education institutions and industry. At the micro level, cluster studies provided a basis for initiating and supporting innovative micro-level cluster projects aiming to increase co-operation among major companies, their (main) suppliers, (semi-) public knowledge institutes, as well as various other bridging institution (engineering companies, innovation centres, and so on).

Box 1. The example of the construction cluster

The construction industry is an important cluster in the Netherlands in terms of value added and employment. It comprises various industry groups (house building and maintenance; industrial buildings; installation of cables/pipes; construction of roads and bridges; dredging and "wet infrastructure" construction; environmental construction works). A complete Porter analysis was performed for six segments[6], including the role of specialised suppliers such as installation services, construction materials producers and wholesalers, architectural and technical engineering services. Although it is difficult to generalise the findings, the studies did reveal that the majority of firms focused mainly on cost-based price competition and much less on product differentiation and product innovation. Instead of offering complete solutions to client needs, most firms still sell construction capacity. Most notably, some smaller construction firms, as well as project developers and some engineering firms, were found to be among the most innovative. In particular, the latter firms transfer knowledge from best practices in other clusters to the construction cluster.

In terms of the balance between competition and co-operation, negative dynamics were also identified. It was observed that a reluctant attitude towards innovation, together with a combination of collusion and cut-throat competition among construction firms, and a lack of trust from clients, contributes to the idea that innovation does not pay. The study made recommendations on all aspects of the diamond to turn these negative cluster dynamics into more positive ones. These suggestions were clearly linked to how construction firms, suppliers and government (both as user and as regulator) can contribute to increasing the competitiveness of the cluster as a whole.

The Dutch government has recently initiated a policy programme "Innovation in Construction" which aims to improve the level of innovation in the construction cluster by stimulating the creation of innovative clusters and consortia (see Section 3 for more details). Moreover, it is expected that the new Dutch Competition Law will enhance the economic dynamism in the construction cluster in a positive way.

Input-output analysis

Due to the high level of aggregation of I/O tables and the complexity of linkages in the economic structure, I/O analysis encounters a number of data and methodological problems. Moreover, I/O analysis does not link actors, but rather statistical and sometimes (depending on the aggregation level) heterogeneous industry groups. The number and character of the identified clusters strongly depend on the aggregation level of the I/O tables. Clusters of economic activities often cross the traditional borders between sectors. Cluster analysis therefore needs data on linkages at a low level of aggregation. The Netherlands studies make use of I/O tables at the aggregation level of 214 x 214 industry groups and of make & use tables drawn up for 650 product categories and 260 categories of economic activities.[7] The I/O and make & use tables were analysed using different techniques.[8]

Basically, in both cases, clusters of related economic activities are identified by linking the main suppliers of goods and services to the main users. Foreign trade linkages were included during the clustering process so that the identified clusters also contain specialised suppliers from outside the Netherlands. The two techniques, although quite different and adopted on different I/O tables with different levels of aggregation, show –(more or less) the same results.[9] Indeed, a fairly simple clustering technique was used, but the findings of the I/O analysis appeared consistent with the cluster chart identified by the qualitative research effort in the case studies.[10] To gain insight into the knowledge flows and the importance of various technology fields, we made use of the Dutch Innovation Survey 1992 (SEO) and the Dutch Research and Development Survey 1995 (CBS), using the information on R&D expenditures to calculate the R&D intensity (partly differentiated by the technology fields electronics, information technology, materials technology and manufacturing technology) of the trade flows (at a three-digit level) in the I/O tables.[11]

1.3. Clusters in the Dutch economy

The I/O analyses identified 12 large conglomerates of interlinked industry groups in the Dutch production structure (Figure 1).[12] Table 1 presents the main economic activities of each cluster, its contribution to the national product and its market orientation.

Figure 1. Clusters in the Dutch economy at the national level

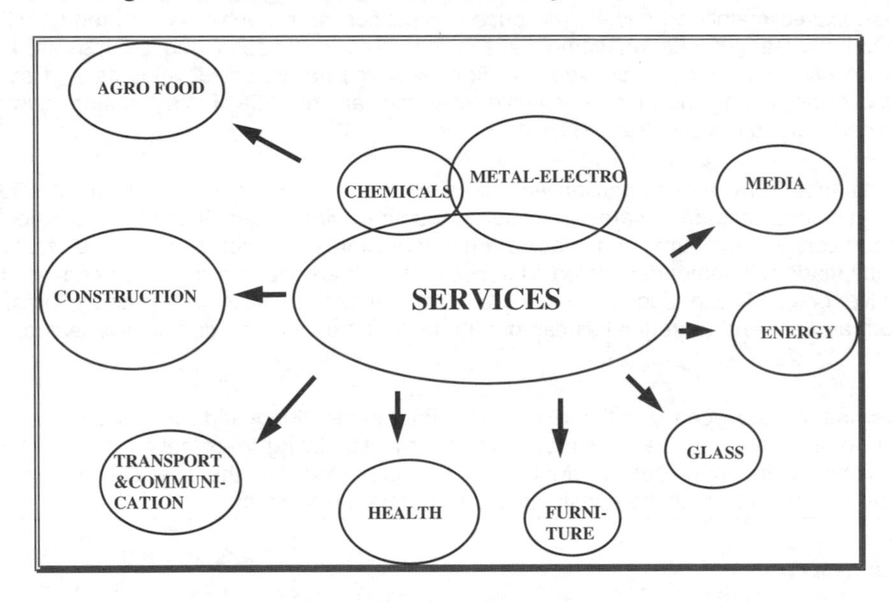

From the table we can conclude that the classification of economic activities into clusters clearly differs from the traditional sectoral classification. The identified clusters cross the borders of traditional sectors. There is no longer a clear distinction between primary, secondary and tertiary industries, but rather a mixture of industry groups in each cluster. The importance of the "commercial services" cluster for the Dutch economy is shown by its relatively high contribution to the national product (nearly 30%). From the viewpoint of employment, the two services clusters are also highly important: some 50% of the total working population is employed in these two clusters.[13] The rapid rise of media-related activities – a recent development in the Dutch economy, – can be deduced from the results. A "media" cluster has been identified, and interdependence is growing between the that cluster and the communication industries in the ports, transport and communication cluster.[14] The manufacturing cluster and the chemical industries cluster are strongly dependent upon the activities of specialised suppliers abroad. The two latter clusters mainly produce for foreign markets, together with the transport, port and communication cluster and the agro-food cluster. The clusters around health and non-commercial services focus relatively strongly on consumer markets. This also applies to the media cluster, although to a lesser extent.

Table 1. Clusters in the Dutch economy

Cluster	Main economic activities	VA	Market orientation
Construction	Manufacture of building materials, pottery, building and installation companies, workmanship, gardeners, estate agencies	87.9	Consumption-oriented; generic delivery of capital goods
Chemical industries	Basic chemicals industry, manufacture of artificial and synthetic yarns and fibres, chemical end-product industry, rubber and synthetic-processing industry, petroleum industry	27.7	Export-oriented; dependent upon specialised foreign suppliers; generic delivery to other clusters
Commercial services	Wholesale and retail, hotels and restaurants, business services, consultancies, cleaning services, computer services	150	Consumption-oriented; generic delivery of specialised services to all other clusters
Non-commercial services	Education, public administration and national insurance, social services, public research institutes	79.1	Consumption-oriented; generic delivery to all other clusters
Energy	Waterworks and production and distribution of electricity and gas, production and exploration of crude petroleum and natural gas	25.5	Consumption-oriented; dependent upon raw materials from abroad; autarkic
Health	Health and veterinary services (hospitals, psychiatric institutions and other medical services, dentists and veterinary surgeons), manufacture of pharmaceutical products and antiseptic dressings	30.4	Highly consumption-oriented
Agro-food	Agriculture, horticulture, forestry, fishing, meat processing industry, grain and flour processing industry, canning, preserving and processing of fruits, vegetables and flowers, manufacture of dairy products, sugar (processing), cacao and chocolate industry	45.8	Export- and consumption-oriented; strong relationships with other clusters
Media	Printing, publishing and related industries, photo studios, computer services, advertising and publicity agencies, theatres, cinemas	14.4	Consumption-oriented
Manufacturing paper	Paper and cardboard industry	3.0	Highly export- and import-oriented
Manufacturing (metal-electro)	Electro-technical industries, basic metals industry, manufacture of metal products, manufacture of instruments and optical products, automobile industry, manufacture of means of transport, machinery	46.8	Highly export-oriented; dependent upon foreign suppliers; generic delivery (of capital goods) to all the other clusters
Manufacturing – other	Wood and furniture industry, manufacture of glass and glass products, textile industry	6.2	Highly export- and import-oriented
Port, transport and communication	Sea and air transport, rail, tram and bus services, travel agencies, communication services	37.7	Export-oriented; generic delivery to other clusters

Note: VA = Value added (in NLG billion): production value minus intermediary use minus value added tax at current prices.
Source: Dutch Ministry of Economic Affairs.

2. Clusters: specialisation patterns and innovation styles

2.1. *Specialisation patterns*

Section 1 showed how cluster studies in the Netherlands have developed during the past decade. All the studies highlight the importance of network relations for innovation (Figure 2).[15] The majority of the innovative firms in clusters co-operate with suppliers, competitors, clients or equipment suppliers. Innovative firms need close relationships with suppliers and clients with complementary technology and competencies. However, the nature of these relations can differ. As shown in Table 2, different specialisation patterns, as identified in Section 1.3, result in different innovation styles and each cluster has its own needs.

Figure 2. Networks of innovation, by cluster[16]

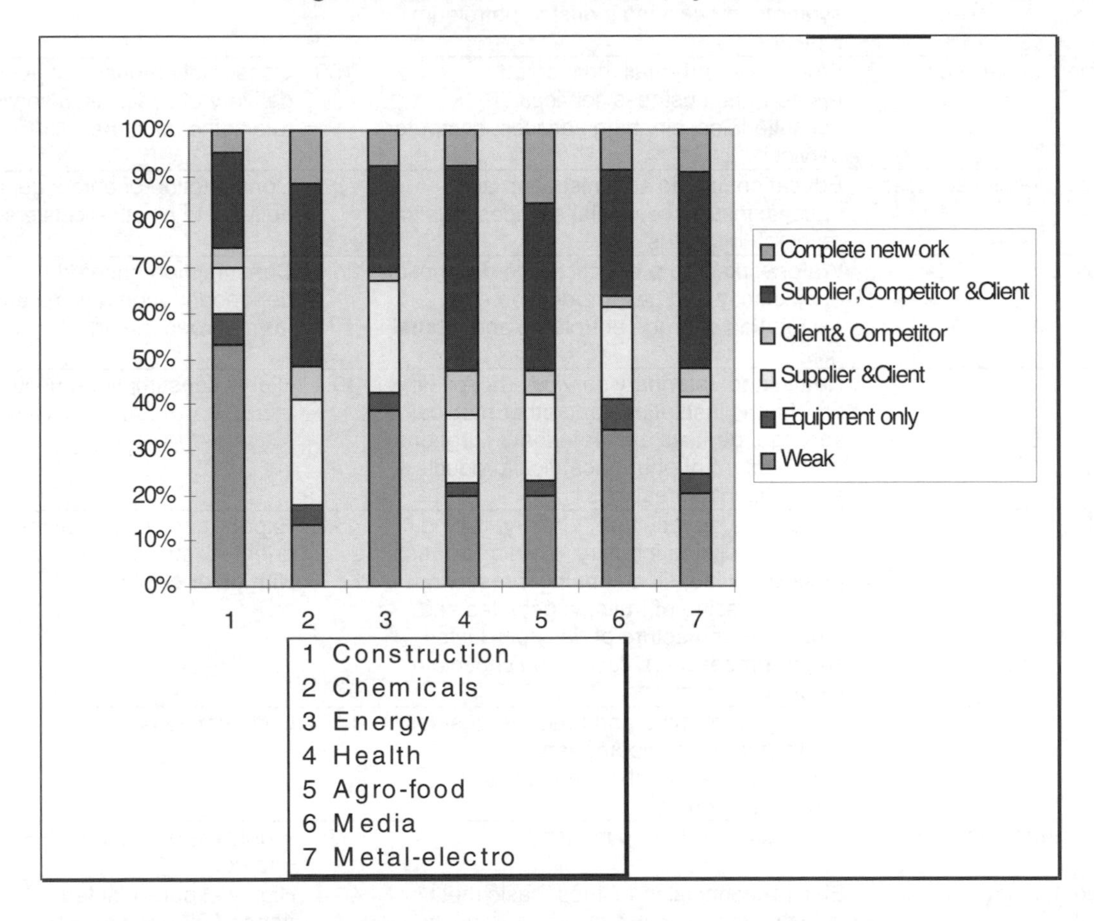

In the construction cluster there is a strongly felt need for product differentiation, which can only be fulfilled by entering into relationships with suppliers outside the cluster. The agro-food cluster is lacking in product innovation; this will be addressed primarily through interaction between the existing collective infrastructure and specialised suppliers from the manufacturing and commercial services clusters. Within the chemical industries cluster, there is a need for upgrading the domestic suppliers. Innovation in the commercial services cluster is mainly realised in co-operation with the manufacturing industries and primary producers in other clusters. Innovation in the media cluster requires the establishment of relations for the development of information and communication technology within the port, transport and communication cluster and the manufacturing industries.

Table 2. Clusters: innovation style, technology fields and their role in the economy

Cluster	Innovation style and technology field
Construction	Acquisition of technology primarily via specialised suppliers; technology follower; high absorptive capacity; strong emphasis on process innovation. *Need*: product innovation and differentiation. *Main technology field*: construction technology and civil engineering.
Chemical industries	High extent of technology development; engine for innovation in other clusters; high extent of process R&D. *Need*: upgrading. *Main technology fields*: high and low molecular materials technology and process technology.
Commercial services	Innovation strongly dependent upon specialised suppliers in the manufacturing industry in particular; large extent of process innovation; engine for innovations in other clusters; interactive relationship with innovation processes in other clusters. *Need*: knowledge-intensive innovation. *Main technology fields*: Information technology and electrotechnical technology.
Non-commercial services	Acquisition of technology via specialised suppliers; engine behind upgrading of knowledge base via public knowledge infrastructure. *Need*: greater spin-off to commercial production clusters. *Main technology field:* fundamental and basic research.
Energy	Relatively low tech. *Main technology fields*: energy technology and mineralogy.
Health	Large extent of product and process innovation; innovations strongly focused upon own cluster (autarkic). *Main technology fields*: medical research and pharmacy and basic research.
Agro-food	Autarkic in technological terms; acquisition of technology via independent collective infrastructure in combination with specialised suppliers in generic clusters; potential innovations, especially in overlapping areas with other clusters. *Need*: product innovation. *Main technology field*: food and process technology.
Media	Technology follower with a high absorptive capacity; for innovation dependent upon specialised suppliers; strong pattern of product innovation. Growing link with communication industries. *Main technology field*: information technology.
Manufacturing (metal-electro)	Significant technological development; engine behind many innovation processes in other clusters; significant process innovation; product innovation, especially in overlapping areas with other clusters. *Main technology fields*: electrotechnical research, manufacturing technology, transport technology and information technology.
Port, transport & communication	Process innovation primarily via specialised suppliers.. *Need*: increased added value, and application of information and communication technologies. *Main technology fields*: logistics, information technology and defence technology.

Clearly, clusters vary widely and each cluster plays its own role in the economy-wide innovation process. Specialisation around core activities and outsourcing of economic activities that are distanced from the own knowledge base are the common characteristics. A typology characterising the innovation pattern of individual clusters is presented below.

2.2. *Patterns of innovation*[17]

Van den Hove *et al.* (1998) identified four patterns of innovation (Figure 3):

- Clusters which are knowledge-intensive and specialised suppliers for innovations in other clusters, such as the *metal and electrical engineering cluster,* the *chemicals cluster* and the *non-commercial services cluster*. These clusters supply the knowledge-intensive services, machinery and equipment, education and materials necessary for almost all the primary production processes in other clusters. Production of these inputs falls outside the scope of the users' technology base and is therefore subcontracted out to specialised suppliers in the services, metal and electrical engineering and chemicals clusters.

- Absorptive clusters with a relatively low R&D effort which are dependent on their suppliers for innovations (*construction* and *media*).

- Clusters which are relatively autonomous and autarkic in their innovation activities, (*agro-food, energy*, and *port, transport and communication*).

- Knowledge-intensifying clusters which absorb knowledge and research from other clusters and enlarge the knowledge intensities of their own products and services, which are then applied elsewhere by others in the production processes of other clusters. The *commercial services* are an example of such a cluster.

These patterns of innovations will be further elaborated below.

Self-creating clusters

The first innovation model was found in large clusters such as metal and electrical engineering, chemicals, and non-commercial services. These clusters can be typified as "self-creating". They carry out substantial amounts of research and absorb relatively little knowledge from other clusters; expenditure on research subcontracted to others is lower than average. The greater part of the knowledge used by these companies in their product and production processes comes from the R&D performed by companies and institutions within the cluster. In addition, the knowledge which is absorbed and incorporated in goods, services and products produced by other companies generally comes from within the same cluster. Furthermore, because of their specialised knowledge, other clusters tap the knowledge of the metal and electrical engineering and chemicals clusters for use in their own production processes and economic activities.

The metal and electrical engineering and chemical clusters are organised around specific core competencies. Van den Hove *et al.* (1998) found that business groups belonging to the metal and electrical engineering cluster have a clear specialisation in electronics. The business groups in the chemicals cluster concentrate their research on materials technology. Apparently, these technologies are essential to product and process innovation in these clusters. Although other clusters, such as the health cluster, may also perform research on materials technology, the chemicals cluster does not absorb this specific kind of materials research, which falls outside its scope in terms of technology. Both the metal and electrical engineering and chemicals clusters are major players in disseminating to the rest of the Dutch economy, know-how in the fields of electronics and materials technology respectively. The importance of these two technologies for the production processes of these clusters, plus the fact that they are outside the technological scope of other clusters, determines the innovation patterns of the metal and electrical engineering and chemicals clusters and their capacity to function as an important source for other clusters.

Figure 3. Four patterns of innovation

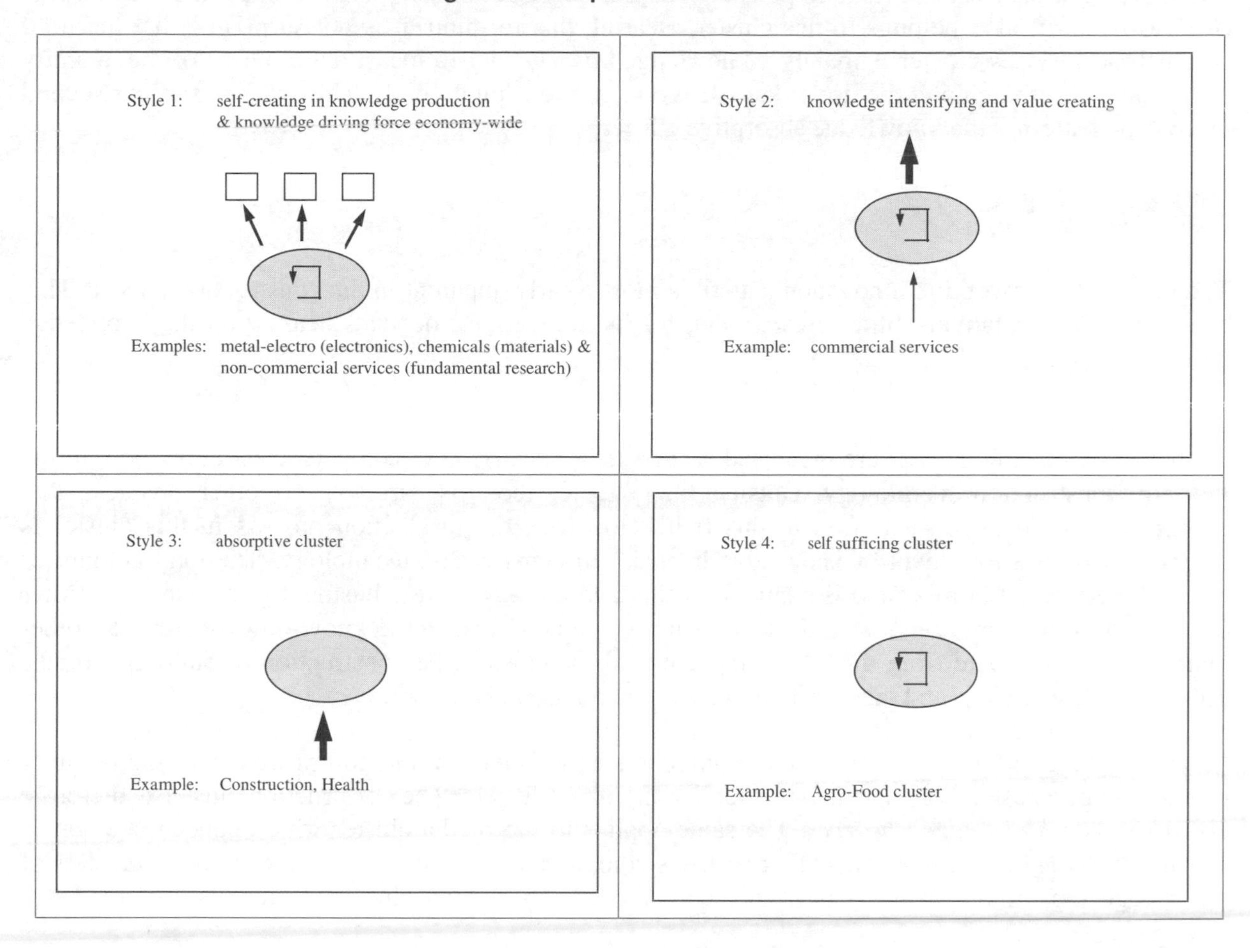

Know-how in other technologies (of which less input is required and which are too remote from the core competencies of the metal and electrical engineering and chemicals clusters) is purchased elsewhere. This is most evident in the chemicals cluster with regard to knowledge on electronics and information technology. The chemical cluster *creates* and *diffuses* know-how on materials but *absorbs* knowledge on electronics and information technology.

The non-commercial services cluster displays the same duality; this sector acts partly as a knowledge *diffuser* and partly as a knowledge *absorber*. To fully understand this, one must distinguish between technology flows in intermediate deliveries, on the one hand, and technology flows in capital stock and subcontracted research, on the other. Non-commercial services mainly comprise government institutes(education and general government policy, social security and other national services, etc.). The Dutch research organisations NWO (Netherlands Organisation for Scientific Research) and KNAW (Royal Netherlands Academy of Arts and Sciences), which perform very broad fundamental research, also belong to this cluster. It will therefore come as no surprise that this cluster is a knowledge *diffuser* when measured in terms of intermediate deliveries: the main outputs of the Dutch research institutes are results from research projects. In addition, there is no single, specific knowledge base for this cluster.[18] It is comprised of sectors specialising in diverse technologies (information technology, technology on transport equipment and technology on pharmaceuticals and energy, etc.). None of these technologies necessarily have strong ties with a specific physical production process. A large number of the research projects carried out in the non-commercial services cluster are performed for other public institutes in the same cluster (national government, for instance).

The Dutch national service unit responsible for the construction of roads, bridges and waterworks (*Rijkswaterstaat*) also belongs to this cluster. Bearing this in mind, it is not surprising that the non-commercial services cluster is mainly a knowledge *absorber* when measured in terms of the quantity of outsourced research and the use of knowledge-intensive capital goods. This brings us to our second innovation pattern – the knowledge absorptive cluster.

Absorptive clusters

The knowledge absorptive innovation pattern is most clearly apparent in the construction cluster. This cluster performs relatively little research and, for its innovations, depends heavily on the knowledge base of its suppliers.

The fact that the construction cluster mainly absorbs knowledge is a consequence of the way in which the businesses in this cluster are organised around their specific core competencies: know-how in the field of construction technology. Outside this cluster, there is virtually no other business that concentrates its research activities in this field. However, the production process in this cluster is characterised by a low level of knowledge intensity in construction technology. The total amount of research performed in this field is relatively small, and although a few business groups in this cluster do carry out research on materials (for instance, cement research), know-how relating to other materials is too remote from the core competence of the cluster (the construction of buildings, roads, railways, tunnels, etc.), and these technologies are purchased elsewhere.

Likewise, the expertise in other fields required for an efficient production process and high-quality products is purchased from specialised suppliers in other clusters. The construction cluster is therefore classified as a *knowledge absorber*. The same applies to the media cluster which concentrates on (a specific type of) IT research. This IT expertise is complemented by the purchase from other clusters of knowledge too remote from its own core competence. The media cluster is thus also a *knowledge absorber*.

In the case of the health cluster, a distinction must be made between the knowledge intensity of the intermediate deliveries and that of the flows of capital goods. The intermediate deliveries made by the health cluster mean that it absorbs more knowledge than it produces. This is not surprising as these supplies consist partly of knowledge-intensive medicine used in hospitals and the health-care sector. Consumers, in particular, benefit from the research generated and absorbed by this cluster.

However, the health cluster is self-sufficient in terms of the knowledge imbedded in its capital stock. This duality in the innovation pattern is caused by the different ways in which the various business groups organise their production process. This cluster also comprises the medical machinery and equipment industry, an industry which is largely responsible for the supply of knowledge-intensive capital goods to businesses and institutes which also belong to the health cluster. This brings us to our third innovation pattern – the self-sufficient cluster.

Self-sufficient clusters

The agro-food cluster provides a good example of a self-sufficient cluster. In 1995, some NLG 1 billion were spent on research, conducted mainly in food and agricultural technology. The companies and institutions in this cluster *absorb* knowledge by virtue of intermediate deliveries (such as food, nutrition and seeds) which originate chiefly within the same cluster. The agricultural sector consists mainly of small and medium-sized enterprises which are not large enough to invest heavily in

R&D. The knowledge in the field of food and agricultural technology absorbed by these enterprises by way of their intermediate deliveries stems from specialised research institutes in the agro-food cluster (the Dutch Research Institute for Advanced Agriculture (DLO), for example). Hence the agro-food cluster is self-sufficient in terms of its agricultural and food technology. It is also the only cluster in which the companies focus some of their research on biotechnology. Because the required know-how on agriculture, food and biotechnology is too far removed from the core competencies of companies in the other clusters, and because the businesses in the agricultural sector are too small to invest heavily in R&D, research activities are organised in concentrated research institutes within the cluster. These research activities provide the required know-how.

On the other hand, knowledge in the fields of information technology, electronics and materials technology is far removed from the core competencies of the R&D institutes in this cluster (development of agricultural machinery, fertilisers and pesticides), and these activities are subcontracted to suppliers in the metal and electrical engineering and the chemicals clusters.

Knowledge intensifying clusters

The fourth innovation model is "knowledge intensifying". These clusters are neither self-creating nor knowledge absorbing, nor are they self-sufficient. The companies in this model make use of research and technologies to increase the knowledge intensity of their products and services. In other words, they intensify the know-how of their throughputs. This innovation pattern was found in the commercial services cluster. In particular, the engineering and computer service agencies play an important role in the diffusion of hardware knowledge by incorporating advanced communication, information and operating systems. These are the key factors for the commercial services cluster. On the other hand, engineering agencies are of paramount importance for the renewal and innovation processes in construction work (buildings and architecture) and other infrastructure (road, bridges, railway and tunnels). The specialised knowledge services of the agencies, consultants and other businesses in the commercial services cluster complement the machinery, equipment and instruments produced in the metal and electrical engineering cluster. The fact that the enterprises in this cluster are specialised knowledge suppliers is one of the reasons for the total lack of a core knowledge base for this cluster (in the same way that there is no core knowledge base for the non-commercial services cluster). The fact that this cluster supplies knowledge to various other clusters explains why it consists of business sectors which specialise in a wide variety of research – transport equipment technology, logistics, energy, food and agriculture, and electronics. Because of complementary aspects, the commercial services cluster plays a significant role in innovation in virtually all other clusters (in a similar way to the self-sufficient metal and electrical engineering and chemicals clusters).

In conclusion, it can be argued that the cluster map is fairly paradoxical in character (see, for instance, the example of the metal-electro cluster in Box 2). On the one hand, a cluster is a closely knit framework of trade linkages among industry groups around a core economic activity (*e.g.* the construction of buildings, the production of machinery, chemicals, multi-media products). The industry groups within a cluster specialise around the primary production processes and the core knowledge base of the cluster. On the other hand, there is a clear division of labour between clusters. Activities that do not fit into the knowledge base of a cluster and that are too remote from the core competencies and activities of the cluster (secondary production process), are outsourced to external and specialised suppliers in other clusters.

Box 2. The role of the metal-electro cluster in the Dutch economy

The metal-electro cluster has a high level of R&D and – together with the services cluster and to a lesser extent the chemicals cluster – is one of the most important engines behind innovation processes economy-wide. The metal-electro cluster undertakes the greater part of research in the Netherlands: in 1995, over a third of total R&D expenditure took place in this cluster. Research in this cluster focuses particularly on the technology field of electronics. More than NLG 1 billion are spent annually on research in electronics (almost 90% of all research in electronics in the Netherlands). In addition to electronics, research and development in this cluster is directed towards information technology and process technology.

The metal-electro cluster absorbs relatively little knowledge from other clusters. It is more or less self-supporting in the production of new knowledge. Expenditure on outsourced research (as a percentage of expenditure on own research) is relatively low (van den Hove *et al.*, 1997). The greater part of the knowledge used by the companies in the metal-electro cluster is produced within the cluster itself, *i.e.* either by performing research or by absorbing knowledge embodied in goods or services of specialised suppliers within the cluster. The metal-electro cluster diffuses knowledge to other clusters, especially in the field of electronics, mostly embedded in machinery, instruments and equipment. Almost all the other clusters absorb electronic technology and the greater part of the knowledge in this field is produced within the metal-electro cluster (van den Hove *et al.*, 1997). For clusters outside the metal-electro cluster, the production of machinery, instruments and equipment is too far removed from their core activities and knowledge base and is outsourced to the specialised suppliers in the metal-electro cluster. The heterogeneous demand for the products of the firms in the metal-electro cluster gives them the opportunity to differentiate their products on several "niche" markets. Apart from knowledge in the field of electronics, this diffusion of embodied technology by the metal-electro cluster also takes place in the field of information technology and fabrication and process technology.

3. Cluster-based policies in the Netherlands

3.1. Introduction

The Dutch Government recently published a White Paper on cluster policy entitled *Opportunities through Synergy. Government and the Emergence of Innovative Clusters in the Market*. This document was sent to Parliament in September 1997. This section will discuss recent policy initiatives. First a few words will be said about cluster policies initiatives before the publication of the White Paper. This will be followed by a brief in-depth presentation of the rationales for cluster-based policy making. Subsequently, three different roles for governments in cluster-based policy making will be discussed. Finally, policy instruments which aim to upgrading the quality of the Dutch economic structure will be reviewed.

During the last five years, Dutch innovation policy has increasingly focused on technological co-operation aimed at improving the competitiveness and innovativeness of firms. Since 1993, various industry-oriented instruments of innovation policy have been launched to stimulate the emergence of innovative clusters and networks. Over the period 1994-97, the available annual budget for facilitating technological co-operation more than doubled (from NLG 71 million in 1994 to NLG 169 million in 1997).

In the public research sector, greater emphasis was placed on co-operation between research institutes and private firms. A larger amount of the government funding of public research organisations was earmarked for programmes performed for and co-financed by private firms. Moreover, four centres of excellence (*Topinstituten*)[19] were established to stimulate co-operation between large industrial firms, academic research groups and research organisations.

3.2. *Rationale for cluster-based policies*

The ability of firms to innovate successfully strongly depends on their capacity to organise complementary knowledge by participating in strategic production networks. Firms that wish to innovate must interact and exchange knowledge with customers, competitors and specialised suppliers of machinery, services and inputs. The emergence of clusters is mainly a market-induced process with little governmental interference. Do governments have a role in strengthening or facilitating the emergence of strategic and innovative clusters? It could be argued that establishing alliances and combining various skills in production chains takes place in the market. Following this line of reasoning, the primary task of government would be to facilitate the dynamic functioning of markets and ensure that co-operation does not lead to collusive behaviour which restricts competition.

Yet governments can do more to facilitate the emergence of innovations and innovative networks in the market. Government should not "pick winners" or "let the winners pick the government": informational complexities and the speed of market developments make it impossible for government planning agencies to successfully create clusters. However, governments can facilitate the smooth and dynamic functioning of markets by creating favourable framework conditions (vigorous competition policy, smooth macroeconomic policy, deregulation), and by reducing market (or systemic) imperfections (informational barriers, organisational failures and externalities, and spillovers).

The Dutch Government distinguishes three rationales for government action, all aiming to remove market and system imperfections:

- The competitiveness of a country's production networks strongly depends upon the synergies that arise from interactions between actors involved in the innovation process. A first rationale for economic policy is the market imperfections which hinder the realisation of these synergies: informational and organisational failures and externalities. These market imperfections can, for instance, result from a lack of strategic information (on market developments as well as on public needs), bottlenecks in dialogue and co-operation between the various actors, or environmental externalities. Many Dutch SMEs, for example, are unaware of the opportunities that co-operation with other firms and knowledge institutes can offer. Correcting these imperfections in the Dutch innovation system can improve the opportunities for firms to respond quickly to changing market needs.

- The social rate of return on investment in R&D is a second argument as to why governments have a role to play. This is the case when the social benefits of R&D investment clearly outrun the private returns on R&D investment; for instance in the fields of energy, the environment, the infrastructure and large-scale innovation projects on the electronic highway.

- A third rationale for government action in this field is increasing the rate of return on investments in public R&D by matching private needs with public funded research. By improving the co-operation between firms, on the one hand, and the public R&D infrastructure (universities, research institutes), on the other, more firms can profit from public R&D efforts and the diffusion of knowledge can increase, especially towards SMEs.

The existence of imperfections in the innovation system does not necessarily imply that government should interfere, as we can learn from past experiences. Government too can fail and should be aware of the pitfalls of cluster policy when designing a cluster-based industrial policy.[20] These pitfalls can be formulated as follows:

- ♦ The creation of clusters should not be a government-driven effort but should be the result of market-induced initiatives.
- ♦ Government policy should not have a strong orientation towards subsidising clusters or to limiting rivalry in the market.
- ♦ Cluster policy should not ignore small and emerging clusters; nor should it focus only on "classic" and existing clusters.
- ♦ While cluster policy requires cluster analysis and cluster studies, the government should not focus on analysis alone. An effective cluster policy means interaction between researchers, captains of industry, policy makers and scientists.
- ♦ Clusters should not be created from "scratch".

3.3. Policy roles and instruments for cluster policies

Following the above rationales for cluster policy, the Dutch government focuses on three tasks for innovation policy: *i) creating favourable framework conditions*; *ii) broker policies*; and *iii)* the government as *demanding customer* in providing for public needs. These three roles are briefly explained below. Table 4 summarises the different roles and policy initiatives.

Creating favourable framework conditions

Chains of value-adding economic activities emerge where markets demand them, and where framework conditions and entrepreneurship stimulate firms to innovate in order to meet the demand of demanding customers. Rivalry creates economic dynamism and the government has a role to play in facilitating the smooth and dynamic functioning of markets to secure a healthy and dynamic private sector. Relevant fields of policy making include vigorous competition policy, deregulation, structural reform, infrastructural policies and the creation of a favourable and stable macroeconomic climate.

Broker policies

The government has the task of removing systemic imperfections in the national innovation system arising from informational and organisational imperfections. Due to these imperfections, some opportunities for innovation are not recognised or can not be realised in the market. Broker policies can help to establish a meeting ground for dialogue between firms and research institutes to identify new market opportunities.

Government can play a role in organising a dialogue on future technological and market developments and providing strategic information. This requires an intensive exchange of information between policy makers, market actors and knowledge-producing agents. Concrete examples in the Netherlands include: technology forecast studies mapping future trends in technology and conducting benchmark studies on clusters.[21] These initiatives are usually joint projects between the public and private sectors.

Table 4. Cluster policy in the Netherlands: rationale and instruments[1]

Policy initiative	Rationale/role of government	Scope/cluster
Framework policy		
• Competition policy	Facilitating dynamic functioning of markets	Generic (all clusters)
• Deregulation	Structural reform of government regulations impeding innovation	Generic (all clusters)
• Macroeconomic policy	Creating favourable macroeconomic conditions	Generic (all clusters)
Broker policy	Supplying strategic information, organising platforms identifying business opportunities	
• Technology forecasts	Information and organisation	Specific fields of technology
• Focus groups, for instance: "Platform Electronic Highway"	Public consultancy	Identifying market opportunities in media, communication & services cluster
Cluster projects, for instance:	Facilitating co-operation at the pre-competitive stage; public consultancy	Upgrading specific clusters:
• "Multimedia on the Electronic Highway"		• developing new multi-media services in the media cluster (tele-learning, tele-working, tele-conferencing)
• New transport systems		• innovation in the transport cluster
• Electromagnetic power technology		• more efficient energy use (energy cluster)
• Life Sciences		• developing biotechnology in the medical and agro-food cluster
• Sub-soil building		• innovation in the construction cluster
• Electronic commerce		• development of electronic business services
• Networks in manufacturing (metal-electro)		• integral design of products and production processes in the manufacturing cluster
• Product Data Interchange		• implementing PDI in the chemicals and energy clusters
• Co-operation facility	Facilitating co-operation in innovative clusters	Generic (all clusters)
Matching public infrastructure to private needs by tendering programmes and co-financing, for instance:	Improving social rate of return on public R&D expenditures by facilitating co-operation between industry and knowledge infrastructure	Generic (all clusters)
• Economy, ecology and technology	• public needs	• stimulating environmental innovations in various clusters
• Centres of excellence (*Topinstituten*)	• increasing social rate of return of public R&D	• telematics, food processing, polymer and metal technology
Demanding customer and innovative procurement policy	High-level provision of public needs by renewal of government procurement policy	Facilitating the creation of innovative networks in:
Programmes, such as:		
• Innovation in construction		• construction cluster
• Sustainable energy		• energy cluster
• Care system of the future		• health cluster
• Optimal energy infrastructure		• energy cluster
• Road pricing		• transport & communication cluster
Horizontal policy making	Innovation in policy making through the integration of functional aspects of policy instruments and institutional renewal	Government agencies

Source: Dutch Ministry of Economic Affairs, 1997.

The government is able to bring actors together by utilising its organisational capabilities. Sometimes market parties express the wish to co-operate, yet require an external party to support the process of co-operation. For example, in the area of the development of chipcards and, more recently, the electronic Superhighway, the Ministry of Economic Affairs brought the major players together on a platform. Since 1994, a total of 24 cluster projects in several industries have emerged as a result of public brokerage.

The government: a demanding and sophisticated customer

Economic history has shown that innovation can be triggered by the presence of a demanding customer. In some areas governments can fulfil the role of a sophisticated buyer. There are several markets in which the government is a large and important player, including construction, health and infrastructure. A well-designed government procurement policy could create incentives to develop innovative solutions in these markets (Dalpe, 1994). This kind of policy should be distinguished from the traditional technology procurement policy, which mainly focused on developing new technology, without actually making the link to market-driven applications or to the realisation of public works. Today, the government is faced with the challenge of providing for public needs in these fields in an innovative and cost-effective way. This can be achieved by initiating and actively organising innovative cluster formation of market parties charged with the provision of public needs. In addition to better provision of public needs, the innovations resulting from this clustering could lead to improved competitiveness of the companies involved: the knowledge gained during the project could be used to move into new markets. The rationale for a more proactive role of government is to benefit from as yet unexploited positive externalities. The following points substantiate this policy role:

- An analysis was made of the changing procurement practices in Dutch industry. The idea was that in industry, especially in those sectors in which large projects are undertaken (*e.g.* chemical industry, offshore industry, energy industry), there are lessons to be learned in relation to, for example, design competition, more integrated project responsibility (turnkey contracting), the sharing of risks and benefits and functional specifications.

- The Ministry of Economic Affairs, in association with various other ministries, examined the possibilities for initiating procurement procedures for a number of large projects that are clearly aimed at stimulating innovation and the formation of innovative clusters. Initially, a total of eight test projects were selected. These include road construction, construction of new hospitals, houses and homes for the elderly, and road pricing. A small budget is available to support the idea of design competition.

- Plans are underway to monitor more systematically in the longer term projects initiated by government which lend themselves to further application of these more proactive public procurement practices.

It is clear that, depending on the type of market (and provided that there is a role for government), one or more of these roles and related instruments can be used to support the (further) development of innovative clusters in the economy.

4. Conclusion

Clusters can be characterised as economic networks of strongly interdependent firms linked in a value-adding production chain. Firms almost never innovate in isolation. Close interaction and knowledge exchange with customers, competitors and suppliers of machinery, services, knowledge, components and materials is a prerequisite for successful innovation. In some cases, clusters encompass strategic alliances with agents in the knowledge infrastructure, such as research institutes, universities, engineering companies and firms of consultants.

The "value chain approach" has been adopted in the Netherlands, analysing linkages and interdependencies between various actors in a value chain. Since the mid-1980s, a number of empirical studies have been conducted using a combination of monographic case study material as well as I/O analysis. The analyses revealed the cluster chart of the Dutch economic structure, its specialisation pattern, its innovation patterns and knowledge flows as well as innovation needs, the quality of the network structure, opportunities to upgrade and the strengths and weaknesses of market positions.

Governments can play a role in facilitating the creation of innovations and innovative networks in the market. Governments should not "pick winners" or "let the winners pick the government", but should first and foremost leave the selection process to the market place. Primarily, the emergence of innovative clusters is a process that takes place in the market and cannot be successfully orchestrated by government planning agencies. Governments can only facilitate the smooth and dynamic functioning of markets by creating favourable framework conditions (vigorous competition policy, smooth macro-economic policy, deregulation) and reduce systemic imperfections (information barriers, organisational failure and externalities). In addition to these imperfections, increasing the social rate of return of investment in private and public R&D is another rationale for policy making. Following these rationales for cluster policy in the Netherlands, three basic government roles can be distinguished: *i)* framework policy; *ii)* broker policy; and *iii)* the government as a sophisticated and demanding customer (innovative procurement policy).

Policy making in this field is an interactive process involving government, firms, knowledge-producing agents and bridging institutions. Indeed, effective policy making requires interaction. In facilitating the formation of innovative clusters, policy makers can:

- Point out business opportunities and provide strategic information to the market.
- Encourage innovation by facilitating the creation of platforms and focus groups.
- Link business development projects in the market to policy instruments.

These activities aim to identify business opportunities in the market and create a flexible mix of policy instruments directed to the needs of the various clusters. Policy making can only fulfil this task if research and cluster analyses (quantitative as well as qualitative) highlight the strengths, the weaknesses and the opportunities of specific clusters.

Table 4 summarised cluster policy roles and instruments in the Netherlands. The review might suggest more systemic action than the real world can handle. Basically, the development of Dutch cluster policy is an incremental and interactive process between market initiatives and public initiatives. Creating incentives for innovative behaviour in the market requires innovations in policy making and institutional renewal of government agencies. There is a growing need for "horizontal policy", integrating the various aspects of functionally organised policy instruments (education policy, science policy, trade policy, competition policy, technology policy, public works, fiscal policy, and so on). Innovative markets must be go hand in hand with innovative policies.

Technical Annex

DESCRIPTION OF THE INPUT-OUTPUT ANALYSIS

The "value chain approach" discussed in Section 2 focuses on trade linkages. These linkages can be quantified using I/O tables.

Input-output analysis

We used I/O tables at the aggregation level of 214 industry groups to describe the I/O method for cluster identification. The method used by van den Hove and Roelandt (1997) and Witteveen (1997) is adapted from that used by Montfort and Dutailly (1983), which introduced the *"filière approach"*, clustering industry groups on the basis of forward or backward trade linkages. The method used by Roelandt *et al.* concentrates on the intermediary use and supply of products and services, and excludes trade flows within industry groups. It includes deliveries from foreign suppliers. The starting point is the intermediary part of the I/O table. Clusters are identified on the basis of intermediary deliveries. The main diagonal of the matrix is assumed to be zero (and subtracted from total deliveries): the focus is exclusively on interindustry-group intermediary deliveries.

Forward linkages

Using the "forward linkages" method, industry groups are clustered on the basis of relatively strong forward trade linkages. In an I/O table row the production of an industry group is differentiated according to its using-industry groups. To identify the main user, a matrix M with the elements $m_{\alpha\beta}$ is drawn up comprising intermediary deliveries between industry groups:

$$m_{\alpha\beta} = \max_j m_{\alpha j} \quad \text{and} \quad m_{\alpha\alpha} = 0$$

The main user (j) of a supplying industry group (i) is identified by determining the maximum of a row. When all the row maxima have been established, we then compute whether it is a substantial delivery, *i.e.* whether this delivery, as a percentage of total (intermediary) deliveries, is higher than a chosen value k.

$$\frac{m_{\alpha\beta}}{t_\alpha} > k \quad \text{and} \quad t_\alpha = \sum_{\alpha=1}^{n} m_{\alpha j} - m_{\alpha\alpha}$$

Following Monfort and Dutailly (1983), the value of k has been assigned by trial and error. If the row maximum is higher than this k-value, then the rows and columns of the industry groups are aggregated and the elements of the main diagonal are again given the value zero. This procedure is repeated until there are no more industry groups to be aggregated.

Backward linkages

To analyse the "backward linkages", industry groups are clustered on the basis of relatively strong backward linkages. Using this method we are able to identify the main *supplier* of an industry group. Aggregation takes place when a delivery is sufficiently large. This procedure is repeated until there are no more industry groups to be aggregated:

$$n_{\alpha\beta} = \max_i n_{i\beta} \quad \text{and} \quad n_{\beta\beta} = 0$$

The main supplier (i) of a using industry group (j) is identified by determining the maximum of a column. When all the column maxima have been determined we then compute whether it is a substantial delivery, *i.e.* whether this delivery, as a percentage of total (intermediary) deliveries, is higher than a chosen value l.

$$\frac{n_{\alpha\beta}}{v_\beta} > l \quad \text{and} \quad v_\beta = \sum_{\beta=1}^{n} n_{i\beta} - n_{\beta\beta}$$

The similar matrices M ($m_{\alpha\beta}$) and N ($n_{\alpha\beta}$) make up the intermediary part of the I/O table, i denotes the row number of an industry group and j the column number.

Besides domestic intermediary deliveries, import, export and investment relationships between domestic and foreign sectors also have a role to play. Because of problems related to statistical registration we have to ignore flows of exports. With the help of import matrices and investment matrices we are able to gain an insight into how flows of imports and investments are structured. By using a similar method, the largest buyers and suppliers of investment and import goods can be identified. The values of k and l are chosen at 0.20 and 0.15, respectively. Determining k and l is a process of trial and error. The lower (higher) k or l, the more (less) aggregation will take place.

Make & use tables and the method of maxima

Make & use tables contain data on the supply and use of intermediary goods and services, final demand categories, imports and exports at a very low level of aggregation. Each table encompasses 650 product categories and 260 industry groups. The make table reviews the production of goods and services by supplying industry groups. The use table shows the use of products and services by different industry groups. For the clustering procedure, we used the simple "method of maxima", implying that we identified the main producer as well as the main user of a specific product category, and subsequently linked both industry groups.

This means:

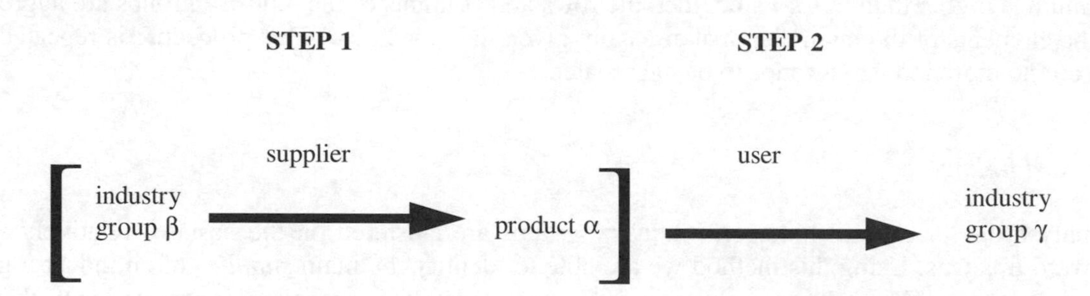

Industry group β is linked to industry group γ if:

$$\underline{Z_{\alpha\beta} = \max_i Z_{i\beta}}\rfloor \text{ and } \underline{N_{\alpha\gamma} = \max_j N_{\alpha j}}\rfloor$$

$\underline{Z_{\alpha\beta}}\rfloor$ = intermediate supply of product category α produced by industry group β, and

$\underline{N_{\alpha\gamma}}\rfloor$ = intermediate use of product category α by industry group γ.

NOTES

1. For a review, see Jacobs and de Man (1996).

2. For a review, see OECD (1992); DeBresson (this volume); and Edquist (1997).

3. In addition, a third cluster approach is developing at the micro level, focusing on identifying small-scale clusters and firm linkages. This approach, which is mainly policy-led, aims at developing collaborative projects and focuses particularly on (regionally) upgrading the quality of constructive co-operation within production networks. This latter approach is excluded here.

4. For an in-depth review and discussion of cluster concepts, cluster dimensions in relation to industrial policy making and firm strategy, refer to Jacobs and de Man (1996) and Jacobs (1997).

5. Jacobs, Boekholt and Zegveld (1990).

6. Jacobs, Kuijper and Roes (1992).

7. The make & use tables can also be used to analyse chains at the meso level, *i.e.* the identification of networks of related branches in a production chain producing one or similar products.

8. A full description of these techniques is given in the Technical Annex.

9. For a detailed description of the differences and similarities, see Witteveen (1997).

10. Some refinements in the algorithm have recently been made to correct for the relative importance of the trade flows from the perspective of the supplying- and using-industry groups. For details and results, please contact the authors.

11. See the Technical Annex for details. A full description of the methodology is given in van den Hove *et al.* (1998).

12. The services cluster actually consists of the commercial services cluster and the non-commercial services cluster.

13. Witteveen (1997).

14. Den Hertog and Maltha (this volume) describe the information and communication cluster as an emerging cluster.

15. Calculated on the base of the Dutch Innovation Survey 1991.

16. For the clusters (Non) Commercial services and Transport no comparable data are available.

17. This section on patterns of innovation draws heavily on van den Hove *et al.* (1998).

18. While Figure 3.5 does show a specialisation pattern on energy for this cluster, technical reasons prohibit inclusion of the research institutes that perform a large part of the research in the non-commercial services cluster. Had research institutes like KNAW and NWO been included in Figure 3.5, a different pattern would have emerged for the non-commercial services cluster.

19. In telematics, food processing, polymers, and metal technology.

20. Held (1996) and Porter (1997).

21. The "Technology Radar" (a technology foresight study) was published recently and various benchmark studies were also performed, including one for the service cluster.

REFERENCES

Dalpe, R. (1994), "Effects of Government Procurement on Industrial Performance", *Technology in Society: An International Journal,* Vol. 16, No. 1, pp. 65-84.

DeBresson, C. (ed.) (1996), *Economic Interdependence and Innovative Activity. An Input-Output Analysis*, Edgar Elgar, Cheltenham.

DeBresson, C. (1999), "Identifying Clusters of Innovative Activity: A New Approach and a Tookbox", this volume.

Edquist, Ch. (ed.) (1997), *Systems of Innovation. Technologies, Institutions and Organisations,* Pinter, London.

Held, J.R. (1996), "Clusters as an Economic Development Tool. Beyond the Pitfalls", *Economic Development Quarterly*, Vol. 10, No. 3, August, pp. 249-261.

Hertog, P. den and S. Maltha (1999), "The Emerging Information and Communication Cluster in the Netherlands", this volume.

Hove, N. van den and Th. Roelandt (1997), "Clusters en technologie", in CBS, *Kennis en Economie 1997*, Voorburg.

Hove, N. van den, Th. Roelandt and Th. Grosfeld (1998), *Cluster Specialisation Patterns and Innovation Styles,* Dutch Ministry of Economic Affairs, The Hague.

Jacobs, D. (1997), "Knowledge-intensive Innovation: The Potential of the Cluster Approach", *The IPTS Report*, 2(16), pp. 22-28.

Jacobs, D., P. Boekholt and W. Zegveld (1990), *De economische Kracht van Nederlan* ("The Economic Strength of the Netherlands"), SMO, The Hague.

Jacobs, D., J. Kuijper and Bert Roes (1992), *De economische kracht van de bouw. Noodzaak van een culturele trendbreuk* ("The Economic Strength of the Construction Industry. The Need for a Cultural Breakthrough"), SMO, The Hague.

Jacobs, D. and A.-P. de Man (1996), "Clusters, Industrial Policy and Firm Strategy: A Menu Approach", *Technology Analysis and Strategic Management*, Vol. 8, No. 4., pp. 425-437.

Ministry of Economic Affairs (1997), *Opportunities through Synergy. Government and the Emergence of Innovative Clusters in the Market*, The Hague.

Montfort, M.J. and J.C. Dutailly (1983), "Les filières de production", *INSEE Archives et Documents*.

OECD (1992), *Technology and the Economy. The Key Relationships,* Paris.

Porter, M.E. (1990), *The Competitive Advantage of Nations,* Macmillan Press, London.

Porter, M.E. (1997), "Knowledge-based Clusters and National Advantages", paper presented at the International Conference on Metropolitan Concentrations of Knowledge-based Industries, 9-12 September.

Witteveen, W. (1997), *Clusters in Nederland. Een verkenning van het bestaan van clusters van toeleveranciers en afnemers en de implicaties voor het overheidsbeleid,* Dutch Ministry of Economic Affairs, The Hague.

Chapter 14

CREATING A COHERENT DESIGN FOR CLUSTER ANALYSIS AND RELATED POLICIES: THE AUSTRIAN "TIP" EXPERIENCE

by

Michael Peneder*
Austrian Institute of Economic Research (WIFO)

1. Introduction

The Austrian innovation research programme ("Technology, Information, Policy consulting" – TIP) is a major vehicle in the promotion of the cluster perspective in Austrian technology policy. Through the provision of a long-term commitment and a stable research environment, considerable experience on cluster analysis has been built up over the past few years. This chapter summarises the basic knowledge and insights gained from this process.

From its beginnings in 1992, the cluster approach has been a constituent part of the TIP programme. Inspired by the popularity of Porter's *Competitive Advantage of Nations* (1990), Austrian policy makers demanded a similar tool. However, Porter's notion of clusters left several conceptual and even more methodological questions unanswered.[1] What followed was a process of integrating alternative approaches and dimensions, among them the work of Jaffe (1986, 1989) on patent activities, and rediscovering the fundamental theoretical sources of Marshall (1920) and their modern revitalisation by Krugman (1991). For a summary see Hutschenreiter and Peneder (1994). Finally, an overall programme design was created, which tried to match the analytical claims as well as the practical limitations of the data in a coherent manner. This led to the combination of a horizontal screening across industries and technology fields, on the one hand, and the vertical focus on a number of selected case studies, on the other.

Since the term "cluster analysis" became popular, it has been used in various contexts. This chapter first sets out the general conceptual background of cluster analysis in the Austrian TIP programme. It then goes on to outline data restrictions and the choice of analytical method. Selected results of the empirical studies are presented to illustrate: *i)* the usefulness of cluster analysis as a policy tool; and *ii)* the kinds of conclusions that can typically be drawn. The final section summarises the main policy implications of the cluster approach.

---

* This chapter has benefited from inputs by colleagues in the TIP team, notably Kurt Bayer, Gernot Hutschenreiter, Leonhard Jörg, Norbert Knoll, Hannes Leo, Fritz Ohler, Wolfgang Polt and Katharina Warta. Nevertheless, the author remains responsible for the overall interpretation and the selection of supporting arguments and illustrative examples. The usual disclaimer applies.

The Austrian "TIP" research programme

TIP is an innovation research and consulting programme based on an initiative of the Austrian Federal Ministry of Science and Transport and the Austrian Federal Ministry of Economic Affairs. It is carried out by the Austrian Institute of Economic Research (WIFO), in co-operation with the Austrian Research Centre Seibersdorf (ARCS). WIFO is in charge of directing and managing the programme.

The overall goal is to produce information and recommendations relevant to Austrian technology policy, based on analyses of technological change and its impact on the national system of innovation at the macro level (enterprises, public and private institutions and their interactions), the meso level (structural analyses of the Austrian economy) and the micro level (analyses of firm behaviour).

Between 1993 and 1995, TIP evolved as a major research and consultancy programme designed to foster the knowledge base of Austrian technology policy through the provision of relevant analysis and the building of capabilities in the field. During this first term, strong emphasis was placed on cluster analysis of various kinds.

Based on the experience gained from the first phase of the programme, the core theme of the current phase (1996-99) is: "Ways into the Knowledge-based and Information Society – Technology Policy in Complex Systems". The current work plan comprises the following modules:

- *National System of Innovation* (topics include a cluster case study on multimedia and cultural content; inter-firm co-operation and innovation; the role of the service sector in the Austrian NIS; the role of the science system in technology development).
- *Productivity, Growth and Employment* (quantitative analysis of embodied interindustry flows; econometric analysis of the relationship between R&D investment and productivity growth).
- *Modern Regulation as an Instrument of Technology Policy* (impact of the regulatory framework on innovative performance in the telecommunication and energy sectors).
- *Consultancy for Technology Policy* (providing stand-by capacity, *e.g.* for the monitoring of current developments in European Union industrial and technology policy).

2. The cluster concept

2.1. Density

Clusters are not unique to economics, and are used increasingly frequently in, for example, statistics, music or the computer sciences. In its literal and most general meaning, a "cluster" is defined by the *Concise Oxford Dictionary* as a "close group of things" (Fowler, 1982). Thus, the synonymous notions of "density", "relative nearness" and "similarity" lie at the heart of the cluster concept. *A priori*, "closeness" or "similarity" is not restricted to any particular dimension (geography, technology, social characteristics, etc.) or limited by any specific scale. Consequently, both have to be chosen exogenously to fit the question under investigation.

2.2. Marshall´s cluster hypothesis

In economics, the cluster concept usually implies more than the literal meaning of density by reference to a hypothesis which states that the geographic agglomeration of economic activity may lead to improved technological or economic performance of the participating units.[2]

The cluster hypothesis, in its economic sense, is based on Marshall (1920), who explained the development of industrial complexes by the existence of *positive externalities* in agglomerations of interrelated firms and industries. These externalities are caused by three major forces: *i)* knowledge spillovers between firms;[3] *ii)* specialised inputs and services from supporting industries;[4] and *iii)* a geographically pooled labour market for specialised skills.[5] Marshall's cluster hypothesis basically states the existence of *dynamic complementarity* within a system of interdependent economic entities that influences specialisation patterns in production: for the reasons given above, innovation and growth in one economic unit can exert positive impulses for innovation and growth in other parts of the system as well. Therefore, a cluster of industrial complexes is expected to perform better than the sum of its individual units in a more scattered distribution. It is worth noting that this idea considerably extended conventional economic wisdom, which relied solely on exogenous *comparative advantages* or *internal economies of scale*, respectively, to explain specialisation and concentration of economic activities.

Marshall's analysis does not stop there, but gives equal weight to two fundamental economic causes which work against too highly specialised industrial locations: *i)* differentiated skills in local labour markets call for a certain spread of skill requirements and associated sectoral structures;[6] and *ii)* a differentiated sectoral composition in a given location offers a greater spread of the risks associated with exogenous shifts in demand or input prices specific to individual industries.[7]

2.3. *"Organic" economic systems*

From the beginning, the Austrian TIP approach was driven by Marshall's hypothesis of positive externalities within dense economic structures. A closer focus on the "organic", evolutionary properties of interrelated units in industrial complexes would be a fruitful area for future research.

Marshall's cluster hypothesis applies to agglomerations of independent decision-making units, *i.e.* there is not a single hierarchical command and control structure between different sub-units within "large" enterprises or conglomerates. The rationale for the latter structure is based on economies of scale and scope as opposed to the external economies associated with clustered firms. Although there are some similarities, the dynamic properties of innovation, diffusion and adaptability to changes in the economic environment will differ considerably across the two types of organisational structure.

Building on the concept of *interdependent, but organisationally independent* decision-making units, industrial clusters can be seen as *"organic" economic systems*, where the principles of evolutionary complexity apply. Thus, the main focus of research will be directed to the manifold feedback mechanisms within the system, and the inherent potential for: *i)* variety creation and innovation; and *ii)* accumulation and growth; as well as the inherent scarcities that cause competitive pressure.

3. Methodology and data problems

The application of quantitative cluster analysis is limited by its demanding data requirements. Use of statistical data is always restricted by the dependence on official classification systems, which tend to reflect more traditional concepts of how industrial activities should be categorised. The opportunities for regrouping the official classifications to suit the specific purpose of cluster research in a significantly better way are hindered by the lack of *interrelational flow* data. In the case of Austria, this situation is further aggravated by the lack of recent input-output tables.

In practice, the most flexible instruments for detecting relevant flow relationships are interviews and questionnaires in the framework of a narrowly targeted cluster case study. However, the results may be seriously distorted by the fact that knowledge spillovers (in the sense of Marshall's cluster hypothesis) depend to a large extent on *implicit knowledge*. Economic agents may benefit from these flows, but in most cases will not be aware of their actual impact or will not be able to communicate them.

In designing the research programme, the following methodological approaches were considered:

- *Input-output analysis:* Because of its direct use of flow data, input-output (I/O) analysis is often regarded as the preferred methodological approach for detecting and quantifying interdependencies among different industries. Although I/O tables are based on material flows, it can plausibly be assumed that these economic interactions also enhance the probability of complementary flows of (embodied) technical knowledge.

  The major obstacle to the use of input-output analysis in the Austrian case has been the lack of recent I/O tables. At the beginning of TIP in 1992, the most recent official Austrian I/O table dated from 1976. Currently, the most up-to-date table was compiled in 1983. An I/O table for 1988 has been compiled at WIFO but is too highly aggregated for the detection of cluster relations. A more recent official table at the 3-digit level (covering some 230 industries) for 1993 has been promised, but is not yet available.

- *Cluster screening:* A wide range of statistical and analytical techniques exists for detecting inherent regularities and similar patterns in multivariate data sets. The following techniques can be particularly useful in the context of research on cluster phenomena:[8]

  - *Principal component analysis* can serve to reduce the number of dimensions by creating new and uncorrelated variables as linear combinations of the original variables, thus facilitating the detection of similarities and relative nearness of observations.

  - Applied for a similar purpose, *factor analysis* additionally identifies the common factors responsible for the correlations among variables in the first instance.

  - *Cluster analysis* produces a classification scheme of individual observations, based on their relative similarity or nearness to an array of different variables. The basic idea is to divide a specific data profile into segments by creating maximum homogeneity within and maximum distance between groups of observations.

  - Finally, *discriminant analysis* can be used to explore the different properties of these groups and to integrate new observations into a classification scheme.

  Within the TIP programme, statistical cluster analysis has been applied most extensively for the screening of patterns and the mapping of patent activities in Hutschenreiter (1994), as well as trade performance by sector in Peneder (1995). Factor analysis has occasionally been applied in cluster case studies such as Jörg *et al.* (1995). Input-output analysis, based on the WIFO I/O table for 1988, is applied in a quantitative analysis of embodied interindustry technology flows (Hutschenreiter *et al.*, 1998).

- *Case studies:* Case studies provide the most flexible approach for linking cluster-related phenomena with particular policy recommendations. Interviews and questionnaires allow the general research design to be tailored to the individual objectives so as to provide valuable micro data (including interrelations within the cluster). The main drawbacks from a methodological point of view are the heterogeneity and the high dependence on intuition, which result in a lack of comparability of the individual studies.

  The major drawback from a practical point of view is the enormous need for research capacity and therefore the high costs involved. In this regard, one of the most significant lessons from the TIP programme has been the rather low "economies of learning by doing" in carrying out the case studies. Because of the heterogeneity of institutional frameworks, technological regimes and strategic settings within different clusters, the economies of learning have been much lower than was anticipated at the outset. Each attempt to analyse some part of the dynamic relationships that take place inside a particular cluster requires huge (and mostly sunk) investment in learning the specific technical and institutional details. However, the accumulated knowledge is of limited use when the commission switches to a different cluster. Although this is common to most case studies, the complex structure of the target (*i.e.* the badly documented interrelations within various kinds of activities) multiplies the fixed investment in comparison with traditional industry studies. If this aspect is neglected in the planning of resources, there is a serious danger of never doing more than scratching the surface.

4. General programme design

Cluster analysis is only part of the TIP programme. Figure 1 summarises the major cluster-related studies and the logic of the programme design. Initially, the lack of current I/O tables forced us to concentrate on the identification of general patterns and "dense locations" within certain dimensions of technological and economic performance. The overall goal of this screening for relevant clusters was to draw a map in order to define priority areas for the case studies that would follow. In the next stage, areas of significant economic or technological activity were selected for case study. The studies began with an analysis of industries having wood and paper as their common resource, followed by the telecommunications sector, the pharmaceutical industry and, currently, the potential cluster at the interface of multimedia and cultural content. With regard to the case studies on "wood and paper" and "telecommunications", a policy review workshop was also organised. Finally, TIP participation in the OECD National Innovation Systems project can be seen as a further extension of the cluster-related research agenda.

Figure 1. The TIP cluster research programme design

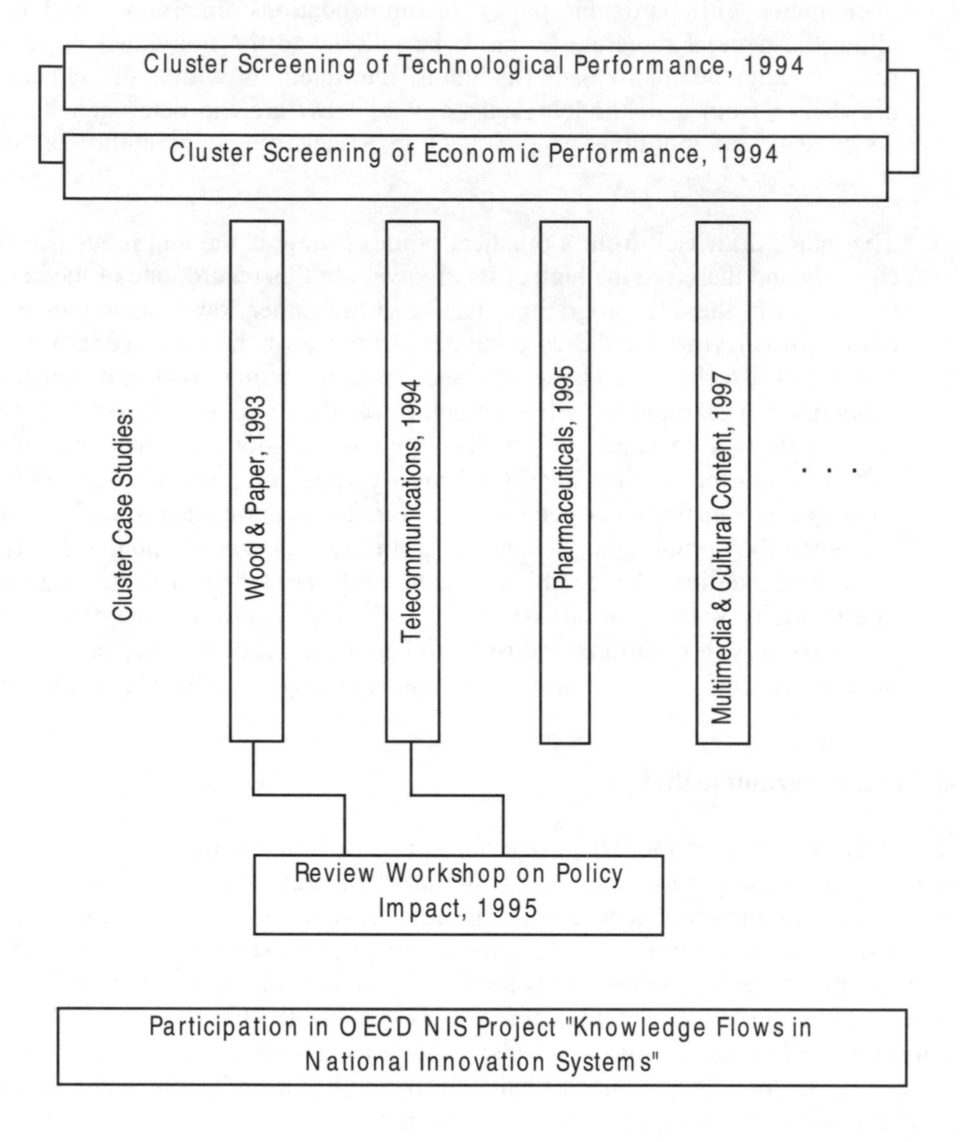

5. Cluster screening

5.1. Technological clusters (Hutschenreiter, 1994)

Technological clusters in Austrian manufacturing were identified on the basis of patent applications from Austrian firms in the years 1987-91, categorised in terms of the patent classes of the International Patent Classification ("technologies"). Using statistical cluster analysis, patenting Austrian firms were combined into groups exhibiting similar patent structures. The externalities which are potentially present in clusters of this kind include pooled labour markets, shared research and training facilities, the division of labour in the core technologies of the cluster, etc.

The findings provided a visualisation of technological clusters including the number of applicant firms and patents. From the bird's eye's perspective, the following groups of technological clusters may be discerned:

- electrical – electronics – telecommunications;
- transportation;
- construction – housing;
- skis – ski boots – sports equipment;
- pharmaceuticals – chemicals.

The cluster analysis resulted in a technological panorama of Austrian manufacturing, including innovative small and medium-sized enterprises, with the following features:

- Comparatively small technological niches play a relatively large role. Given the industrial structure of Austria, niche strategies appear to be rational. It has to be noted, however, that the technological and *a fortiori* the economic value of patents is virtually unknown.
- Larger enterprises with diversified patenting activities tend to be isolated and are not attracted by any cluster. This may reflect both overly diversified activities as well as simply their own narrow home market.
- Clusters typically consist of just a handful of enterprises. The opportunities to reap positive externalities are therefore limited. Strengthening the evolution of technological clusters may be a good basis for technology policy.

5.2. Economic performance (Peneder, 1994, 1995)

Porter's influential study on the competitive advantage of nations inspired a methodologically extended analysis of Austrian data. However, in contrast to Porter's analysis, competitiveness has been determined endogenously by means of statistical cluster techniques. Avoiding his "cut-off" approach, which *a priori* eliminates a large majority of industries that perform below an exogenously pre-specified level, industries that perform either "well" or "badly" have been the objects of analysis. The resulting cluster centres constitute the typical pattern of competitiveness for the chosen trade indicators, while the classifications produce a "map" of Austrian industrial export performance. In a nutshell, the results: *i)* showed that "clustered" industries are generally rare in the case of Austria; *ii)* indicated that a considerable share of these industries are located in declining sectors; and *iii)* underline the importance of transnational links (as opposed to narrow national boundaries) for the creation of successful industries.

To account for the quantitative (volume-dependent) as well as the qualitative (price-dependent) dimension of competitiveness, trade performance has been measured by four variables: international market shares (MAS); revealed comparative advantage (TSP); relative position in export prices (RUV); and revealed comparative price advantage (CPA). The overall profile of Austrian industrial performance for 1992 is shown in Figure 2. The panorama results from a three-step clustering process, which aggregated 208 SITC product classifications up to 21 clusters and grouped them according to their relative nearness in the performance variables. Using this "map", the position of any single industry relative to others can be identified. The only effort required is to look up the appropriate number of the 21 clusters with which it is classified in the relevant tables of the original publication.[9]

Figure 2. Panorama of Austrian industrial performance 1992

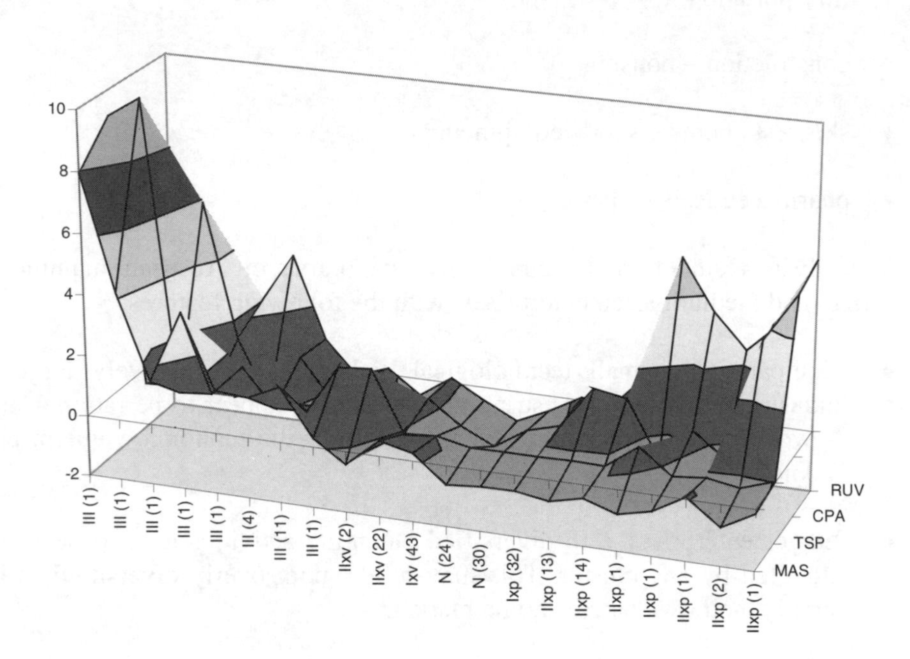

Source: Peneder, 1995.

The performance levels resulting from the clustering procedure, are interpreted as follows:

- Level N: "non-performers".
- Level Ixp: badly performing industries with an average position in export prices.
- Level IIxp: industries competing in small niches of the product spectrum, characterised by high price levels but low export volumes.
- Level Ixv: industries competing on their "low cost/low price" position, and which have modest success in terms of export volumes.
- Level IIxv: industries performing well, mainly in terms of market shares.
- Level III: "champions", performing best in terms of quantity as well as quality.

Industries belonging to Level III are mainly spread around the *materials and metals*, *forest products*, *transport* and *textiles* sectors. In the area of basic industries, an internationally successful cluster has grown out of rich endowments of wood and covers *simply worked wood*, *chipboards* and *wood manufacturers* as well as *paper and paperboard*. The production of *internal combustion piston engines* is one of the most successful areas of Austrian industrial activity; in addition to the innovative capacity of a number of Austrian manufacturers, this extraordinary performance is mainly due to successful transnational links with the international automotive industry.

6. Cluster case studies

The final selection of case studies within the research programme was partly linked to the cluster screening of patent activities ("telecommunications", "pharmaceuticals") and of economic performance ("wood and paper"). However, it also reflected the more immediate need for more prospective policy advice in a particularly fast-moving industry ("multimedia and cultural content").

6.1. Wood and paper (Bayer et al., 1993)

The "wood and paper" cluster occupies a special position within the Austrian economy as it covers a complete value chain ranging from the raw material to the highly processed final product. With the sustainable raw material – wood – available locally, a tradition of crafts and industries has developed over the centuries to provide the foundation for modern production processes and product developments. The "wood and paper" cluster is interlinked in terms of products and processes, but not with regard to organisation and property rights.

The subject of this study is not a single technology, company or business line, but rather an industrial complex in the sense of horizontally related or vertically linked companies, characterised by a range of interactions. The goal is to observe the intricately interwoven parts of the cluster in its processes, in terms of its production, services, customers and public regulation bodies, and in this way to highlight the numerous interdependencies and spillovers between its various parts as a focus of research and economic policy. Taking this holistic view should lead to political decisions that offer sector-crossing integral solutions. In order to indicate economic interdependencies, an input-output analysis was used, and a material flow analysis, performed by the Austrian Central Statistical Office, introduced.

The "wood and paper" cluster figures among those sectors of the Austrian industry that are low on research and innovation input. In the sawmills, innovation concentrates entirely on process innovation, while product innovation dominates in the woodworking and paper industries. Both sectors purchase technology primarily from outside (mostly from the machine and chemicals industries). An essential cluster-internal adaptation activity is then performed by the intelligent design of the production layout. Technological progress is primarily incremental and determined mainly by the development of generic technologies (chiefly microelectronic applications such as sensor and CIM technologies). In recent years, the impetus for technological innovation has come mainly from environmental standards. In many areas, production processes are characterised by a high degree of mechanisation and automation. As a consequence, labour costs tend to decrease in importance, while industry concentration – fuelled by increasing capital intensity – is becoming more significant (especially among sawmills). Summarising the situation, the study describes this cluster's innovation system as largely in "equilibrium" – albeit at a low level.

The analysis produced a number of starting-points for economic policy measures to strengthen the innovation and market performance of this cluster. In this chapter, we are limited to sketching the general approaches (Table 1): technology, structure, ecology, standardisation, education and training. To achieve the maximum possible effect, these fields need to be co-ordinated.

Table 1. Recommendations for strengthening the industrial complex "wood and paper"

	Umbrella project "Wood processing" Initiative on Energy-Wood-Paper Awareness, programme management and marketing, management of research and development					
	Technology	Structure	Ecology	Standards	Education & training	Innovation
Sawmills		Promote forward integration	Energy-generating disposal	Fire protection standards	College course on "sawmill engineering"	
Wood construction	Strengthening C technologies	Public demand	Environmentally compatible construction materials and binders	Building codes, fire protection standards, quality standards	Courses for wood construction environmental balance	Compound materials, wood
Production of boards	Co-operative R&D		Binders, coatings, disposal	Environmental standards, quality standards		Boards made of wood substitutes, bend-resistant boards
Furniture production	C technologies, design	Increase production depth, public procurement	Disposal design	Quality	Training for "wood engineers", reskilling	Improve system idea
Paper production and processing	Co-operative R&D	New business segment: energy supply	Chlorine-free bleach, wastewater, de-inking	Emission standards		New paper grades

Source: Bayer *et al.*, 1993.

6.2. Telecommunications (Leo et al., 1994)

The cluster case study on telecommunications provides a comprehensive exposition of the evolution of major technical trajectories, changes in international regulatory regimes and the Austrian policy network. Moreover, the strategic options for the Austrian telecommunications cluster – comprised of hardware suppliers, the national PTO and new value-added services – are examined in view of the new legal, technical and competitive environment these actors have to face.

Technical and regulatory innovations have led to a process of profound and global change in the field of telecommunications. In the short and medium term, the required changes will give rise to major adjustment costs, mainly for the PTO. In the long run, however, rising productivity and higher utility levels for users are expected to lead to considerable welfare gains for the economy as a whole.

Figure 3. Impulses for innovation and growth within the telecommunications cluster

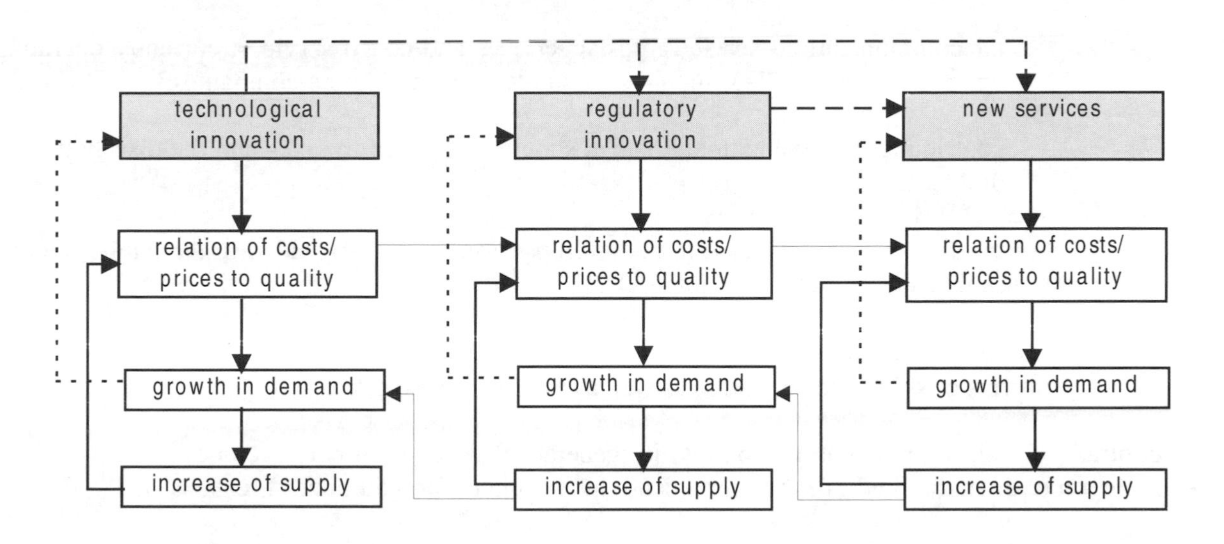

Source: Peneder, 1995.

In order to define priority areas for policy measures, the inherent dynamic complementarities and associated mutual impulses for innovation and growth within the "telecommunications" cluster have been analysed schematically. Figure 3 illustrates the presence of vertical *pecuniary external economies* via dynamic forward and backward linkages between and within the three cluster components. These linkages arise from a circular relationship in which the decision to invest in large-scale production depends on the size of the market, while at the same time the size of the market depends on the amount of investment. The *forward linkages* stem from cost reductions or quality improvements for potential downstream users because of innovation and growth in an upstream industry. The *backward linkages* stem from economies of scale which are enabled by growth of the downstream industries. Thus economies of scale are the essential criterion for the existence and economic relevance of these pecuniary external economies.[10] In the case of the telecommunications cluster, economies of scale are particularly significant because of the large outlays for R&D as fixed investment (apparatus and equipment) as well as considerable network externalities (basic network services and value-added services).

As technological innovation in the field of hardware suppliers was considered to be the major driving force for the dynamic development of this sector, lack of complementary regulatory innovation and organisational adaptation was identified as the major barrier to further growth.

The report stated that Austria still lacks well-established specific telecommunications policies. Policy measures directed at the telecommunications sector are largely determined by other areas (*e.g.* fiscal policy or accession to the European Union). Pressures for reform are being exerted from outside but implementation is slow. A detailed examination led to the following recommendations for Austrian telecommunications policy; these recommendations have now been partially fulfilled under pressure from EU legislation:

- Gradual increase of competition through an accelerated opening of markets.

- Liberalisation requires accompanying organisational changes in the Austrian PTO:

  - separation of the Austrian PTO from the public sector;

  - telecommunications services, postal services and coach services, currently operating under the national PTO, should be split off to form separate companies;

  - contributions to the national budget should be limited to taxes and the distribution of dividends;

  - funding instruments need to be developed for services provided in the public interest (universal service obligations).

- To rise to these challenges, the regulatory bodies will require additional resources in terms of skills and economic as well as managerial expertise.

In contrast to telecommunications policy in general, the scope for a sector-specific, national technology policy in the field of telecommunications is limited, for a number of reasons:

- The majority of firms supplying telecommunications equipment, systems and components are affiliates of multinational enterprises, and for that reason have only limited capacity to decide autonomously on matters of overall strategic importance.

- The development and manufacturing activities of Austrian suppliers are widely dispersed, making it almost impossible to identify common inter-firm priority areas in R&D.

- The globalisation of competition enhances the importance of R&D projects conducted at the European rather than at the national level.

Under these conditions, technology policy must mainly focus on the provision of a *supportive framework for R&D activities*. Specific recommendations for technology policy measures directed at the telecommunications sector are summarised as follows:

- An investigation into the R&D potential of research institutions (universities, etc.) and their complementarity with industrial production, service providers and specific applications in Austria is necessary.

- National telecommunications policy projects must focus on the upgrading of networks and services to enable better provision of user-specific applications.

- Public support through subsidies or other measures should focus on general applications rather than on specific technologies or firms.

6.3. *Pharmaceuticals (Jörg et al., 1995)*

The production of pharmaceuticals in Austria accounted for only approximately 2% of overall industrial value added in 1992. However, the growth of this sector has been impressive: its share in both total value added and the total number of industrial employees has doubled since 1980. The

Austrian pharmaceutical sector is exposed to two major global changes in its economic and technological environment:

- Competitive pressures on pharmaceutical companies have increased due to the escalating cost of developing new drugs in recent years accompanied by the increased market shares of "generic" drugs (*i.e.* drugs manufactured by other companies once their patent protection has expired). Reducing the economic appropriability of successful innovations, this development has made it harder for research-intensive companies to recuperate their initial R&D outlays.

- A new technological window has opened in the form of biotechnologies, which have become one of the pharmaceuticals industry's most promising sources of innovation.

Within the Austrian pharmaceutical industry, two types of companies can be identified:

- The first group includes a small number of large science-based companies with a strong international focus. Nearly all of these firms are owned by foreign multinational enterprises based in Germany and Switzerland.

- The second group – mainly comprising firms of Austrian ownership – is made up of small and medium-sized companies which concentrate their innovation activities on the improvement of established products and which focus largely on the domestic market.

Innovation in the Austrian pharmaceutical complex is concentrated among a small number of players. As far as the private sector is concerned, three-quarters of all research expenditure in the pharmaceutical sector is spent by the three largest companies. Furthermore, the five companies with the largest innovation capabilities are all subsidiaries of multinational pharmaceutical companies.

In tracking knowledge flows within the pharmaceutical industry, the study identified a largely "insular" attitude to R&D. Although strong links exist between universities and the larger firms, inter-firm co-operation and knowledge spillovers through personnel mobility are poor. One explanation for this lack of information flows within the NIS is that the Austrian "pharmaceutical" cluster can at best be labelled "incomplete". Particularly in the new area of biotechnology, the start-up group is almost completely lacking. Access to competence in biotechnology can only be achieved by cross-border co-operation. In addition, knowledge spillovers are hindered by the fact that the co-operation strategies of the major players in the Austrian field are determined and constrained by their foreign-based headquarters.

Finally, the study put forward a number of policy recommendations that could be implemented to strengthen knowledge-related ties in the sector. Among others, attention was directed on the following policy variables:

- There are shortcomings in the public discussion of the pharmaceutical and biotechnology sector. In order to provide momentum and strengthen the debate, a dialogue should be initiated on the following themes:

  – the current and future role of pharmaceutical therapies compared with other treatments available in the health-care system;

  – the role of biotechnology as a specific branch of R&D in the pharmaceutical industry.

Figure 4. Intensity of R&D co-operation within the incomplete Austrian "pharmaceutical" cluster

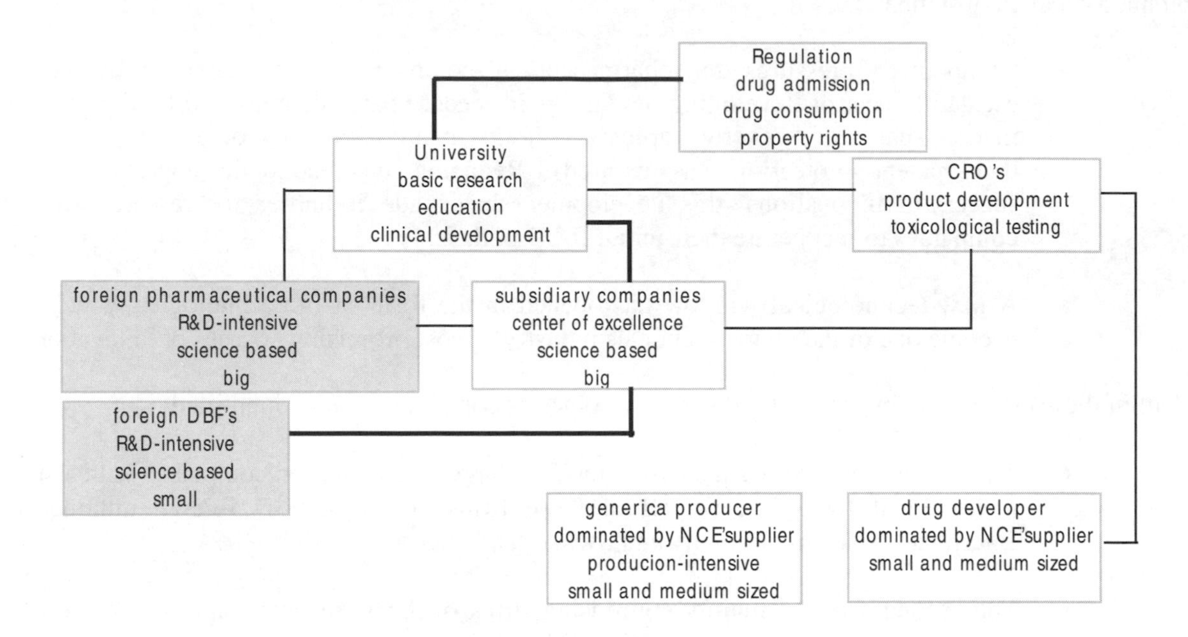

Note: CRC = Contract research organisations, DBF = Dedicated biotechnology firms, NCE = New chemical entities.
Source: Jörg *et al.*, 1995.

- Many researchers at universities and – to an even greater degree – in companies in the pharmaceuticals field are recruited from abroad. The hiring of international specialists should not be hindered by problems with residence and work permits.

- In Austria, the right to exploit patents from university research belongs to the public authorities rather than to the scientists or their departments. Offering scientists a share in these rights would generate a strong incentive to improve the patent rate achieved by university research.

- To increase the transfer of knowledge from universities to the business sector, a special focus on biotechnology and pharmaceuticals within a "seed-financing" programme was proposed along with complementary measures to enhance access to venture capital. The rationale for this focus stems from the fact that: *i)*a strong scientific competence base was identified in these fields; and *ii)* new spin-off firms could enhance the dynamic performance within the "cluster" well beyond the current incentives provided by existing co-operative activities.

- A specific marketing scheme to attract pharmaceutical and biotechnology companies is proposed. As a business location, Vienna offers considerable capacities for complementary research in biotechnology and pharmaceuticals.

6.4. Multimedia and cultural content (Warta et al., 1997)

At the present stage of development, it is too early to speak of an Austrian "multimedia" cluster. The investigation therefore concentrated on the question of whether and under what conditions a new cluster could emerge from the combination of multimedia technologies and cultural content. The study starts with the hypothesis that, in order to position itself successfully in "electronic space", Austria needs to bundle its resources under common objectives. The combination of multimedia and cultural content is perceived as an opportunity to furnish Austrian suppliers with a separate and internationally identifiable profile.

The empirical investigation included interviews with experts and a complementary survey among companies engaged in multimedia services and products. It is generally acknowledged that competitive advantages stem from the assumed "relative abundance" of cultural content as well as the geographic proximity between customers, suppliers and partners for co-operation. However, this clearly will not suffice to turn Austria into an attractive location for multimedia production, as long as the following barriers, identified in the questionnaire, remain prohibitively high: excessive telecommunication rates, inadequacies in the telecommunications infrastructure and, as a consequence, a lack of international networking in general. Additional problems are encountered by multimedia suppliers with regard to tapping sources for R&D support and venture capital.

A potentially powerful leverage for creating competitive advantage is the large pool of qualified labour. Unlimited and free access to universities, in addition to other factors such as the numerous cultural activities of international renown in Vienna, Salzburg, Graz and Linz, has produced a surplus of labour interested in working in the arts. This potential has so far been largely ignored because the labour market is not able to offer sufficient professional activities to complement such qualifications. Offering complementary education directed at multimedia technologies and skills may activate at least some of these human capital resources, at the same time offering employment to this abundant source of human capital.

Figure 5. Functional components within the "multimedia and cultural content" cluster

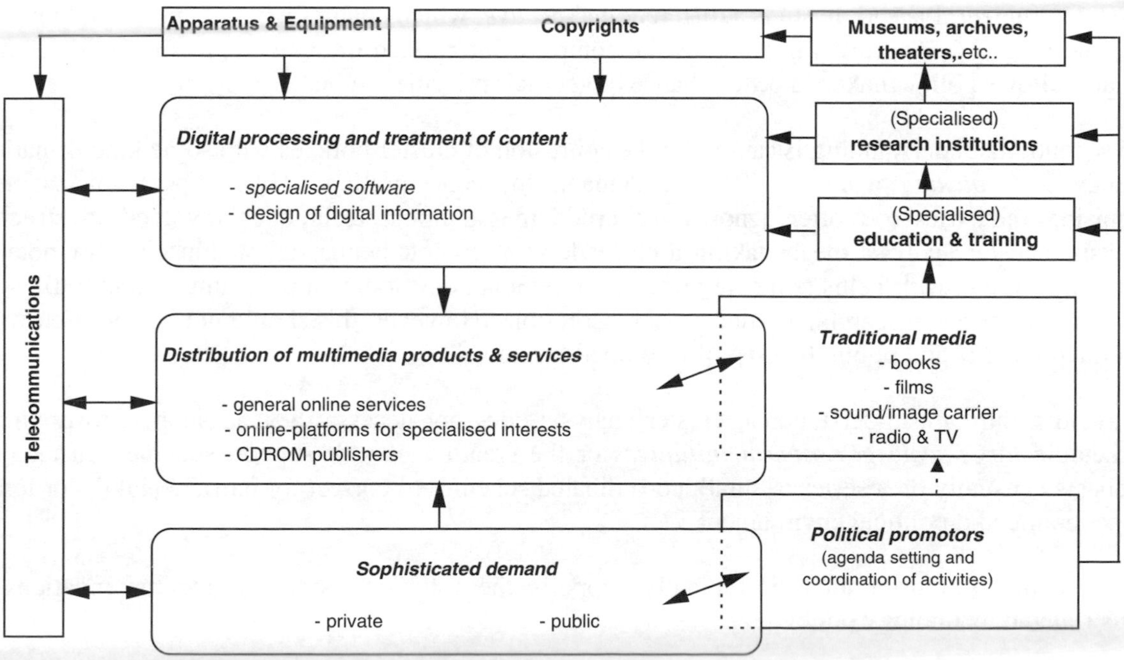

Source: Peneder, 1998.

The results of an analysis of the potential and preconditions for an industrial cluster which currently only exists in rather small fragments are necessarily more vague than for the other studies undertaken, where the targets were more visible. The basic message for policy makers can be summarised as follows.

Although its excellent supply of human resources for multimedia professions provides Austria with the main prerequisite of utilising existing market potential, it is not expected to realise that potential, given the current terms of reference. In particular, the continuing lack of competition in the telecommunications and TV markets is a substantial barrier to the generation of sophisticated demand for multimedia applications. Any policy that attempts to nurture the sector exogenously by subsidies and government contracts would necessarily fail, given the current situation.

The study also emphasizes the fact that these barriers to growth are parameters that can be shaped by policy. However, in the questionnaire, large economies of scale have been identified. This signals that the time factor plays a major role in the dynamic evolution of a new cluster: if the market starts to consolidate after the initial innovatory phase, the current opportunities for endogenous growth of an internationally competitive Austrian multimedia cluster could be lost within a few years.

7. Cluster-oriented policy

As noted in the Introduction, the notion of clusters and cluster policy became popular in the Austrian policy debate following the publication of Porter's work at the beginning of the 1990s. In sharp contrast to its popularity, its implications for policy are much less clear. The following points, mainly based on popular misconceptions, may help to summarise the major difficulties encountered in translating refined and generally appreciated analysis into operational policy.

The first misunderstanding is based on the erroneous belief that the cluster approach offers a *specific policy instrument*. Instead, it offers a different *perspective*, which enables a better understanding of the dynamic potential for growth in an industrial complex. Interpreted in this way, its major contribution is that it allows policy makers a better chance to get their priorities right.

The second misunderstanding is caused by the confusion of cluster policies with some kind of national strategy for *"picking winners"*. Popular demand for concentrating public support on the most promising industries too often ignores the crude measurement techniques involved in structural analysis. Cluster analysis means taking a closer look at complementarities within given or potential economic structures. It helps to define priorities in regional development programmes and to fine-tune the supply of public research and educational institutions. However, this should not be confused with a general "picking" and public funding of presumed "winners".

To avoid a third misunderstanding, cluster analysis does not necessarily lend support to politically marketable *large-scale government initiatives* or the creation of new support schemes. Instead, the emphasis is usually on a series of small, co-ordinated schemes to correct for barriers and distortions in the economic and political environment.

The outcome of cluster analysis typically supports the following kinds of recommendations for industrial and technology policy:

- improving the design of regulatory frameworks (*e.g.* through the elimination of regulatory distortions which tend to favour well-established industries to the detriment of new, and therefore smaller, innovative industries);

- filling institutional gaps by launching or supporting "Coasean" institutions which trim transaction costs through the provision of institutional platforms for experimentation and co-operation as public goods;

- creating complementary human capital through government spending on education;

- raising public awareness of potential opportunities, especially with regard to new technologies through the dissemination of relevant information;

- triggering demand-pull effects by means of public procurement (critical mass, high quality standards);

- setting priorities for focused R&D support schemes; and finally

- using the cluster notion as an instrument for the focused marketing of business locations, thus providing an internationally visible profile for potential investors and further strengthening the cluster by attracting an inflow of foreign direct investment.

While some of the above recommendations refer to the common set of innovation policy instruments, the final argument of "clusters" as a useful instrument for the *focused marketing of business locations* deserves special attention. The "Styria Automobile Cluster" is a particularly instructive example in this regard.[11] Building on high-performing local enterprises in motor technology and gear units, a successful agglomeration of related companies (covering assembling as well as special automotive components) has built up in this south-eastern province of Austria. The cluster currently groups more than 120 individual enterprises, research institutes and technical colleges. Using the cluster notion to create a "brand name" for the location proved to be a helpful instrument in attracting further foreign direct investment, *e.g.* for automotive parts and accessories. The cluster idea also proved useful in fostering co-operation among local suppliers, ranging from transport logistics to the new development of particular automotive components and, finally, to joint research on noise-reducing motor technologies.

A second illustration of the combination of traditional cluster motives such as the fostering of co-operation and knowledge spillovers, on the one hand, and the creation of marketable brands, on the other, is a recent initiative called "Austrian Water". This project comprises a number of firms engaged in different areas of water technology (supply, processing, purification, electricity production). Based on the presumed high level of Austrian water technology, the prime target is to co-ordinate the activities of over 20 individual, highly specialised firms with a particular focus on export promotion and the ability to provide complete systems for the international market.

Cluster-oriented policy in Austria has also played an important role in the creation of a new type of technical college, specialising in a range of activities from information and communication technologies, through engineering skills or training for wood manufacturers to the professional management of tourism. The impressive growth of these *Fachhochschulen* over the past few years has served to strengthen the links between the educational system, on the one hand, and the requirements of local businesses for specialised skills, on the other.

Without doubt, the most ambitious plan for translating the cluster idea into new policy is the "Competence Centre Programme", publicly announced by the federal government in December 1997. Still in its pilot phase, the target of this initiative is to improve the interaction between academic research and private industrial R&D through the establishment of collaborative research institutions, called "competence centres". Set up as independent legal entities, they will be selected by competitive tender based on the double criteria of scientific excellence and economic relevance. Funding by federal and other public bodies can cover up to 60% of the budgets, while a minimum of five strategic private partners must cover at least 40% of total expenditure.[12]

Another example of Austrian cluster-oriented policies is an initiative of the Austrian Industrial Research Promotion Fund focusing on the industrial wood processing complex. In the telecommunications field, the cluster approach with its emphasis on systemic feedback mechanisms served to strengthen the call for liberalisation and abolition of the many regulatory distortions that handicap new services and technologies. In addition to these specific examples, general awareness of the cluster idea has considerably grown over the past few years, fuelling and partly steering the many less spectacular day-to-day decisions affecting industrial and innovation policy.

NOTES

1. See, for example, Peneder (1994).

2. See, for example, Hutschenreiter (1994).

3. "When an industry has thus chosen a locality for itself, it is likely to stay there long: so great are the advantages which people following the same skilled trade get from near neighbourhood to one another. The mysteries of trade become no mysteries; but are as it were in the air, and children learn many of them unconsciously. Good work is rightly appreciated, inventions and improvements in machinery, in process and the general organisation of the business have their merits promptly discussed: if one man starts a new idea, it is taken up by others and combined with suggestions of their own; and thus it becomes the source of further ideas." (Marshall, 1920, IV, x, 3)

4. "Again, the economic use of expensive machinery can sometimes be attained in a very high degree in a district in which there is a large aggregate production of the same kind, even though no individual capital employed in the trade be very large. For subsidiary industries devoting themselves each to one small branch of the process of production, and working it for a great many of their neighbours, are able to keep in constant use machinery of the most highly specialised character, and to make it pay its expenses, though its original cost may have been high, and its rate of depreciation very rapid." (Marshall, 1920, IV, x, 3)

5. "Again, in all but the earliest stages of development a localised industry gains a great advantage from the fact that it offers a constant market for skill. Employers are apt to resort to any place where they are likely to find a good choice of workers with the special skill which they require; while men seeking employment naturally go to places where there are many employers who need such skill as theirs and where therefore it is likely to find a good market." (Marshall, 1920, IV, x, 3)

6. "On the other hand a localised industry has some disadvantage as a market for labour if the work done in it is chiefly of one kind, such for instance as can be done only by strong men. ..[T]he remedy.. is found in the growth in the same neighbourhood of industries of a supplementary character." (Marshall, 1920, IV, x, 3)

7. "A district which is dependent chiefly on one industry is liable to extreme depression, in case of a falling-off in the demand for its produce, or of a failure in the supply of the raw material which it uses. This evil again is in a great measure avoided by those large towns or large industrial districts in which several distinct industries are strongly developed. If one of them fails for a time, the others are likely to support it indirectly; and they enable local shopkeepers to continue their assistance to workpeople in it." (Marshall, 1920, IV, x, 4)

8. See, for example, Sharma (1996).

9. For the current purpose, Figure 2 can only illustrate how the screening process works. A full list of all the 208 industries and their respective cluster membership, which would identify their position on the map, is available in a separate Working Paper (Peneder, 1995).

10. Krugman (1995).

11. See, for example, Steiner *et al.* (1996).

12. Detailed and continuously updated information is provided at http://www.bmwf.gv.at/4fte/k-plus/ausw.htm.

REFERENCES

Bayer, K., M. Peneder, F. Ohler and W. Polt (1993), "Zwischen Rohstoff und Finalprodukt: Die wirtschaftliche Wettbewerbsfähigkeit des Wirtschaftsbereiches Holz-Papier", TIP, Vienna.

Fowler, H.W. (ed.) (1982), *The Concise Oxford Dictionary of Current English* (7th ed.), Oxford.

Hutschenreiter, G. (1994), "Cluster innovativer Aktivitäten in der österreichischen Wirtschaft", TIP, Vienna.

Hutschenreiter, G., S. Kaniovski and K. Kratena (1998), "Embodied Technology Diffusion in the Austrian Economy", mimeo, TIP, Vienna.

Hutschenreiter, G. and M. Peneder (1994), "Ziele und Methoden der Clusteranalyse wirtschaftlicher und innovativer Aktivitäten", *WIFO-Monatsberichte*, Vol. 67, No. 11, pp. 617-623.

Jaffe, A.B. (1986), "Technological Opportunity and Spillovers of R&D: Evidence from Firms' Patents, Profits, and Market Value", *American Economic Review*, Vol. 76, No. 5, pp. 984-1001.

Jaffe, A.B. (1989), "Characterizing the 'Technological Position' of Firms, with Application to Quantifying Technological Opportunity and Research Spillovers", *Research Policy* 18, pp. 87-97.

Jörg, L., K. Bayer, G. Hutschenreiter, and W. Polt (1995), "Spezialisierung und Diversität. Die wirtschaftliche und technologische Wettbewerbsfähigkeit der österreichischen pharmazeutischen Industrie im internationalen Umfeld", TIP, Vienna.

Knoll, N. (1998), "Sectoral Analyses and Case Studies on the Diffusion and Utilisation of ICT in Austria: Banking, Mechanical Engineering, Textiles and Clothing", mimeo, WIFO.

Krugman, P. (1991), *Geography and Trade*, MIT, Cambridge, Massachusetts.

Krugman, P. (1995), *Development, Geography and Economic Theory*, MIT, Cambridge, Massachusetts.

Leo, H., M. Peneder, N. Knoll, F. Ohler and M. Latzer (1994), "Telekommunikation im Umbruch. Innovation – Regulierung – Wettbewerb", TIP, Vienna.

Marshall, A. (1920), *Principles of Economics* (8th ed.), London.

Peneder, M. (1994), "Clusteranalyse und sektorale Wettbewerbsfähigkeit der österreichischen Industrie", WIFO, Vienna.

Peneder, M. (1995), "Cluster Techniques as a Method to Analyse Industrial Competitiveness", International Advances in Economic Research (IAER), Vol. 1, No. 3, pp. 295-303 (also available with extended tables as WIFO Working Paper No. 80).

Peneder, M., (1995), "Technologiepolitische Herausforderungen in der Telekommunikation", *WIFO-Monatsberichte*, 1995, 6, pp. 435-442.

Peneder, M. (1998), "Industrial Location and Sectoral Specialisation. An Extension to Krugman's Model of Cluster Formation by Pooled Labour Markets", paper presented at the 25th EARIE Conference, Copenhagen, 27-30 August.

Peneder, M. (1998), "Evolutionäre Ökonomie und Clusterbildung. Dargestellt am Beispiel Multimedia", *Wirtschaftspolitische Blätter*, 2-3, pp. 160-167.

Porter, M. (1990), *The Competitive Advantage of Nations*, New York.

Sharma, S. (1996), *Applied Multivariate Techniques*, New York.

Steiner, M., T. Jud, A. Pöschl and D. Sturn (1996), *Technologiepolitisches Konzept Steiermark*, Graz.

Warta, K., N. Knoll and M. Peneder (1997), "Multimedia, Kultur und Konvergenz: Perspektiven einer Clusterbildung in Österreich", TIP, Vienna.

Chapter 15

FINNISH CLUSTER STUDIES AND NEW INDUSTRIAL POLICY MAKING

by

Petri Rouvinen and Pekka Ylä-Anttila
The Research Institute of the Finnish Economy (ETLA), Helsinki

1. Introduction

This chapter presents the "meso" clusters identified in the Finnish economy, discusses recent cluster developments and sheds some new light on "cluster-based" industrial policy making in Finland.

1.1. Why did we need something new?

Although Porter's (1990) book aroused considerable discussion in Finland, it was some time before cluster analysis was recognised as being of possible use in investigating the origins of competitive advantage and in redesigning industrial and technology policies. Despite the fact that Finland was a late starter in cluster studies, the intermediate studies and the final report (Hernesniemi *et al.*, 1995) have, nevertheless, had a profound effect on the Finnish economy and on Finnish policy making. In the early 1990s, new ideas were in great demand as it became increasingly evident that macroeconomic policies were simply unable to cope with the profound structural changes taking place in the economy.

"Advantage Finland", the Finnish cluster study co-ordinated by ETLA (The Research Institute of the Finnish Economy), was from the outset designed to examine the future prospects for industry and to produce information essential for a reshaping of industrial policy. The preliminary results of this work were exploited in 1992–93 when the Ministry of Trade and Industry prepared new policy guidelines. The *National Industrial Strategy* (Ministry of Trade, 1993) was well received upon its publication.[1]

Much of the policy discussion, as well as policy making since 1993 has been based on the guidelines outlined in the *Industrial Strategy*. There has been a clear shift in policy thinking away from *old-style policies* – subsidising ailing industries, restricting competition, sheltering strategic industries, and/or backing national champions – towards *new policies* providing favourable framework conditions and promoting the better functioning of markets.

European integration and increased global competition have highlighted the importance of national competitive advantage. Any deterioration in competitive advantage is rapidly felt through financial markets and investment flows. In this new "borderless" world, capital is more responsive to changes in operational environment and firms are constantly seeking better locations to earn returns on firm-specific knowledge and capital investments.

Although it is firms that compete on the marketplace, national and local governments have a major role in creating attractive locations for business. Indeed, the policy emphasis, in Finland as in most other developed countries, has shifted away from macro policies towards industrial and technology – or competitiveness – policies. Cluster analysis has clarified the respective roles of private enterprises and the public sector and has shown that, despite some differences, the interests of the two run largely parallel.

Why is cluster analysis such a useful policy tool?

- The cluster concept shifts attention away from individual industries and the outcomes of the competitive process, and focuses instead on linkages between industries and firms and on the pre-conditions for competitiveness, *i.e.* upgrading the pool of production factors and enhancing competition. Thus, policy making is given a *positive role*. Cluster analysis emphasizes productivity as a key source of firm competitiveness and economic growth. The clusters system brings about reductions in transaction costs, boosts efficiency and, above all, creates positive external economies due to information and knowledge flows among firms and industries.

- The cluster concept brings together researchers, policy makers and the business community. It has opened a new dialogue between these three major actors, thanks to both the positive view offered by the cluster approach and the familiarity of the concepts used in the analysis.

- The diamond framework helps to define the roles of the different actors in improving the competitiveness of economies. This point is clarified in Section 4.

1.2. The context of the Finnish cluster studies – a clear sense of urgency

In the early 1990s, the Finnish economy was in the middle of its deepest recession of the century. GDP fell by more than 10% in 1991–93. The unemployment rate was rocketing; by 1994 it had risen to almost 20%. The economy was in a deep slump, accompanied by a severe structural crisis. The slump revealed weaknesses in production and export structures. The open sector proved to be simply too small and insufficiently competitive to support the prevailing standard of living and continuous growth. There was an urgent need to study competitiveness and its origins, to predict the factors that would lead to future competitive edge and industrial structure. The Finnish cluster study responded to these needs.

It is our belief that the project contributed in many ways to the rapid structural changes that took place in the 1990s (Steinbock, 1998). The results of the research programme had an impact not only on industrial and technology policies, but also on science, educational and regional policies (Helander, 1997). Many of the project reports have since been used as text books in universities and other educational institutions. The results have also benefited business firms in their strategic decision making.

2. Using statistical analysis and expert opinions to define clusters

In the definition of Finnish clusters, we attempted to remain as faithful as possible to Porter's (1990) original work. We also studied carefully the, considerable, "Porter critique" literature in order to overcome some of the approach's shortcomings.[2]

Substantial and sustained exports to a range of countries and/or significant outward foreign investment were taken as signs of international competitive advantage of an industry (Porter, 1990, p. 25). However, since comprehensive and detailed statistics are only available for the former, our initial statistical analysis relied heavily on export statistics.

Following Porter, we evaluated "narrowly defined" industries[3] at three points in time: 1980, 1985 and 1990. This analysis gave us a list of potentially interesting branches.[4] We combined these into somewhat larger groups based on our prior knowledge of industries' operating logic, to obtain a number of internationally competitive industries or "cluster skeletons".

For each of these "skeletons", we drew a cluster map, and met with experts in the particular field. With their help, we evaluated whether the suggested entity was indeed a reasonable starting-point for analysis and outlined the characteristics of the potential cluster. Once we had a good idea of the entity in question, we undertook the tedious task of studying each actor in the cluster in detail, with a particular focus on its interlinkages with other actors in the operating environment. Throughout the analysis, we continuously re-evaluated our cluster definitions – on occasion new interlinkages were revealed and sometimes our initial thoughts on the cluster's structure were proved to be wrong.

As of 1992, the strictest definition of clusters would have left only the forestry cluster to be studied. Furthermore, we also wanted to include a few less-successful clusters as a "control group" and to study a number of clusters that were considered to have particularly bright future prospects. Therefore, we supplemented Porter's approach with a classification of clusters according to their relative strengths. Clusters were classified as "strong", "fairly strong", "potential" or "latent (defensive)".[5]

Compared to Porter's work, our clusters were somewhat more broadly defined. In our opinion, the Porter method laid too little emphasis on international aspects in the case of a small open economy – our desire for a more global perspective led us to include *international business activities* (IBA) as the third outside force of the diamond. Although the initial cluster definition was largely based on statistical evidence, our method was essentially case-driven. The following section presents the clusters identified and reviews a few illustrative examples in detail.

3. Finnish clusters

Table 1 lists the clusters identified in the "Advantage" project. Only one of the clusters, forestry, was characterised as "strong". Note that in 1992, telecommunications were still included in the "potential clusters" group. The case of the telecommunication cluster illustrates one of the shortcomings of Porter's methodology: had we solely relied on our original export data of 1980, 1985 and 1990, ignoring expert opinions and our own intuition, this cluster would not have been included.[6]

Table 1. Finnish industrial clusters: development and growth prospects

Type	Cluster	Export value in 1996 (billion USD)	Average annual growth in exports 1980-94 (%)	Average annual growth potential up to 2010 (%)
Strong clusters	Forestry	13.6	3.5	3.0
Fairly strong clusters	Base metals	3.7	8.5	6.0
	Energy	2.4	6.0	7.0
Potential clusters	Telecommunications	4.1	13.5	15.0
	Environment	--[1]	--[1]	10.0
	Well-being	0.7	7.0	10.0
	Transport	3.0[2]	2.5	4.0
	Chemicals	4.0	6.0	4.0
Latent or defensive clusters	Construction	5.5	7.0	2.0
	Foodstuffs	7.0	7.0	1.0

1. Hard to quantify; often embodies in other clusters' output.
2. Estimate.
3. This cluster has a 50% overlap with the forestry cluster, and a 25% overlap with the base metals cluster.

Source: Hernesniemi *et al.;* 1995; Mäkinen, 1998, and author's own estimates.

Despite the fact that the forestry cluster is characterised as "strong", *i.e.* it has identifiable strengths in all corners of the diamond, none of the Finnish clusters is fully developed. In fact, a cluster hopefully never reaches its final form – because if it does, it loses its ability to adapt to changes in its operating environment and is thus doomed to failure.

In our opinion, an understanding of the dynamic nature of the cluster concept is essential. The development of a cluster from a few companies to a major branch can take decades. Cluster development requires strengths in all the corners of the diamond. The Finnish forestry cluster has existed for over 500 years, but has only recently entered the innovation-driven stage. Exports by the telecommunications cluster have expanded during the last decade, but the Finnish root – competitive telephone service – is over 100 years old.

Table 2 lists past, current, and what we see as future, sources of competitive advantage of the Finnish clusters. The example of the forestry cluster illustrates that, historically speaking, comparative advantages in factor conditions are often the initial impetus behind the development of a cluster. The development of the energy cluster is an example of how selective disadvantage may contribute to a cluster's success in the long run.

In the fastest growing industries today, however, the created and advanced factors are the sources of competitive strength. This shows particularly in the growth of the telecommunications cluster.

Table 2. Main sources of competitive advantage of the Finnish clusters

Cluster	Previous sources of competitive advantage	Current sources of competitive advantage	Predicted future sources of competitive advantage
Forestry	*Factor conditions:* abundant raw material (wood), waterways suitable for transportation.	*Related industries:* national innovation system supporting the industry.	*Related industries:* national innovation system. *Strategy:* customer orientation.
Base metals	*Factor conditions:* ore deposits, demanding conditions in mining and metallurgy.	*Strategy:* specialisation, process and logistics know-how.	*Demand:* potential in regional markets. *Factor conditions:* potential of Russian ore deposits.
Energy	*Factor conditions:* harsh climate, lack of domestic energy sources. *Demand:* needs of the two clusters listed above. *Competition* in power generation.	*Factor conditions:* technological know-how. *Strategy:* product integration.	*Strategy:* exports of services, increasing service content of existing products.
Telecommunications	*Competition* in operation. *Government:* Scandinavia-wide Nordic Mobile Telephone (NMT) standard.	*Competition* in operation (mobile in particular). *Factor conditions:* supply of skilled labour.	*Related industries:* applications in health care, integration of telephones, TV and PC, etc.
Environment	*Demand:* the needs of the Finnish processing industry (end-of-pipe technologies).	*Demand:* the needs of the Finnish processing industry – environmentally friendly production technologies.	*Demand:* the needs of the Finnish processing industry – environmentally friendly consumer products.
Well-being	*Factor conditions:* combining medical and technological know-how.	*Factor conditions:* know-how.	*Demand:* responding to the "double ageing" problem while curbing health-care costs.
Transport	*Demand:* Russian need for vessels, domestic need for ice-breakers.	*Factor conditions:* project management. *Related industries:* specialised supplies for components of luxury vessels. *Demand:* transit.	*Strategy:* making Finland into a logistics hub between East and West.
Chemicals	*Demand:* domestic needs in process industries.	*Factor conditions:* processing know-how. *Demand:* needs of domestic processing industry.	*Strategy:* environmental orientation, customer orientation.
Construction	*Strategy:* standardised production. *Factor conditions:* excellence in architecture.	--	*Strategy:* customer orientation. *Related industries:* environmentally friendly construction.
Foodstuffs	--	*Demand:* expanding regional markets.	Increased *competition.*

Source: Hernesniemi *et al.*, 1995, slightly modified.

New branches are often spin-offs of existing ones – thus, in the policy section, we suggest building on existing strengths. This does not mean that one should not attempt to discover new businesses – we simply suggest that new businesses may be founded near existing "centres of excellence". For instance, there is great potential for exploiting telecommunications cluster expertise in health care.

In the early stages of an industry, the pioneers' personal qualities are crucial – there is an inherent randomness in where new growth takes place. The current success of Nokia is, in part, the result of the "stubbornness" of a few key managers, who for years were willing to pour money to a sector that at times must have seem like a black hole (Lemola and Lovio, 1997). Despite the random nature of cluster development, the public sector can play a role in guaranteeing the pre-conditions for entrepreneurial activity, thus increasing the likelihood that these "accidents" will happen.

Recent statistics (Table 1 above) show that, with exports of USD 13.6 billion in 1996, the forestry cluster is the biggest in Finland. In the past decade or so, however, telecommunications is the only cluster to sport double-digit average annual growth rates. In 1998, exports of telecommunication equipment were larger than Finnish exports of paper and, among the OECD countries Finland and Sweden are the most export-specialised in this particular branch.[7] The telecommunications cluster is likely to have the brightest growth prospects in a ten-year horizon.

In the following section, we present three Finnish clusters as illustrative examples. While all three are fairly innovative, the characteristics of their knowledge creation and upgrading processes are quite different. The forestry cluster has accumulated an enormous amount of tacit knowledge over hundreds of years. Products markets are, however, mature and are characterised by "incremental" product and process innovations. Internationally, the Finnish patenting intensity is unparalleled, but on an interindustry basis, it is nevertheless modest. There is much more turmoil in the fairly mature energy technology markets. Furthermore, this cluster is quite heterogeneous, and some segments are only now emerging. Thus there is plenty of room for innovation. The Finns have been at the cutting edge, *e.g.* in combustion technologies. The telecommunications cluster has the most rapidly evolving markets and, in some senses, the highest rate of innovation. Product cycles are short and new technological possibilities emerge almost daily. Nokia alone accounts for roughly one-third of the country's total industrial expenditure on R&D. According to Vuori (1997), this cluster (or what she calls the ICT cluster) accounts for around half of the interindustry R&D spillovers in the Finnish manufacturing industry.

The three clusters also shed light on the quite different roles the public sector has played. Policies related to the forestry cluster had some of the characteristics of the *old-style* policy making (see above). Since the pulp and paper industry was the single most important industry in the country for decades, its competitiveness was backed by an occasional devaluation. Other characteristics of the old policies were also used, along with more positive technological and educational policies. Although the policy stance has since changed, some implications of the old policy practices can still be seen in the structure of the branch.

As producers of infrastructure services, the energy and telecommunications clusters have been subject to government regulation in all industrialised countries. In Finland, however, these clusters have been less regulated and are more open to international competition than most of their counterparts abroad. This is one of the key explanations for the good performance of these clusters. These two examples show that an appropriate combination of competition and regulation can be an important source of competitive advantage.

Overall, the Finnish telecommunications (or ICT) cluster is a shining example of *positive* policy making. The public sector has taken an active role in setting standards and acting as a demanding customer, as well as creating and improving framework conditions by investing in R&D and education. The branch is one of the focal points of science and technology policy.

3.1. *The forestry cluster – 500 years old and still swinging*

The forestry cluster evolves around the conversion of wood. We thus contradict Porter by arguing that a cluster can indeed be based on a shared resource base.

The cluster originated in wood-tar burning over 500 years ago, although currently sophisticated printing papers are at centre-stage. Development of forest industry machinery and equipment has been carried out near or within the parent branch. The majority of forestry industry chemicals were imported until the 1970s; since then domestic companies have gained a foothold.

Today, strength in the core products of the cluster, *i.e.* pulp, paper, paperboard and sawn wood, is accompanied by perhaps even stronger Finnish presence in virtually all related machinery and equipment segments. Finnish companies frequently provide related inputs, such as consulting, power generation, automation, etc. Universities and research organisations are important components of the industrial network. Strong interactions among the participants have made the forestry cluster a prosperous one.

The forestry cluster has continuously invested in state-of-the-art production facilities – possibly to an excessive extent. Strategically, the goal has been to increase the value-added content of production. Close interaction within the cluster is most explicit between paper mills and engineering workshops. The bond between the mechanical forestry industry and its suppliers is somewhat weaker.

The greatest pressures within the cluster are faced by the pulp and paper industry: deciding the location of future production plants is problematic, and the usage of recycled fibre may reshape the competitive arena.

Customers in the fine paper market often require fast shipments of small amounts of tailored products. This adds to the already high logistics costs faced by Finnish manufacturers. One solution could be to deliver domestically produced pulp to paper factories in Central Europe.

Relatively speaking, the weakest link in the forest cluster diamond is the connection to the end-user. This may be partly due to geographical distance.

The technological superiority of the Finnish forestry cluster supports production in Finland. The existing human capital could be further fine-tuned by investing in university education and by clarifying the missions of the various educational and research units. The maturing-forest-related chemicals industry promotes the competitiveness of the Finnish forestry cluster.

The key question in the mechanical forestry industry is the value-added content of the products. The share of price-sensitive bulk products should be reduced, while market and delivery channels need to be trimmed.

Finnish paper companies have been among the country's forerunners in outward foreign direct investment. Currently, they own production units in all major markets and one-third of production is abroad. Finnish paper companies have attracted considerable portfolio investment – for example, 49% of UPM-Kymmene's stock is held by foreigners (The Finnish Central Securities Depository, (15 September 1998). In particular, the forestry-related chemicals business has attracted considerable foreign direct investment. There are some signs that foreign companies in the field want to be present in Finland in order to benefit from the dynamism of the cluster.

Figure 1. The diamond of the Finnish forestry cluster

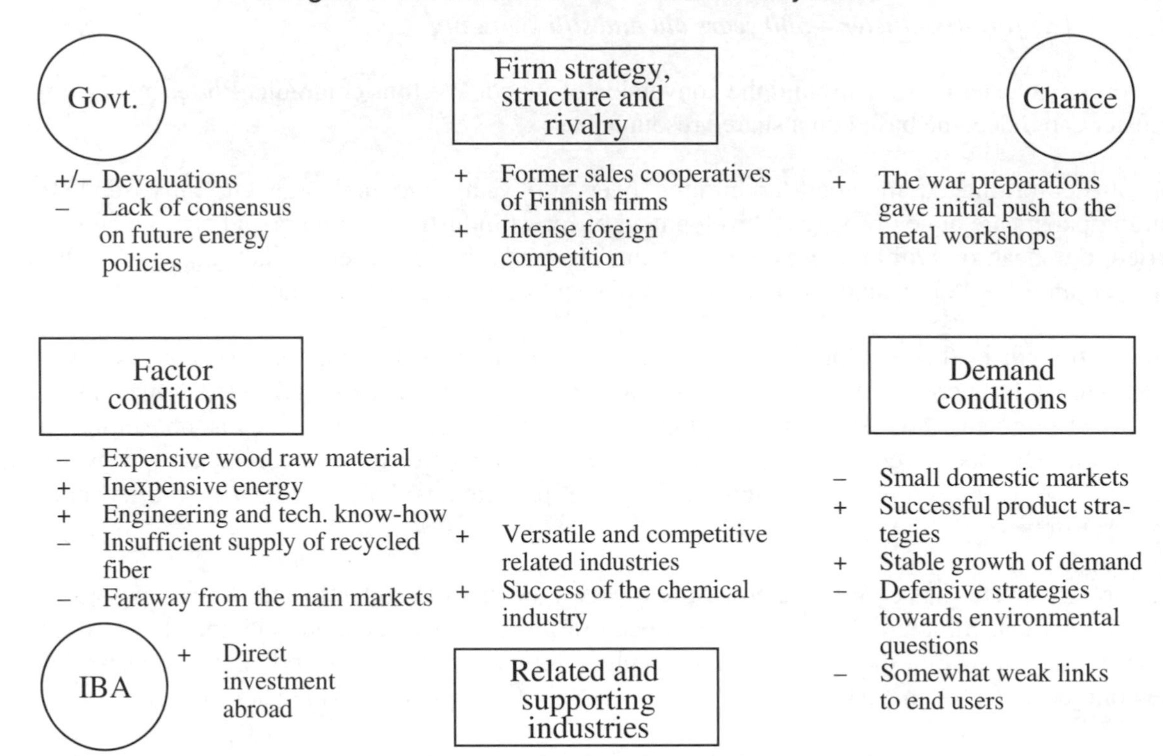

Source: Hernesniemi *et al.*, 1995, slightly modified.

Although the perspective of the "Advantage Finland" project was considerably more international than that of Porter, in retrospect it seems that the global aspects were nevertheless understated. There is no reason for a cluster to respect national borders and today the forest cluster may be seen as a Scandinavian rather than a Finnish phenomenon.[8] Markku Lammi and Colin Hazley of ETLA are currently studying the European forestry industry using the cluster methodology.

3.2. The energy cluster – from scarcity of energy to exports of technology

Historically, the two main clusters in Finland, forestry and metals, are fairly energy-intensive. This, the arctic climate, long distances and a low population density are factors underlying Finland's high per capita energy consumption. With the exception of some hydropower, Finland lacks domestic energy sources. This has led to great desire to develop a low-cost and efficient source of energy. The Finns have been largely successful in their efforts here and domestic energy prices are extremely competitive.

The utility companies and large-scale energy users have been central to the development of the Finnish energy cluster. In developing the Finnish power generation and distribution system, many unique technological solutions had to be found to overcome the harsh conditions and limited financial resources. Domestic demand encouraged the growth of Finnish equipment suppliers.

The equipment used in energy production and distribution was identified as central to the energy technology cluster – thus the definition of the cluster is somewhat more "Porterian" than that of the forestry cluster.

Unlike in some other countries, the demand for electric lighting and other electrical appliances was not the main driving force behind technological development in Finland. In fact, it was the main industries, forestry and metals, that set the strict requirements which the power companies struggled to fulfil. The power producers turned to domestic metal workshops for equipment. In the field of energy technology, this co-operation in minimising expenditures on energy has engendered one of the most efficient and technologically progressive energy systems in the world.

Energy technology provides an example of fast technological adaptation and diffusion, followed by innovation. The cluster is relatively young. In the early years of the century, most equipment was imported. Later, it was manufactured in Finland under foreign licence; not until the 1970s were there significant exports based on own R&D. The energy cluster, as defined here, is rather heterogeneous but in our opinion the suggested definition is nevertheless justified by the close interlinkages among participants. At the turn of the decade, Finland was less specialised in exports of energy technology than were the OECD countries on average. Today, Finland is one the most export-specialised OECD countries in this field. Even so, the Finns are rarely active in the biggest segments – rather they tend to carve niche markets in which they seek worldwide dominance.

The case of Strömberg (currently a subsidiary of ABB) shows how the right owner, regardless of nationality, can promote a business by providing marketing and distribution networks and efficiently allocating activities across the globe. As other major players are foreign controlled, its fair to say that, of the branches studied here, the energy cluster is the most influenced by foreign ownership.

Figure 2. The diamond of the Finnish energy cluster

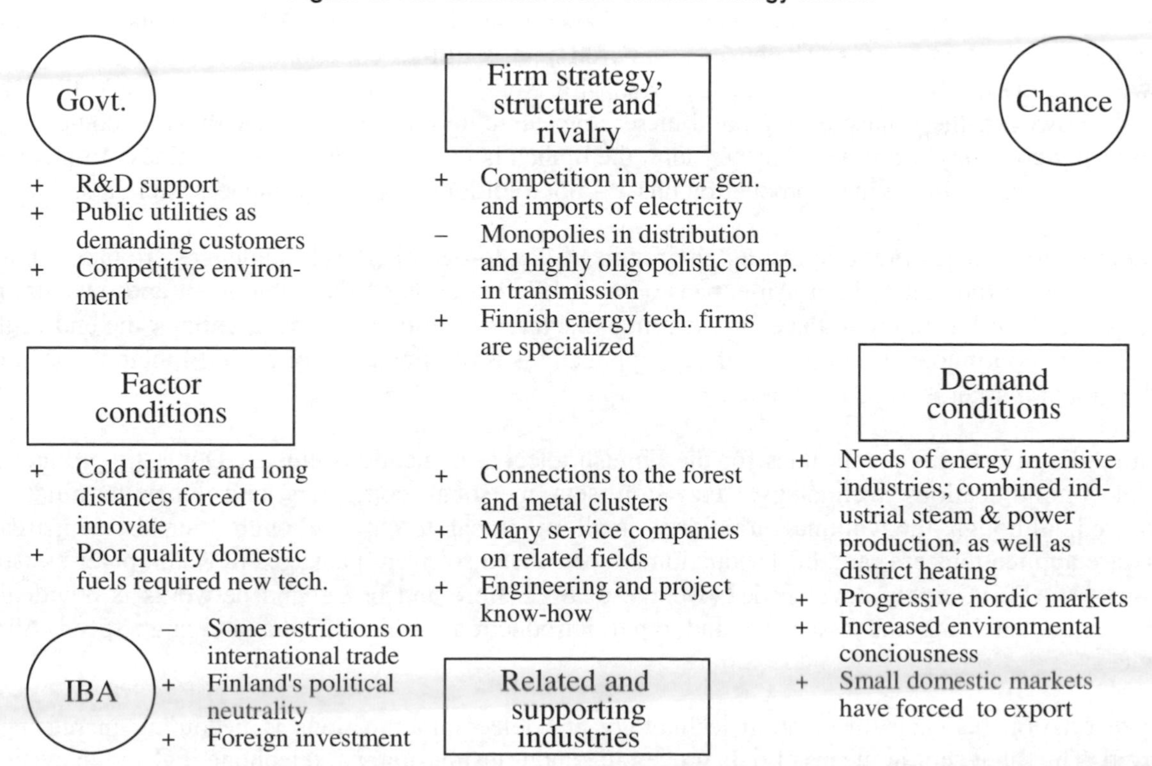

Source: Hernesniemi *et al.*, 1995, slightly modified.

Energy technology is a politically sensitive subject. Thus, implicit trade restrictions remain, although the trend is towards more open competition. A wave of mergers and acquisitions has been going on in the industry for quite some time, and maintaining a competitive environment should be high on the political agenda.

As discussed above, our analysis concentrated on energy technology. Esa Viitamo and Hannu Hernesniemi of ETLA are currently considering a broader cluster definition, with energy raw materials, power generation and distribution, etc., at one end of the spectrum, and energy usage, at the other.

3.3. The telecommunications cluster – making all the right moves

The telecommunications cluster is the first significant Finnish industrial cluster in which raw materials are less important while know-how plays a central role. The strengthening of the telecommunications cluster has been supported by phenomenal growth in global demand and trade that is expanding, on average, faster than production.

The key products of the telecommunications cluster are defined as equipment manufacturing,[9] operation,[10] and value-added network services (VANS).[11] This somewhat unorthodox definition was favoured because of the high and increasing service content of physical equipment and the myriad roles of the cluster participants.

The role of human capital in the development of the cluster can hardly be overstated. The cluster has indeed experienced some growing pains as, despite significant private and public efforts, qualified labour has been in short supply.

Business in the telecommunications sector is characterised by huge R&D investments and rapidly evolving product generations. The business environment is further complicated by the lack of global network standards. So far, Finnish companies, with Nokia as the locomotive, have made the right moves. However, the stakes are high and increasing; jumping on the wrong bandwagon could easily change any company's fortunes. Furthermore, the branch is moving away from a business-to-business and pioneer market into a mass production phase – not traditionally one of Finland's strengths.

In addition to changes in the operating environment, day-to-day operative challenges are taking place at both ends of the value chain. An expansion and differentiation of the component and outsourcing sectors could further improve the equipment manufacturers' competitive edge, while at the end of the value chain, continuous development of new applications is of utmost importance if Finland's position in the global market is to be maintained.

Multimedia offers vast possibilities for the Finnish telecommunications cluster. Domestic production of telecommunications technology, television sets, personal computers and other terminals is advanced, although the volumes are fairly small in global terms. Although Finnish mainstream software applications are rare, the production of specialised software packages (*e.g.* computer assisted design – CAD) is highly developed. Coverage of optic fibre and broadband networks is practically nation-wide, enabling the extensive and rapid introduction of multimedia and high-speed ATM services.

The distinction between information technology and telecommunications is becoming increasingly blurred. The three central means of data processing and transmission, *i.e.* telephone, PC and television, are converging, offering tremendous opportunities in applications. Exploiting these possibilities requires knowledge in all the respective fields. Thus, the building of suitable industrial networks through mergers and acquisitions, joint projects, co-operation and partnerships is crucial for future success.

In addition to investing in the multimedia market, there is vast potential in more extensive applications – education, inter-organisational communication and health care are examples of domains that could benefit greatly from new telecommunications applications. Finnish industries, such as forestry and basic metals, are already taking advantage of highly developed intelligent machines – products that could also become successful export items.

Figure 3. The diamond of the Finnish telecommunications cluster

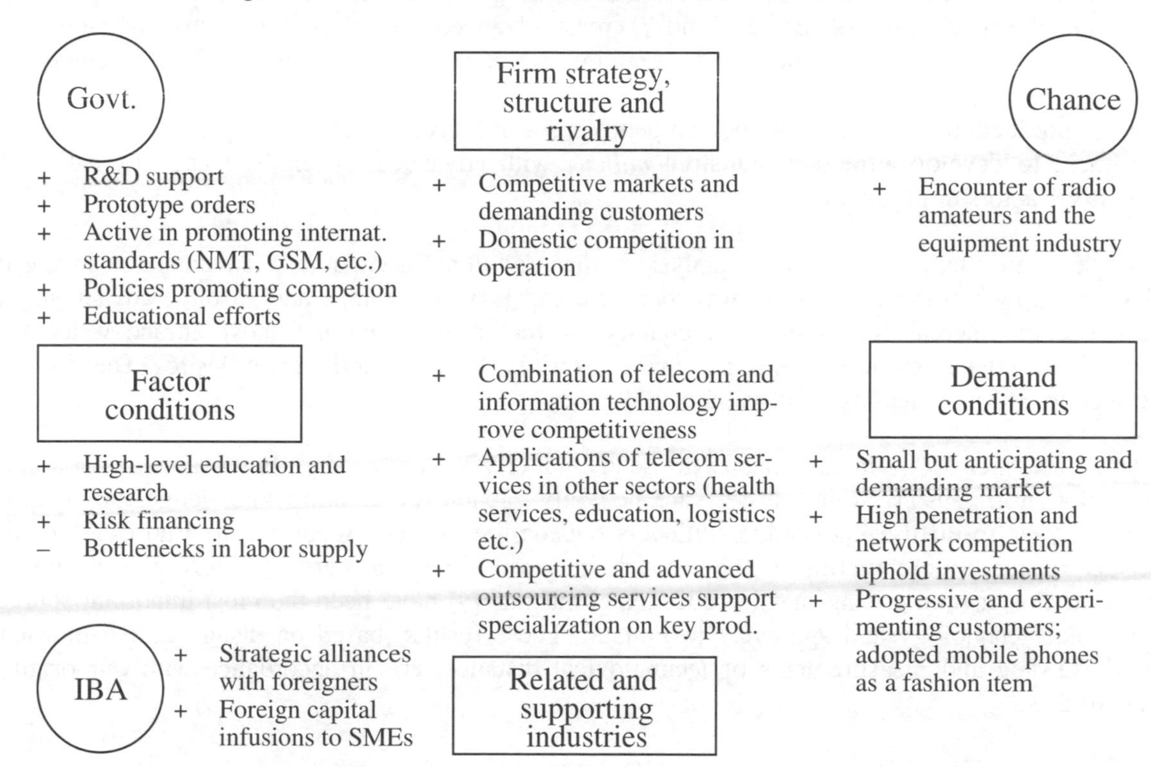

Source: Hernesniemi *et al.*, 1995, slightly modified.

Based on her ongoing work, Heli Koski of ETLA suggests considering the information and communication technology (ICT) cluster, rather than the telecommunications cluster, in order to capture the changing marketplace dynamics brought about by technology convergence.

4. New industrial policy making

4.1. Changing patterns of industrial and technology policies

Industrial and technology policy making increasingly recognises the systemic nature of innovation and economic growth. This is reflected in several related system approaches to competitiveness and policy analysis. Smith (1995) distinguishes three main approaches: *i)* technological systems; *ii)* national systems of innovation; and *iii)* industrial clusters (see also Vuori, 1997; Vartia and Ylä-Anttila, 1996). All three approaches aim at modelling interaction among system participants and recognise the importance of externalities. All three have been considered in Finland, although the cluster approach has been most widely applied as a practical policy tool.[12]

In addition to the systemic nature of technological change, globalisation has been a focal points in policy thinking. In a world where productive assets move freely, the basic policy issue is how to make a country or a region attractive to internationally competitive firms. This issue is in accordance with, and can be inferred from, cluster analysis. This was the other major starting-point for the design of a new industrial policy for Finland (Ministry of Trade and Industry, 1993a).

However, subsidies and compensatory policies are not appropriate tools for increasing the attractiveness of certain locations. Instead, the major goals of the new type of policy are to: *i)* guarantee the functioning of markets; and *ii)* create advanced and specialised factors of production and sustain high-level technological and social infrastructures. It is recognised that, together with increased specialisation and product differentiation, it is the firm-specific capabilities and created factors of production that determine the competitiveness of a given country or region. It is the task of government to develop attractive industrial *milieux* with advanced, specialised and internationally competitive factors of production.

One of the main messages of cluster analysis is that traditional comparative advantage is losing its explanatory power at the expense of firm-specific competitive advantage and absolute advantage. As factors become increasingly mobile, a country or region must be the most attractive location worldwide in order to attract the multinational enterprises' business activities it desires. This fact has to be accounted for in shaping modern industrial policy.

Clusters vary greatly in their capacity to send and receive knowledge flows and technological spillovers (Vuori, 1997). Some clusters are not only capable of creating knowledge, but – more importantly – also diffuse knowledge spillovers outside the cluster. According to Vuori (1997), the materials cluster (a combination of the forestry and base metals clusters), the ICT cluster and the fabrication cluster (comprising metal products and machinery) have been the most important sources of embodied technology and spillovers in Finland. These results, based on disaggregated industry-level R&D data and measurements of technological distance, are in accordance with our original cluster studies.

4.2. *Broadening the policy scope*

The above discussion has two major policy implications:

- Industrial policy must be broad in scope. It should involve not only industries and business firms, and the reallocation existing resources, but should also focus on the creation of future factor conditions.

- Public expenditure on R&D and education are perhaps the most important channels through which the public sector can influence national competitiveness. Taking into account the abilities of different clusters to generate and distribute knowledge throughout the economy, the ways in which public expenditure on R&D and education is allocated are not insignificant.

Figure 4 clarifies the role of public policy and its broad scope, emphasizing the indirect role played by policy. The main economic, industrial and technology policy blocks have been added to the diamond model. Each policy block influences competitiveness via one or more facets of the diamond.

Figure 4. Determinants of competitive advantage and the components of economic and industrial policies

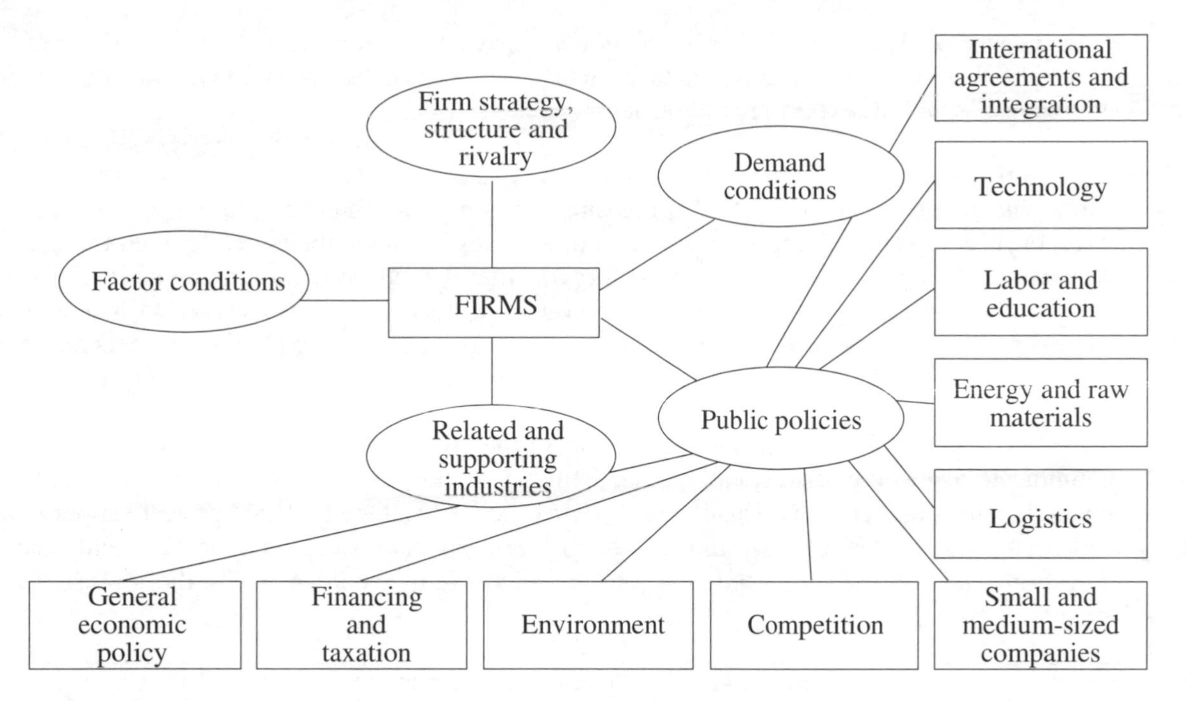

Source: Hernesniemi *et al.*, 1995, slightly modified.

Education and technological policies create the pool of advanced and specialised factors which are the main sources of sustainable long-term growth. Competition policy is used to establish a competitive environment in which companies formulate their own strategies and which affects firm and industry structure. Financial institutions, together with tax policy, affect the forms of commercial activity, co-operative networks, etc.

Figure 4 shows that virtually *all* government decisions matter and have implications for competitiveness. The diamond model helps to better understand the mechanisms through which the decisions affect – directly or indirectly – the competitive advantages of firms.

Cluster and networks can be seen as collective assets reducing transaction costs by internalising transactions involving positive externalities. Policy measures should attempt to strengthen the common knowledge base within the cluster, and thus correct market failures implied by the existence of external economies. Thus, industrial clusters – including both private and public agents – can be seen as entities with built-in mechanisms for correcting market failures.

In addition to technical advances and innovative activities, research on economic growth emphasizes the importance of another fundamental source of growth: specialisation associated with a deeper domestic and international division of labour. Deepening division of labour, expansion of world trade, and the consequent economic growth are possible only if well-functioning economic institutions exist at the national and international levels. These institutions – broadly taken to mean trust between different parties – reduce the inherent risks involved in specialisation and thus encourage firms to invest. Industrial networks are often, in practice, the organisational forms of these collective risk-sharing mechanisms. Hence, public policies should encourage business networking.

4.3. *Cluster-based policies in Finland – experiences and recent developments*

In Finland, cluster analysis and cluster-based policies have been applied at both the national and regional levels. Fields of application include industrial, science & technology, educational and regional policies, as well as export promotion activities.

Many observers (Steinbock, 1998) argue that the "Advantage Finland" project provided a much needed new vision for the economy, reshaping thinking and policy practices nation-wide. Although the final report highlighted the importance of structural change, it nevertheless underlined the need to build on existing strengths and capabilities. It foresaw that the future growth areas often lie in the interface of traditional strong clusters and new emerging clusters. It was also suggested that some of the new fast-growing clusters could improve their growth prospects by exploiting the technological know-how of other clusters, as was the case for the telecommunications and well-being clusters.

In his recent writings, Porter has emphasized that one of the important roles of government is to create and communicate a positive, distinctive, and challenging *economic vision* for the nation, mobilising government, business and citizens. This is exactly what the "Advantage Finland" project set out to do. In a sense, argues Porter (1998), the competitiveness agenda of a country is really about communication. The policy recommendations of the project are in accordance with those presented in Box 1 below.

Box 1. Appropriate roles for government

Establish a stable and predictable economic and political environment.

Improve the availability, quality and efficiency of general-purpose inputs and institutions.

Create a context that encourages innovation and upgrading.

Reinforce cluster formation and upgrading throughout the economy.

Adopt an action/problem-solving orientation.

Create and communicate a positive, distinctive and challenging economic vision for the nation.

Source: Porter, 1998.

Although our cluster study concentrated on national "mega"- or "meso"-level clusters, dozens of regional studies and development programmes use our work as a starting-point. Cluster studies have encouraged regional and local industrial associations and private businesses to take a more proactive role in improving their operating environment.

Perhaps the most worthwhile policy innovation associated with cluster analysis is the increased understanding that the creation and diffusion of technological knowledge occur through interactions between different economic agents and institutions. This has been the starting-point for recent government decisions to increase funding for R&D and allocate part of this funding to the sectoral ministries for various cluster programmes.[13] There are cluster programmes in telecommunications, foodstuffs, transportation (logistics), environment, forestry and health care. The programmes aim to encourage new forms of co-operation among businesses, scientific and technological organisations, and government bodies. The goal of this co-operation is: "…to further strengthen the links between innovation policy and other relevant policy sectors, providing new opportunities, for example, for developing regulatory frameworks so that they become more conducive to innovation and generating demand for new innovations within the public sector" (Ormala, 1998, p. 282).

5. Concluding remarks – small country strategies in global competition[14]

The competitive advantage of the Finnish economy and the firms in it has changed significantly as the Finnish industrial structure has shifted away from slow-growth industries towards knowledge-driven industries and clusters. This move has made the country less dependent on world markets for wood-based products.

The sensitivity of the Finnish economy to world market fluctuations has also been reduced as a consequence of the rapid internationalisation of business, due to the worldwide diversification of exports and production.

From the perspective of a small open economy, overcoming the relative disadvantage of size means specialising in niche markets and exploiting company networks; this is particularly true in the case of small and medium-sized enterprises.

It was believed that economic integration would strengthen industries which enjoy comparative advantage based on (relatively abundant) factors of production, such as labour, raw materials and energy. However, the significance of comparative advantage in the traditional sense, as a determinant of the location of production, has changed as a consequence of the increased mobility of production factors. Furthermore, together with increased specialisation and product differentiation, it is firm-specific capabilities and created – rather than inherited – factors of production that now determine the competitiveness of a country or a region.

The comparative advantage of a country and the competitive advantage of a firm can no longer be equated. Hence, policy thinking has changed. The 1990s have seen a shift away from "picking winners" towards "letting the market pick winners".[15] This concurs with the internationalisation of firms and the changing mechanisms for creating competitive advantages. Industrial policy aims at improving framework conditions or the operational environment of firms. Direct subsidies are harmful in the long run since they distort competition and reduce incentives to innovate and upgrade. However, as discussed above, industrial policies have an important role to play. The main goals of these policies are, on the one hand, to ensure the efficient functioning of markets and, on the other hand, to create advanced and specialised factors of production. Industrial policies becoming broader in scope in modern policy thinking: educational, trade, energy, environmental and competition policies overlap, to a large extent, with the areas covered by industrial policy.

To summarise, industrial policies are becoming competitiveness policies. Governments are trying to create attractive locations for internationally competitive firms by developing high-level technological infrastructures and advanced factors of production.

Clusters provide a coherent framework for policy design, enhance the dialogue between various partners in the economy, clarify the roles of different agents in policy formulation, and provide an insight into how sustainable and productivity-based competitiveness emerges. For small open economies, knowledge-based growth and continuous productivity improvements are the only route to high and improving standards of living. Productivity growth and competitive advantage inevitably require specialisation. Specialisation, in turn, often leads to higher risks. Industrial networking and co-operation of the various agents within clusters provides a means to cope with such risks. Clustering and networking are, to some extent, collective goods with various external economies. This situation calls for, and justifies, active public policies.

Globalisation and the volatility of the global economy are here to stay; the only way to survive is through co-ordination of appropriate macro and competition-enhancing micro policies. Industrial clusters and cluster-based policies provide a consistent micro-oriented framework to respond to this need for co-ordination.

Let us conclude with a word of warning. As discussed above, we bent Porter's methodology to better suit our needs. Nevertheless, we attempted to remain true to the original ideas and kept in mind what the diamond model was intended for. There have been numerous examples where the diamond framework has been used to disguise a re-marketing of older approaches. Likewise, we are aware of a few examples where the diamond model has been used for something for which it was not intended, *e.g.* as a justification of cartel-like co-operative behaviour. As stated above, in our opinion the diamond model is a medium- to long-term model of structural competitiveness and should be treated as such. In that model, government is one of the *outside* forces indirectly affecting the facets of the diamond. The consideration of short-term and macroeconomic policies requires a different approach.

NOTES

1. The *New Outlook on Industrial Policies* was published in 1996. It continued the agenda set forth in the previous policy document, but, as suggested by Porter himself in the Foreword of Steinbock's (1998) book, some of the lessons learned from cluster studies were perhaps forgotten.

2. Penttinen (1994) divided the critique into nine categories: *i)* competitiveness may also be found outside clusters (we conducted a study of successful "lone stars"); *ii)* the diamond model does not properly account for foreign direct investment and multinational enterprises (we explicitly considered international business activities); *iii)* the model may not be suited to small open economies (in many cases we were forced to use somewhat broader cluster definitions than those used by Porter); *iv)* the model may not be applicable to resource-based industries (we successfully applied the model to two resource-based clusters); *v)* cultural variation should be given more emphasis (this is not an issue in a one-country study); *vi)* the methodology is not valid – Porter combines old ideas with a seemingly theoretical framework (we agree that the framework is to some extent a rewrap of old ideas – however, after carefully studying Porter's connections to economic theories and his previous work, we found the diamond internally consistent and in line with the mainstream competitiveness literature), *vii)* the role of macroeconomic variables in Porter's model is unclear (this is indeed true. In our opinion the model should be considered as a medium- to long-term model of structural competitiveness); *viii)* it is unclear whether the model is dynamic or static (the evolution of a cluster – if understood correctly – is a dynamic process; however, Porter's model is too static and backward-looking. We made great effort to predict the future developments of Finnish clusters, which can only be done by understanding technological trajectories and business dynamics of individual clusters); and *ix)* studies may not be conducted with sufficient rigor (the loosely defined theory offers possibilities for misuse).

3. In practice, we used the United Nations' SITC (Standard International Trade Classification), Revision 2 and Revision 3 five-digit (or four-digit when the five-digit level was not available) commodity classes as our starting-point. Export data came from OECD's Foreign Trade database.

4. Practical criteria in constructing the lists of successful industries were: *i)* Finland's share of world market exports in the commodity in question exceeded the average Finnish share of world trade; *ii)* industries where the trade balance was negative were eliminated; and *iii)* if a commodity was traded almost exclusively with neighbouring nations, the industry was excluded.

5. In order for a cluster to be defined as "strong", all parts of the cluster's diamond should be strong and well-balanced, competition among domestic firms should be fierce, Finnish R&D should be significant even on a global scale, and there should be a considerable network of supporting and related industries and organisations. In the case of "fairly strong" clusters, all the above conditions are fulfilled, but to a somewhat lesser extent. "Potential" clusters are still fragile and their industrial diamonds are incomplete, although many positive factors support their growth. "Defensive" clusters have some cluster characteristics, but their development trends have been negative.

6. From the policy point of view, it is perhaps more important to consider "potential" than "currently dominant" clusters.

7. By export specialisation, we refer to the so-called revealed comparative advantage (RCA) index.

8. The recent merger of the Finnish Enso and the Swedish Stora may be interpreted as anecdotal evidence of this.

9. Including, *e.g.* switching and transmission systems, terminals, data communications and mobile communication equipment.

10. Operation involves planning, construction, and maintenance of networks, as well as running of telephony and data services therein.

11. VANS utilise operator services as inputs.

12. In technology policy, the innovation systems approach has caught most attention. However, the two approaches in no way in conflict (Ormala, 1998).

13. The new funding originates from the privatisation of state-owned companies.

14. Partly based on Pajarinen, Rouvinen and Ylä-Anttila (1998).

15. See Jacobs and de Man (1996) for a discussion of the main phases of industrial policy making.

REFERENCES

Helander, E. (1997), “Finland's Research Clusters – Important Assets for a Member of the European Union”, *ETLA Discussion Papers*, No. 617.

Hernesniemi, H., M. Lammi, P. Ylä-Anttila and P. Rouvinen (eds.) (1995), *Advantage Finland – The Future of Finnish Industries*, ETLA, The Research Institute of the Finnish Economy B:113, and SITRA, The Finnish National Fund for Research and Development 149. Distributed by Taloustieto Oy, Helsinki, Finland. Also published in Finnish.

Jacobs, D. and A. de Man (1996), “Clusters, Industrial Policy and Firm Strategy – A Menu Approach”, *Technology Analysis & Strategic Management*, Vol. 8, No. 4, pp. 425-437.

Lemola, T. and R. Lovio (eds.) (1997), *Miksi Nokia, Finland* (*Why Nokia, Finland*), Juva, WSOY, Finland.

Ministry of Trade and Industry (1993), *National Industrial Strategy for Finland*, M. Pietarinen and R. Ranki, Industry Department, Publication 3/1993, Helsinki, Finland. Also published in Finnish.

Mäkinen, M. (1998), “Suomen viennin rakennemuutos ja klustereiden vientimenestys 1990-luvulla” (“The Structural Change of Finnish Exports and the Export Performance of Industrial Clusters in the 1990s”), *ETLA keskusteluaiheita (Discussion Papers)*, No. 643.

Ormala, E. (1998), “New Approaches in Technology Policy – The Finnish Example”, *STI Review*, No. 22, Special Issue on “New Rationale and Approaches in Technology and Innovation Policy”, pp. 278-282, OECD, Paris.

Pajarinen, M., P. Rouvinen, P. Ylä-Anttila (1998), *Small Country Strategies in Global Competition – Benchmarking the Finnish Case*, ETLA, The Research Institute of the Finnish Economy B:(forthcoming), and SITRA, The Finnish National Fund for Research and Development (forthcoming). Distributed by Taloustieto Oy, Helsinki, Finland.

Penttinen, R. (1994), “Summary of the Critique on Porter’s Diamond Model – Porter’s Diamond Model Modified to Suit the Finnish Paper and Board Machine Industry”, *ETLA Discussion Papers*, No. 462.

Porter, M.E. (1990/98), *The Competitive Advantage of Nations – With a New Introduction*, The Free Press, New York.

Porter, M.E. (1998), “The Competitive Advantage of Nations – The Finnish Case”, Foreword in D. Steinbock (ed.), *The Competitive Advantage of Finland –From Cartels to Competition*, Taloustieto Oy, Helsinki, Finland.

Smith, K. (1995), "Interactions in Knowledge Systems – Foundations, Policy Implications and Empirical Methods", *STI Review*, Special Issue on Innovation and Standards, No. 16, pp. 69-102, OECD, Paris.

Steinbock, D. (1998), *The Competitive Advantage of Finland – From Cartels to Competition*, Taloustieto Oy, Helsinki, Finland. With a Foreword by M.E. Porter.

Vartia, P. and P. Ylä-Anttila (1996), "Technology Policy and Industrial Clusters in a Small Open Economy – The Case of Finland", *ETLA Discussion Papers*, No. 550.

Vuori, S. (1997), *Technology Sources and Competitiveness – An Analysis of Finnish Industries,* ETLA Elinkeinoelämän Tutkimuslaitos B:138. Distributed by Taloustieto Oy, Helsinki, Finland.

Chapter 16

PUBLIC POLICIES TO FACILITATE CLUSTERS: BACKGROUND, RATIONALE AND POLICY PRACTICES IN INTERNATIONAL PERSPECTIVE

by

Patries Boekholt*
Technopolis, Amsterdam

Ben Thuriaux*
Technopolis, Brighton

1. Introduction

Cluster-based policies have been implemented by a number of governments in both industrialised and developing countries. A broad set of initiatives, ranging from cluster-mapping studies to inter-firm network brokerage, have been launched by international organisations, national ministries, regional development agencies, local and regional governments, business support organisations and businesses themselves. This chapter describes the policy practices implemented by a number of countries that have pursued these initiatives, and discusses their rationale and objectives. The chapter is based on an ongoing study, conducted on behalf of the Dutch Ministry of Economic Affairs, in collaboration with the OECD Focus Group on Cluster Mapping and Cluster Policy. It focuses on the cluster policies implemented in a number of industrialised OECD countries, mainly at the national level, and in some cases at the regional level.[1]

The definition of cluster policy in this study closely follows the working definition of clusters outlined in the Synthesis Report of the OECD Focus Group. Clusters are characterised as networks of production of strongly interdependent firms (including specialised suppliers), knowledge producing agents (universities, research institutes, engineering companies), bridging institutions (brokers, consultants) and customers, linked to each other in a value-adding production chain. Cluster policies comprise the set of policy activities that aim to: stimulate and support the emergence of these networks; strengthen the inter-linkages between the different parts of the networks; and increase the value added of their actions.

This international review looks at the various "cluster policy models", which form the overall national policy visions for the development of activities in favour of a "cluster approach" in industrial, S&T, regional, educational policies, etc. The implementation of these models can be achieved through:

---

* This chapter is partly based on the paper, "The Public Sector at Arm's Length or in Charge? Towards a Typology of Cluster Policies", by Patries Boekholt, presented at the OECD Cluster Focus Group Workshop in Amsterdam, 9-10 October 1997.

- *Policy programmes*, *i.e.* mechanisms for allocating support to collaborative projects with the objective of improving the competitiveness of clusters and inter-firm networks, or fostering business links with the S&T system.

- Influencing the *framework conditions* for specific clusters, particularly through standards, tax regimes, etc., or alternatively by the provision of specialised facilities and assets for a particular cluster (R&D, training, technology centres) with the objective of altering the competitive conditions for this cluster.

- *Strategic action* for clusters through the provision of codified information (market knowledge, benchmarking), mapping exercises, or by facilitating communication within and outside the cluster through platforms such as cluster associations, foresight activities, external promotion, etc.

This review has identified four broad "cluster policy models" and a wide range of instruments for the implementation of cluster policy. The four models are, respectively, *national advantage, inter-firm networking*, *regional development* and *industry-research*. Before describing these models, the chapter discusses the emergence of cluster-based policies and the shift in the rationale for public intervention in this area. Some of the confusion surrounding the wider concept of clusters needs to be clarified, in addition to discussing the pitfalls of a policy focus on clusters.

As mentioned above, we do not claim to have made a comprehensive overview of all cluster policy activities, not even in the OECD countries participating in the Focus Group. Our method has been to focus on a number of countries that are known to be developing cluster policies and policy instruments, and to study their general approach to the emerging phenomenon. We asked the OECD network to provide complementary information on the countries identified for study: Belgium (Flanders), Canada, Denmark, Finland, Norway, the Netherlands, Sweden, the United Kingdom, the United States. We later added some cluster policy initiatives in Australia, Austria, Germany and New Zealand. The research project is now nearing completion. This chapter describes some of the results obtained so far.

2. Clusters as a policy mechanism for interactive learning

The underlying question in this chapter is why governments – national, regional and local – are increasingly engaging in the facilitation and support of cluster-based policies? A critical observer of economic policy could well ask: "Isn't this a process which should be brought about by the marketplace?" What is the rationale for the role of government in cluster policy? And, why has it shown such explosive growth in the past decade?

The recent studies of networks and industrial clusters have their origin in the literature on industrial districts, based on empirical research in regions mostly with mature industries (Piore and Sabel, 1984; Morgan and Cooke, 1991). Europe's best-known examples of clusters with a long-standing tradition of industrial activity are the textile industry in Emilia-Romagna, and the engineering, machine tool building and automotive sectors in Baden-Württemberg. More recently, studies have been conducted on the factors for success or failure of the high-technology poles such as Silicon Valley, Boston Route 128, Sophia-Antipolis and the numerous science parks that have tried to repeat these success stories.[2] These examples have inspired policy makers to create similar interlinked systems, which should function as growth poles in their own economies.

Another influential contribution that emphasized the importance of networking and clusters of related industries was Porter's (1990), *The Competitive Advantage of Nations*. In his view, industries in a particular nation or region have a competitive advantage if they are embedded in a wide and deep network. A well-developed and dense network consists not only of inter-firm links within a particular industry, but also with related industries, specialised knowledge centres, education facilities, innovation support agencies and direct links with clients. Firms are likely to perform better in an institutional context where co-operative public and private relations create an environment in which firms find stimuli to upgrade their activities through access to training, finance, business services and sources of knowledge.

The evolutionary economics literature, in particular studies on the sources of innovation, stresses that in many industries, interactions with suppliers and customer firms are a key source for innovative ideas, more so than formal R&D.

A crucial element of clusters is inter-firm networks, where companies co-operate directly with each other. Co-operation can be formal, *i.e.* with an explicit contract (*e.g.* a supplier contract, a joint venture), or informal. In the latter case, we may think of forms of informal knowledge transfer, relations with "related" industries (*i.e.* industries which have no direct supplier relationships, but which may share some "economies of scope", *e.g.* similar technologies, similar markets).

Strategic alliances between large multinational companies are now common practice in international business. These alliances are not necessarily clusters, but mostly project-related bilateral arrangements between companies and possibly R&D laboratories. Heightened global competition and rising costs of R&D underlay this change in co-operation patterns. Small and medium-sized enterprises, even when operating in regional markets, feel the challenges of global competition to the same extent as the large companies operating internationally. This is evidenced, for instance, in the change in subcontractor relationships: contractors require higher quality, greater flexibility and more complex products from their suppliers. SMEs find it increasingly difficult to face these challenges in isolation. Co-operation with other firms and the creation of networks can offer:

- *More channels for learning and creating expertise.* Companies rely heavily on other firms – their competitors, suppliers and customers – and on further knowledge carriers for innovative ideas. Learning from others is an effective way of channelling the huge amounts of knowledge that firms need for innovation.

- *Economies of scale.* Networks aimed at joint purchasing, distribution or sharing of (knowledge) facilities can reduce costs for individual firms. If the results of R&D can be shared by a large set of companies, diffusion of knowledge will speed up.

- *Economies of scope.* Combined expertise can open new market niches for high-end, value-added products. A network of firms with complementary expertise will be better equipped to deal with the increased demand for high-quality products and services.

- *Heightened flexibility and shared risk.* The creation of a pool of expertise in a flexible network increases firms' ability to respond to market requirements.

Thus, there is a clear business logic for sectoral clusters and inter-firm networks, for large companies as well as for small companies operating on regional markets. Why then should governments intervene in what seems to be a market-driven process?

Cluster policies are situated at the boundaries of industrial policy (including SME policy), regional development policy, and science and technology (S&T) policy. Cluster policy differs from traditional industrial policy, which aimed at (financially) supporting particular sectors or firms thought to be vital for the economy. The objective of industrial policy was to subsidise national industries facing restructuring due to increased competition from abroad. Sustaining employment was the key rationale for these "defensive" support actions. The classic industrial policies of the 1970s have since been abandoned due to pressure from international fair trade and competition agreements. In Europe, the European Commission exercised a strong influence on its member states to abolish these "protective" measures. Rules on state aid no longer allow direct subsidies to particular firms or sectors. In addition, many of these protective policies in support of declining sectors failed due to increased competition from the newly industrialised countries. In the 1980s, most OECD countries shifted the focus of industrial policy towards the more strategic and technology-oriented industries such as aerospace, nuclear energy and defence-related industries. Due to international restrictions on state aid, public intervention increasingly focused on subsidising S&T, both in firms and in semi-public organisations.

At the same time, more macro-oriented industrial policies were used to improve the framework conditions for competition, through deregulation, tax and labour regulations, all of which required less "hands-on" action from government. In Europe, national governments' control over many macroeconomic parameters was reduced through European integration agreements. Therefore, a certain interest in influencing the meso level came back on the policy agenda in the early 1990s. The difference between traditional industrial policy and current cluster policy is that the target group has shifted away from horizontal sectors to wider "value chains", and the support mechanism has changed from direct financial support to indirect facilitation. However, in some instances, the borderline between the "old" and the "new" approaches is not so clear.

The rationale for *science and technology policy* was for a long time based on the notion of imperfect markets, justifying public intervention. The guiding principle was "the need to provide subsidies, and to strengthen intellectual property protection for industrial R&D in order to overcome the so-called "appropriability problem" – the weakening of private incentives for such investments, as a result of informational "spillovers" from innovations" (David, 1996). This classic approach to S&T policy had the side-effect of focusing on knowledge creation and R&D investments in either public research organisations or in private sector R&D laboratories. The result was the build-up of large stocks of knowledge, with limited flows between the different actors performing the R&D.

In the 1980s, the notion of the interactive innovation model, first described by Kline and Rosenberg (1986), and the theories of the evolutionary economists became more commonly accepted and inspired scholars and eventually policy makers to adopt the "national systems of innovation approach". Similarly S&T policy makers embraced a wider notion of modernisation which included both knowledge creation and absorption, commercial as well as organisational skills, human capital resources as well as "hard" technologies, *i.e.* innovation policy. Innovation policy broadened from public investment in R&D to a concern about how these investments could be most effectively used to increase the competitiveness of our industry and services.

The "interactive learning model" incited the debate on the European paradox: Europe is strong in basic research but weak in innovation. One of the key deficiencies of European systems of innovation was considered to be their inability to turn research-based knowledge into commercialised innovations. The cultural gap between industry and publicly funded research was fingered as the main culprit.

Cluster policy is not solely about S&T policy; in fact, in many policy programmes S&T is not a key issue at all. Nevertheless, cluster-based activities are inspired by the notion that the linkages between industry and research need to improve in order to sustain competitiveness. In a national system of innovation (NIS) approach, policy interventions can be seen as different ways of encouraging interactive learning between the sub-elements of the system: the company system, the research and technological development (RTD) or knowledge supply system; and the public sector. Acceptance of this model means a shift in the rationale for science and technology policy. Public action is justified to overcome imperfections in the "innovation system"; either where essential elements in the system are missing or where the linkages and flows in the system are not functioning well. The NIS approach emphasizes that interactions between organisations and people are needed to increase the learning effect, which in turn stimulates innovation. The findings of many empirical studies confirm that firms do not innovate in isolation and require a broad set of external information and knowledge to improve their processes, products and services. This is where science and technology policy and cluster policy come together; both aim to increase the learning capabilities of individual firms through intensified interaction with other firms, organisations and individuals.

3. The different rationales for cluster policy and their main policy tools

Thus, the "protectionist" rationale of "old-style" industrial policy has been replaced by a more market-oriented economic policy, where the framework conditions are shaped for more competitive industries. At the same time, the sole "market imperfection" rationale for science and technology policy cannot narrow the gap that exists between the creation of knowledge and the commercialisation of that knowledge. Today's economic development policies need a rationale that fully incorporates the interactive element of innovation as well as the market-oriented approach. Cluster policy can combine these two elements, as we have seen in several countries.

Combining the NIS perspective with a more market-oriented industrial policy approach, as most OECD countries have done, the *rationales* for cluster policy could be described as follows:

- Government regulations hamper innovation or competitiveness in a particular cluster.
- SMEs do not grasp the opportunities for collaboration with other firms that could increase their interactive learning and resource basis.
- Firms, particularly SMEs, cannot access strategic knowledge when operating in isolation.
- Firms do not utilise the expertise of knowledge suppliers, while knowledge suppliers are not sufficiently equipped to market their knowledge.
- Existing or potential clusters lack identity and self-awareness.
- Existing or potential clusters lack crucial elements which would increase synergy.

Each of these "system deficiencies" requires different policy initiatives to address the issues which arise. In the practice of cluster and networking policies, we can identify the following roles of public policy to counter the system failures described above:

- Identify clusters and their strengths and weaknesses through mapping exercises and analyses.
- Provide strategic information to clusters through benchmarking and foresight studies.
- Amend regulations or legislation that hamper the emergence of clusters.

- Raise awareness of the benefits of networking.
- Facilitate informal contacts through "industrial circles", sector groups or cluster platforms, etc.
- Help to bring firms together by acting as broker or encouraging brokerage by other stakeholders.
- Provide (financial) support for collaborative (knowledge) facilities and technical services.
- Provide financial support for the (launch of) networks and inter-firm co-operation (feasibility projects, management support, etc.).
- Act as "launching customer", bringing together various partners to develop new technologies, products or services in areas where the public sector is the main client (infrastructure, ICT, defence, etc.).
- Attract missing elements to a cluster.

Table 1 sums up the different rationales, the corresponding policy actions and the tools identified in this international review. The authorities of the countries reviewed explicitly or implicitly used one or more of these rationales for specific policy action. There is a strong element of selection in the proposed rationales: policies are addressed to specific subsets of the economy which need attention either because they have a strong presence in the economy or because they are seen as strategic for future growth and employment.

In general, the support of clusters followed one of the following general "policy models":

- *To improve the "national advantage" of certain sectors or value chains.* This means identifying clusters which are of importance (in terms of number of firms, share of employment, historical strengths, etc.) for the country in question, and ensuring that favourable conditions are put in place to sustain or develop this position. Denmark and Finland, and to a lesser degree Canada at the federal level, are examples of this type of cluster-based policy. A national advantage approach is also emerging in industrial and innovation policy in the Netherlands.
- *To improve SME competitiveness.* The lack of ability of small firms to innovate and learn from others has prompted public action to increase their interactions with external knowledge carriers, *i.e.* other firms. Networking schemes designed for this purpose are not necessarily embedded in cluster policy if the networks do not cover a particular subset of the value chain. Many US state-level initiatives, as well as cluster-based initiatives in Australia, New Zealand and Norway have a strong SME support element.

Table 1. Cluster policy rationales, initiatives and tools

Policy rationales	Cluster-oriented policy action	Tools
Lack of cluster identity and awareness	• Identification and public marketing of clusters	• Mapping exercises • External promotion of regional clusters • External/internal promotion of cluster member's competencies
Government regulations hamper innovation or competitiveness	• Organise cluster specific fora to identify regulative bottlenecks and take actions to improve them	• Cluster platforms and focus groups • Tax reform • Regulation reform (environment, labour markets, financial markets)
Firms do not take up opportunities for collaboration with other firms	• Encourage and facilitate inter-firm networking • Purchase innovative products through collaborative tender procedures	• Networking programmes • Brokerage training • Public procurement for consortia
Firms, particularly SMEs, cannot access strategic knowledge	• Support cluster based retrieval and spread of information • Organise dialogue on strategic cluster issues	• Set up cluster specific information and technology centres • Platforms to explore market opportunities • Foresight exercises
Firms do not utilise the expertise of knowledge suppliers	• Collaborative R&D actions and cluster specific R&D facilities	• Set up cluster specific technology and research centres/initiatives • Subsidise collaborative R&D and technology transfer
Lack of crucial elements in a cluster	• Attract or promote growth of firms in cluster • Attract major R&D facilities	• Targeted inward investment • Support start-up firms in particular cluster

- *To improve the attractiveness and the economic performance and development of a region.* Many development agencies, intermediaries and policy makers at regional level have taken up this approach. Some regions with an active cluster policy, for example Wales, Steiermark, the Basque Country, use a mix of policy instruments such as inward investment, supply chain development, SME networking and support of emerging technologies. In Norway, the national SME networking policy also has a strong regional development aspect.

- *To intensify industry-research collaboration in particular technologies or types of firms.* Although most countries' efforts to improve industry-research links are based on non-cluster-related policy initiatives, there is potential for encouraging industry-research networks and centres of excellence with the aim of stimulating more user-oriented research in a particular field. This approach is applied in a concentrated geographic area (cities, regions), with the objective of kick-starting economic strength in emerging technologies. It is assumed that firms specialising in emerging technologies will develop more rapidly if they can share complementary assets with other firms and knowledge carriers. Public action is launched with the aim of creating "critical mass" in newly emerging fields of technology by attracting research facilities and funding, large investors with R&D capabilities and new technology-based firms. Recent initiatives in Germany to encourage competence centres aim to help small regions become world-class "technopoles". Austria, Germany and the Netherlands can also be placed in this category, although a number of the regional policy initiatives of these countries have the objectives of the second model described above. Sweden also takes a research-industry approach to its networking activities.

As can be seen from Table 1, the term cluster policy is used to describe a large variety of policy measures, programmes and initiatives. Some clarification is required to allow the issues to be more clearly understood.

First, there is widespread confusion due to the fact that the term "cluster" is used to describe very distinct types of interlinked systems. Since Porter's *Competitive Advantage of Nations*, policy makers have adopted the concept of clustering to include initiatives ranging from traditional industrial sector policy to the financing of RTD projects for a group of two or three companies. In addition, the terms "clusters" and "network" are often used to describe different phenomena. Rosenfeld (1996) states that the crucial difference between the two is that the latter refers to a group with more formal, often contractual collaborations between members. In networks, membership is restricted, whereas membership of a cluster is open.

Compared to clusters, business networks comprise a group of firms with restricted membership sharing specific business objectives. Networking practitioners often refer to them as *business opportunity networks*. In principle, these networks can be part of a broader cluster but this is not necessarily the case. They can be both "soft networks" aiming at sharing information and contacts and "hard networks" aiming at doing joint business. Network members usually draw up a plan how to achieve these common goals, which is subsequently formalised. In terms of cluster-based policy instruments, the facilitation of horizontal and vertical networks is the most frequent goal. Many case studies of these business opportunity networks report an increase in economic performance among members. Nevertheless, the evaluation of the Danish Network Programme showed that many of these networks are dissolved after several years of existence.

For the aims of the OECD Working Group, it would be premature to impose strict demarcation lines on the use of clusters. Many initiatives labelled "cluster policy" are, in fact, "networking". Networking can be seen as one of the tools used to foster clusters.

A next step to clarify the situation is to distinguish between the levels of aggregation on which cluster policies focus: networks can be mega-level (*i.e.* agro-food), meso-level (*i.e.* the machine tool building or yacht building sector), or micro-level (*i.e.* a collaborative network of individual firms) (Figure 1). Each level of aggregation requires a different policy approach using different policy tools. The brokerage type of policy instrument typically operates at the micro-cluster level, the development of specialised factor conditions such as the establishment of technology centres concentrates on the meso level, while changing environmental regulations for certain product groups affects the mega level.

Figure 1. Level of aggregation of clusters

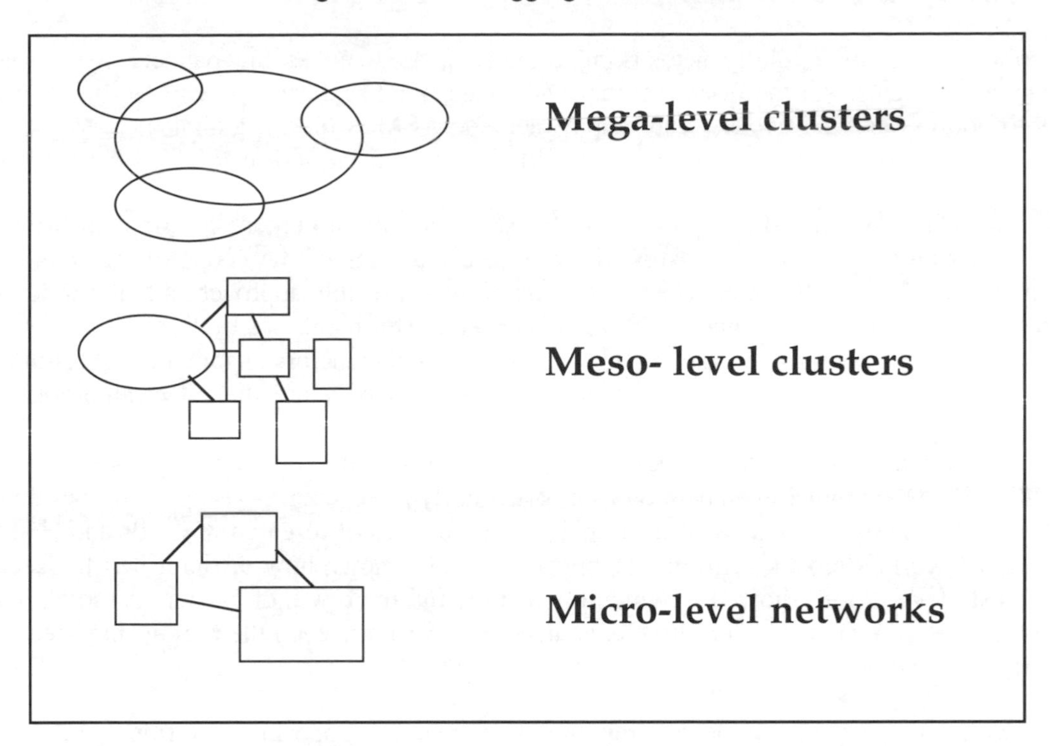

In most academic and policy debates on clusters, the perspective is on the meso level, following Porter's sectoral approach. However, the concept used here is broader than the traditional sector-based categorisation of industries since clusters include laterally related sectors and institutional infrastructures.

An examination of a large set of European policy initiatives that are labelled as cluster policy reveals that many policy support initiatives operate on the micro level, with the aim of bringing together a restricted group of partners to achieve a common purpose.

Cluster policy at the mega level is geared towards providing the infrastructure and general conditions for innovative clusters. If we consider only those activities geared towards intensifying the knowledge base, the main efforts can be found in the areas of science, education and general innovation support. Prioritisation in science and innovation in favour of clusters is an issue of debate in many industrialised countries. Those in favour hold that governments have limited financial resources which should be used as efficiently as possible. Critics argue that this is a new form of "picking the winners", ostracising those companies and researchers that fall outside the dominant clusters. This could lead to a built-in conservatism in the national innovation system since radical new approaches can come from unexpected angles. In addition, governments are notoriously bad at picking winners. The counter argument is that clustering is about choosing based, not on future potential but on present strengths (Rosenfeld, 1996). This is one reason why developing good tools to map and analyse clusters which are (potentially) strong and strategic is so important.

3.2. *The pitfalls of cluster policy*

Cluster policies, and particularly networking initiatives, are seen as the panacea to all economic development problems. All too often regional and local public sector actors call for new networking projects to address competitiveness and to put their local SMEs on the global map. What is easily forgotten is that networking can also have some pitfalls for the firms and sectors involved.

Clustering can have several drawbacks on the meso level. One argument for implementing cluster policies is that they allow policy makers to target selected parts of the economy in order to spend public investment more efficiently. The most common pitfall of this approach is that public agencies develop their policy strategy around "high-tech" clusters, even if the preconditions for such clusters are not present in the country or region. A large number of policy actors engage in wishful thinking in the identification of the growth areas in their economy. Not only are they over-ambitious in their targeting of high-tech sectors; they show little imagination in their choice of growth poles. The same clusters – multimedia, medical technology or biotechnology – feature on many regional strategy plans. The creation of "technopoles" can be seen as a particular type of science-based cluster building. Very often these are attempts to create high-tech poles around centres of research. Castells and Hall (1994), comparing the world's most significant "technopoles", have shown how difficult it is to successfully pursue this strategy. Porter showed convincingly that the future growth of clusters is rooted in present strengths and determinants of competitive advantage. Furthermore, even the famous high-tech clusters needed time to develop.

Focusing exclusively on the traditional strengths of an area, *i.e.* supporting traditional and established clusters to perform more efficiently, could lead to "path dependency" and lock-in effects in a regional or national economy. Public support for existing trade and products can delay a more radical reorientation of the economy towards new and more competitive industries. Even Baden-Württemberg, the role model for many regional clusters, is encountering difficulties in adapting its dense institutionalised engineering clusters to the changing demands of the international market (Cooke *et al.*, 1998, forthcoming). Cluster policy approaches should, therefore, be open to new developments in technologies and international markets and constantly strive to make companies aware of these international developments.

Another pitfall is that public brokers tend to underestimate the risks and efforts involved for companies to engage in inter-firm networking. All to easily, initiatives are launched to bring together firms for collaboration on very strategic parts of their business. Setting up a firm-to-firm network is a complicated task requiring time and professional mentoring. Companies, and particularly SMEs:

- Need to build up trust with their potential partners; this takes time and requires (informal) contacts.
- Are reluctant to spend valuable time and effort on a network if the objectives and potential benefits are not clear; the initial (human) investments are high compared to uncertain outcomes.
- Fear losing strategic assets and information to other network members, particularly if these are larger companies.
- Have varying needs and expectations of networking depending on their own (technological) capabilities.
- Will be disillusioned if their first experiences with networking are negative.

- Are more likely to start off with less strategic alliances before entering into complicated R&D collaborative efforts.

Thus, public agents setting up network activities should have experience in considering networks from the business perspective, while public brokers should be sufficiently trained or experienced to deal with the multifaceted aspects of inter-firm collaboration. Given the "cluster hype", these pitfalls are all too often neglected by enthusiasts who believe they have discovered the ultimate tool for economic development.

4. Cluster policy models in practice

In Section 3.1 we briefly described the four policy models that can be distinguished on the basis of the dominant "rationale" for cluster policy. Each of these models is described in more detail in the following sections.

4.1. The national advantage model

The essential features of this model are:

- Clusters at the mega and meso levels are supported by improving framework conditions through adapted regulation, access to RTD, access to foreign markets, education, etc.
- In order to identify the strengths and weaknesses of national clusters, policy agencies, or research organisations acting on behalf of policy agencies, conduct mapping exercises and studies.
- Dialogue and discussion platforms are organised to identify the needs and bottlenecks of the firms and organisations in the cluster, in order to develop further public action.
- Policy tools are mostly generic and aim to improve the framework conditions of the clusters.

Few countries have implemented any very explicit policy consisting of a cluster approach to enhance the *national advantage* of its economy for its strong clusters. In this policy model, clusters, mainly at the mega and meso levels, are supported through strategic mapping studies, adaptation of regulations, specialised support infrastructures, etc., with an emphasis on providing good framework conditions.

The country that takes the most explicit "national advantage" approach is *Denmark* (Drejer *et al.*, this volume) with its identification of "resource areas" as the main value chains in the Danish economy. *Finland* (Rouvinen and Ylä-Anttila, this volume) has also conducted cluster studies to identify how Finland's competitive advantage could be improved. This has been translated into general policy measures rather than specific cluster-focused initiatives. The *Netherlands* (Roelandt *et al.*, this volume) is gradually moving towards a more pronounced "national advantage" approach.

In Denmark, the concept of resource areas has been central to the technology and industrial policies of the Danish Ministry of Business and Industry since 1993, although attempts to identify production clusters go back to the early 1980s.

The theoretical foundation for cluster studies in Denmark is largely the same as in other countries. Porter's *The Competitive Advantage of Nations* (1990) influenced much of the thinking about and policy development for clusters, both in Denmark and elsewhere. When the Danish *Erhvervsutviklingsradet* (Council for Industrial Development – EUR), an advisory board for the government, initiated the cluster support programme, preparatory work had already been carried out by industry associations and the EUR in analysing cluster-specific challenges and bottlenecks. The new government saw in this approach a way of conducting economic policy on the basis of closer consultations between the government, the administration and industry. A major difference between the Danish cluster mapping compared with mapping exercises in other countries was that the clusters covered almost the entire economy, rather than only selected parts.

The "resource areas" approach has enabled Danish industrial policy to move from support for individual business (for instance, through subsidy schemes for certain sectors or industries), to a more systemic approach, enabling business to "maintain or increase their competitive position". Six broad economic areas, or mega-clusters, were defined as key areas for the Danish economy:

- food products;
- leisure/consumer goods;
- construction/housing;
- communication;
- transport/environment/energy; and
- medical/health.

A resource area "consists of those firms and institutions which together make, produce and distribute a set of products with homogenous market characteristics."

This is a vertical view of the market structure, where the value chain and its end-products define the resource area. The Danish Ministry states that this constitutes the main difference with "traditional" sector policy:

> "The traditional way of looking at the economy was based on a division into *sectors,* which distinguished between primary producers, manufacturers and service industries. In practice though, relations of interdependence go across the sector boundaries."[3]

Within each of these resource areas, the Danish Ministry of Business and Industry set up a *dialogue* with a group of key people from industry, research, business associations and other related organisations. Such a dialogue should, together with an analysis of the challenges and opportunities of each business sector, lead to the tailoring of "framework conditions" suitable for each resource area. The resource area approach has made it vital to distinguish between the general framework conditions and those that apply to particular areas of the economy – the *proximate framework conditions*.

The underlying basis for the Danish cluster support programme is that macroeconomic framework conditions (interest rates, exchange rates, etc.) have always been, and remain, important for competitiveness. These are increasingly outside the control of national governments and it was made clear that several macroeconomic issues, such as the general tax legislation, were to be kept outside the resource area dialogue. Five *core areas* for potential policy action were defined:

- government regulation;
- access to knowledge;
- access to capital;
- interaction between the public and the private sector;
- conditions for international competition.

The dialogue process is organised as follows:

- A reference group with representatives from the resource area identifies the challenges and critical framework conditions and sets up working groups.
- The working groups make specific proposals to improve sector-specific framework conditions.
- The Danish Ministry of Business and Industry monitors the implementation of the proposals and keeps the relevant parties and the public informed.

Initiatives resulting from such dialogue include the creation of "centres without walls" (virtual centres) for government and private fundamental and strategic research. These centres are established in areas where Danish public and private research are of world-class standard. The public and private sectors both contribute funding and research staff, thus stimulating face-to-face collaboration between public and private R&D. Virtual centres have been created in the areas of health, food, tourism and information and communication technology (ICT). Action has also been undertaken in standardisation, for instance in transportation, construction, and environment/housing. In the area of public procurement, the Danish government used its political influence to alter the interpretation of EU rules relative to calls for tender, to the benefit, for instance, of Danish ICT firms.

A key to the success of the resource area approach is that it requires the involvement of a number of Ministries in the implementation of dialogue outcomes. The Ministry of Business and Trade is officially responsible for co-ordinating the implementation of the outcomes since the issues relate to a broad set of policy measures, ranging from education and research, legislation on intellectual property rights and biotechnology, etc. One of the key challenges of the exercise was to break down the walls between administrative bodies who had got used to working on their own territory.

Finland has also engaged in extensive mapping and research exercises to define its strengths and weaknesses. Porter's *Competitive Advantage of Nations* had a major influence on Finnish industrial policy. Rouvinen and Ylä-Antilla (1999) report that almost every industrial branch in Finland has been analysed through the *Competitive Advantage Finland* project, which was co-funded by the government agencies. The project led to a vast array of publications which distinguish both Finland's traditional (paper and pulp, forestry, chemistry) and emerging clusters. It provided the input for the 1993 Government White Paper, *A National Industrial Strategy for Finland,* a strategy paper by the Ministry of Trade and Industry announcing a new industrial policy.[4] As Finnish analysts state: "*A National Industrial Strategy for Finland* introduced the notion of the cluster into the Finnish discourse on industrial policy and competitiveness – not only did it popularise the notion, it made the "cluster" *the* catchword of the Finnish policy institutes and think-tanks in the mid-1990s." (Steinbock, 1998).

However, contrary to the approach taken by Denmark, the Finnish *national advantage approach* did not have such an explicit cluster support focus as the resource area activities.

Finland's industrial policy is characterised as having "moved from an interventionistic policy to a pro-market approach".[5] Direct subsidies to Finnish companies have been reduced and the policy focus had shifted to the creation of "advanced and specialised" factor conditions. Finland's main policy approach can be characterised as an increased market orientation through stricter competition policy and the strengthening of cluster factor conditions, in particular on the R&D supply side. Analysts conclude that the main impact on S&T policy has been the reorientation of the science and technology centres to focus on Finland's emerging technological clusters.

A similar approach can be found in *Sweden*. Although Sweden undertook various cluster mapping exercises, including the original Porter study, these have not had a deep impact on its industrial or innovation policy.

Canada's policy has similarities with the Finnish approach: the main focus is on reorienting the S&T system to make it more user- (*i.e.* business-) oriented. The focus of S&T on economic strengths implicitly means taking account of Canada's strong industrial sectors. The government's role is shifting away from that of a provider of subsidies into that of a catalyst of better co-ordinated action between academia, business and government. This means support for research networks with a specific focus on certain technological fields and/or sectors, such as wood pulp and robotics. However, an explicit cluster approach is not taken in industrial and S&T policy at the federal level. The Canadian telecommunication and broadcasting sector has benefited most from intervention related to regulation, through the introduction of more vigorous competition rules. Proactive cluster initiatives have taken place in a number of Canadian states or state groupings, such as Quebec and West Canada.

In the *Netherlands* the emergence of a national advantage approach is more pronounced than in Sweden. In 1990, the Ministry of Economic Affairs commissioned a Dutch "Porter" study, assessing the economic strength of the economy in general and of a selection of clusters (Jacobs *et al.*, 1990). Since then mapping and cluster analysis studies have been conducted or supported by the Ministry. In sectors such as telematics and telecom, discussion platforms have been organised with industry to discuss common bottlenecks and opportunities for the future, as well as foresight exercises in particular technological and industrial fields. In 1997, the Ministry of Economic Affairs published its first White Paper on cluster policy: *Opportunities Through Synergy: The Public Sector and Innovative Clustering in the Market.* The White Paper distinguishes three public roles for clusters:

- To create favourable framework conditions for industry and services in general.
- To act as broker in identifying cluster opportunities by bringing together supply and demand and providing strategic intelligence.
- To operate as demanding customer in the provision of societal needs.

In terms of implementing policy tools, however, the industry-research linkage approach has received greater emphasis in the Netherlands. Most of the measures put into place over the last five years either apply to industry in general or are aimed at stimulating the development of technological innovation projects by consortia of industry, research and end-users. The traditional interventionist industrial policy was convincingly dismissed during the mid-1980s after bad experiences with this approach in the preceding years. Therefore, a more hands-on inter-firm networking approach was unlikely to be adopted by national policy makers.

To summarise, the national advantage policy model is typically based on extensive mapping exercises which indicate the key clusters on which to focus attention and the main bottlenecks to innovation. The general competitive framework conditions are adapted to address the specific needs of the key clusters, through cluster panels and other consultation mechanisms. Countries with a national advantage model, particularly Canada, Denmark and Finland, have opted not to develop more hands-on brokerage and direct policy interventions to initiate and support clusters.

4.2. The SME networking model

In terms of policy tools the most established form of cluster-based policy is the multitude of inter-firm networking programmes set up around the world. In many countries, "cluster policy" equals the encouragement of networking.

The essential features of this policy model are:

- Networks are mainly regional in scope, thus allowing the emergence of people-to-people networks.
- Support mechanisms focus on encouraging and supporting collaborative networks between firms, and between firms and knowledge providers.
- Public or private agencies adopt an active brokerage role to initiate and manage collaborative networks.
- Part of the support activities focus on improving the brokerage skills of intermediaries.

In this model, the main objective of cluster policy is to establish, sustain and encourage collaborative networks between firms. There are three typical structures of network initiatives:

- *Network programmes*, where the participating firms share and upgrade common resources (training, technologies, managerial experiences, strategic information).
- *Supply chain networks*, where groups of suppliers are encouraged to collaborate with each other and their (potential) contractors.
- *Horizontal networks*, where companies from a similar sector join forces to improve their competitive position (joint marketing, joint purchases) *vis-à-vis* their (potential) clients or suppliers.

Collaboration between members of a cluster or network can involve a large variety of business functions including:

- purchasing;
- logistics;
- research and development;
- prototype development and adaptation;

- technology search;
- market research;
- marketing;
- quality management;
- certification and standardisation;
- systems supply development;
- distribution;
- informal exchange of know how (trends in technologies, markets, life styles patterns, etc.).

The OECD Working Group focuses on innovative clusters, which often translates into clusters with a high R&D intensity. In practice, many network programmes and projects do not explicitly focus on R&D as a necessary element in the linkages formed. Nor are research organisations a necessary partner in the network configurations. Of course, many countries have a tradition of collaborative research programmes which create small-scale networks. However, these are aimed at short-term project-related collaborations. Figures 2a and 2b show examples of network initiatives in various European countries. These reveal that the firm-to-firm component is the dominant pattern.

When setting up university-industry linkage initiatives for a particular cluster, policy makers should consider whether their target groups have the capabilities and needs to form networks with knowledge suppliers such as universities. There are ample examples of where these type of initiatives have failed simply because the targeted members from industry and local universities were not able to "speak the same language". In these cases, the inclusion of technology centres in the cluster's innovation system, with an orientation on diffusion and the provision of support and information on applied technologies would be more appropriate.

We can roughly distinguish between horizontal (firms having an equal position in the supply chain of a sector), vertical (firms having a client-supplier relation to one another), or lateral (firms from different sectors sharing common skills, technologies or other business needs) cluster initiatives.

Figure 2a. Examples of network models (1)

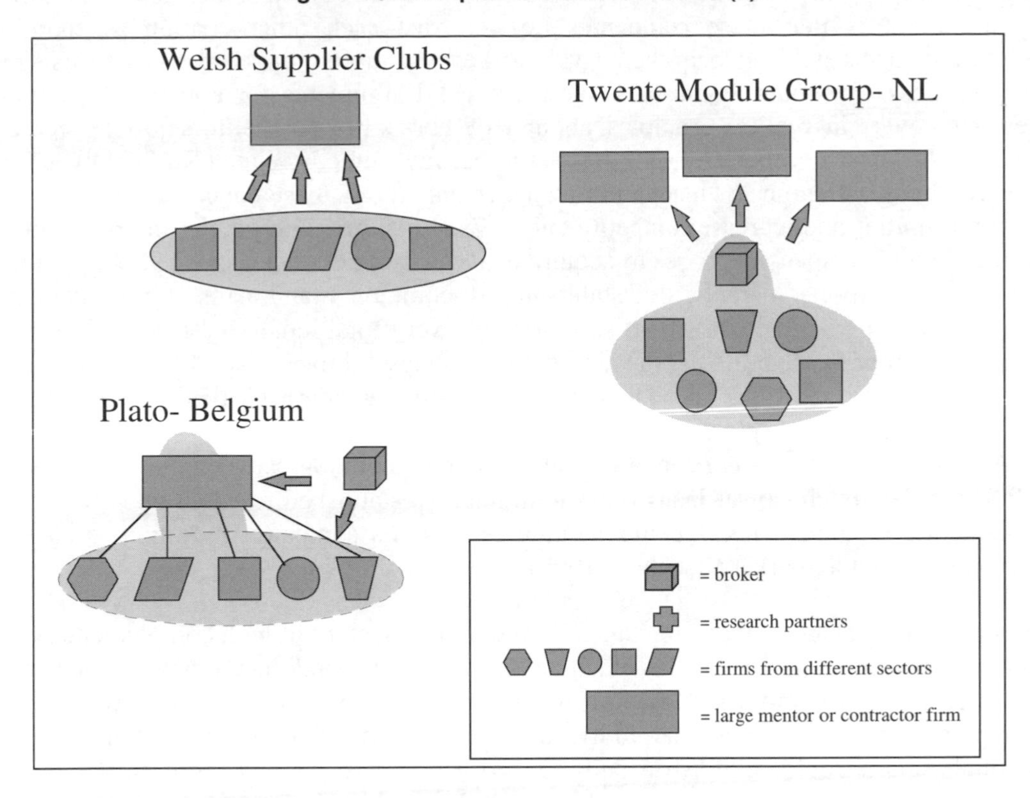

Figure 2b. Examples of network models (2)

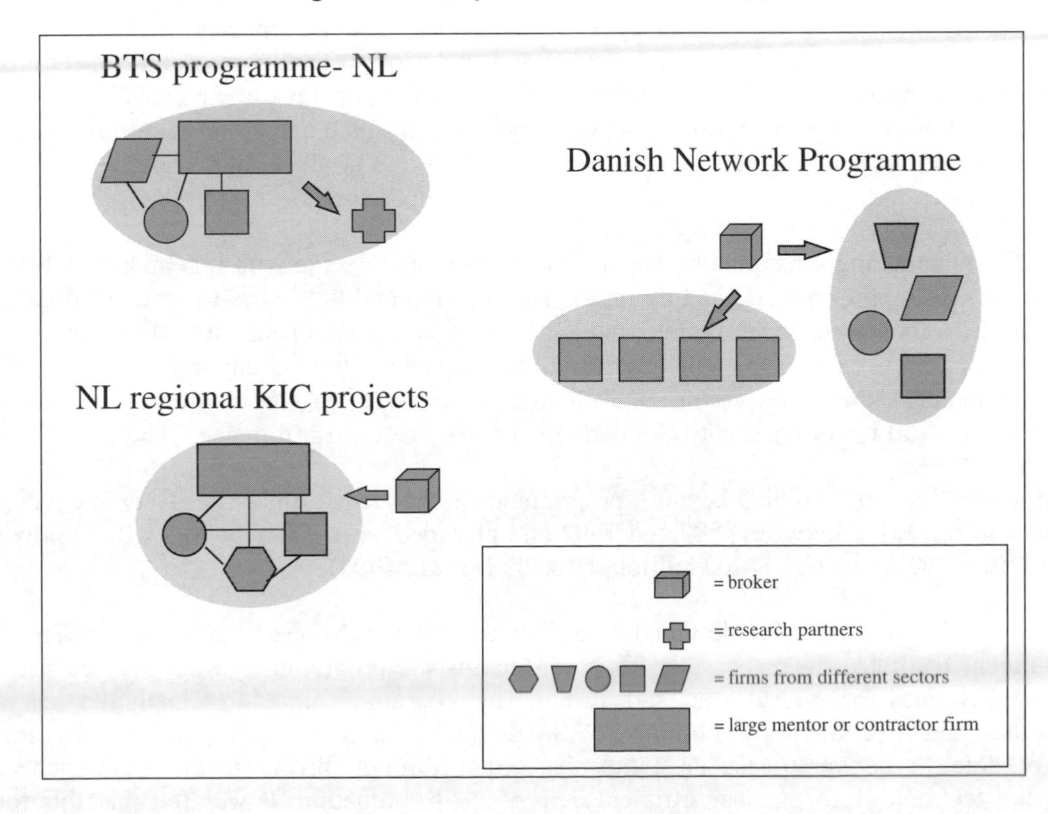

It is often argued that it is not possible to create clusters between direct competitors in the same sector. The key argument is that these companies do not trust each other enough to share strategic information, and firms will not help their rivals to become more competitive since this would hurt their own business. For many companies, these are valid arguments for not entering network-type initiatives. However, in reality, examples abound of horizontal competitors joining forces in the common interest. These co-operative activities are not always directly related R&D or even innovation in the narrow sense, although in "innovation management" terms these concepts certainly create new business opportunities and increase competitiveness. Examples are joint purchasing of commodities to reduce prices, multi-company exercises to acquire quality standards such as ISO 9000, joint marketing to access distant (export) markets, or establishing a common brand name for a product from a particular region. An example of such an approach in a very traditional sector is the lace industry in the Ayrshire region of Scotland. After the almost total collapse of the sector, the remaining companies got together and developed value-added product targets. Among their activities were the establishment of a joint brand name, the exploration of new products, and the setting up of databases with design patterns. Another example of a horizontal cluster where collaboration between direct competitors does include R&D is the Dutch flower industry. A traditionally established research infrastructure ranging from fundamental to applied R&D, demonstration centres and company study groups co-exists with fierce competition on the market (Boekholt, 1997).

Horizontal networks can consist of companies from the same sector but with complementary expertise. In the Twente region, suppliers in the machine-building and metal sector have joined together as system suppliers to large contractors. Even if these companies' expertise can overlap in certain areas, offering expertise as a group rather than individually offers greater business opportunities for all. This type of collaboration could reinforce a shift towards core business and increase the efficiency of companies' investments. The Twente Modulen Group has attracted interest from some 60 companies and has won contracts from several large companies.

Vertical clusters are those where several layers of the supply chain are represented. These clusters aim to increase the level of the subcontractor's capabilities and product quality in order to meet increasing customer requirements. In the longer term, the effect could be to raise their capability for product development, thus creating new business opportunities. In the Dutch agro-food sector, regional cluster activities have been set up to increase hygiene standards in the product chain, ranging from primary product, transport, industrial food processing to storage in supermarkets.

Policy makers launching cluster initiatives need to explore and decide with potential members which model is the most appropriate depending on the target group and the issues to be tackled. Horizontal cluster activities for the purpose of joint strategic R&D have a good chance to fail unless a culture of trust and co-operation exists among the participating companies. Vertical clusters are more difficult to launch and manage since they include stakeholders from different communities. Lateral networks require greater effort to convince potential partners that they can learn from each other.

The inspiration for most existing network programmes is the Danish Industrial Network programme, which was in operation between 1989 and 1992 and provided over 300 networks with approximately USD 25 million in grants (DKK 160 million) (Boekholt *et al.*, 1998).

The Network Programme was designed and initiated by the Ministry of Industry. The Danish Technological Institute was responsible for co-ordinating the programme, which was carried out by the National Agency for Industry and Trade (AIT). The programme aimed to remedy the lack of large firms in the Danish economy by "bulking up" SMEs through networks. The rationale was that large firms are able to bootstrap smaller companies either through procurement requirements or by transferring technology to facilitate efficient delivery and production. It was felt that this force for change was missing in the Danish economy. The aim of the Network Programme was to change the structure of Danish industry towards greater co-operation, both nationally and internationally.

The programme provided financial support to SMEs through grants to help them implement co-operative projects that would improve their capabilities in important strategic issues. Support was based on a three-phase network development model with support for each stage dependent on success in the previous stage. To be eligible for support, the network had to consist of a minimum of three firms, which could apply for subsidies to cover the external expenses incurred in finding partners and to carry out a feasibility study of co-operation possibilities.

The Network Programme also covered the training of 40 network brokers who were to assist the participating companies. Their role was to identify the right companies and to provide the networks with secretarial and administrative support. Their role in Phases 1 and 2 are fundamental in that they deal with the initial administrative problems of the network; in Phase 3, however, the relationships between firms are strong and established and the role of the broker is minimal (in many cases they were no longer needed). In addition, the programme financed sector analyses and the services of a group of lawyers to develop standard contracts addressing the needs of different types of network.

A significant number of companies were encouraged to establish a new legal entity (a new firm). This new entity served as the network centre or hub, controlling the joint (and jointly owned) resources of the network. The programme was biased towards manufacturing companies and the establishment of a network allowed these smaller firms to replicate the behaviour of a larger company:

- Firms were able to purchase or resource joint solutions to common problems, such as monitoring markets, competitors and technology; purchasing advanced equipment; joint R&D; joint finance and credit line guarantees.

- Firms could specialise within the network to exploit complementarity in much the same way as large firms have divisions specialising in different areas.

- Firms were able to leverage their subcontracting links both downstream to pooled contractors, and upstream through access to markets that had previously been denied them because of their small size.

The programme was extended to other sectors in the form of the Tourism Network Programme, the Environmental Technology Network programme and the Export Network Programme. The programme as a whole was very costly in proportion to its achievements. The evaluation of the programme has not been very positive in terms of value for money, although other experts say that the evaluation was unable to properly analyse the effects because of the time lapse between the study and the participation in the programme.

Following the Danish example, other countries and regions copied its basic structure and philosophy. One of the most comprehensive networking programmes is the *Australian* AusIndustry's Business Network Programme (BNP). The programme was one of a number of industrial policy initiatives announced in the Australian Federal Government's 1994 *Working Nation* statement. Although federally funded, the programme is being implemented in conjunction with industry associations, regional development authorities and private consultants, as well as agencies of federal state and local government. The programme has been allocated USD 38 million and has created some 200 networks, of which approximately 40% are in the service sector.

Other examples of network programmes which aim to establish micro-level networks are the Joint Action Groups and the Hard Network Programmes in *New Zealand*. Tradenz, the New Zealand government's networking sponsor, has been set up to manage and implement these activities. Although the Danish structure was followed, a distinctive characteristic of the "hard network" programme is its use of private consultants who are trained in facilitation skills and expected to perform the role of network brokers. They will act as a secretariat and deal with any administrative tasks related to the network, including developing a network business plan and managing the grant application process. The brokers receive no funding from Tradenz and pay for their training. It is expected that they will use networking as one of the tools they can offer their client firms, and that firms will pay for these services. In its first years, the main objective was increased exports by New Zealand firms, and gradually developing local clusters has become a major activity of this organisation.

Another example of a networking policy programme is the USNet Programme. Again, this contains similar elements to the Danish Networking Programme: strong emphasis on the training of brokers to start up and manage inter-firm collaborative networks. The programme is jointly funded by the Department of Defence (the Technology Reinvestment Project) and 15 partner states. The National Institute of Standards is also involved as a partner. The managing partner, Regional Technology Strategies Inc., is responsible for administering USNet and delivering additional services. The programme is not aimed at setting up inter-firm collaboration networks directly, but rather at training and providing services to the organisations responsible for economic development policy and innovation in order to build their capacity to foster networking. Thus, the programme is targeted at the intermediaries who encourage and facilitate networks, rather than at the networks themselves.

Two evaluation studies of the networks supported through USNet show that, overall, the mainly small firms in the networks are satisfied with the achievements of the networks and have increased their performance as a result. Information sharing is the primary objective of their participation.

The *Belgian* PLATO programme takes a completely different approach to regional networking. The programme is managed by the Regional Development Agency in the Kempen region. It aims to increase the transfer of knowledge between SMEs and large international companies through "mentoring". The programme is funded by the European Union and the Flemish Government. The concept has been exported to a number of European regions in Denmark, Germany, Ireland, the Netherlands, Slovenia and Sweden.

PLATO brings together a group of SMEs from different sectors in thematic sessions, with a mentor from a large (international) company. A group of some 12 SME entrepreneurs and two mentors from a large company participate in these groups for a period of two years. The emphasis is on sharing knowledge on management issues. Guest speakers can be invited to the monthly sessions. Bilateral sessions with the mentor and a small-company manager take place if the need arises. A pact of confidentiality is agreed between the group members regarding all information shared. These meetings serve to stimulate the exchange of knowledge between the SME managers in the group. The objective is to encourage the development of more formal networks linking "allied enterprises".

The task of the management is to find and select mentor companies and SMEs to participate in the programme. They also support and train the mentors from the large companies. Companies that have taken up the role of mentors are well-established large to medium-sized businesses, ranging from multinationals such as Philips, Agfa-Gevaert and Siemens, to more regional medium-sized firms.

This initiative has been in operation since 1988 and has been copied in several other European countries. In the Kempen region alone, more than 600 SMEs have participated over the last ten years.

An evaluation conducted in 1995 found that SMEs which had participated in the programme performed significantly better (growth in turnover and employment) than those which had not. The entrepreneurs surveyed saw a direct relationship between their improved performance and their PLATO activities. It could be questioned whether this type of SME networking should be seen as a "cluster" initiative since the firms brought together are not in direct business with each other. In the first instance, they share a common learning process to increase their internal capabilities rather than increasing their interaction. One could argue, however, that doing this in a network architecture increases their awareness of the learning opportunities to be had from working with other firms. The typical "SME isolation" problem and the lack of trust in exchanging knowledge with other firms create an important barrier to cluster building in general.

In addition to the examples described here, many similar national and regional initiatives have been set up to encourage inter-firm networking. The main policy target group is SMEs since small firms suffer most from lack of access to strategic market information. Experience with these initiatives has shown that involving large firms in these networks can augment the capabilities of SMEs as suppliers, co-developers or, in general, companies that are better equipped to take up new market opportunities.

Although we have established that inter-firm networking does not equal clustering, the creation of networks can be one of the tools for policies addressing clusters at a higher level of aggregation. The political decision that needs to be taken when designing these programmes is to what extent public sector agencies should act as "network managers". The majority of the programmes described above involved a far-reaching role for public agencies in taking up activities to match, mentor, manage and even administrate the networks. A proactive role for government can be expected in less-favoured regions with many isolated small firms with little exposure to markets outside their own area. The Danish case shows that the "over-pampering" of networks can lead to perverted behaviour, in which case the networks do not survive long beyond their subsidised phase. Experience from several of these initiatives confirms that if network participants are not committed to a common business opportunity, public brokerage will not have much effect.

4.3. The regional cluster development approach

The most explicit hands-on and comprehensive initiatives to stimulate clusters can be found at the sub-national policy level. Depending on the region's size and the empowerment of regional actors *vis-à-vis* national authorities, independent activities to promote local clusters can be quite successful. Often this serves to promote the attractions of the region to potential investors and to upgrade the capabilities of indigenous companies.

Essential elements of the regional development model are:

- The active involvement of regional public agencies (government bodies, development agencies, innovation centres), focusing on strong sectors in their local economy, either in traditional areas or in areas which are emerging through knowledge strengths or inward investment.
- The use of a mix of policy tools to stimulate this development consisting of a combination of policy "disciplines" (industrial policy, S&T, education, foreign acquisition, etc.).
- Encouragement of informal face-to-face contacts between cluster members.

In particular, regions are adopting clusters and networks as a key tool in their economic development and innovation policies. Several developments have led to the emergence of the region as the initiating policy level in innovation and cluster policy:

- The growing understanding of the innovation process as an interactive learning process. Following this approach, the direct environment in which – particularly small – firms operate, is vital for the development of learning capabilities. An institutional environment where firms have easy access to innovation support services, informal and formal contacts with other firms, availability of skilled labour, are all aspects of the firm's learning environment. For these sources of innovation, proximity has a role to play. Therefore, a large part of regional innovation action is aimed at the establishment of inter-firm links and links between firms, knowledge suppliers and intermediaries.

- Building networks is essentially a person-to person affair. Many interlinkages start with informal contacts where trust can build up, making more strategic alliances easier to establish. Here again, proximity makes a difference as it can help increase the frequency of meetings.

- In recent years, policy makers have been more realistic with regard to the economic potential of attracting (foreign) investment from high-tech industries and "building high-tech fantasies" around science parks. At the same time, the authorities have reawakened to the importance of innovation in indigenous firms. Regional authorities, and especially those in less-favoured regions, are increasingly aware that offering financial incentives for industrial investment alone is not sufficient for sustainable economic development. Embedding these firms in the local economy and creating synergy between indigenous firms and investors is now on many regional policy agenda. Cluster policy is a suitable tool for that task.

- In Europe, the European Structural Funds have shifted their focus away from infrastructural and general economic activities to developing strategies for regional innovation. Regions have gained more opportunities to surpass their national governments and develop their own policies and activities. Through this empowerment many cluster initiatives have been launched in the last five years. More action-oriented EU initiatives, such as the Regional Innovation and Technology Transfer Strategy (RITTS) projects and Regional Technology Plans (RTP), are making a clear contribution to the dissemination of these policy lessons throughout Europe.

- Policy discussions on subsidiarity have reinforced the necessity of designing and implementing policies at the most appropriate level. The regional policy level is increasingly seen as the appropriate level for diffusion-type innovation activities.

The discussion should not be a question of "*either* national *or* regional *or* international". The appropriate combination of supporting clusters at all these levels will lead to greater success. An example could be an agro-food cluster project in a Dutch region, bringing together local firms to increase the quality of their supply chain, which can rely on the extensive national infrastructure of expertise and skills in this area. At the same time, a regional cluster could have great potential on the international market, but not feature at all on the national map of clusters. Porter (1990) argued that, given the concentration of so many strong clusters, there are good arguments for studying competitive advantage at the regional level since "the conditions that underlie competitive advantage are indeed often localised within a nation, though at different locations for different industries". However, he concluded that it is the combination of national and intensely local conditions that fosters competitive advantage.

The case described in this study is the *United Kingdom* (Lagendijk and Charles, this volume), which has no national-level cluster-based policies.

Regional cluster policy in the United Kingdom is limited to the two regions, Wales and Scotland, that were given a certain degree of autonomy and budgets to develop their own regional policies. Both regions have developed a "package" of clustering initiatives, including:

- Supply chain development by networking suppliers in a certain sector (such as Source Wales).

- Activities to cluster industry with the research capacities in the region (the appointment of Centres of Excellence in Wales, activities of the Scottish Electronics Forum).

- Inward investment strategies aimed at attracting foreign investors from certain industrial sectors.

Wales has set up a number of initiatives to encourage clustering. The Welsh Supplier Associations (UK-Welsh Development Agency) bring together, in an informal setting, subcontractors which share the same large contractors. The Sources Wales Programme has initiated 31 supplier associations, some of which have formed the basis for long-term relationships between suppliers. One such association that has been in operation for some time is the Automotive Suppliers Association. Subcontractors hold network meetings, inviting the large – often international – automotive companies in Wales and the rest of the United Kingdom. The idea is that, together, subcontractors can increase their business opportunities in relation to their clients. The network is organised on an informal basis. Other networking activities in Wales are the Technology Clubs, a meeting-place for firms, academics, funding bodies and legislators. A third type of clustering in Wales is the Welsh Medical Technology Forum, which was founded as a collaborative venture between industry, The Welsh Office, the Welsh Development Agency, the National Health Service and academia. Its mission is to improve the competitiveness of the medical sector in Wales. The Forum has an industrial chairman and representatives from industry on its steering group. More than 90% of all health-care companies in Wales have participated in seminars and workshops since its launch in 1992. The UK Government White Paper, *Competitiveness: Helping Business to Win* highlighted the Forum as a model for best practice for regionally based technology associations.

In Scotland, Scottish Enterprise, the Development Agency has set up a number of initiatives to support the clustering of firms in various parts of the region. The most well known are aimed to support the image of Silicon Glen, the Scottish electronics cluster. An illustrative example is the Scottish Electronics Forum (SEF).

Scottish Enterprise helped to create this industry-led group, the membership of which encompasses all the Original Equipment Manufacturers (OEMs) based in Scotland. It has four main themes:

- Development and upgrading of skills, both short-term (retraining, etc.) and the identification of possible long-term shortages and appropriate responses.

- Commercialisation of university R&D – in particular to act as a catalyst through the provision of advice and informal evaluation of university projects.

- Encouragement of R&D, closely associated with the goal of commercialisation.

- To establish suppliers' needs of microelectronics companies and to encourage the establishment of local supply sources, either from existing enterprises or by attracting new companies.

Scotland has been less successful than Wales in embedding foreign companies into the regional economy. Analyses show that the ties between electronics companies and local suppliers, the labour market, and science and technology centres are not very strongly developed. The recent financial crisis in Asia has brought back fears that assembly plants in the region could be relocated or shut down, since the region does not offer sufficient value added to the electronics cluster.

The federal structure and sheer size of the *United States* mean that cluster activities take place at the state and sub-state level rather than at the federal level. Nevertheless, some national cluster-based initiatives have been taken by different agencies. The National Institute of Standards and Technology, and the Technology Administration of the US Department of Commerce have, as part of their Manufacturing Extension Partnership, prepared an "Entrepreneurial Guide to Co-operative Strategies for Manufacturing Competitiveness". This guide is aimed at brokers to help entrepreneurs set up firm collaboration networks as part of general business support services. The Manufacturing Extension Centres are located throughout the country and operate as regional intermediaries. Networking activities focus on supply-chain development to help smaller manufacturers meet the needs of the supply chains to which they belong. USNet, described above, fits into this type of networking policy approach.

The most prominent and most frequently quoted high-technology cluster in the United States, Silicon Valley, was not the result of explicit policy initiatives. The provision of large-scale defence-related R&D contracts from as early as the 1940s strongly influenced its emergence (Saxenian, 1996), but it was not foreseen that such a cluster of IT-related industries would spin out. Silicon Valley has become a successful model that many policy makers have tried to copy.

The main industrial policy approach in many US states focuses on attracting inward investment from other states or from abroad. Potential investors are offered facilities, tax credits, etc., to persuade them to locate in a particular area. Job creation is the main objective of this approach. Few states relate this acquisition policy to an explicit cluster-oriented policy. Those states that did develop a cluster policy approach started by conducting studies with the aim of improving strategic decision making (Held, 1996). A study commissioned by the US Department of Commerce on Cluster-based Economic Development describes 17 state-led cluster initiatives, mainly relating to emerging high-tech clusters. The list is far from complete since other studies have identified further initiatives; for example, New York State has developed cluster policy initiatives in Information Technologies and Distribution (Held, 1996).

The overall picture emerging from this regional development and SME capability model is that:

- Regional development actors tend to use instruments that are more hands-on in terms of intervening in the emergence and sustaining of inter-firm networks. Compared to national measures, regional policy measures, particularly in less-favoured areas, come under less scrutiny by international competition policy rules.

- The role of brokers in performing this role is a crucial element in the policy initiatives and, in many cases, the training of these brokers takes up much of the resources of the policy initiative.

- Regions that are actively involved in "making" clusters happen use a mixed package of policy instruments, ranging from inward investment through SME networking to technology transfer activities.

4.4. Cluster policy to improve research-industry relations

A less explicit cluster orientation in national policies can be found in those countries where networks and inter-firm collaboration are stimulated in order to make better use of (public) knowledge resources or to conduct joint research.

The key characteristics of this approach are:

- The partners in the clusters are firms and research establishments, therefore the clusters are not defined as a product value chain but rather as a knowledge value chain, ranging from basic research to commercialisation.

- The focus is on creating micro-level networks (a limited set of industrial and research organisations) and/or spatially concentrated centres of excellence.

- The firms involved are technologically competent and often have internal R&D facilities.

- Typical tools are financial support for collaborative R&D, support for technology-based start-up firms, and the provision of strategic information on science and technology developments for particular sectors.

The policy philosophy is that groups of firms are better equipped than individual companies to make use of available knowledge resources. This could be the case for SMEs, which do not have sufficient financial resources to conduct R&D, or for groups of companies benefiting from joint pre-competitive research in different fields of application. These groups of firms can be *ad hoc* working arrangements for the duration of an R&D project, or longer-term arrangements, for instance on behalf of a certain sector. Increasingly, we see initiatives where joint industry-research organisations are set up in which industry has a large say in determining the research agenda. Examples include the Swedish Competence Centres, the Dutch Technological Top Institutes and the Interdisciplinary Research Centres in the United Kingdom. Nevertheless, in general, the objective to these initiatives is to increase the access to research of a particular sector or type of firm, and not a combination of firms linked through the value chain.

Germany has a long tradition of collaborative and "network" research, without having an explicit cluster focus in the sense of the definition used here.[6] Many of their RTD support mechanisms contain an element of collaborative research. Examples are programmes (*Forschungskooperation in der Mittelständischen Wirtschaft*) to subsidise R&D personnel of a group of SMEs engaged in either national or international R&D co-operation, or groups contracting out R&D to a research organisation. Other sector-based R&D facilities are provided by the German Federation of Industrial Co-operative Research Associations (*Arbeitsgemeinschaft industrieller Forschungsvereiningungen* – AIF). These are industry-led initiatives for collaborative research in particular industrial sectors. The AIF links SMEs by co-ordinating their R&D requirements. If a sufficient number of firms have an interest in a particular research project, their association prepares a detailed plan and costing of the research to be performed. This is then carried out either in the research association's own institutes or in another scientific establishment.

Despite this long-standing experience with collaborative research, this activity can not be labelled as cluster-based policy since it is not aimed at creating partnerships between firms.

More recently, cluster-type initiatives are being developed by the German Federal Government to stimulate regional "centres of excellence", *i.e.* clusters of research and industry in a particular technological field. The German Government (BMBF) has launched a series of debates and studies on the issue of supporting regional high-technology clusters or *Kompetenzzentren*. The aim is to select regions that have the potential to develop into internationally acclaimed centres of expertise. A pilot project was launched in 1996 through a competition for the selection of "Bio-Regions". The regions selected were Munich, Rheinland, Rhein-Neckar-Triangle and Jena.

The aim of the project is to develop German's biotechnology RTD activities into a commercially successful and powerful sector on the world market. To achieve this goal, experience, potential and network activities were called upon. The selected regions had a strong existing potential of firm start-ups, research facilities and organisational experience in this field. The scheme supports:

- Co-operation between research institutions and medical centres in the area of research and testing.
- Help in management issues for start-up biotechnology firms.
- A network to exchange experiences within workshops and conferences.
- Privileged funding from the BMBF-biotechnology promotion programme during 1997-2001.

Each region has a co-ordination team or agency to manage the operational issues relating to the federal Bio-Region programme. One of the main tasks is to obtain access to risk capital from private firms or banks. The federal government (BMBF) initially controlled the main financial resources. Once the four regions had been selected, this responsibility was transferred to their management offices. Semi-public organisations or companies co-ordinate the work, the financial flows and the co-operation activities. Thus, the initiative combines a technology cluster perspective with a spatial concentration perspective, with the objective of kick-starting an emerging technology into a commercially successful production cluster. At the time of writing, the initiative is too young to report on its impacts. However, it has become the model for future initiatives which are now being put into place, for example on the theme of mobility (transport technologies, logistics, etc.).

Two smaller countries that have placed the emphasis on the industry-research perspective are the Netherlands and Austria.

The *Netherlands*, as stated above, aims to move to a more national advantage approach, applying a wide range of policy tools to enhance collaborative networking between firms, and between firms and knowledge centres. At present, most of its policy tools are focused on research collaboration. It has recently restructured its R&D subsidy approach: whereas in the past individual companies could apply for R&D grants, now only a group of firms linked by (possibly) a research centre can apply for funding. This scheme (*Bedrijfsgerichte Technologie Stimulering* – BTS), launched in 1997 by the Ministry of Economic Affairs, is representative of the new approach: public support is given to a network of firms, even if this is at the most micro level.

Austria, inspired by the German "competence centre" approach, has launched a similar *Kompetenzzentren+* initiative. The main difference is that these activities focus on research centres and their partners, rather than on agglomerations of firms around a number of research centres in a concentrated geographical area.

5. Conclusions

As this review has shown, there is no unambiguous or "best practice" definition of precisely what cluster policy is and where its boundaries lie. We have seen very different interpretations of public action to stimulate clustering and networking. At the national level, the characteristics and content of cluster policy are interpreted in such a way as to fit the general policy visions and traditions in industrial policy, science and technology policy, and regional development policy. At the regional level, more experimentation has been accepted, particularly on the micro-networking level. Many bottom-up inter-firm networking models have been developed with some support from the public sector. Nevertheless, many countries have some kind of networking initiatives, which form part of their efforts to increase SME competitiveness, encourage regional development, improve industry-research collaboration, or a combination of these objectives.

Few countries have an *explicit* cluster policy at the national mega level, where a choice for strong clustered sectors is made and for which a set of focused policy initiatives are developed. For very large countries with strong sub-national governance levels, such as Germany and the United States, focusing on nationally strong clusters is politically precarious, since this could lead to rivalry among states. Although a number of countries (Austria, Canada, the Netherlands, Sweden) have conducted extensive mapping exercises to identify the nation's clusters and their issues, only Denmark has so far translated this into a comprehensive cluster policy approach that includes regulation, standards, education, R&D, procurement policy and networking. Denmark and Finland apply the clearest "national advantage" model, where the one country (Denmark) has chosen to directly support its clusters through focused investment and policy initiatives, while the other (Finland) has chosen to adopt a more general "competition and framework conditions policy". In some countries which do not have an explicit cluster policy at the national or federal level, specific regions have been very active (Canada, Germany, the United Kingdom, the United States) in developing regional clusters. Other countries have adopted a regional development model operated on the national level (Norway and, to lesser degree, Sweden).

Table 3 links the dominant policy models identified in this review with the tools used to implement them.

Table 3. Policy models and their main instruments and public roles

	Mega level	Meso level	Micro level
1. National advantage	• Mapping • Competitive markets • Regulations and standardisation	• Foresights • Specialised RTD facilities	• Collaborative RTD programmes
2. Inter-firm (SME) networking		• Supply chain development	• Brokerage • Networking programmes • Awareness raising
3. Regional development	• Regional Competence Centre development	• Focused Inward investment • Supply-chain associations • Specialised technology transfer • Marketing clusters	• Brokerage • Networking programmes • Awareness raising
4. Industry-RTD clustering	• Incentives RTD-industry collaboration (IPR, financial, etc.)	• Collaborative RTD centres programmes in specific areas • Prioritisation of R&D expertise	• Technology circles • NTBF support • Procurement policy

The review of countries has provided a typology of different cluster models and policy instruments that could be used in a cluster policy model. The following figure places the different countries in these four models (Table 4).

Table 4. Dominant cluster policy focus per country

	Mega level	Meso level	Micro level
1. National advantage	Denmark Finland		
2. Inter-firm (SME) networking			Australia New Zealand Ireland Norway
3. Regional development		US states UK regions Canada	Netherlands regions
4. Industry-RTD clustering		Germany Netherlands	Austria Sweden

An intriguing question for policy makers is whether clusters can be created or designed by policy. Can public efforts and investment trigger the emergence of clusters – preferably high-tech and fast-growing clusters – for instance by large investments in research and technology? The ambition of many economic developers is to create a new Silicon Valley. With the exception of the ITC and biotechnology clusters, the targeted clusters in the countries under review all have a long historical development behind them. There are very few examples of high-technology clusters that had been successfully "planned" by government and those that can be pointed to, tend to be regionally concentrated. High-tech boom cities, such as Austin (United States) in ICT and multimedia, and Olou (Finland) in mobile telecommunication, have their own factors of success in which targeted and sustained public investment played a major role. The public investment functioned as a leverage mechanism to attract even larger amounts of private investment in technological capabilities.

However, for each success story, there are even more examples of "technopoles" that failed. The policy tools described in this chapter mainly function as facilitators of market-led developments.

Although it was not possible to look at the regional level in great detail in this international review, there are some examples of regional governments that have been successful in establishing new clusters. One of the success factors was an integrated approach to cluster support. This included traditional industrial policy mechanisms such as an active inward investment policy, tax and infrastructure facilities as well as initiatives in the area of specialised training and education, and collaboration between university and industry.

The main conclusion from the international review is that no single policy instrument can support the emergence of value-adding and innovative clusters. Given:

- the multiple levels and geographical locations in which clusters are positioned;
- the need for horizontal, vertical and lateral linkages in a dynamic cluster;
- the role of knowledge carriers to bring in new ideas;
- the constant changes in international markets and technologies; and
- the influence of economic, financial and other regulations on clusters;

cluster policy strategies should aim at a broad set of policy tools that do not conflict which each other. Since, in most countries, policies for macroeconomic development, industry, financial markets, science and technology, education and regional development are divided across various administrative bodies, the creation of integrated cluster policy "packages" will be a major challenge in terms of changing the policy process as well as changing its content.

NOTES

1. Nevertheless, the authors are aware of interesting activities taking place in countries such as Brazil, India, Mexico and many developing countries around the world. Within the scope of this OECD-related study, it was not possible to make a comprehensive overview of all cluster-related policies.

2. Examples include Castells and Hall (1994) and Saxenian (1994).

3. Danish Ministry of Business and Industry, "Dialogue with Industry", undated.

4. For an extensive discussion on these matters and the influence of Porter's study, see Steinbock (1998).

5. Quoted from Pertti Monto, editor–in-chief of *Talouselämä* (Economic Life) in Steinbock (1998), p. 94.

6 The authors would like to thank Katrin Vopel and Alfred Spielkamp of ZEW, partners in the OECD Focus Group on Clusters, for their contribution to the parts of the text which refer to Germany.

REFERENCES

Boekholt, P. (1997), "Innovative Networks in Regional Economies: Enhancing Infrastructures for Technology Creation and Transfer", in J. Mitra and P. Formica (eds.), *Innovation and Economic Development: University-Enterprise Partnerships in Action*, Oak Tree Press.

Boekholt, P., J. Clark, P. Sowden and J. Niehoff (1998), "An International Comparative Study on Initiatives to Build, Develop and Support 'Kompetenzzentren'", study for BMBF, Bonn, Technopolis, Amsterdam.

Castells, M. and P. Hall (1994), *Technopoles of the World: The Making of 21st Century Industrial Complexes*, Routledge, London.

Cooke, P., P. Boekholt and F. Tödtling (1998), *Regional Innovation Systems: Designing for the Future*, TSER Report for the European Commission, forthcoming, Pinter, London.

Danish Ministry of Business and Industry, "Dialogue with Industry", undated.

David, Paul A. (1996), "Accessing and Expanding the S&T Knowledge Base: Public Policy and Private Action", presentation to the Six Countries Programme on "R&D Subsidies at Stake? In Search of a Rationale for Public Funding of Industrial R&D", Gent.

Drejer, I., F. Skov Kristensen and K. Laursen (1990), "Studies of Clusters as a Basis for Industrial and Technology Policy in the Danish Economy", this volume.

Held, J.R. (1996), "Clusters as an Economic Development Tool: Beyond the Pitfalls", *Economic Development Quarterly*, Vol. 10, No. 3, August, pp. 249-261.

Jacobs, D., P. Boekholt and W. Zegveld (1990), *De economische kracht van Nederland*, SMO, The Hague.

Kline, S.J. and N. Rosenberg (1986), "An Overview of Innovation", in R. Landau and N. Rosenberg (eds), *The Positive Sum Strategy: Harnessing Technology for Economic Growth*, The National Academy Press, Washington D.C.

Lagendijk, A. and D. Charles (1999), "Clustering as a New Growth Strategy for Regional Economies? A Discussion of New Forms of Regional Industrial Policy in the United Kingdom", this volume.

Morgan, K. and P. Cooke (1991), *The Intelligent Region – Industrial and Institutional Innovation in Emilia-Romagna*, Regional Industrial Research Report, Wales Development Agency.

Piore, M.J. and C.F. Sabel (1984), *The Second Industrial Divide: Possibilities for Prosperity*, Basic Books, New York.

Porter, M. (1990), *The Competitive Advantage of Nations*, Macmillan, London.

Roelandt, T., P. den Hertog, J. van Sinderen and N. van den Hove (1999), "Cluster Analysis and Cluster Policy in the Netherlands", this volume.

Rosenfeld, S.A. (1996), *Achievers and Over-achievers*, RTS, Chapel Hill.

Rouvinen, P. and P. Ylä-Anttila (1999), "Finnish Cluster Studies and New Industrial Policy Making", this volume.

Saxenian, Annalee (1994), *Regional Advantage, Culture and Competition in Silicon Valley and Route 128*, Harvard University Press.

Steinbock, D. (1998), *The Competitive Advantage of Finland: From Cartels to Competition*, Taloustieto Oy, Helsinki.

Chapter 17

CLUSTER ANALYSIS AND CLUSTER-BASED POLICY MAKING: THE STATE OF THE ART

by

Theo J.A. Roelandt
Dutch Ministry of Economic Affairs

Pim den Hertog
Dialogic, Utrecht

1. Introduction

The contributions presented in the preceding chapters illustrate that both cluster analysis and cluster policies are practised worldwide. The *cluster approach* is a viable alternative to the traditional sectoral approach. The cluster approach is valued in many countries because it offers useful insights into the linkages and interdependencies among networked actors in the production of goods and services and in innovation (Box 1). It has revealed systemic imperfections in innovation systems as well as policy responses aimed at improving the efficient and dynamic functioning of innovation systems. However, this book has pointed to the wide variety of cluster approaches adopted in different countries. In fact, not only do levels of analysis and methodologies differ across countries, but also the degree to which cluster-based policies have been implicitly or explicitly implemented as well as their form in terms of instruments used. This chapter discusses and summarises the similarities and differences between the contributions to this book. What are the lessons that have been learned so far? And how can the cluster perspective be further developed in the near future?

Box 1. Innovation, interdependency and the cluster approach

Innovation is not usually a single-firm activity, it increasingly requires an active search process in order to tap new sources of knowledge and technology and apply these in products and production processes. Systems of innovation approaches give shape to the idea that companies in their quest for competitiveness are becoming more dependent upon complementary knowledge in firms and institutions other than their own. The cluster approach focuses on the linkages and interdependencies among networked actors in the production of goods and services and in innovation. In so doing, the cluster approach offers an alternative to the traditional sectoral approach.

The following research questions were formulated in Chapter 1:

1. Which clusters can be identified economy-wide?
2. How do clusters innovate? Which innovation styles are most successful in which clusters?
3. How do the same clusters in different countries vary in their economic and innovation performance and how can the differences in performance be explained?

4. What are the lessons to be learnt from the above for policy making?

5. Which policy instruments have been used in the various countries and what is the role of cluster analysis?

6. What are the key instruments and pitfalls of cluster-based policy making?

These questions are addressed below, synthesising the results on cluster analysis (mainly questions 1-3) and cluster-based policies (questions 4-6) and using some overview tables.

Box 2. Clusters as reduced-scale national innovation systems

Economic clusters can be characterised as networks of production of strongly interdependent firms (including specialised suppliers) linked to each other in a value-adding production chain. In some cases, clusters also encompass strategic alliances with universities, research institutes, knowledge-intensive business services, bridging institutions (brokers, consultants) and customers. Clusters are usually cross-sectoral (vertical and/or lateral) networks and contain dissimilar and complementary firms specialised around a specific link or knowledge base in the value chain. The cluster concept is, in fact, a specific type of a much larger family of "systems of innovation" approaches which have systems analysis as their common-starting point but which differ in the object and level of analysis (supranational, regional, sectoral or technological systems of innovation, clusters). Clusters can be interpreted as reduced-scale national innovation systems: The dynamics, system characteristics and interdependencies are similar to those for national innovation systems.

2. The scope of cluster analysis

In the countries in which cluster analysis is practised, it has produced useful information on the actors involved in clusters, value-chain relations of firms, and innovation interaction linkages as well as the institutional setting for clusters' innovation systems and the imperfections of these cluster-based innovation systems. Many of the country contributions have in common that they describe *networks of strongly interdependent firms or industry groups*:

- In some cases based on *trade linkages* (Hauknes, this volume; Roelandt *et al.*, this volume; Bergman and Feser, this volume).
- Sometimes on *innovation linkages* (DeBresson and Hu, this volume).
- Sometimes on *knowledge flow linkages* (Viori, 1995; Poti, 1997; Roelandt *et al.*, this volume; van den Hove and Roelandt, 1997).
- Sometimes based on a *common knowledge base or common factor conditions* (Drejer *et al.*, this volume).

In all events, the common starting-point of these perspectives is the assumption that, in order to innovate successfully, firms need a network of suppliers, customers and knowledge-producing agents. Most cluster analyses use a combination of different techniques at different levels of aggregation. Table 1 shows how the level of analysis, cluster techniques and cluster concept used varies among countries. Most countries combine various techniques to overcome the limitations of a single technique; different methodologies can be used to answer different questions and to provide different sorts of information.

Table 1. Level of analysis, cluster technique and cluster concept adopted in various countries

Country	Level of analysis			Cluster technique					Cluster concept
	Micro	Meso	Macro	I/O	Graph	Corresp.	Case	Other	
AUS		X	X	X		X	X		Networks of production, networks of innovation, networks of interaction.
AUT		X	X			X	X	Patent data and trade performance	Marshallian industrial districts.
BEL	X				X			Scientometrics	Networks or chains of production, innovation and co-operation.
CAN		X	X	X			X		Systems of innovation.
DK	X	X		X	X		X		Resource areas.
FNL	X	X					X		Clusters as unique combinations of firms tied together by knowledge.
GER	X	X		X		X			Similar firms and innovation styles.
IT		X		X					Interindustry knowledge flows.
MEX		X	X				X		Systems of innovation.
NL		X	X	X			X		Value chains and networks of production.
NOR		X	X	X			X		Value chains and networks of production.
SP		X		X			X		Systems of innovation
SWE		X					X		Systems of interdependent firms in different industries.
SWI	X	X				X	X	Patent data	Networks of innovation.
UK	X	X					X		Regional systems of innovation.
USA		X		X			X		Chains and networks of production.

What can we learn from cluster analysis?

The cluster analyses conducted in the various countries reveal the *value added of using cluster analysis*.[1] The advantages of cluster analysis highlighted in this book include the following:

- Cluster analysis offers a new way of thinking about the economy and organising economic development efforts; it overcomes some of the limitations of traditional sectoral analysis.

- Cluster analysis accounts better for the changed nature of competition and market-based innovation systems and the main sources of competitive advantage. It captures important linkages in terms of technology, skills, information, marketing and customer needs that cut across firms and industries. Such linkages and interdependencies are fundamental to the direction and pace of innovation.

- Studies of clusters, as reduced-scale innovation systems, have improved our understanding of innovation systems, including systemic imperfections and policy options.

- Cluster studies are now the cornerstone of industrial policy making in many countries. Cluster studies not only provide an analytical tool for studying systems of innovation, they can also be used as a working method for policy making in this area and as an economic development tool for strategic business development, in industrialised as well as developing countries (Ceglie *et al.*, this volume).

- Cluster analysis provides the possibility to recast the role of the private sector, government, trade associations and educational and research institutions, and presents business development opportunities of firms of all sizes, crossing traditional industry lines.

- Cluster analysis provides a starting-point for a forum for constructive business-government dialogue. Not only have common problems been identified, cluster analyses can serve to identify common development opportunities and highlight attractive public and private investment opportunities.

Methodological bottlenecks

However, a number of methodological bottlenecks and complications seriously hamper the international comparability (both quantitative and qualitative) of cluster studies performed in individual (national, regional, cluster) innovation systems. Countries' experiences with cluster analysis revealed the following methodological bottlenecks:

- The use of existing official national and international data sources for cluster analysis is limited by conventions on official classification systems of economic activities and industries. These sources were not designed to cover flow relations between different industries (Peneder, this volume) or to measure dynamic interactions and linkages between industries and firms. Some countries (especially Canada, Denmark and Finland) have decided to improve the statistical information on clusters by establishing statistical groups and research teams to produce data in line with the needs of cluster analysis and cluster-based policies (Drejer *et al.*, this volume; Sulzenko, 1997). Other countries (*e.g.* Belgium, the Netherlands, Sweden) would also like to improve their statistics for cluster analysis.

- Using input-output (I/O) tables to identify clusters or technology flows has considerable methodological limitations (Drejer *et al.*, this volume). Identifying networks of production requires a very fine level of aggregation of the I/O tables, and cluster analysis needs data at very low levels of aggregation (the three- or four-digit industry code level). Some countries (Canada, Denmark, the Netherlands, the United States) have very detailed and useful I/O tables (as well as make & use tables at the product level); other countries only have tables at a relatively high level of aggregation (two-digit) (Germany, Spain), while yet another group of countries have severe data shortages in this field (Austria, Belgium, Sweden, Switzerland). The data for OECD countries contained in the official OECD I/O database is too aggregated for internationally comparable cluster analysis. Countries that do have access to very detailed I/O tables have been able to produce stable and useful results in identifying networks of production and innovation. Countries with severe data problems are currently seeking to improve their I/O data sources (*e.g.* Belgium, Germany, Sweden).

- The use of innovation interaction matrices describing the flows of innovations from suppliers to users is promising, but is limited to the flows of major innovations of using and supplying industries. The main advantage of these tables is their focus on innovation interdependency and interaction between industry groups when innovating. However, the tables suffer from their relatively high level of aggregation. In future, the availability of this type of data could be improved with the addition of questions on the main users and producers of innovations to Eurostat's Community Innovation Survey (CIS) questionnaire.

- In addition to statistical analysis, most countries combine their statistical cluster analysis with qualitative and monographic cluster case studies. One of the major disadvantages of case studies is that the approach is intrinsically qualitative. A quantitative approach is needed to map production relations, innovative networks and clusters of economic activity. Combining the more qualitative cluster studies with input-output analysis can considerably reinforce the results. The dynamics in the clusters identified statistically can only be interpreted sensibly in combination with the more qualitative insights gained through monographic case studies.

- A final complication concerning an international comparison of the same clusters in different countries is changing specialisation patterns worldwide.[2] A trend towards growing specialisation among OECD countries, and among the same industry groups and clusters in different countries has been observed (OECD, 1997). This implies that the growing importance of networking between dissimilar and complementary firms with different specialisation patterns discussed above has an important international dimension. As a consequence, the innovation systems and specialisation patterns of the same clusters (operating in value chains producing products and services for the same end-product markets) within different countries can differ significantly in institutional setting and innovation performance. This makes identifying "best practices" or "optimal incentive structures in innovation systems" almost tautological. International comparative research in this field can reveal the critical factors of these diverging strategies.

Table 2. Clusters identified in the participating countries

	1	2	3	4	5	6	7	8	9	10	11	12	13	14	15	16	17	18	19	20	21	22
AUS			X				X			X	X	X	X									
AUT									X	X				X		X			X		X	
BEL					X								X		X	X						
CAN																						
DK	X				X	X	X				X		X				X	X				
FNL	X	X			X	X	X			X	X		X	X					X			
GER		X			X			X		X			X		X							
NL	X	X	X	X	X	X	X	X	X	X	X											
NOR	X				X		X		X		X											
SP													X									
SWE	X				X		X			X			X	X		X				X	X	
UK																						
USA	X	X			X		X	X		X			X	X	X						X	X

Note: 1. Construction, 2. Chemicals, 3. Commercial services, 4. Non-commercial services, 5. Energy, 6. Health, 7. Agro-Food, 8. Media, 9. Paper, 10. Metal-electro, 11. Transport & communication, 12. Bio-medical, 13. Information & communication technology, 14. Wood & paper, 15. Biotechnology, 16. Materials, 17. General supplier business, 18. Consumer goods & leisure, 19. Environmental, 20. Machinery, 21. Transport (vehicles), 22. Aerospace.

Table 2 illustrates the clusters identified in the participating countries. Due to differences in cluster methodology (see Table 1), this table should be interpreted with caution when comparing countries' cluster maps internationally.[3]

3. Countries' strategies in cluster-based policy

Clustering/networking is basically a bottom-up, market-induced and market-led process. Following the classical line of reasoning, the primary task of government should be to facilitate the dynamic functioning of markets and ensure that co-operation does not lead to collusive behaviour which restricts competition. This classical approach can be criticised for its limited scope and for the fact that it has not evolved in line with the changing character of market-based innovation systems, growing understanding of the functioning of market-based innovation systems and insights derived from modern innovation theory. Nevertheless, cluster studies have also revealed the need to redefine the role of the government as a facilitator of networking, as a catalyst of dynamic comparative advantage and as an institution builder, creating an efficient incentive structure to remove systemic inefficiencies in systems of innovation.

Box 3. The systemic imperfections argument and systemic cluster-based innovation and industrial policies

Four rationales for innovation and industrial policy making are reported in the literature: *i)* creating favourable framework conditions for an efficient and dynamic functioning of markets; *ii)* externalities associated with investments in knowledge; *iii)* the economic role of government as a demanding customer; and *iv)* systemic imperfections. In the majority of countries, industrial and innovation policy making actually focus on improving the efficient functioning of their systems of innovation. The latter rationale is increasingly seen as the key rationale for systemic innovation and industrial policies.

In practice, countries' cluster policy approaches differ. One fundamental difference relates to the distinction between a *bottom-up approach,* on the one hand, and a more or less *top-down approach,* on the other (Boekholt *et al.*, this volume). The first approach focuses on fostering dynamic market functioning and removing market imperfections; the starting-point lies in market-induced initiatives, with the government acting as a facilitator and moderator but with no setting of national priorities (the Netherlands, the United States). In the second approach, government (in consultation with industry and research agencies) sets national priorities, formulates a challenging vision for the future and – prior to initiating the dialogue process – decides on the actors to be involved in the dialogue (this is the case in some of the Nordic countries). Once national priorities have been set and the dialogue groups implemented, the clustering process becomes a market-led process, with little government intervention.

What country strategies can be discerned in cluster-based policy? Policy researchers[4] point to various roles for government in cluster-based policy, for example:

- Establishing a stable and predictable economic and political climate.
- Creating favourable framework conditions for the smooth and dynamic functioning of markets (infrastructure, competition policy and regulatory reform, provision of strategic information).
- Creating a context that encourages innovation and upgrading by setting a challenging economic vision for the nation or region.
- Raising awareness of the benefits of knowledge exchange and networking.
- Providing support and appropriate incentive schemes for collaboration and initiating network brokers and intermediaries to bring actors together.

- Acting as a facilitator and moderator of networking and knowledge exchange.
- Acting as a demanding and launching customer when addressing needs.
- Facilitating the informal and formal exchange of knowledge.
- Setting up competitive programmes and projects for collaborative research and development.
- Providing strategic information (technology foresight studies, strategic cluster studies).
- Ensuring that (public) institutions (especially schools, universities, research institutes) cultivate industry ties.
- Ensuring that rules and regulations maximise flexible adaptation to changed market conditions and stimulate innovation and upgrading processes.

In most countries with cluster-based policies, these initiatives have originated in a trend towards designing governance forms and incentive structures to reduce systemic imperfections in national systems of innovation. These policy responses to systemic imperfections can be categorised as follows:

- Establishing a stable and predictable economic and political climate.
- Creating favourable framework conditions for the efficient and dynamic functioning of free markets.
- Stimulating interactions and knowledge exchange between the various actors in systems of innovation.
- Removing informational failures by providing strategic information.
- Removing institutional mismatches and organisational failures in systems of innovation, *i.e.* mismatches between the (public) knowledge infrastructure and private needs in the market or a missing demanding customer in the value chain.
- Removing government failures and government regulations that hinder the clustering and innovation process.

Table 3 summarises countries' cluster-based policy responses to systemic imperfections.

In many countries, the clustering process has been initiated through the establishment of forums, platforms and regular meetings of firms and organisations related to a particular network of production in the value chain. Strategic information (technology foresight studies, strategic cluster studies) is often used as an input to the dialogue process. The organisation of the clustering process differs across countries, depending on policy culture, the way in which dialogue among industry, research and governments is institutionalised, the size of the economy, the level of government intervention and the degree of industrial and technological specialisation.

Box 4. Market-induced cluster formation and new forms of governance

Cluster initiatives originate in a trend towards governance forms and incentive structures based on networks and partnerships. In practice, this means a shift away from direct intervention towards indirect inducement. The main task of policy makers is to facilitate the networking process and to create an institutional setting which provides incentives for market-induced cluster formation.

Table 3. Systemic and cluster-based policy response

Systemic and market failures	Policy response	Countries' focus in cluster-based policy making
Inefficient functioning of markets	▪ Competition policy and regulatory reform.	▪ Most countries.
Informational failures	▪ Technology foresight. ▪ Strategic market information and strategic cluster studies.	▪ Netherlands, Sweden. ▪ Canada, Denmark, Finland, Netherlands, United States.
Limited interaction between actors in innovation systems	▪ Broker and networking agencies and schemes. ▪ Provision of platforms for constructive dialogue. ▪ Facilitating co-operation in networks (cluster development schemes).	▪ Australia, Denmark, Netherlands. ▪ Austria, Denmark, Finland, Germany, Netherlands, Sweden, United Kingdom, United States. ▪ Belgium, Finland, Netherlands, United Kingdom, United States.
Institutional mismatches between (public) knowledge infrastructure and market needs	▪ Joint industry-research centres of excellence. ▪ Facilitating joint industry-research co-operation. ▪ Human capital development. ▪ Technology transfer programmes.	▪ Belgium, Denmark, Finland, Netherlands, Spain, Sweden, Switzerland. ▪ Finland, Spain, Sweden. ▪ Denmark, Sweden. ▪ Spain, Switzerland.
Missing demanding customer	▪ Public procurement policy.	▪ Austria, Netherlands, Sweden, Denmark
Government failure	▪ Privatisation. ▪ Rationalise business. ▪ Horizontal policy making. ▪ Public consultancy. ▪ Reduce government interference.	▪ Most countries. ▪ Canada. ▪ Canada, Denmark, Finland. ▪ Canada, Netherlands. ▪ Canada, United Kingdom, United States.

Our review of experiences with clusters in OECD countries has highlighted some of the *pitfalls* of cluster-based industrial policy making:

- The creation of clusters should not be government-driven but rather should result from market-induced and market-led initiatives.
- Government policy should not be strongly oriented to directly subsidising industries and firms or to limiting rivalry in the marketplace.
- Government policy should shift away from direct intervention towards indirect inducement. Public interference in the marketplace only can be justified in the presence of a clear market or systemic failure. Even if clear market and systemic imperfections exist, it cannot necessarily be concluded that government intervention will improve the situation.

- Government should not try to take the direct lead or ownership in cluster initiatives, but should work as a catalyst and broker, bringing actors together and supplying support structures and incentives to facilitate the clustering and innovation process.

- Cluster policy should not ignore small and emerging clusters; nor should it focus only on "classic", existing clusters.

- Clusters should not be created from "scratch". The cluster notion has sometimes been appropriated by (industrial) policy makers and used as an excuse to continue more or less traditional ways of defensive industrial policy making

An awareness of these pitfalls can be helpful in designing the leading policy principles of a comprehensive cluster-based policy.[5]

Table 4 summarises the strategies used in the cluster analysis and cluster-based policy initiatives of a number of countries. The most common features of cluster-based policy include:

- Vigorous competition and regulatory reform policy (almost all countries).

- Providing strategic information through technology foresight studies (*e.g.* the Netherlands, Sweden,), cluster studies (*e.g.* Austria, Denmark, Finland, Italy, the Netherlands, Sweden, the United Kingdom, the United States), special research groups (*e.g.* the Austrian TIP research programme, Denmark, the German Delphi report), or special Web sites (*e.g.* STRATEGIS in Canada).

- Broker and network agencies and schemes (*e.g.* the Danish network programme and the Dutch Innovation Centres).

- Cluster development programmes (*e.g.* cluster programmes in Finland and the Netherlands, regional development agencies in Germany, the United Kingdom and the United States, and Flemish R&D support to clusters).

- Initiating joint industry-research centres of excellence (*e.g.* Belgium, Denmark, Finland, Germany, the Netherlands, Spain, Sweden, Switzerland).

- Public procurement policy (*e.g.* Austria, Denmark, the Netherlands).

- Institutional renewal in industrial policy making (*e.g.* Canada, Finland).

- Providing platforms for constructive dialogue (*e.g.* the Danish reference groups, Dutch broker policy, the Finnish National Industrial Strategy, the German Council for Research, Technology and Innovation, the Swedish industrial system approach, the UK regional development agencies, and the US focus groups).

Table 4. Countries' strategies in cluster-based policy

Country	Approach	Cluster analysis	Policy initiatives/Policy principles
Austria	Systems of interdependent economic entities.	▪ Improving I/O tables. ▪ Traditional statistical cluster analysis screening for patterns of innovative activities. ▪ Case studies.	▪ Cluster policy under construction. ▪ Framework conditions (regulatory reform, human capital development). ▪ Providing platforms for co-operation and experimentation. ▪ Raising public awareness of technologies. ▪ Demand pull by public procurement.
Australia	Networks of economic activity.	▪ Case studies of industrial districts (geographical propinquity) and resource based clusters. ▪ I/O analysis of interindustry linkages.	▪ No comprehensive cluster-based policy. ▪ Networking schemes encouraging the emergence of inter-firm networks.
Belgium (Flanders)	Networks or chains of production, innovation and co-operation.	▪ Graph analysis and case study work. ▪ Improving I/O statistics. ▪ Technology flows. ▪ Technology clubs (similar collaboration patterns).	▪ Cluster-based policy under construction. ▪ Market induced cluster initiatives. ▪ Government facilitating co-operation. ▪ Subsidies and firms' co-financing in cluster programmes (in metal processing industry, plastics, space industry, SMEs, furniture). ▪ Stimulating cross-sectoral technology diffusion ▪ Supporting supplier-producer networks. ▪ Centres of excellence around newly emerging technologies.
Denmark	Resource areas	▪ Industrial districts/ development blocks. ▪ Porter-like cluster studies. ▪ Improving statistics. ▪ Cluster analysis as an input to the process of dialogue.	▪ Dialogue in reference groups. ▪ Centres of excellence in specific areas. ▪ New educational programmes in specific areas ▪ Development centres in specific areas ▪ Top down approach (selected priority fields) ▪ Institutional reform in policy making (co-ordination between ministries)
Finland	Clusters as a unique combination of firms tied together by knowledge and production flows.	▪ Porter-based cluster studies.	▪ Clusters as an economic development tool. ▪ Identifying sources of competitive advantages in Finnish economy. ▪ Competition policy and structural reform. ▪ Creating advanced factors of production (basically creating favourable framework conditions). ▪ Cluster programmes, strategic research, centres of excellence.
Netherlands	Value chain approach.	▪ Porter-like cluster studies. ▪ Cluster benchmark studies. ▪ Input-output analysis.	▪ Dialogue in specific platforms. ▪ Brokerage and network policy. ▪ Public consultancy. ▪ Providing strategic information (a/o. technology foresight studies). ▪ Renewal in procurement policy. ▪ Deregulation and competition policy.
Spain	Inter-sectoral linkages and dependency.	▪ Technology and innovation flow analysis.	▪ Framework policy. ▪ Stimulating R&D co-operation and R&D networks. ▪ Research centres (mixed private and public participation) and science parks.

Country	Approach	Cluster analysis	Policy initiatives/Policy principles
Sweden	Interdependencies between firms in different sectors.	▪ Development blocks (1950s). ▪ Technological systems (late 1980s). ▪ Network approach (since the 1970s). ▪ Porter studies (since the mid-1980s).	▪ Cluster-based policy under construction. ▪ General framework conditions. ▪ Technology procurement. ▪ Stimulating R&D co-operation. ▪ Research centres. ▪ Industrial systems project (is being set up) to stimulate strategic dialogue. ▪ Technology foresight studies identifying actual or potential innovative clusters.
Switzerland	Networks of innovation.	▪ Case study work on restructuring system of production and innovation (Swiss Jura arc). ▪ Analysing technological spillovers and innovation styles.	▪ Action programme for diffusion of specific technology (Computer Integrated Manufacturing). ▪ Setting up competence centres integrated in regional networks.
United Kingdom	Regional systems of innovation.	▪ Cluster case studies focus on identifying actors and development opportunities for the region.	▪ Clusters as a regional development tool. ▪ Government as catalyst and broker. ▪ Regional cluster programmes.
United States	Clusters (chains of production) as a regional development tool.	▪ Cluster analysis focusing on the strengths and weaknesses of the local economic structure and identifying business opportunities. ▪ Cluster analysis used as an input to the consultation process. ▪ I/O analysis combined with insight information from business.	▪ Dialogue in regional focus groups. ▪ Regional development plans.

Most countries use the cluster approach to organise a market-led economic development strategy by initiating dialogue between the various actors in their relevant systems of innovation. In the majority of countries reported in this book (Finland, Denmark, the Netherlands, Sweden, the United Kingdom, the United States), cluster-based policy is seen as a market-led business development strategy to bring together actors and organisations and to foster knowledge exchange and transfer. One common lesson from the cluster-based policy review is that cluster studies not only provide an analytical tool for studying systems of innovation at the reduced-scale level of networks, but in practice can also be used as a working method for policy making and as an economic development tool for strategic business development. Cluster policy making, in this sense, is – as are many policy processes – a policy-learning process and thus requires a willingness on the part of policy makers to see cluster policy making as a continuous learning process.

Cluster-based systemic innovation policies aimed at increasing the competitiveness of clusters from a systems of innovation perspective offer a powerful alternative to partial and rather "old fashioned" interventionist technology and industrial policy making. Cluster analysis is increasingly perceived as a useful working method for systemic innovation policy making, as it serves not only to link cluster analysis to cluster-based policy making, but also greatly facilitates cluster policy learning.

Creating incentives for innovative behaviour in the market requires innovations in policy making and institutional renewal of government agencies (Ormala, 1998; Sulzenko, 1997; Roelandt *et al.*, this volume). There is a strong and growing need for "horizontal policy", integrating the various aspects of functionally organised policy instruments (*e.g.* education policy, science policy, trade policy, competition policy, technology policy, public works, fiscal policy, etc.). As stated by Ormala (1998), governments are not necessarily organised to manage innovation policy in the best possible way. Ministries usually have sectoral and functional responsibilities. Innovation policy demands horizontal policies, requiring a co-ordinated contribution from a number of different sectors. Governments have a key role to play not only in managing knowledge in their ministries and agencies, but also in improving the acquisition and application of knowledge on an economy-wide basis. One solution could be to encourage the mobility of personnel between the public sector and business (Ormala, 1998).

NOTES

1. See, in particular, the contributions of Drejer *et al.* (this volume); Roelandt *et al.* (this volume); Rouvinen and Ylä-Anttila (this volume); DeBresson and Hu (this volume); DeBresson (1996); Porter (1997).

2. It goes without saying that this bottleneck also holds for sectoral analyses.

3. The OECD Focus Group is currently working on developing a common cluster methodology and some pilot studies adopting a common methodology. For preliminary results, please contact the authors.

4. See, for example, Boekholt *et al.* (this volume); Heath (this volume); Rouvinen *et al.* (this volume); Roelandt *et al.* (this volume); Lagendijk and Charles (this volume); Ormala (1998); Held (1996); Porter (1997).

5. See also Held (1996); Porter (1997); Roelandt *et al.* (this volume); Rouvinen *et al.* (this volume); Dunning (1997).

REFERENCES

Arvanitis, S. and H. Hollenstein (1997), "Innovative Activity and Firm Characteristics: An Exploration of Clustering at Firm Level in Swiss Manufacturing", paper presented at the OECD Workshop on Cluster Analysis and Cluster-based Policy, Amsterdam, 10-11 October.

Boekholt, P. and B. Thuriaux (1998), "Overview of Cluster Policies in International Perspective", report prepared for the Dutch Ministry of Economic Affairs, Technopolis, Amsterdam.

Boekholt, P. and B. Thuriaux (1999), "Public Policies to Facilitate Clusters. Background, Rationales and Policy Practices in International Perspective", this volume.

Chaminade, C. (1999), "Innovation Processes and Knowledge Flows in the Information and Communication Technologies (ICT) Cluster in Spain", this volume.

Cimoli, M. (1997), "Methodologies for the Study of NIS: A Cluster-based Approach for the Mexican Case", paper presented at the OECD Workshop on Cluster Analysis and Cluster-based Policy, Amsterdam, 10-11 October.

Debackere, K. (1997), "Cluster-based Innovation Policies: A Reflection on Definitions and Methods", paper presented at the OECD Workshop on Cluster Analysis and Cluster-based Policy, Amsterdam, 10-11 October.

DeBresson, Ch. (ed.) (1996), *Economic Interdependence and Innovative Activity,* Edward Elgar.

DeBresson, Ch. and X. Hu (1999), "Identifying Clusters of Innovative Activity: A New Approach and a Toolbox", this volume.

Drejer, I., F.S. Kristensen and K. Laursen (1999), "Studies of Clusters as a Basis for Industrial and Technology Policy in the Danish Economy", this volume.

Dunning, J.H. (1997), *Alliance Capitalism and Global Business,* Routledge, London.

Feser, E.J. and E.M. Bergman (1997), "National Industry Clusters: Frameworks for State and Regional Development Policy", University of North Carolina (prepared for *Regional Studies).*

Heath, R. (1998), "The Policy Utility of Systemic Analysis", draft internal report.

Held, J.R. (1996), "Clusters as an Economic Development Tool: Beyond the Pitfalls", *Economic Development Quarterly*, Vol. 10, No. 3, August, pp. 249-261.

Hove, N. van den, T. Roelandt and T. Grosfeld (1998), *Cluster Specialisation Patterns and Innovation Styles,* Ministry of Economic Affairs, The Hague.

Jacobs, D. and A.-P. de Man (1996), "Clusters, Industrial Policy and Firm Strategy: A Menu Approach", *Technology Analysis and Strategic Management,* Vol. 8, No. 4, pp. 425-437.

Lagendijk, A. and D. Charles (1999), "Clustering as a New Growth Strategy for Regional Economies? A discussion of New Forms of Regional Industrial Policy in the United Kingdom", this volume.

Marceau, J. (1999), "The Disappearing Trick: Clusters in the Australian Economy", this volume.

Meeuwsen, W. and M. Dumont (1997), "Some Results on the Graph-theoretical Identification of Micro-clusters in the Belgian National Innovation System", paper presented at the OECD Workshop on Cluster Analysis and Cluster-based Policy, Amsterdam, 10-11 October.

Morgan, K. (1996), "Learning by Interacting. Inter-firm Networks and Enterprise Support, *Local Systems of Small Firms and Job Creation*, OECD, Paris.

Ormala, E. (1998), "New Approaches in Technology Policy – The Finnish Example", *STI Review*, No. 22, Special Issue on "New Rationale and Approaches in Technology and Innovation Policy", OECD, Paris, pp. 277-283.

Peneder, M. (1999), "Creating a Coherent Design for Cluster Analysis and Related Policies. The Austrian 'TIP' Experience", this volume.

Porter, M.E. (1997), "Knowledge-based Clusters and National Competitive Advantage", paper presented at Technopolis 97, 12 September, Ottawa.

Poti, B. (1997), "The Interindustrial Distribution of Knowledge: The Example of Italy", paper presented at the OECD Workshop on Cluster Analysis and Cluster-based Policy, Amsterdam, 10-11 October.

Rouvinen, P. (ed.) (1996), "'Advantage Finland'. The Future of Finnish Industries", ETLA, The Research Institute of the Finnish Economy, Helsinki.

Rouvinen, P. and P. Ylä-Antilla (1999), "Finnish Cluster Studies and New Industrial Policy Making", this volume.

Spielkamp, A. and K. Vopel (1999), "Mapping Innovative Clusters in National Innovation Systems", this volume.

Stenberg, L. and A.-C. Strandell (1997), "An Overview of Cluster-related Studies and Policies in Sweden", paper presented at the OECD Workshop on Cluster Analysis and Cluster-based Policy, Amsterdam, 10-11 October.

Sweeney, S.H. and E.J. Feser (1997), "Plant Size and Clustering of Manufacturing Activity", University of North Carolina (forthcoming in *Geographical Analysis*).

Vock, P. (1979), "Swiss Position Paper for the Focus Group on Mapping Innovative Clusters of the OECD NIS Project", paper presented at the OECD Workshop on Cluster Analysis and Cluster-based Policy, Amsterdam, 10-11 October.

Vuori, S. (1995), "Technology Sources in Finnish manufacturing", Series B 108, ETLA, The Research Institute of the Finnish Economy, Helsinki.

Vuori, S. (1997), "Technology Sources and Competitiveness – An Analysis of Finnish Industries", ETLA, The Research Institute of the Finnish Economy, Helsinki.

OECD PUBLICATIONS, 2, rue André-Pascal, 75775 PARIS CEDEX 16
PRINTED IN FRANCE
(92 1999 04 1 P) ISBN 92-64-17080-4 – No. 50745 1999